Spon's
External Works
and Landscape
Price Book

2011

Spon's External Works and Landscape Price Book

Edited by
DAVIS LANGDON LLP

in association with
LandPro Ltd
Landscape Surveyors

2011

Thirtieth edition

Spon Press
an imprint of Taylor & Francis
LONDON AND NEW YORK

First edition 1978
Thirtieth edition published 2010
by Spon Press
2 Park Square, Milton Park, Abingdon, Oxon, OX14 4RN

Simultaneously published in the USA and Canada
by Spon Press
270 Madison Avenue, New York, NY 10016

Spon Press is an imprint of the Taylor & Francis Group, an informa business

The right of Davis Langdon to be identified as the Author of this Work has been asserted by them in accordance with the Copyright, Designs and Patents Act 1988

Typeset in Arial by Taylor & Francis Books

Printed and bound in Great Britain by
TJ International Ltd, Padstow, Cornwall

British Library Cataloguing in Publication Data
A catalogue record for this book is available from the British Library

ISBN 13: 978-0-415-58849-2 (hardback)
ISBN 13: 978-0-203-84610-0 (ebook)

ISSN: 0267-4181

Contents

Preface to the Thirtieth Edition

Market Conditions

Landscape contractors still report an active market in the commercial and public tender sectors. High end domestic work is still unaffected but smaller contractors dealing with small domestic projects have experienced cutbacks earlier in the year. They now report a resurgence which may be seasonal at the time of the writing of this preface. In general the market is more competitive with tighter margins but there is plenty of work around for the well established contractors. The effects of the 2012 London Olympics are also beginning to have an impact. Tender prices are stable and steady compared to last year's market. Analysis of tenders received by Davis Langdon in the first three months of 2010 has shown that prices were 0.5% higher than in the last quarter of 2009.

Supplier Prices

During the course of our price enquiries for this edition we have noted the following main trends: the main industry drivers vary, steel prices are lower and cement prices have increased slightly. Many of the suppliers who have contributed to this edition have asked us to keep their material costs the same. Those that have raised their prices have stated that in most cases pressure could be brought to bear to maintain 2009 prices.

Labour Rates

Last year we reported that the impact of Eastern European labour returning to their home countries had stabilized the labour rates. Currently we note an increase of about 2% due to employers wishing to ensure that they retain their skilled labour. Spon's External Works and Landscape rates reflect these increases in both Major Works (£18.50 per hour) and Minor Works (£20.50 per hour).

Plant Rates

Most of the plant suppliers referenced in this publication have advised us during our update to keep their prices the same. We have kept the price of red diesel the same as last year's rate. The crane hire rates which we use in this book have remained the same for three years running.

Tables and Memoranda

We remind readers of this section at the back of the book. It is often overlooked but contains a wealth of information useful to external works consultants and contractors.

Profit and Overhead

Spon's External Works and Landscape prices do not allow for profit or site overhead. Readers should evaluate the market conditions in the sector or environment in which they operate and apply a percentage to these rates. Company overhead is allowed for within the labour rates used. Please refer to 'Part 1 General' for an explanation of the build-up of this year's labour rates.

Prices for Suppliers and Services

We acknowledge the support afforded to us by the suppliers of products and services who issue us with the base information used. Their contact details are published in the directory section at the front of this book. We advise that readers wishing to evaluate the cost of a product or service should approach these suppliers directly should they wish to confirm prices prior to submission of tenders or quotations.

Whilst every effort is made to ensure the accuracy of the information given in this publication, neither the editors nor the publishers in any way accept liability for loss of any kind resulting from the use made by any person of such information.

We remind readers that we would be grateful to hear from them during the course of the year with suggestions or comments on any aspect of the contents of this year's book and suggestions for improvements. Our contact details are shown below.

DAVIS LANGDON   SAM HASSALL
MidCity Place   LANDPRO LTD
71 High Holborn   Landscape Surveyors
London   14 Upper Bourne Lane
WC1V 6QS   Farnham
                                               Surrey
                                               GU10 4RQ

Tel: 02070 617000   Tel: 01252 795030
e-mail: spons@davislangdon.com   e-mail: info@landpro.co.uk

Acknowledgements

This list has been compiled from the latest information available but as firms frequently change their names, addresses and telephone numbers due to reorganization, users are advised to check this information before placing orders.

ABG Environmental Geosynthetics Ltd
Unit E7
Meltham Mills Road
Meltham
West Yorkshire HD9 4DS
Geosynthetics
Tel: 01484 852096
Fax: 01484 851562
Website: www.abg-geosynthetics.com
E-mail: sales@abgltd.com

Aco Technologies Plc
Aco Business Park
Hitchin Road
Shefford
Bedfordshire SG17 5TE
Linear drainage
Tel: 01462 816666
Fax: 01462 815895
Website: www.aco-online.co.uk
E-mail: technologies@aco.co.uk

Addagrip Surface Treatments UK Ltd
Addagrip House
Bell Lane Industrial Estate
Uckfield
East Sussex TN22 1QL
Epoxy bound surfaces
Tel: 01825 761333
Fax: 01825 768566
Website: www.addagrip.co.uk
E-mail: sales@addagrip.co.uk

Agripower Ltd
Broomfield Farm
Rignall Road
Great Missenden
Bucks HP16 9PE
Drainage
Tel: 01494 866776
Fax: 01494 866779
Website: www.agripower.co.uk
E-mail: info@agripower.co.uk

Alumasc Exterior Building Products Ltd
White House Works
Bold Road
Sutton
St Helens
Merseyside WA9 4JG
Green roof systems
Tel: 01744 648400
Fax: 01744 648401
Website: www.alumasc-exterior-building-products.co.uk
E-mail: info@alumasc-exteriors.co.uk

Amenity and Horticultural Services Ltd
Coppards Lane
Northiam
East Sussex TN31 6QP
Horticultural composts and fertilizers
Tel: 01797 252728
Fax: 01797 252724
Website: www.ahsdirect.co.uk
E-mail: info@ahsdirect.co.uk

Amenity Land Services Ltd (ALS)
Unit 2/3
Allscott
Telford
Shropshire TF6 5DY
Horticultural products supplier
Tel: 01952 641949
Fax: 01952 247369
Website: www.amenity.co.uk
E-mail: sales@amenity.co.uk

Anderton Concrete Products Ltd
Unit 1&2 Cosgrove Business Park
Soot Hill
Anderton
Northwich
Cheshire CW9 6AA
Concrete fencing
Tel: 01606 595300
Fax: 01606 75905
Website: www.andertonconcrete.co.uk
E-mail: civils@andertonconcrete.co.uk

Anglo Aquarium Plant Co Ltd
Strayfield Road
Enfield
Middlesex EN2 9JE
Aquatic plants
Tel: 020 8363 8548
Fax: 020 8363 8547
Website: www.anglo-aquarium.co.uk
E-mail: sales@anglo-aquarium.co.uk

Architectural Heritage Ltd
Taddington Manor
Cutsdean
Cheltenham
Gloucestershire GL54 5RY
Ornamental stone buildings and water features
Tel: 01386 584414
Fax: 01386 584236
Website: www.architectural-heritage.co.uk
E-mail: puddy@architectural-heritage.co.uk

Autopa Ltd
Cottage Leap off Butlers Leap
Rugby
Warwickshire CV21 3XP
Street furniture
Tel: 01788 550556
Fax: 01788 550265
Website: www.autopa.co.uk
E-mail: info@autopa.co.uk

AVS Fencing Supplies Ltd
Unit 1 AVS Trading Park
Chapel Lane
Milford
Surrey GU8 5HU
Fencing
Tel: 01483 410960
Fax: 01483 860867
Website: www.avsfencing.co.uk
E-mail: sales@avsfencing.co.uk

Bauder Ltd
Broughton House
Broughton Road
Ipswich
Suffolk IP1 3QR
Green roof systems
Tel: 01473 257671
Fax: 01473 230761
Website: www.bauder.co.uk
E-mail: info@bauder.co.uk

Bituchem
Laymore
Forest Vale Industrial Estate
Cinderford
Gloucestershire GL14 2YH
Macadam contractor
Tel: 01594 826768
Fax: 01594 826948
Website: www.bituchem.com
E-mail: info@bituchem.com

Boddingtons Ltd
Blackwater Trading Estate
The Causeway
Maldon
Essex
Erosion control, soil stabilization, plant protection
Tel: 01621 874200
Fax: 01621 874299
Website: www.boddintons-ltd.com
E-mail: sales@boddingtons-ltd.com

Breedon Special Aggregates
(A division of Ennstone Johnston Ltd)
Breedon-on-the-Hill
Derby
Derbyshire DE73 8AP
Specialist gravels
Tel: 01332 694001
Fax: 01332 695159
Website: www.ennstone.co.uk
E-mail: sales@ennstonejohnston.co.uk

Britannia Rainwater Recycling Systems
Studio House
Delamere Road
Cheshunt
Herts EN8 9SH
Rainwater tanks
Tel: 01992 633211
Fax: 01992 633212
E-mail: britanniaservices@btopenworld.com

British Seed Houses Ltd
Camp Road
Witham St Hughs
Lincoln LN6 9QJ
Grass and wildflower seed
Tel: 01522 868714
Fax: 01522 868095
Website: www.bshamenity.com
E-mail: seeds@bshlincoln.co.uk

Broxap Street Furniture
Rowhurst Industrial Estate
Chesterton
Newcastle-under-Lyme
Staffordshire ST5 6BD
Street furniture
Tel: 0844 800 4085
Fax: 01782 565357
Website: www.broxap.com
E-mail: sales@broxap.com

Capital Garden Products Ltd
Gibbs Reed Barn
Pashley Road
Ticehurst
East Sussex TN5 7HE
Plant containers
Tel: 01580 201092
Fax: 01580 201093
Website: www.capital-garden.com
E-mail: sales@capital-garden.com

CED Ltd
728 London Road
West Thurrock
Grays
Essex RM20 3LU
Natural stone
Tel: 01708 867237
Fax: 01708 867230
Website: www.ced.ltd.uk
E-mail: sales@ced.ltd.uk

Charcon Hard Landscaping
Hulland Ward
Ashbourne
Derbyshire DE6 3ET
Paving and street furniture
Tel: 01335 372222
Fax: 01335 370074
Website: www.aggregate.com
E-mail: ukenquiries@aggregate.com

City Electrical Factors
Eastern Road
North Lane
Aldershot
Hampshire GU12 4YD
Electrical wholesaler
Tel: 01252 327661
Fax: 01252 343089
Website: www.cef.co.uk

Cooper Clarke Civils and Lintels
Special Products Division
Bloomfield Road
Farnworth
Bolton BL4 9LP
Wall drainage and erosion control
Tel: 01204 862222
Fax: 01204 793856
Website: www.heitonuk.com
E-mail: marketing@cooperclarke.co.uk

Crowders Nurseries
Lincoln Road
Horncastle
Lincolnshire LN9 5LZ
Plant protection
Tel: 01507 525000
Fax: 01507 524000
Website: www.crowders.co.uk
E-mail: sales@crowders.co.uk

CU Phosco Ltd
Charles House
Lower Road
Great Amwell
Ware
Hertfordshire SG12 9TA
Lighting
Tel: 01920 860600
Fax: 01920 860635
Website: www.cuphosco.co.uk
E-mail: sales@cuphosco.co.uk

Deepdale Trees Ltd
Tithe Farm
Hatley Road
Sandy
Bedfordshire SG19 2DX
Trees, multi-stem and hedging
Tel: 01767 262636
Fax: 01767 262288
Website: www.deepdale-trees.co.uk
E-mail: mail@deepdale-trees.co.uk

DLF Trifolium Ltd
Thorn Farm
Evesham Road
Inkberrow
Worcestershire WR7 4LJ
Grass and wildflower seed
Tel: 01386 791102
Fax: 01386 792715
Website: www.dlf.co.uk
E-mail: amenity@dlf.co.uk

Duracourt (Spadeoak) Ltd
Town Lane
Wooburn Green
High Wycombe
Buckinghamshire HP10 0PD
Tennis courts
Tel: 01628 529421
Fax: 01628 810509
Website: www.duracourt.co.uk
E-mail: info@duracourt.co.uk

Earth Anchors Ltd
15 Campbell Road
Croydon
Surrey CR0 2SQ
Anchors for site furniture
Tel: 020 8684 9601
Fax: 020 8684 2230
Website: www.earth-anchors.com
E-mail: sales@earth-anchors.com

EasyMix Concrete Ltd
Unit 12
Enterprise Close
Croydon CR0 3RZ
Concrete
Tel: 020 8643 3795
Website: www.easymixconcrete.com
E-mail: info@easymixconcrete.com

Elliott Hire
The Fen
Baston
Peterborough
Cambridgeshire PE6 9PT
Site offices
Tel: 01778 560891
Fax: 01778 560881
Website: www.elliotthire.co.uk
E-mail: hirediv@elliott-algeco.com

English Woodlands
Burrow Nursery
Cross in Hand
Heathfield
East Sussex TN21 0UG
Plant protection
Tel: 01435 862992
Fax: 01435 867742
Website: www.ewburrownursery.co.uk
E-mail: sales@ewburrownursery.co.uk

Eura Conservation Ltd
Unit H10, Halesfield 19
Telford
Shropshire TF7 4QT
Metalwork restoration
Tel: 01952 680 218
Fax: 01952 585 044
E-mail: enquiries@eura.co.uk

Eve Trakway
Bramley Vale
Chesterfield
Derbyshire S44 5GA
Portable roads
Tel: 08700 767676
Fax: 08700 737373
Website: www.evetrakway.co.uk
E-mail: marketing@evetrakway.co.uk

ExcelEdge
Kinley Systems
Haywood Way
Hastings
East Sussex TN35 4PL
Aluminium edgings
Tel: 01424 201111
Fax: 01424 533004
Website: www.exceledge.co.uk

Exclusive Leisure Ltd
28 Cannock Street
Leicester LE4 9HR
Artificial sports surfaces
Tel: 0116 233 2255
Fax: 0116 246 1561
Website: www.exclusiveleisure.co.uk
E-mail: info@exclusiveleisure.co.uk

Fairwater Water Garden Design Consultants Ltd
Lodge Farm
Malthouse Lane
Ashington
West Sussex RH20 3BU
Water feature contractors
Tel: 01903 892228
Fax: 01903 892522
Website: www.fairwater.co.uk
E-mail: info@fairwater.co.uk

Farmura Environmental Products Ltd
Stone Hill
Egerton
Ashford
Kent TN27 9DU
Organic fertilizer suppliers
Tel: 01233 756241
Fax: 01233 756419
Website: www.farmura.com
E-mail: info@farmura.com

Fleet (Line Markers) Ltd
Spring Lane
Malvern
Worcestershire WR14 1AT
Sports line marking
Tel: 01684 573535
Fax: 01684 892784
Website: www.fleetlinemarkers.com
E-mail: sales@fleetlinemarkers.com

Forticrete Ltd
Anstone Works
Kiverton Park Station
Kiverton Park
Sheffield S26 6NP
Retaining wall systems
Tel: 0870 9034015
Fax: 01909 775043
Website: www.forticrete.com
E-mail: enq@forticrete.com

FP McCann Ltd
Whitehill Road
Coalville
Leicester LE67 1ET
Precast concrete pipes
Tel: 01530 240056
Fax: 01530 240015
Website: www.fpmccann.co.uk
E-mail: info@fpmccann.co.uk

Furnitubes International Ltd
Meridian House
Royal Hill
Greenwich
London SE10 8RT
Street furniture
Tel: 020 8378 3200
Fax: 020 8378 3250
Website: www.furnitubes.com
E-mail: sales@furnitubes.com

The Garden Trellis Company
Unit 1 Brunel Road
Gorse Lane Industrial Estate
Great Clacton
Essex CO15 4LU
Garden joinery specialists
Tel: 01255 688361
Fax: 01255 688362
Website: www.gardentrellis.co.uk
E-mail: info@gardentrellis.co.uk

Grace Construction Products
Ajax Avenue
Slough
Berkshire SL1 4BH
Bituthene tanking
Tel: 01753 692929
Fax: 01753 691623
Website: www.uk.graceconstruction.com
E-mail: uksales@grace.com

Grass Concrete Ltd
Duncan House
142 Thornes Lane
Thornes
West Yorkshire WF2 7RE
Grass block paving
Tel: 01924 379443
Fax: 01924 290289
Website: www.grasscrete.com
E-mail: info@grasscrete.com

Greenfix Soil Stabilization and Erosion Control Ltd
Allens West
Durham Lane
Eaglescliffe
Stockton On Tees TS16 ORW
Seeded erosion control mats
Tel: 01642 888693
Fax: 01642 888699
Website: www.greenfix.co.uk
E-mail: stockton@greenfix.co.uk

Greenleaf Horticulture
Ivyhouse Industrial Estate
Haywood Way
Hastings TN35 4PL
Root directors
Tel: 01424 717797
Fax: 01424 205240
E-mail: info@greenleafhorticulture.net

Grundon Waste Management Ltd
Goulds Grove
Ewelme
Wallingford OX10 6PJ
Gravel
Tel: 0870 4438278
Website: www.grundon.com
E-mail: sales@grundon.com

HSS Hire
Unit 3
14 Wates Way
Mitcham
Surrey CR4 4HR
Tool and plant hire
Tel: 020 8685 9500
Fax: 020 8685 9600
Website: www.hss.com
E-mail: hire@hss.com

Haddonstone Ltd
The Forge House
Church Lane
East Haddon
Northampton NN6 8DB
Architectural stonework
Tel: 01604 770711
Fax: 01604 770027
Website: www.haddonstone.co.uk
E-mail: info@haddonstone.co.uk

Harrison External Display Systems
Borough Road
Darlington
Co Durham DL1 1SW
Flagpoles
Tel: 01325 355433
Fax: 01325 461726
Website: www.harrisoneds.com
E-mail: sales@harrisoneds.com

Havells Sylvania Fixtures UK Ltd
Avis Way
Newhaven
East Sussex BN9 0ED
Lighting
Tel: 0870 6062030
Fax: 01273 512688
Website: www.havells-sylvania.com
E-mail: info.concord@havells-sylvania.com

Headland Amenity Ltd
1010 Cambourne Business Park
Cambourne
Cambridgeshire CB23 6DP
Tel: 01223 597834
Fax: 01223 598052
Chemicals
Website: www.headlandamenity.com
E-mail: info@headlandamenity.com

Heicom UK
4 Frog Lane
Tunbridge Wells
Kent TN1 1YT
Tel: 01892 522360
Fax: 01892 522767
Tree sand
Website: www.treesand.co.uk
E-mail: mike.q@treesand.co.uk

Hepworth
Hazelhead
Crow Edge
Sheffield S36 4HG
Drainage
Tel: 0870 4436000
Fax: 0870 4438000
Website: www.hepworthdrainage.co.uk
E-mail: info@hepworthdrainage.co.uk

Icopal UK Ltd
Lyon Way
St Albans
Hertfordshire AL4 0LB
Lake liner
Tel: 01727 830116
Fax: 01727 868045
Website: www.monarflex.icopal.co.uk
E-mail: enq@monarflex.co.uk

Inturf
The Chestnuts
Wilberfoss
York YO41 5NT
Turf
Tel: 01759 321000
Fax: 01759 380130
Website: www.inturf.com
E-mail: info@inturf.co.uk

J Toms Ltd
7 Marley Farm
Headcorn Road
Smarden
Kent TN27 8PJ
Plant protection
Tel: 01233 770066
Fax: 01233 770055
Website: www.jtoms.co.uk
E-mail: jtoms@btopenworld.com

Jacksons Fencing
Stowting Common
Ashford
Kent TN25 6BN
Fencing
Tel: 01233 750393
Fax: 01233 750403
Website: www.jacksons-fencing.co.uk
E-mail: sales@jacksons-fencing.co.uk

James Coles & Sons (Nurseries) Ltd
The Nurseries
Uppingham Road
Thurnby
Leicester LE7 9QB
Nurseries
Tel: 01162 412115
Fax: 01162 432311
Website: www.colesnurseries.co.uk
E-mail: sales@colesnurseries.co.uk

John Anderson Hire
Unit 5 Smallford Works
Smallford Lane
St Albans
Herts AL4 0SA
Mobile toilet facilities
Tel: 01727 822485
Fax: 01727 822886
Website: www.superloo.co.uk
E-mail: sales@superloo.co.uk

Johnsons Wellfield Quarries Ltd
Crosland Hill
Huddersfield
West Yorkshire HD4 7AB
Natural Yorkstone pavings
Tel: 01484 652311
Fax: 01484 460007
Website: www.johnsons-wellfield.co.uk
E-mail: sales@johnsons-wellfield.co.uk

Jones of Oswestry
Whittington Road
Oswestry
Shropshire SY11 1HZ
Channels, gulleys, manhole covers
Tel: 01691 653251
Fax: 01691 658222
Website: www.jonesofoswestry.com
E-mail: sales@jonesofoswestry.com

KAR UK
Invar Road
Swinton
Manchester M27 9HF
Irrigation
Tel: 01617 939703
Fax: 01617 945145
E-mail: sales@karuk.com
Website: www.karuk.co.uk

Kompan Ltd
20 Denbigh Hall
Bletchley
Milton Keynes
Bucks MK3 7QT
Play equipment
Tel: 01908 642466
Fax: 01908 270137
Website: www.kompan.com
E-mail: kompan.uk@kompan.com

Land & Water Group Ltd
3 Weston Yard
The Street
Albury
Guildford
Surrey GU5 9AF
Specialist plant hire
Tel: 01483 202733
Fax: 01483 202510
Website: www.land-water.co.uk
E-mail: enquiries@land-water.co.uk

Landline Ltd
1 Bluebridge Industrial Estate
Halstead
Essex CO9 2EX
Pond and lake installation
Tel: 01787 476699
Fax: 01787 472507
Website: www.landline.co.uk
E-mail: sales@landline.co.uk

Lappset UK Ltd
Lappset House
Henson Way
Kettering
Northants NN16 8PX
Play equipment
Tel: 01536 412612
Fax: 01536 521703
Website: www.lappset.com
E-mail: customerservicesuk@lappset.com

LDC Limited
The Charcoal House
Blacksmith Lane
Guildford
Surrey GU4 8NQ
Willow walling
Tel: 01483 573817
Fax: 01483 532078
Website: www.ldclandscape.co.uk
E-mail: info@ldc.co.uk

Leaky Pipe Systems Ltd
Frith Farm
Dean Street
East Farleigh
Maidstone
Kent ME15 0PR
Irrigation systems
Tel: 01622 746495
Fax: 01622 745118
Website: www.leakypipe.co.uk
E-mail: sales@leakypipe.co.uk

Lister Lutyens Co Ltd
6 Alder Close
Eastbourne
East Sussex BN23 6QF
Street furniture
Tel: 01323 431177
Fax: 01323 639314
Website: www.listerteak.com
E-mail: sales@listerteak.com

Lorenz von Ehren
Maldfeldstrasse 4
D-21077 Hamburg
Germany
Topiary
Tel: 0049 40 761080
Fax: 0049 40 761081
Website: www.lve.de
E-mail: sales@lve.de

Maccaferri Ltd
7400 The Quorum
Oxford Business Park North
Garsington Road
Oxford OX4 2JZ
Gabions
Tel: 01865 770555
Fax: 01865 774550
Website: www.maccaferri.co.uk
E-mail: oxford@maccaferri.co.uk

Marshalls
Landscape House
Premier Way
Lowfields Business Park
Elland HX5 9HT
Hard landscape materials and street furniture
Tel: 01422 312000
Fax: 01422 330185
Website: www.marshalls.co.uk

Maxit Building Products Ltd
Heath Business Park
Runcorn
Cheshire WA7 4QX
Expanded clay aggregate
Tel: 01928 565656
Website: www.maxit-uk.co.uk/2367
E- mail: sales@maxit-uk.co.uk

McArthur Group Ltd
Foundry Lane
Bristol BS5 7UE
Security fencing
Tel: 0117 943 0500
Fax: 0117 943 0577
Website: www.mcarthur-group.com
E-mail: marketing@mcarthur-group.com

Melcourt Industries Ltd
Boldridge Brake
Long Newton
Tetbury
Gloucestershire GL8 8RT
Mulch and compost
Tel: 01666 502711
Fax: 01666 504398
Website: www.melcourt.co.uk
E-mail: mail@melcourt.co.uk

Milton Pipes Ltd
Cooks Lane
Milton Regis
Sittingbourne
Kent ME10 2QF
Soakaway rings
Tel: 01795 425191
Fax: 01795 478232
Website: www.miltonpipes.com
E-mail: sales@miltonpipes.com

Neptune Outdoor Furniture Ltd
Thompsons Lane
Marwell
Winchester
Hampshire SO21 1JH
Street furniture
Tel: 01962 777799
Fax: 01962 777723
Website: www.nofl.co.uk
E-mail: info@nofl.co.uk

Nomix Enviro
A Division of Frontier Agriculture Ltd
The Grain Silos
Weyhill Road
Andover
Hants SP10 3NT
Herbicides
Tel: 01264 388050
Fax: 01264 337642
Website: www.nomixenviro.co.uk
E-mail: nomixenviro@frontierag.co.uk

Norris and Gardiner Ltd
Lime Croft Road
Knaphill
Woking
Surrey GU21 2TH
Grounds maintenance
Tel: 01483 289111
Fax: 01483 289112
E-mail: rich@norg.co.uk

Oakover Nurseries Ltd
Maidstone Road
Hothfield
Ashford
Kent TN26 1AR
Native plant nurseries
Tel: 01233 713016
Fax: 01233 713510
Website: www.oakovernurseries.co.uk
E-mail: enquiries@oakovernurseries.co.uk

Orchard Street Furniture Ltd
Whistler House
51 The Green North
Warborough
Oxfordshire OX10 7DW
Street furniture
Tel: 01491 642123
Fax: 01491 642126
Website: www.orchardstreet.co.uk
E-mail: sales@orchardstreet.co.uk

PBA Solutions
Bryn
Rake Road
Liss
Hants GU33 7HB
Japanese Knotweed consultants
Tel: 01730 893460
Website: www.pba-solutions.com
E-mail: info@pba-solutions.com

Phi Group
Harcourt House
13 Royal Crescent
Cheltenham
Gloucestershire GL50 3DA
Retaining wall specialists
Tel: 01242 707600
Fax: 01242 254342
Website: www.phigroup.co.uk
E-mail: southern@phigroup.co.uk

Platipus Anchors Ltd
Kingsfield Business Centre
Philanthropic Road
Redhill
Surrey RH1 4DP
Tree anchors
Tel: 01737 762300
Fax: 01737 773395
Website: www.platipus-anchors.com
E-mail: info@platipus-anchors.com

Recycled Materials Ltd
PO Box 519
Surbiton
Surrey KT6 4YL
Disposal contractors
Tel: 020 8390 7010
Fax: 020 8390 7020
Website: www.recycledmaterialsltd.co.uk
E-mail: mail@recycledmaterialsltd.co.uk

Rigby Taylor Ltd
The Riverway Estate
Portsmouth Road
Peasmarsh
Guildford
Surrey GU3 1LZ
Horticultural supply
Tel: 01483 446900
Fax: 01483 534058
Website: www.rigbytaylor.com
E-mail: sales@rigbytaylor.com

RIW Ltd
Arc House
Terrace Road South
Binfield
Bracknell
Berkshire RG42 4PZ
Waterproofing products
Tel: 01344 397777
Fax: 01344 862010
Website: www.riw.co.uk
E-mail: enquiries@riw.co.uk

Road Equipment Ltd
28–34 Feltham Road
Ashford
Middlesex TW15 1DL
Plant hire
Tel: 01784 256565
Fax: 01784 240398
E-mail: roadequipment@aol.com

Rolawn Ltd
York Road
Elvington
York YO41 4XR
Industrial turf
Tel: 01904 608661
Fax: 01904 608272
Website: www.rolawn.co.uk
E-mail: info@rolawn.co.uk

RTS Ltd
UK Sales
Daisy Dene
Inglewhite Road
Goosnargh
Preston PR3 2EB
Permaloc edging
Tel: 01772 780234
Fax: 01772 780234
Website: www.rtslimited.uk.com
E-mail: rtslimited@hotmail.co.uk

Scotts UK Ltd
Paper Mill Lane
Bramford
Ipswich
Suffolk IP8 4BZ
Fertilizers and chemicals
Tel: 01473 830492
Fax: 01473 830046
Website: www.scottsprofessional.co.uk
E-mail: prof.sales@scotts.com

Sleeper Supplies Ltd
PO Box 1377
Kirk Sandall
Doncaster DN3 1XT
Sleeper supplier
Tel: 0845 230 8866
Fax: 0845 230 8877
Website: www.sleeper-supplies.co.uk
E-mail: sales@sleeper-supplies.co.uk

SMP Playgrounds Ltd
Ten Acre Lane
Thorpe
Egham
Surrey TW20 8RJ
Playground equipment
Tel: 01784 489100
Fax: 01784 431079
Website: www.smp.co.uk
E-mail: sales@smp.co.uk

Southern Conveyors
Unit 2, Denton Slipways Site
Wharf Road
Gravesend DAI2 2RU
Conveyors
Tel: 01474 564145
Fax: 01474 568036

Spadeoak Construction Co Ltd
Town Lane
Wooburn Green
High Wycombe
Bucks HP10 0PD
Macadam contractors
Tel: 01628 529421
Fax: 01628 810509
Website: www.spadeoak.co.uk
E-mail: email@spadeoak.co.uk

Steelway Brickhouse
Brickhouse Lane
West Bromwich
West Midlands B70 0DY
Access covers
Tel: 01215 214500
Fax: 01215 214551
Website: www.steelway.co.uk
E-mail: sales@steelwaybrickhose.co.uk

Steelway Fensecure
Queensgate Works
Bilston Road
Wolverhampton
West Midlands WV2 2NJ
Fencing
Tel: 01902 490919
Fax: 01902 4909296
Website: www.steelway.co.uk
E-mail: sales@steelway.co.uk

SteinTec UK Ltd
PO Box 381
Grays
Essex RM17 94B
Specialized mortars
Tel: 08707 555448
Fax: 08707 500218
Website: www.steintec.co.uk
E-mail: info@steintec.co.uk

Sugg Lighting Ltd
Sussex Manor Business Park
Gatwick Road
Crawley
West Sussex RH10 9GD
Lighting
Tel: 01293 540111
Fax: 01293 540114
Website: www.sugglighting.co.uk
E-mail: sales@sugglighting.co.uk

Targetti Poulsen UK Ltd
Unit C 44 Barwell Business Park
Leatherhead Road
Chessington KT9 2NY
Outdoor lighting
Tel: 020 8397 4400
Fax: 020 8397 4455
Website: www.targetti.com
Or: www.louispoulsen.com
E-mail: nak-uk@lpmail.com

Tarmac Precast Concrete (RCC)
Barholm Road
Tallington
Stamford
Lincolnshire PE9 4RL
Precast concrete retaining units
Tel: 01778 381000
Fax: 01778 348041
Website: www.tarmac.co.uk
E-mail: enquiries@tarmac.co.uk

Tensar International
Cunningham Court
Shadsworth Business Park
Blackburn
Lancashire BB2 4PJ
Erosion control, soil stabilization
Tel: 01254 262431
Fax: 01254 266868
Website: www.tensar-international.com
E-mail: sales@tensar.co.uk

Terranova Lifting Ltd
Terranova House
Bennett Road
Reading
Berks RG2 OQX
Crane hire
Tel: 0118 931 2345
Fax: 0118 931 4114
Website: www.terranova-lifting.co.uk
E-mail: cranes@terranovagroup.co.uk

Thompson Landscapes
610 Goffs Lane
Goffs Oak
Herts EN7 5EP
Swimming pools
Tel: 01707 873444
Fax: 01707 873447
Website: www.thompsonlandscapes.com
E-mail: info@thompsonlandscapes.com

Townscape Products Ltd
Fulwood Road South
Sutton-in-Ashfield
Nottinghamshire NG17 2JZ
Hard landscaping
Tel: 01623 513355
Fax: 01623 440267
Website: www.townscape-products.co.uk
E-mail: sales@townscape-products.co.uk

Tubex Ltd
Aberaman Park
Aberaman
South Wales CF44 6DA
Plant protection, tree guards
Tel: 01685 888000
Fax: 01685 888001
Website: www.tubex.com
E-mail: plantcare@tubex.com

Wade International Ltd
Third Avenue
Halstead
Essex CO9 2SX
Stainless steel drainage
Tel: 01787 475151
Fax: 01787 475579
Website: www.wadedrainage.co.uk
E-mail: sales@wade.eu

Wavin Plastics Ltd
Parsonage Way
Chippenham
Wiltshire SN15 5PN
Drainage products
Tel: 01249 766600
Fax: 01249 443286
Website: www.wavin.co.uk
E-mail: info@wavin.co.uk

Wicksteed Leisure Ltd
Digby Street
Kettering
Northamptonshire NN16 8YJ
Play equipment
Tel: 01536 517028
Fax: 01536 410633
Website: www.wicksteed.co.uk
E-mail: sales@wicksteed.co.uk

Woodscape Ltd
Church Works
Church Street
Church
Lancashire BB5 4JT
Street furniture
Tel: 01254 383322
Fax: 01254 381166
Website: www.woodscape.co.uk
E-mail: sales@woodscape.co.uk

Wybone Ltd
Mason Way
Platts Common Industrial Estate
Barnsley
South Yorkshire S74 9TF
Street furniture
Tel: 01226 744010
Fax: 01226 350105
Website: www.wybone.co.uk
E-mail: sales@wybone.co.uk

Yeoman Aggregates Ltd
Stone Terminal
Horn Lane
Acton
London W3 9EH
Aggregates
Tel: 020 8896 6820
Fax: 020 8896 6829
Website: www.foster-yeoman.co.uk
E-mail: sales@yeoman-aggregates.co.uk

How to Use This Book

INTRODUCTION

First-time users of *Spon's External Works and Landscape Price Book* and others who may not be familiar with the way in which prices are compiled may find it helpful to read this section before starting to calculate the costs of landscape works.

The cost of an item of landscape construction or planting is made up of many components:

* the cost of the product
* the labour and additional materials needed to carry out the job
* the cost of running the contractor's business

These are described more fully below.

IMPORTANT NOTES ON THE PROFIT ELEMENT OF RATES IN THIS BOOK

The rates shown in the Approximate Estimates and Measured Works sections of this book do not contain a profit element unless the rate has been provided by a subcontractor. Prices are shown at cost. This is the cost including the costs of company overhead required to perform this task or project. For reference please see the tables on pages 6–10.

Analysed Rates Versus Subcontractor Rates

As a general rule if a rate is shown as an analysed rate in the Measured Works section, i.e. it has figures shown in the columns other than the 'Unit' and 'Total Rate' column, it can be assumed that this has no profit or overhead element and that this calculation is shown as a direct labour/material supply/plant task being performed by a contractor.

On the other hand, if a rate is shown as a Total Rate only, this would normally be a subcontractor's rate and would contain the profit and overhead for the subcontractor.

The foregoing applies for the most part to the Approximate Estimates section, however in some items there may be an element of subcontractor rates within direct works build-ups.

As an example of this, to excavate, lay a base and place a macadam surface, a general landscape contractor would normally perform the earthworks and base installation. The macadam surfacing would be performed by a specialist subcontractor to the landscape contractor.

The Approximate Estimate for this item uses rates from the Measured Works section to combine the excavation, disposal, base material supply and installation with all associated plant. There is no profit included on these elements. The cost of the surfacing, however, as supplied to us in the course of our annual price enquiry from the macadam subcontractor, would include the subcontractor's profit element.

The landscape contractor would add this to his section of the work but would normally apply a mark up at a lower percentage to the subcontract element of the composite task. Users of this book should therefore allow for the profit element at the prevailing rates. Please see the worked examples below and the notes on overheads for further clarification.

INTRODUCTORY NOTES ON PRICING CALCULATIONS USED IN THIS BOOK

There are two pricing sections to this book:

Major Works – Average contract value £100,000.00–£300,000.00

Minor Works – Average contract value £10,000.00–£70,000.00

Typical Project Profiles Used in This Book

	Major Works	Minor Works
Contract value	£100,000.00–£300,000.00	£10,000.00–£70,000.00
Labour rate (see page 6)	£18.50 per hour	£20.50 per hour
Labour rate for maintenance contracts	£15.50 per hour	–
Number of site staff	30	6–9
Project area	6000 m²	1200 m²
Project location	Outer London	Outer London
Project components	50% hard landscape 50% soft landscape and planting	20% hard landscape 80% soft landscape and planting
Access to works areas	Very good	Very good
Contract	Main contract	Main contract
Delivery of materials	Full loads	Part loads

As explained in more detail later the prices are generally based on wage rates and material costs current at spring 2010. They do not allow for preliminary items, which are dealt with in the Preliminaries section of this book, or for any Value Added Tax (VAT) which may be payable.

Adjustments should be made to standard rates for time, location, local conditions, site constraints and any other factors likely to affect the costs of a specific scheme.

Term contracts for general maintenance of large areas should be executed at rates somewhat lower than those given in this section.

There is now a facility available to readers that enables a comparison to be made between the level of prices given and those for projects carried out in regions other than outer London; this is dealt with on page 20.

Units of Measurement

The units of measurement have been varied to suit the type of work and care should be taken when using any prices to ascertain the basis of measurement adopted.

The prices per unit of area for executing various mechanical operations are for work in the following areas and under the following conditions:

Prices per m² relate to areas not exceeding 100 m²
    (any plan configuration)

Prices per 100 m² relate to areas exceeding 100 m² but not exceeding ¼ ha
    (generally clear areas but with some subdivision)

Prices per ha relate to areas over ¼ ha
    (clear areas suitable for the use of tractors and tractor-operated equipment)

The prices per unit area for executing various operations by hand generally vary in direct proportion to the change in unit area.

Approximate Estimates

These are combined Measured Work prices which give an approximate cost for a complete section of landscape work. For example, the construction of a car park comprises excavation, levelling, road-base, surfacing and marking parking bays. Each of these jobs is priced separately in the Measured Works section, but a comprehensive price is given in the Approximate Estimates section, which is intended to provide a quick guide to the cost of the job. It will be seen that the more items that go to make up an approximate estimate, the more possibilities there are for variations in the PC prices and the user should ensure that any PC price included in the estimate corresponds to his specification.

Worked example

In many instances a modular format has been use in order to enable readers to build up a rate for a required task. The following table describes the trench excavation, pipe laying and backfilling operations as contained in section R12 of this book.

Pipe laying 100 m

	£
Remove topsoil 150 mm × 300 wide	40.95
Excavate for drain 150 × 450 mm; excavated material to spoil heaps on site by machine	436.38
Lay flexible plastic pipe 100 mm wide	155.90
Backfilling with gravel rejects, blinding with sand and topping with 150 mm topsoil	356.16
Total	**989.39**

The figures given in Spon's External Works and Landscape Price Book are intended for general guidance, and if a significant variation appears in any one of the cost groups, the Price for Measured Work should be recalculated.

Measured Works

Prime Cost: Commonly known as the 'PC'. Prime Cost is the actual price of the material item being addressed such as paving, shrubs, bollards or turves, as sold by the supplier. Prime Cost is given 'per square metre', 'per 100 bags' or 'each' according to the way the supplier sells his product. In researching the material prices for the book we requested that the suppliers price for full loads of their product delivered to a site close to the M25 in London. Spon's rates do not include VAT. Some companies may be able to obtain greater discounts on the list prices than those shown in the book. Prime Cost prices for those products and plants which have a wide cost range will be found under the heading of Market Prices in the main sections of this book, so that the user may select the product most closely related to his specification.

Materials: The PC material plus the additional materials required to fix the PC material. Every job needs materials for its completion besides the product bought from the supplier. Paving needs sand for bedding, expansion joint strips and cement pointing; fencing needs concrete for post setting and nails or bolts; tree planting needs manure or fertilizer in the pit, tree stakes, guards and ties. If these items were to be priced out separately, Spon's External Works and Landscape Price Book (and the Bill of Quantities) would be impossibly unwieldy, so they are put together under the heading of Materials.

Labour: This figure covers the cost of planting shrubs or trees, laying paving, erecting fencing etc. and is calculated on the wage rate (skilled or unskilled) and the time needed for the job. Extras such as highly skilled craft work, difficult access, intermittent working and the need for labourers to back up the craftsman all add to the cost. Large regular areas of planting or paving are cheaper to install than smaller intricate areas, since less labour time is wasted moving from one area to another.

Plant, consumable stores and services: This rather impressive heading covers all the work required to carry out the job which cannot be attributed exactly to any one item. It covers the use of machinery ranging from JCBs to shovels and compactors, fuel, static plant, water supply (which is metered on a construction site), electricity and rubbish disposal. The cost of transport to site is deemed to be included elsewhere and should be allowed for elsewhere or as a preliminary item. Hired plant is calculated on an average of 36 hours working time per week.

Subcontract rates: Where there is no analysis against an item, this is deemed to be a rate supplied by a subcontractor. In most cases these are specialist items where most external works contractors would not have the expertise or the equipment to carry out the task described. It should be assumed that subcontractor rates include for the subcontractor's profit. An example of this may be found for the tennis court rates in section Q26.

Labour Rates Used in This Edition

The rates for labour used in this edition have been based on surveys carried out on a cross section of external works contractors. These rates include for company overheads such as employee administration, transport insurance, and on costs such as National Insurance. Please see the calculation of these rates shown on pages 6–10. The rates do not include for profit.

Overheads: An allowance for this is included in the labour rates which are described above. The general overheads of the contract such as insurance, site huts, security, temporary roads and the statutory health and welfare of the labour force are not directly assignable to each item, so they are distributed as a percentage on each, or as a separate preliminary cost item. The contractor's and subcontractor's profits are not included in this group of costs. Site overheads, which will vary from contract to contract according to the difficulties of the site, labour shortages, inclement weather or involvement with other contractors, have not been taken into account in the build up of these rates, while overhead (or profit) may have to take into account losses on other jobs and the cost to the contractor of unsuccessful tendering.

ADJUSTMENT AND VARIATION OF THE RATES IN THIS BOOK

It will be appreciated that a variation in any one item in any group will affect the final Measured Works price. Any cost variation must be weighed against the total cost of the contract. A small variation in Prime Cost where the items are ordered in thousands may have more effect on the total cost than a large variation on a few items. A change in design that necessitates the use of earth moving equipment which must be brought to the site for that one job will cause a dramatic rise in the contract cost. Similarly, a small saving on multiple items will provide a useful reserve to cover unforeseen extras.

Worked examples

A variation in the Prime Cost of an item can arise from a specific quotation.

For example:
The PC of 900 × 600 × 50 mm precast concrete flags is given as £3.78 each which equates to £7.00/m², and to which the costs of bedding and pointing are added to give a total material cost of £10.88/m².
Further costs for labour and mechanical plant give a resultant price of £18.28/m².
If a quotation of £8.00/m² was received from a supplier of the flags the resultant price would be calculated as £18.28 less the original cost (£7.00) plus the revised cost (£8.00) to give £19.28/m².

A variation will also occur if, for example, the specification changes from 50 light standard trees to 50 extra heavy standard trees. In this case the Prime Cost will increase due to having to buy extra heavy standard trees instead of light standard stock.

Original price: The prime cost of an Acer platanoides 8–10 cm bare root tree is given as £12.00, to which the cost of handling and planting, etc. is added to give an overall cost of £19.40. Further costs for labour and mechanical plant for mechanical excavation of a 600 × 600 × 600 mm pit (£3.37), a single tree stake (£5.17), importing 'Top-grow' compost in 75 litre bags (£1.88), backfilling with imported topsoil (£8.81) and disposing of the excavated material (£3.86), give a resultant price of £42.49. Therefore the total cost of 50 light standard trees is £2124.71

Revised price: The prime cost of an Acer platanoides root ball 14–16 cm is given as £71.00, however, taking into account, extra labour for handling and planting (£22.20), a larger tree pit 900 × 900 × 600 mm (£7.56), double staking (£7.17), more topsoil (£19.82), compost (5.55) and the disposal (£8.69), this cost gives a total unit price of £141.99. This is significantly more than the light standard version. Therefore the total cost of 50 extra heavy trees is £7099.50 and the additional cost is £4974.79.

This example of the effects of changing the tree size illustrates that caution is needed when revising tender prices, as merely altering the prime cost of an item will not accurately reflect the total cost of the revised item.

HOW THIS BOOK IS UPDATED EACH YEAR

The basis for this book is a database of Material, Labour, Plant and Subcontractor resources each with its own area of the database.

Material, Plant and Subcontractor Resources

Each year the suppliers of each material, plant or subcontract item are approached and asked to update their prices to those that will prevail in September of that year.

These resource prices are individually updated in the database. Each resource is then linked to one or many tasks. The tasks in the task library section of the database is automatically updated by changes in the resource library. A quantity of the resource is calculated against the task. The calculation is generally performed once and the links remain in the database.

On occasions where new information or method or technology are discovered or suggested, these calculations would be revisited. A further source of information is simple time and production observations made during the course of the last year.

Labour Resource Update

Most tasks, except those shown as subcontractor rates (see above), employ an element of labour. The Data Department at Davis Langdon conducts ongoing research into the costs of labour in various parts of the country. Tasks or entire sections are then re-examined and recalculated. Comments on the rates published in this book are welcomed and may be submitted to the contact addresses shown in the Preface.

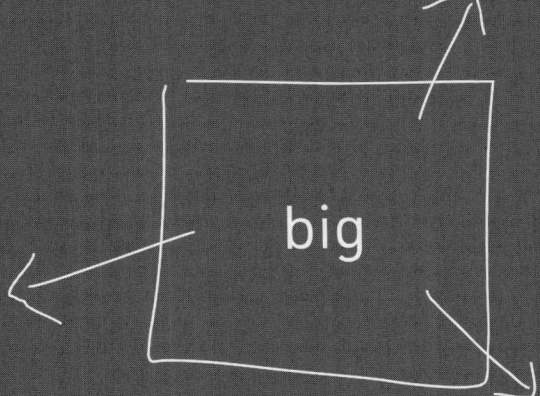

'Thinking Big'

Davis Langdon ✪

Global construction consultants

Cost Management | Project Management | Program Management | Banking Tax & Finance | Building Surveying | Design Project Management | Engineering Services | Health & Safety Services | Legal Support | Management Consulting | Mixed-use Masterplanning | Specification Consulting | Value Planning & Risk

www.davislangdon.com

PART 1

General

This part of the book contains the following sections:

Spon's Asia-Pacific Construction Costs Handbook

4th Edition

Edited by **Davis Langdon & Seah**

Spon's Asia Pacific Construction Costs Handbook includes construction cost data for 20 countries. This new edition has been extended to include Pakistan and Cambodia. Australia, UK and America are also included, to facilitate comparison with construction costs elsewhere. Information is presented for each country in the same way, as follows:

• key data on the main economic and construction indicators.

• an outline of the national construction industry, covering structure, tendering and contract procedures, materials cost data, regulations and standards

• labour and materials cost data

• Measured rates for a range of standard construction work items

• Approximate estimating costs per unit area for a range of building types

• price index data and exchange rate movements against £ sterling, $US and Japanese Yen

The book also includes a Comparative Data section to facilitate country-to-country comparisons. Figures from the national sections are grouped in tables according to national indicators, construction output, input costs and costs per square metre for factories, offices, warehouses, hospitals, schools, theatres, sports halls, hotels and housing.

This unique handbook will be an essential reference for all construction professionals involved in work outside their own country and for all developers or multinational companies assessing comparative development costs.

April 2010: 234x156: 480pp
Hb: 978-0-415-46565-6: **£120.00**

To Order: Tel: +44 (0) 1235 400524 **Fax:** +44 (0) 1235 400525
or Post: Taylor and Francis Customer Services,
Bookpoint Ltd, Unit T1, 200 Milton Park, Abingdon, Oxon, OX14 4TA UK
Email: book.orders@tandf.co.uk

For a complete listing of all our titles visit:
www.tandf.co.uk

Taylor & Francis
Taylor & Francis Group

Common Arrangement of Work Sections

The main work sections relevant to landscape work and their grouping:

A   **Preliminaries/general conditions**

A10	Project particulars
A11	Tender and contract documents
A12	The site/existing buildings
A13	Description of the work
A20	The contract/subcontract
A30	Employer's requirements: Tendering/Subletting/Supply
A31	Employer's requirements: Provision, content and use of documents
A32	Employer's requirements: Management of the Works
A33	Employer's requirements: Quality standards/control
A34	Employer's requirements: Security/Safety/Protection
A35	Employer's requirements: Specific limitations on method/sequence/timing/use of site
A36	Employer's requirements: facilities/temporary works/services
A37	Employer's requirements: operation/maintenance of the finished building
A40	Contractor's general cost items: Management and staff
A41	Contractor's general cost items: Site accommodation
A42	Contractor's general cost items: Services and facilities
A43	Contractor's general cost items: Mechanical plant
A44	Contractor's general cost items: Temporary works
A50	Works/Products by/on behalf of the Employer
A51	Nominated subcontractors
A52	Nominated suppliers
A53	Work by statutory authorities/undertakers
A54	Provisional work
A55	Dayworks
A60	Preliminaries/General conditions for demolition contract
A61	Preliminaries/General conditions for investigation/survey contract
A62	Preliminaries/General conditions for piling/embedded retaining wall contract
A63	Preliminaries/General conditions for landscape contract
A70	General specification requirements for work package

B   **Complete buildings/structures/units**

B10	Prefabricated buildings/structures
B11	Prefabricated building units

C   **Existing site/buildings/services**

C10	Site survey
C11	Ground investigation
C12	Underground services survey
C13	Building fabric survey
C20	Demolition
C50	Repairing/Renovating/Conserving metal

D   **Groundwork**

D11	Soil stabilization
D20	Excavating and filling
D41	Crib walls/gabions/reinforced earth

E   **In situ concrete/large precast concrete**

E05	In situ concrete construction generally
E10	Mixing/Casting/Curing in situ concrete
E20	Formwork for in situ concrete
E30	Reinforcement for in situ concrete

F   **Masonry**

F10	Brick/Block walling
F20	Natural stone rubble walling

F21   Natural stone ashlar walling/dressings
F22   Cast stone ashlar walling/dressings
F30   Accessories/Sundry items for brick/
      block/stone walling
F31   Precast concrete sills/lintels/copings/
      features

G   Structural/carcassing metal/timber

G31   Prefabricated timber unit decking

H   Cladding/covering

H51   Natural stone slab cladding/features
H52   Cast stone slab cladding/features

J   Waterproofing

J10   Specialist waterproof rendering
J20   Mastic asphalt tanking/damp-proofing
J21   Mastic asphalt roofing/insulation/finishes
J22   Proprietary roof decking with asphalt finish
J30   Liquid applied tanking/damp-proofing
J31   Liquid applied waterproof roof coatings
J40   Flexible sheet tanking/damp-proofing
J44   Sheet linings for pools/lakes/waterways
J50   Green roof systems

M   Surface finishes

M10   Cement:sand/Concrete screeds
M20   Plastered/Rendered/Roughcast coatings
M40   Stone/Concrete/Quarry/Ceramic tiling/
      Mosaic
M60   Painting/clear finishing

P   Building fabric sundries

P30   Trenches/Pipeways/Pits for buried
      engineering services

Q   Paving/planting/fencing/site furniture

Q10   Kerbs/Edgings/Channels/paving
      accessories
Q20   Granular subbases to roads/pavings
Q21   In situ concrete roads/pavings
Q22   Coated macadam/Asphalt roads/pavings
Q23   Gravel/Hoggin/Woodchip roads/pavings
Q24   Interlocking brick/block roads/pavings
Q25   Slab/Brick/Sett/Cobble pavings
Q26   Special surfacings/pavings for sport/
      general amenity
Q30   Seeding/Turfing
Q31   Planting
Q32   Planting in Special environments
Q35   Landscape maintenance
Q40   Fencing
Q50   Site/Street furniture/equipment

R   Disposal systems

R12   Drainage below ground
R13   Land drainage

S   Piped supply systems

S10   Cold water
S14   Irrigation
S15   Fountains/Water features

V   Electrical supply/power/lighting systems

V41   Street/Area/Flood lighting

Labour Rates Used in This Edition

Based on surveys carried out on a cross section of external works contractors, the following rates for labour have been used in this edition.

These rates include for company overheads such as employee administration, transport insurance, and on-costs such as National Insurance.

The rates used do not include for profit.

The rates used may include a subcontractor's profit where subcontractors rates are used to build up the costs of an item in the book.

The rates for all labour used in this edition are as follows:

Major Works

General contracting	£18.50/hour
Maintenance contracting	£15.50/hour

Minor Works

General contracting	£20.50/hour

Computation of Labour Rates Used in This Edition

Different organizations will have varying views on rates and costs, which will in any event be affected by the type of job, availability of labour and the extent to which mechanical plant can be used. However this information should assist the reader to:

(1)   compare the prices to those used in his own organization
(2)   calculate the effect of changes in wage rates or prices of materials
(3)   calculate prices for work similar to the examples given

From 30 June 2010 – assumed basic weekly rates of pay for craft and general operatives are £409.73 and £308.30 respectively; to these rates have been added allowances for the items below in accordance with the recommended procedure of the Chartered Institute of Building in its 'Code of Estimating Practice'. The resultant hourly rates are £10.51 and £7.91 for craft operatives and general operatives respectively.

The items for which allowances have been made are:

* Lost time
* Construction Industry Training Board Levy
* Holidays with pay
* Accidental injury, retirement and death benefits scheme
* Sick pay
* National Insurance
* Severance pay and sundry costs
* Employer's liability and third party insurance

The tables that follow illustrate how the above hourly rates have been calculated. Productive time has been based on a total of 1802 hours worked per year for daywork calculations above and for 1953.54 hours worked per year for the all-in labour rates (including 5 hours per week average overtime) for the all-in labour rates below.

How the Labour Rate has been Calculated

A survey of typical landscape/external works companies indicates that they are currently paying above average wages for multi-skilled operatives regardless of specialist or supervisory capability. In our current overhaul of labour constants and costs we have departed from our previous policy of labour rates based on national awards and used instead a gross rate per hour for all rate calculations. Estimators can readily adjust the rate if they feel it is inappropriate for their work.

The productive labour of any organization must return their salary plus the cost of the administration which supports the labour force.

The labour rate used in this edition is calculated as follows:

Team size

A three man team with annual salaries as shown and allowances for National Insurance, uniforms, site tools and a company vehicle, returns a basic labour rate of £15.05 per hour to which company overhead is added as per the following tables.

Basic labour rate calculation for Spon's 2011 Major and Minor Works

Standard Working Hours per year (2010/11) = 1802
Less allowance for sick leave = −39
Less allowance for down time = −35
Actual Working Hours per year (20010/11) = 1728

	Number of	Labour team	NI	Nett Cost	Clothing	Site Tools	TOTAL 3 Man Team
Foreman	1	22440.00	2872.32	25312.32	200.00	150.00	66631.92
Craftsman	1	19890.00	2545.92	22435.92	200.00	150.00	
Labourer	1	15810.00	2023.68	17833.68	200.00	150.00	
	3.00	58140.00	7441.92	65581.92	600.00	450.00	
Overtime Hours							
nominal hours	40.00	19.48	2.49	878.90			
/annum	40.00	17.27	2.21	779.03			
at 1.5 times normal	40.00	13.72	1.76	619.23			
standard rate							
Total staff in Team	3.00			67859.07	600.00	450.00	68909.07
Vehicle Costs Inclusive of Fuel Insurances etc	Working Days	£ /Day					
	240.00	38.00					£9,120.00
						TOTAL	£78029.07

The basic average labour rate excluding overhead costs (below) is

$$\frac{£78029.07}{\text{(Total staff in team)} \times \text{(Working hours per year)}}$$

$$= £15.05$$

Add to the above the overhead costs as per the table on the next page.

Overhead costs
Add to the above basic rate the company overhead costs for a small company as per the table below. These costs are absorbed by the number of working men multiplied by the number of working hours supplied in the table below. This then generates an hourly overhead rate which is added to the Nett cost rate above.

Illustrative overhead for small company employing 12–30 landscape operatives

Cost Centre	Number of	Cost	Total
MD: Salary only: excludes profits; include vehicle	1	43000.00	45000.00
Senior contracts managers; include vehicle	1	37500.00	40000.00
Other contracts managers; include vehicle	1	28000.00	30000.00
Secretary	1.00	17000.00	17000.00
Bookkeeper	1.00	19000.00	19000.00
Rental	12.00	1000.00	12000.00
Insurances	1.00	5000.00	5000.00
Telephone and mobiles	12.00	250.00	3000.00
Office equipment	1.00	1000.00	1000.00
Stationary	12.00	50.00	600.00
Advertising	12.00	200.00	2400.00
Other vehicles not allocated to contract teams	1.00	9000.00	9000.00
Other consultants	1.00	6000.00	6000.00
Accountancy	1.00	2500.00	2500.00
Lights heating water	12.00	200.00	2400.00
Other expenses	1.00	8000.00	10000.00
TOTAL OFFICE OVERHEAD			**198400.00**

The final labour rate used is then generated as follows:

Nett labour rate   + $$\frac{\text{Total Office Overhead}}{(\text{Working hours per year} \times \text{Labour resources employed})}$$

Total nr of site staff	Admin cost per hour	Total rate per man hour
12	9.57	24.62
15	7.65	22.71
18	6.38	21.43
20	5.74	20.79
25	4.59	19.64
30	3.83	18.88
40	2.87	17.92

Hourly labour rates used for Spon's External Works and Landscape 2011 (Major Works)
This year's rates for Major Works are calculated on the rounded value of a 30 man working team

General contracting	£18.50
Maintenance contracting	£15.50

Minor Works

- The minor works labour calculation is based on the above but the administration cost of the organization is £96840.00 per annum
- The owner of the business does not carry out site works and is assumed to be paid a salary of £38,000.00 per annum; this figure constitutes part of the £94,840.00 shown
- There are between 6 and 15 site operatives

Illustrative overhead for small company employing 6–15 landscape operatives

Cost Centre	Number of	Cost	Total
MD: Salary only: excludes profits; include vehicle	1	38000.00	38000.00
Secretary	1.00	17000.00	17000.00
Bookkeeper	1.00	8000.00	8000.00
Rental	12.00	300.00	3600.00
Insurances	1.00	4000.00	4000.00
Telephone and mobiles	12.00	200.00	2400.00
Office equipment	1.00	1000.00	1000.00
Stationary	12.00	20.00	240.00
Advertising	12.00	200.00	2400.00
Other vehicles not allocated to contract teams	1.00	4000.00	4000.00
Other consultants	1.00	3000.00	3000.00
Accountancy	1.00	2000.00	2000.00
Lights, heating, water	12.00	100.00	1200.00
Other expenses	1.00	8000.00	10000.00
TOTAL OFFICE OVERHEAD			94840.00

The final labour rate used is then generated as follows:

$$\text{Nett labour rate} \quad + \quad \frac{\text{Total Office Overhead}}{(\text{Working hours per year} \times \text{Labour resources employed})}$$

Total nr of site staff	Admin cost per hour	Total rate per man hour
3	18.06	33.22
6	9.03	24.08
9	6.02	21.07
12	4.52	19.57
15	3.61	18.66

Hourly labour rates used for Spon's External Works and Landscape 2011 (Minor Works)
This year's rates for Minor Works are calculated on the rounded value of a 9 man working team

General contracting	£20.50

Computation of Labour Rates Used in This Edition

Calculation of Annual Hours Worked

Normal hours worked

Normal hours worked per week		39 hours
Normal hours worked per year	39 hrs × 52 wks	2028 hours

Less holidays

Annual holidays	39 hrs × 4.2 wks	−163.8 hours
Public holidays	39 hrs × 1.6 wks	−62.4 hours

Less sick   39 hrs × 1 wk   −39 hours

Less time lost for inclement weather @ 2%   −35.26 hours

Actual hours worked per annum   **1727.54 hours**

Computation of the Cost of Materials

Percentages of default waste are placed against material resources within the supplier database. These range from 2.5% (for bricks) to 20% (for topsoil). An allowance for the cost of unloading, stacking etc. should be added to the cost of materials.

The following are typical hours of labour for unloading and stacking some of the more common building materials.

Material	Unit	Labourer hours
Cement	tonne	0.67
Lime	tonne	0.67
Common bricks	1000	1.70
Light engineering bricks	1000	2.00
Heavy engineering bricks	1000	2.40

Computation of Mechanical Plant Costs

Plant used within resource calculations in this book have been used according to the following principles:

- Plant which is non-pedestrian is fuelled by the plant hire company.
- Operators of excavators above the size of 5 tonnes are provided by the plant hire company and no labour calculation is allowed for these operators within the rate.
- Three to five tonne excavators and dumpers are operated by the landscape contractor and additional labour is shown within the calculation of the rate.
- Small plant is operated by the landscape contractor.
- No allowance for delivery or collection has been made within the rates shown.
- Downtime is shown against each plant type either for non-productive time or mechanical downtime.

The following tables list the plant resources used in this year's book.

Prices shown reflect the cost of hire of the plant equipment however with the following adjustments for non productive time.
Major works – The weekly rate is divided by 32 working hours per week.
Minor works – The weekly rate is divided by 28 working hours per week.

Plant Supplied by Road Equipment Ltd, Tel: 01784 256565

Description – Major Works – 32 hours productive time/week	Unit	Supply quantity	Rate used £	Down-time %
Dumper 3 tonne Thwaites self drive	hour	1	3.19	20
Dumper 5 ton Thwaites self drive	hour	1	3.83	20
Dumper 6 ton Thwaites self drive	hour	1	4.14	20
Excavator 360 Tracked 21 ton fueled operated	hour	1	42.08	20
Excavator 360 Tracked 7 ton self drive	hour	1	12.11	20
Excavator Tracked 5 ton self drive	hour	1	11.16	20
Fork lift telehandler self drive	hour	1	14.88	60
JCB 3CX 4 × 4 Sitemaster + breaker fueled operated	hour	1	31.88	20
JCB 3CX 4 × 4 Sitemaster fueled operated	hour	1	30.60	20
Mini Excavator 1.5 tonne self drive	hour	1	4.78	20
Mini Excavator JCB 803 Rubber tracks self drive	hour	1	7.33	20
Mini Excavator JCB 803 Steel tracks self drive	hour	1	7.33	20
Skip loader 1 tonne	hour	1	2.87	40
Fuel charge Red diesel	litres	1	0.75	20

Plant Supplied by Road Equipment Ltd, Tel: 01784 256565

Description – Minor Works – 28 hours productive time/week	Unit	Supply quantity	Rate used £	Down-time %
Dumper 3 tonne Thwaites self drive	hour	1	3.64	30
Dumper 6 ton Thwaites self drive	hour	1	4.74	30
Excavator Tracked 5 ton self drive	hour	1	12.75	30
Fork lift telehandler self drive	hour	1	18.06	30
Fuel charge Red diesel	hour	1	0.75	30
JCB 3CX 4 × 4 Sitemaster fueled operated	hour	1	34.97	30
Mini Excavator 1.5 tonne self drive	hour	1	5.46	30
Mini Excavator JCB 803 Rubber tracks self drive	hour	1	8.38	30
Mini Excavator JCB 803 Steel tracks self drive	hour	1	8.38	30
Mini Excavator JCB 803 Steel tracks self drive	hour	1	8.38	30
Skip loader 1 tonne	hour	1	3.28	40

Plant Supplied by HSS Hire

Resource Description	Unit	Supply Quantity	Supply Cost	Rate used £	Down-time %
Access tower; alloy; 5.2 m	day	5.00	165.00	33.00	–
Cultivator; 110 kg	hour	25.00	199.00	7.96	37.5
Diamond blade consumable; 450 mm; concrete	mm	1.00	18.72	18.72	37.5
Heavy-duty breaker; 110 v; 2200 w	hour	25.00	95.50	3.82	37.5
Heavy-duty breaker; petrol; 5 hrs/day; single tool	hour	25.00	158.50	6.34	37.5
Mesh fence; temporary security fencing; 2.85 × 2.00 m high	lm/wk	2.85	6.00	2.11	37.5
Oxy-acetylene cutting kit	hour	25.00	105.50	4.22	37.5
Petrol masonry saw bench; 350 mm	hour	25.00	128.50	5.14	37.5
Petrol poker vibrator + 50 mm head	hour	25.00	49.50	1.98	37.5
Post hole borer; 1 man; weekly rate	hour	25.00	98.50	3.94	37.5
Vibrating plate compactor	hour	25.00	69.00	2.76	37.5
Vibrating roller; 136 kg; 10.1 kN; 4 hrs/day	hour	25.00	139.00	5.56	37.5
Vibration damped breaker (light); 1600 w; 110 v	hour	25.00	64.00	2.56	37.5

Description – Minor Works	Unit	Supply quantity	Supply cost £	Rate used £	Down-time %
Access tower; alloy; 5.2 m	day	1.00	33.00	33.00	-
Cultivator; 110 kg	hour	4.00	99.50	24.88	50.00
Diamond blade consumable; 450 mm; concrete	mm	1.00	18.72	18.72	50.00
Heavy-duty breaker; 110 v; 2200 w	hour	4.00	47.75	11.94	50.00
Mesh fence delivery charge per panel	nr	1.00	0.80	0.80	-
Mesh fence; (heras fence) 2.85 × 2.0 m high	lm/wk	1.00	6.00	6.00	-
Oxy-acetylene cutting kit	hour	4.00	52.75	13.19	50
Petrol masonry saw bench; 350 mm	hour	4.00	64.25	16.06	50
Petrol poker vibrator + 50 mm head	hour	4.00	24.75	6.19	50
Post hole borer; 1 man; daily rate	hour	4.00	49.25	12.31	50
Vibrating plate compactor	hour	4.00	34.50	8.63	50
Access tower; alloy; 5.2 m	day	1.00	33.00	33.00	-
Cultivator; 110 kg	hour	4.00	99.50	24.88	50.00

Landfill Tax

Inert Waste Liable at the Lower Rate

Group	Description of material	Conditions	Notes
1	Rocks and soils	Naturally occurring	Includes clay, sand, gravel, sandstone, limestone, crushed stone, china clay, construction stone, stone from the demolition of buildings or structures, slate, topsoil, peat, silt and dredgings. Glass includes fritted enamel, but excludes glass fibre and glass reinforced plastics.
2	Ceramic or concrete materials		Ceramics includes bricks, bricks and mortar, tiles, clayware, pottery, china and refractories. Concrete includes reinforced concrete, concrete blocks, breeze blocks and aircrete blocks, but excludes concrete plant washings.
3	Minerals	Processed or prepared, not used	Moulding sands excludes sands containing organic binders. Clays includes moulding clays and clay absorbents, including Fuller's earth and bentonite. Man-made mineral fibres includes glass fibres, but excludes glass-reinforced plastic and asbestos. Silica, mica and mineral abrasives.
4	Furnace slags		Vitrified wastes and residues from thermal processing of minerals where, in either case, the residue is both fused and insoluble slag from waste incineration.
5	Ash		Comprises only bottom ash and fly ash from wood, coal or waste combustion but excludes fly ash from municipal, clinical and hazardous waste incinerators and sewage sludge incinerators.
6	Low activity inorganic compound		Comprises only titanium dioxide, calcium carbonate, magnesium carbonate, magnesium oxide, magnesium hydroxide, iron oxide, ferric hydroxide, aluminium oxide, aluminium hydroxide and zirconium dioxide.
7	Calcium sulphate	Disposed of either at a site not licensed to take putrescible waste or in a containment cell which takes only calcium sulphate	Includes gypsum and calcium sulphate-based plasters, but excludes plasterboard.
8	Calcium hydroxide and brine	Deposited in brine cavity	
9	Water	Containing other qualifying material in suspension	

Volume to Weight Conversion Factors

Waste category	Typical waste types	Cubic metres to tonnes – multiply by:	Cubic yards to tonnes – multiply by:
Inactive or inert waste	Largely water insoluble and non or very slowly biodegradable: e.g. sand, subsoil, concrete, bricks, mineral fibres, fibreglass etc.	1.5	1.15

Notes:
The Landfill tax came into operation on 1 October 1996. It is levied on operators of licensed landfill sites at the following rates with effect from 1 April 2010:

Inert or inactive wastes
£2.50 per tonne (frozen until 2010/11) – inactive or inert wastes
Included are soil, stones, brick, plain and reinforced concrete, plaster and glass.

All other taxable wastes
£48.00 per tonne (increasing by £8.00 per tonne per year until at least 2013) – all other taxable wastes
Included are timber, paint and other organic wastes generally found in demolition work, builders skips, etc.

Mixtures containing wastes not classified as inactive or inert will not qualify for the lower rate of tax unless the amount of non-qualifying material is small and there is no potential for pollution. Water can be ignored and the weight discounted.

Calculating the Weight of Waste
There are two options:

- If licensed sites have a weighbridge, tax will be levied on the actual weight of waste.
- If licensed sites do not have a weighbridge, tax will be levied on the permitted weight of the lorry based on an alternative method of calculation based on volume to weight factors for various categories of waste.

For further information contact the National Advisory Service, Telephone: 0845 010 9000

Spon's Irish Construction Price Book

Third Edition

Franklin + Andrews

This new edition of *Spon's Irish Construction Price Book*, edited by Franklin + Andrews, is the only complete and up-to-date source of cost data for this important market.

- All the materials costs, labour rates, labour constants and cost per square metre are based on current conditions in Ireland

- Structured according to the new Agreed Rules of Measurement (second edition)

- 30 pages of Approximate Estimating Rates for quick pricing

This price book is an essential aid to profitable contracting for all those operating in Ireland's booming construction industry.

Franklin + Andrews, Construction Economists, have offices in 100 countries and in-depth experience and expertise in all sectors of the construction industry.

April 2008: 246x174 mm: 510 pages
Hb: 978-0-415-45637-1: **£150.00**

**To Order: Tel: +44 (0) 1235 400524 Fax: +44 (0) 1235 400525
or Post:** Taylor and Francis Customer Services,
Bookpoint Ltd, Unit T1, 200 Milton Park, Abingdon, Oxon, OX14 4TA UK
Email: book.orders@tandf.co.uk

**For a complete listing of all our titles visit:
www.tandf.co.uk**

Taylor & Francis
Taylor & Francis Group

Design for Outdoor Recreation

Second Edition

Simon Bell

A manual for planners, designers and managers of outdoor recreation destinations, this book works through the processes of design and provides the tools to find the most appropriate balance between visitor needs and the capacity of the landscape.

A range of different aspects are covered including car parking, information signing, hiking, waterside activities, wildlife watching and camping.

This second edition incorporates new examples from overseas, including Australia, New Zealand, Japan and Eastern Europe as well as focusing on more current issues such as accessibility and the changing demands for recreational use.

July 2008: 276x219: 272pp
Pb: 978-0-415-44172-8 **£47.99**

To Order: Tel: +44 (0) 1235 400524 **Fax:** +44 (0) 1235 400525
or Post: Taylor and Francis Customer Services,
Bookpoint Ltd, Unit T1, 200 Milton Park, Abingdon, Oxon, OX14 4TA UK
Email: book.orders@tandf.co.uk

For a complete listing of all our titles visit:
www.tandf.co.uk

PART 2

Approximate Estimating – Major Works

This part of the book contains the following sections:

Cost Indices

The purpose of this section is to show changes in the cost of carrying out landscape work (hard surfacing and planting) since 1990. It is important to distinguish between costs and tender prices: the following table reflects the change in cost to contractors but does not necessarily reflect changes in tender prices. In addition to changes in labour and material costs, which are reflected in the indices given below, tender prices are also affected by factors such as the degree of competition at the time of tender and in the particular area where the work is to be carried out, the availability of labour and materials, and the general economic situation. This can mean that in a period when work is scarce, tender prices may fall despite the fact that costs are rising, and when there is plenty of work available, tender prices may increase at a faster rate than costs.

The Constructed Cost Index

A Constructed Cost Index based on PSA Price Adjustment Formulae for Construction Contracts (Series 2). Cost indices for the various trades employed in a building contract are published monthly by HMSO and are reproduced in the technical press.

The indices comprise 49 Building Work indices plus seven 'Appendices' and other specialist indices. The Building Work indices are compiled by monitoring the cost of labour and materials for each category and applying a weighting to these to calculate a single index.

Although the PSA indices are prepared for use with price-adjustment formulae for calculating reimbursement of increased costs during the course of a contract, they also present a time series of cost indices for the main components of landscaping projects. They can therefore be used as the basis of an index for landscaping costs.

The method used here is to construct a composite index by allocating weightings to the indices representing the usual work categories found in a landscaping contract, the weightings being established from an analysis of actual projects. These weightings totalled 100 in 1976 and the composite index is calculated by applying the appropriate weightings to the appropriate PSA indices on a monthly basis, which is then compiled into a quarterly index and rebased to 1976 = 100.

Constructed Landscaping (Hard Surfacing and Planting) Cost Index

Based on approximately 50% soft landscaping area and 50% hard external works.
1976 = 100

Year	First Quarter	Second Quarter	Third Quarter	Fourth Quarter	Annual Average
1990	330	334	353	357	344
1991	356	356	364	364	360
1992	364	365	376	378	371
1993	380	381	383	384	382
1994	387	388	394	397	391
1995	399	404	413	413	407
1996	417	419	428	430	424
1997	430	432	435	442	435
1998	442	445	466	466	455
1999	466	468	488	490	478
2000	493	496	513	514	504
2001	512	514	530	530	522
2002	531	540	573	574	555
2003	577	582	603	602	591
2004	602	610	639	639	623
2005	641	648	684	683	664
2006	688	692	710	710	710
2007	714	718	737	742	728
2008	752	771	803	794	780
2009	789	792	796	802	795
2010	804	807*			

Provisional

This index is updated every quarter in Spon's Price Book Update. The updating service is available, free of charge, to all purchasers of Spon's Price Books (complete the reply-paid card enclosed).

Regional Variations

Prices in *Spon's External Works and Landscape Price Book* are based upon conditions prevailing for a competitive tender in the outer London area. For the benefit of readers, this edition includes regional variation adjustment factors which can be used for an assessment of price levels in other regions.

Special further adjustment may be necessary when considering city centre or very isolated locations.

Region	Adjustment Factor
Outer London	1.00
Inner London	1.07
South East	0.94
South West	0.90
East Midlands	0.84
West Midlands	0.86
East Anglia	0.84
Yorkshire & Humberside	0.86
Northern	0.86
North West	0.84
Scotland	0.92
Wales	0.88
Northern Ireland	0.68
Channel Islands	1.15

The following example illustrates the adjustment of prices for regions other than Outer London, by use of regional variation adjustment factors.

A. Value of items priced using *Spon's External Works and Landscape Price Book* for Outer London   £100,000

B. Adjustment to value of A. to reflect Northern Region price level 100,000 × 0.86   £ 86,000

Approximate Estimating Rates

Prices in this section are based upon the Prices for Measured Works, but allow for incidentals which would normally be measured separately in a Bill of Quantities. They do not include for Preliminaries which are priced elsewhere in this book.

Items shown as subcontract or specialist rates would normally include the specialist's overhead and profit. All other items which could fall within the scope of works of general landscape and external works contractors would not include profit.

Based on current commercial rates, profits of 15% to 35% may be added to these rates to indicate the likely 'with profit' values of the tasks below. The variation quoted above is dependent on the sector in which the works are taking place – domestic, public or commercial.

PRELIMINARIES

Item Excluding site overheads and profit	Unit	Total rate £
CONTRACT ADMINISTRATION AND MANAGEMENT		
Prepare tender bid for external works or landscape project; measured works contract; inclusive of bid preparation, provisional program and method statements		
Measured works bid; measured bills provided by the employer; project value		
£30,000.00	nr	340.00
£50,000.00	nr	540.00
£100,000.00	nr	1125.00
£200,000.00	nr	1750.00
£500,000.00	nr	2200.00
£1,000,000.00	nr	3050.00
Lump sum bid; scope document, drawings and specification provided by the employer; project value		
£30,000.00	nr	640.00
£50,000.00	nr	840.00
£100,000.00	nr	1600.00
£200,000.00	nr	2700.00
£500,000.00	nr	3650.00
£1,000,000.00	nr	5400.00
Prepare and maintain Health and Safety file; prepare risk and COSHH assessments, method statements, works programmes before the start of works on site; project value		
£35,000.00	nr	280.00
£75,000.00	nr	420.00
£100,000.00	nr	560.00
£200,000.00–£500,000.00	nr	1400.00
Contract management		
Site management by site surveyor carrying out supervision and administration duties only; inclusive of on costs		
full time management	week	1200.00
managing one other contract	week	600.00
Parking		
Parking expenses where vehicles do not park on the site area; per vehicle		
metropolitan area; city centre	week	200.00
metropolitan area; outer areas	week	160.00
suburban restricted parking areas	week	40.00
Congestion charging		
London only	week	40.00
SITE SETUP AND SITE ESTABLISHMENT		
Site fencing; supply and erect temporary protective fencing and remove at completion of works		
Cleft chestnut paling		
1.20 m high 75 mm larch posts at 3 m centres	100 m	800.00
1.50 m high 75 mm larch posts at 3 m centres	100 m	960.00

PRELIMINARIES

Item Excluding site overheads and profit	Unit	Total rate £
Heras fencing		
2 week hire period	100 m	870.00
4 week hire period	100 m	1275.00
8 week hire period	100 m	2125.00
12 week hire period	100 m	3000.00
16 week hire period	100 m	3800.00
24 week hire period	100 m	5500.00
Site compounds		
Establish site administration area on reduced compacted Type 1 base; 150 mm thick; allow for initial setup of site office and welfare facilities for site manager and maximum 8 site operatives		
temporary base or hard standing; 20 × 10 m	200 m²	1875.00
To hardstanding area above; erect office and welfare accommodation; allow for furniture, telephone connection and power; restrict access using temporary safety fencing to the perimeter of the site administration area		
Establishment; delivery and collection costs only		
site office and chemical toilet; 20 m of site fencing	nr	770.00
site office and chemical toilet; 40 m of site fencing	nr	830.00
site office and chemical toilet; 60 m of site fencing	nr	900.00
Weekly hire rates of site compound equipment		
site office 2.4 × 3.6 m; chemical toilet; 20 m of site fencing	week	105.00
site office 4.8 × 2.4 m; chemical toilet; 20 m of site fencing; armoured store 2.4 × 3.6 m	week	120.00
site office 2.4 × 3.6 m; chemical toilet; 40 m of site fencing	week	150.00
site office 2.4 × 3.6 m; chemical toilet; 60 m of site fencing	week	190.00
Removal of site compound		
Remove base; fill with topsoil and reinstate to turf on completion of site works	200 m²	1375.00
Surveying and setting out		
Note: In all instances a quotation should be obtained from a surveyor. The following cost projections are based on the costs of a surveyor using EDM (electronic distance measuring equipment) where an average of 400 different points on site are recorded in a day. The surveyed points are drawn in a CAD application and the assumption used that there is one day of drawing up for each day of on-site surveying.		
Survey existing site using laser survey equipment and produce plans of existing site conditions		
site 1,000 to 10,000 m² with sparse detail and features; boundary and level information only; average five surveying stations	nr	750.00
site up to 1,000 m² with dense detail such as buildings, paths, walls and existing vegetation; average five survey stations	nr	1500.00
site up to 4,000 m² with dense detail such as buildings, paths, walls and existing vegetation; average 10 survey stations	nr	2250.00

DEMOLITION AND SITE CLEARANCE

Item Excluding site overheads and profit	Unit	Total rate £
DEMOLITION AND SITE CLEARANCE		
Clear existing scrub vegetation including shrubs and hedges and stack on site		
By machine		
light scrub and grasses	100 m²	17.50
undergrowth, brambles and heavy weed growth	100 m²	39.00
small shrubs	100 m²	39.00
As above but remove vegetation to licensed tip	100 m²	34.00
By hand	100 m²	81.00
As above but remove vegetation to licensed tip	100 m²	55.00
Fell trees on site; grub up roots; all by machine; remove debris to licensed tip		
trees 600 mm girth	each	215.00
trees 1.5–3.00 m girth	each	610.00
trees over 3.00 m girth	each	1400.00
Demolish existing surfaces		
Break up plain concrete slab; remove to licensed tip		
150 mm thick	m²	7.65
200 mm thick	m²	10.20
300 mm thick	m²	15.30
Break up plain concrete slab and preserve arisings for hardcore; break down to maximum size of 200 × 200 mm and transport to location; maximum distance 50 m		
150 mm thick	m²	5.20
200 mm thick	m²	6.90
300 mm thick	m²	10.30
Break up reinforced concrete slab and remove to licensed tip		
150 mm thick	m²	11.50
200 mm thick	m²	15.30
300 mm thick	m²	23.00
Break out existing surface and associated 150 mm thick granular base load to 20 tonne muck away vehicle for removal off site		
macadam surface; 70 mm thick	m²	6.60
macadam surface; 150 mm thick	m²	8.80
block paving; 50 mm thick	m²	6.40
block paving; 80 mm thick	m²	6.95
Demolish existing free standing walls; grub out foundations; remove arisings to tip		
Brick wall; 112 mm thick		
300 mm high	m	10.50
500 mm high	m	15.40
Brick wall; 225 mm thick		
300 mm high	m	13.30
500 mm high	m	17.00
1.00 m high	m	27.50
1.20 m high	m	29.50
1.50 m high	m	36.00
1.80 m high	m	41.00
Site clearance – General		
Clear away light fencing and gates (chain link, chestnut paling, light boarded fence or similar) and remove to licensed tip	100 m	110.00

DEMOLITION AND SITE CLEARANCE

Item Excluding site overheads and profit	Unit	Total rate £
Demolish existing structures Timber sheds on concrete base 150 mm thick inclusive of disposal and backfilling with excavated material; plan area of building		
10 m²	nr	190.00
15 m²	nr	275.00
Building of cavity wall construction; insulated with tiled roof; inclusive of foundations; allow for disconnection of electrical water and other services; grub out and remove 10 m of electrical cable, water and waste pipes		
10 m²	nr	1450.00
20 m²	nr	1950.00

GROUNDWORK

Item Excluding site overheads and profit	Unit	Total rate £
GROUNDWORK		
Cut and strip by machine; turves 50 mm thick; turf to be lifted for preservation		
Load to dumper or palette by hand; stack on site not exceeding 100 m travel to stack	100 m²	17.80
As above but all by hand; including barrowing	100 m²	430.00
Prices for excavating and reducing to levels are for work in light or medium soils; multiplying factors for other soils are as follows		
clay	1.5	–
compact gravel	1.2	–
soft chalk	2.0	–
hard rock	3.0	–
Excavating topsoil for preservation		
Remove topsoil average depth 300 mm; deposit in spoil heaps not exceeding 100 m travel to heap; turn heap once during storage		
by machine	m³	11.90
by hand	m³	110.00
Excavate to reduce levels; remove spoil to dump not exceeding 100 m travel; all by machine		
Average 0.25 m deep	m²	1.80
Average 0.25 m deep	m³	7.20
Average 0.30 m deep	m²	2.20
Average 0.30 m deep	m³	6.65
Average 1.0 m deep; using 21 tonne 360 tracked excavator	m³	4.20
Average 1.0 m deep; using JCB	m³	6.65
Excavate to reduce levels; remove spoil to dump not exceeding 100 m travel; all by hand		
Average 0.10 m deep	m²	10.00
Average 0.20 m deep	m²	20.00
Average 0.30 m deep	m²	30.00
Average 100–300 mm deep	m³	100.00
Average 1.0 m deep	m³	110.00
Extra for carting spoil to licensed tip off site (20 tonnes)	m³	17.90
Extra for carting spoil to licensed tip off site (by skip; machine loaded)	m³	48.50
Extra for carting spoil to licensed tip off site (by skip; hand loaded)	m³	100.00
Spread excavated material to levels in layers not exceeding 150 mm; using scraper blade or bucket		
Average thickness 100 mm	m²	0.40
Average thickness 100 mm but with imported topsoil	m²	3.00
Average thickness 200 mm	m²	0.70
Average thickness 200 mm but with imported topsoil	m²	5.95
Average thickness 250 mm	m²	0.90
Average thickness 200–250 mm	m³	3.60
Average thickness 250 mm but with imported topsoil	m²	7.40
Average thickness 250 mm but with imported topsoil	m³	30.00
Extra for work to banks exceeding 30° slope	30%	–

GROUNDWORK

Item Excluding site overheads and profit	Unit	Total rate £
Rip subsoil using approved subsoiling machine to a depth of 600 mm below topsoil at 600 mm centres; all tree roots and debris over 150 × 150 mm to be removed; cultivate ground where shown on drawings to depths as shown		
100 mm	100 m²	7.90
200 mm	100 m²	8.00
300 mm	100 m²	8.10
400 mm	100 m²	9.05
Form landform from imported material; mounds to falls and grades to receive turf or seeding treatments; material dumped by 20 tonne lorry to location; volume 100 m³ spread over a 200 m² area; thickness varies from 150–500 mm; prices to not include for surface preparations		
Imported recycled topsoil		
by large excavator	m³	22.00
by 5 tonne excavator	m³	24.00
by 3 tonne excavator	m³	25.00
Form landform from excavated material; mounds to falls and grades to receive turf or seeding treatments; material moved by dumper to location maximum 50 m; volume 100 m³ spread over a 200 m² area; thickness varies from 150–500 mm; prices to not include for surface preparations		
Imported recycled topsoil		
by large excavator	m³	2.55
by 5 tonne excavator	m³	4.45
by 3 tonne excavator	m³	5.70
Surface preparations to recently formed soil areas to receive seeding turf or planting treatments (not included)		
TRENCHES		
Excavate trenches for foundations; trenches 225 mm deeper than specified foundation thickness; pour plain concrete foundations GEN 1 10 N/mm and to thickness as described; disposal of excavated material off site; dimensions of concrete foundations		
250 mm wide × 250 mm deep	m	10.70
300 mm wide × 300 mm deep	m	12.80
400 mm wide × 600 mm deep	m	31.50
600 mm wide × 400 mm deep	m	33.00
Excavate trenches for foundations; trenches 225 mm deeper than specified foundation thickness; pour plain concrete foundations GEN 1 10 N/mm and to thickness as described; disposal of excavated material off site; dimensions of concrete foundations		
250 mm wide × 250 mm deep	m³	200.00
300 mm wide × 300 mm deep	m³	140.00
600 mm wide × 400 mm deep	m³	135.00
400 mm wide × 600 mm deep	m³	120.00

GROUNDWORK

Item Excluding site overheads and profit	Unit	Total rate £
TRENCHES – cont		
Excavate trenches for services; grade bottoms of excavations to required falls; remove 20% of excavated material off site		
Trenches 300 mm wide		
600 mm deep	m	5.30
750 mm deep	m	6.80
900 mm deep	m	8.15
1.00 m deep	m	9.00
Ha-ha		
Excavate ditch 1200 mm deep × 900 mm wide at bottom battered one side 45° slope; excavate for foundation to wall and place GEN 1 concrete foundation 150 × 500 mm; construct one and a half brick wall (brick PC £300.00/1000) battered 10° from vertical; laid in cement: lime: sand (1:1:6) mortar in English garden wall bond; precast concrete coping weathered and throated set 150 mm above ground on high side; rake bottom and sides of ditch and seed with low maintenance grass at 35 g/m²		
wall 600 mm high	m	240.00
wall 900 mm high	m	300.00
wall 1200 mm high	m	370.00
Excavate ditch 1200 mm deep × 900 mm wide at bottom battered one side 45° slope; place deer or cattle fence 1.20 m high at centre of excavated trench; rake bottom and sides of ditch and seed with low maintenance grass at 35 g/m²		
fence 1200 mm high	m	18.70

GROUND STABILIZATION

Item Excluding site overheads and profit	Unit	Total rate £
GROUND STABILIZATION		
Excavate and grade banks to grade to receive stabilization treatments below; removal of excavated material not included		
By 21 tonne excavator	m³	0.30
By 7 tonne excavator	m³	2.00
By 5 tonne excavator	m³	2.20
By 3 tonne excavator	m³	3.40
Excavate sloped bank to vertical to receive retaining or stabilization treatment priced below; allow 1.00 m working space at top of bank; backfill working space and remove balance of arisings off site		
Bank height 1.00 m; excavation by 5 tonne excavator		
10°	m	79.00
30°	m	28.00
45°	m	17.20
60°	m	10.90
Bank height 1.00 m; excavation by 5 tonne excavator; arisings moved to stockpile on site		
10°	m	18.40
30°	m	9.40
45°	m	6.40
60°	m	5.95
Bank height 2.00 m; excavation by 5 tonne excavator		
10°	m	305.00
30°	m	100.00
45°	m	63.00
60°	m	40.00
Bank height 2.00 m; excavation by 5 tonne excavator but arisings moved to stockpile on site		
10°	m	63.00
30°	m	27.00
45°	m	20.00
60°	m	16.20
Bank height 3.00 m; excavation by 21 tonne excavator		
20°	m	305.00
30°	m	190.00
45°	m	110.00
60°	m	64.00
As above but arisings moved to stockpile on site		
20°	m	40.00
30°	m	27.00
45°	m	12.40
60°	m	8.10
Bank stabilization; erosion control mats		
Excavate fixing trench for erosion control mat 300 × 300 mm; backfill after placing mat selected below	m	5.80

GROUND STABILIZATION

Item Excluding site overheads and profit	Unit	Total rate £
GROUND STABILIZATION – cont		
To anchor trench above; place mat to slope to be retained; allow extra over to the slope for the area of matting required to the anchor trench Rake only bank to final grade; lay erosion control solution; sow with low maintenance grass at 35 g/m²; spread imported topsoil 25 mm thick incorporating medium grade sedge peat at 3 kg/m² and fertilizer at 30 g/m²; water lightly		
unseeded Eromat Light	m²	2.20
seeded Covamat Standard	m²	2.80
lay Greenfix biodegradable pre-seeded erosion control mat; fix with 6 × 300 mm steel pegs at 1.0 m centres	m²	3.30
Tensar open textured erosion mat	m²	4.50
Excavate trench and lay foundation concrete 1:3:6; 600 × 300 mm deep		
By machine	m	28.00
By hand	m	38.00
Concrete block retaining walls; Stepoc Excavate trench 750 mm deep and lay concrete foundation 600 mm wide × 600 mm deep; construct Forticrete precast hollow concrete block wall with 450 mm below ground laid all in accordance with manufacturer's instructions; fix reinforcing bar 12 mm as work proceeds; fill blocks with concrete 1:3:6 as work proceeds		
Walls 1.00 m high		
type 190; 400 × 225 × 200 mm	m	175.00
type 256; 400 × 225 × 256 mm	m	200.00
As above but 1.50 m high		
type 190; 400 × 225 × 200 mm	m	230.00
type 256; 400 × 225 × 256 mm	m	260.00
Walls 1.50 m high as above but with foundation 1.80 m wide × 600 mm deep		
type 190; 400 × 225 × 200 mm	m	275.00
type 256; 400 × 225 × 256 mm	m	310.00
Walls 1.80 m high		
type 190; 400 × 225 × 200 mm	m	300.00
type 256; 400 × 225 × 256 mm	m	340.00
On foundation measured above, supply and RCC precast concrete L shaped units, constructed all in accordance with manufacturer's instructions; backfill with approved excavated material compacted as the work proceeds		
1250 mm high × 1000 mm wide	m	200.00
2400 mm high × 1000 mm wide	m	335.00
3750 mm high × 1000 mm wide	m	630.00
Grass concrete; bank revetments; excluding bulk earthworks Bring bank to final grade by machine; lay regulating layer of Type 1; lay filter membrane; lay 100 mm drainage layer of broken stone or approved hardcore 28–10 mm size; blind with 25 mm sharp sand; lay grass concrete surface (price does not include edgings or toe beams); fill with approved topsoil and fertilizer at 35 g/m²; seed with dwarf rye based grass seed mix		
Grasscrete in situ reinforced concrete surfacing GC 1; 100 mm thick	m²	43.50
Grasscrete in situ reinforced concrete surfacing GC 2; 150 mm thick	m²	51.00
Grassblock 103 mm open matrix blocks; 406 × 406 × 103 mm	m²	45.00

GROUND STABILIZATION

Item Excluding site overheads and profit	Unit	Total rate £
Gabion walls		
Gabions 500 mm high; inclusive of excavation to sloped bank to vertical; allowance of 1.00 m working space at top of bank; backfill working space and remove balance of arisings off site; lay concrete footing; 200 mm deep × 1.00 m wide		
Existing bank slope		
10°	m	**165.00**
30°	m	**115.00**
45°	m	**105.00**
60°	m	**99.00**
Gabions 1.00 m high; inclusive of excavation to sloped bank to vertical; allowance of 1.00 m working space at top of bank; backfill working space and remove balance of arisings off site; lay concrete footing; 200 mm deep × 1.00 m wide		
Existing bank slope; excavation by 5 tonne excavator		
10°	m	**225.00**
30°	m	**170.00**
45°	m	**160.00**
60°	m	**160.00**
Construct revetment of Maccaferri Ltd Reno mattress gabions laid on firm level ground; tightly packed with broken stone or concrete and securely wired; all in accordance with manufacturer's instructions		
gabions 6 × 2 × 0.17 m; one course	m²	**33.50**
gabions 6 × 2 × 0.17 m; two courses	m²	**59.00**
Timber log retaining walls		
Excavate trench 300 mm wide to one third of the finished height of the retaining walls below; lay 100 mm hardcore; fix machine rounded logs set in concrete 1:3:6; remove excavated material from site; fix geofabric to rear of timber logs; backfill with previously excavated material set aside in position; all works by machine		
100 mm diameter logs		
500 mm high (constructed from 1.80 m lengths)	m	**14.50**
1.20 mm high (constructed from 1.80 m lengths)	m	**28.00**
1.60 mm high (constructed from 2.40 m lengths)	m	**37.00**
2.00 mm high (constructed from 3.00 m lengths)	m	**46.00**
As above but 150 mm diameter logs		
1.20 mm high (constructed from 1.80 m lengths)	m	**28.00**
1.60 mm high (constructed from 2.40 m lengths)	m	**37.00**
2.00 mm high (constructed from 3.00 m lengths)	m	**46.00**
2.40 mm high (constructed from 3.60 m lengths)	m	**55.00**
Timber crib wall		
Excavate trench to receive foundation 300 mm deep; place plain concrete foundation 150 mm thick in 11.50 N/mm² concrete (sulphate-resisting cement); construct timber crib retaining wall and backfill with excavated spoil behind units		
timber crib wall system; average 4.2 m high	m	**175.00**
timber crib wall system; average 2.2 m high	m	**175.00**

GROUND STABILIZATION

Item Excluding site overheads and profit	Unit	Total rate £
GROUND STABILIZATION – cont		
Geogrid soil reinforcement (Note: measured per metre run at the top of the bank)		
Backfill excavated bank; lay Tensar geogrid; cover with excavated material as work proceeds		
2.00 m high bank; geogrid at 1.00 m vertical lifts		
10° slope	m	**160.00**
20° slope	m	**85.00**
30° slope	m	**61.00**
2.00 m high bank; geogrid at 0.50 m vertical lifts		
45° slope	m	**57.00**
60° slope	m	**45.00**
3.00 m high bank		
10° slope	m	**160.00**
20° slope	m	**85.00**
30° slope	m	**61.00**
45° slope	m	**140.00**

IN SITU CONCRETE

Item Excluding site overheads and profit	Unit	Total rate £
IN SITU CONCRETE		
Mix concrete on site; aggregates delivered in 20 tonne loads; deliver mixed concrete to location by mechanical dumper distance 25 m		
1:3:6	m³	79.00
1:2:4	m³	91.00
As above but ready mixed concrete		
10 N/mm²	m³	92.00
15 N/mm²	m³	97.00
Mix concrete on site; aggregates delivered in 20 tonne loads; deliver mixed concrete to location by barrow distance 25 m		
1:3:6	m³	120.00
1:2:4	m³	130.00
As above but aggregates delivered in 1 tonne bags		
1:3:6	m³	160.00
1:2:4	m³	165.00
As above but ready mixed concrete		
10 N/mm²	m³	135.00
15 N/mm²	m³	140.00
As above but concrete discharged directly from ready mix lorry to required location		
10 N/mm²	m³	91.00
15 N/mm²	m³	95.00
Excavate foundation trench mechanically; remove spoil off site; lay 1:3:6 site mixed concrete foundations; distance from mixer 25 m; depth of trench to be 225 mm deeper than foundation to allow for three underground brick courses priced separately		
Foundation size		
200 mm deep × 400 mm wide	m	17.70
300 mm deep × 500 mm wide	m	26.50
400 mm deep × 400 mm wide	m	26.00
400 mm deep × 600 mm wide	m	39.00
600 mm deep × 600 mm wide	m	56.00
As above but hand excavation and disposal to spoil heap 25 m by barrow and off site by grab lorry		
200 mm deep × 400 mm wide	m	36.00
300 mm deep × 500 mm wide	m	59.00
400 mm deep × 400 mm wide	m	59.00
400 mm deep × 600 mm wide	m	66.00
600 mm deep × 600 mm wide	m	125.00
Excavate foundation trench mechanically; remove spoil off site; lay ready mixed concrete GEN1 discharged directly from delivery lorry to location; depth of trench to be 225 mm deeper than foundation to allow for three underground brick courses priced separately; foundation size		
200 mm deep × 400 mm wide	m	16.20
300 mm deep × 500 mm wide	m	23.50
400 mm deep × 400 mm wide	m	23.00
400 mm deep × 600 mm wide	m	34.50
600 mm deep × 600 mm wide	m	49.50

IN SITU CONCRETE

Item Excluding site overheads and profit	Unit	Total rate £
IN SITU CONCRETE – cont		
Reinforced concrete wall to foundations above (site mixed concrete)		
Up to 1.00 m high × 200 mm thick	m²	93.00
Up to 1.00 m high × 300 mm thick	m²	93.00
Reinforced concrete wall to foundations above (ready mix concrete RC35)		
1.00 m high × 200 mm thick	m²	120.00
1.00 m high × 300 mm thick	m²	130.00
Grading; excavate to reduce levels for concrete slab; lay waterproof membrane, 100 mm hardcore and form concrete slab reinforced with A142 mesh in 1:2:4 site mixed concrete to thickness; remove excavated material off site		
Concrete 1:2:4 site mixed		
100 mm thick	m²	62.00
150 mm thick	m²	68.00
250 mm thick	m²	81.00
300 mm thick	m²	88.00
Concrete ready mixed GEN 2		
100 mm thick	m²	60.00
150 mm thick	m²	66.00
250 mm thick	m²	77.00
300 mm thick	m²	83.00

BRICK/BLOCK WALLING

Item Excluding site overheads and profit	Unit	Total rate £
BRICK WALLING		
One brick thick wall (225 mm thick); excavation 400 mm deep; remove arisings off site; lay GEN 1 concrete foundations 450 mm wide × 250 mm thick; all in English Garden Wall bond; laid in cement: lime: sand (1:1:6) mortar with flush joints, fair face one side; DPC two courses engineering brick in cement: sand (1:3) mortar; precast concrete coping 152 × 75 mm		
Wall 900 mm high above DPC		
engineering brick (class B) PC £239.00/1000	m	130.00
brick PC £300.00/1000	m	140.00
brick PC £800.00/1000	m	190.00
Wall 1200 mm high above DPC		
engineering brick (class B) PC £239.00/1000	m	170.00
brick PC £300.00/1000	m	210.00
brick PC £800.00/1000	m	370.00
Wall 1800 mm high above DPC		
engineering brick (class B) PC £239.00/1000	m	225.00
brick PC £300.00/1000	m	300.00
brick PC £800.00/1000	m	600.00
One and a half brick wall; excavate 450 mm deep; remove arisings off site; lay GEN 1 concrete foundations 600 × 300 mm thick; two thick brick piers at 3.0 m centres; all in English Garden Wall bond; laid in cement: lime: sand (1:1:6) mortar with flush joints; fair face one side; DPC two courses engineering brick in cement: sand (1:3) mortar; coping of headers on edge		
Wall 900 mm high above DPC		
engineering brick (class B) PC £239.00/1000	m	530.00
brick PC £300.00/1000	m	180.00
Wall 1200 mm high above DPC		
engineering brick (class B) PC £239.00/1000	m	300.00
brick PC £300.00/1000	m	215.00
Wall 1800 mm high above DPC		
engineering brick (class B) PC £239.00/1000	m	370.00
brick PC £300.00/1000	m	330.00
brick PC £800.00/1000	m	910.00
BLOCK WALLING		
Notes: Measurements allow for works above ground only		
Concrete block walls; including excavation of foundation trench 450 mm deep; remove spoil off site; lay GEN 1 concrete foundations 600 × 300 mm thick; 1 block below ground		
Solid blocks 7 N/mm^2; 100 mm thick		
500 mm high	m^2	47.00
750 mm high	m^2	55.00
1.00 m high	m^2	63.00
1.25 m high	m^2	70.00
1.50 m high	m^2	78.00
1.80 m high	m^2	87.00

BRICK/BLOCK WALLING

Item Excluding site overheads and profit	Unit	Total rate £
BLOCK WALLING – cont		
Concrete block walls – cont		
140 mm thick		
140 mm thick	m²	63.00
100 mm blocks; laid on flat		
500 mm high	m²	72.00
750 mm high	m²	92.00
100 mm blocks laid flat; 2 courses underground		
1.00 m high	m²	120.00
1.20 m high	m²	130.00
1.50 m high	m²	155.00
1.80 m high	m²	180.00
Hollow blocks filled with concrete; 440 × 215 × 215 mm		
500 mm high	m²	71.00
750 mm high	m²	87.00
1.00 m high	m²	105.00
1.25 m high	m²	115.00
1.50 m high	m²	135.00
1.80 m high	m²	155.00
Concrete block retaining walls; 2 courses underground; including foundation trench 450 mm deep; remove spoil off site; lay GEN 1 concrete foundations 1200 × 400 mm thick		
Solid blocks 7 N/mm²; 100 mm thick		
1.00 m high	m	120.00
1.50 m high	m	140.00
1.50 m high	m	150.00
100 mm blocks; laid on flat		
1.00 m high	m²	180.00
1.50 m high	m²	220.00
1.80 m high	m²	240.00
Hollow blocks filled with concrete; 215 mm thick		
1.00 m high	m²	195.00
1.50 m high	m²	240.00
1.80 m high	m²	255.00
Hollow blocks with steel bar cast into the foundation		
1.00 m high	m²	220.00
1.50 m high	m²	270.00
1.80 m high	m²	285.00
Concrete block wall with brick face 112.5 mm thick to stretcher bond; including excavation of foundation trench 450 mm deep; remove spoil off site; lay GEN 1 concrete foundations 600 × 300 mm thick; place stainless steel ties at 4 nr/m² of wall face		
Solid blocks 7 N/mm² with stretcher bond brick face; reclaimed bricks PC £800.00/1000		
100 mm thick	m²	170.00
140 mm thick	m²	180.00

BRICK/BLOCK WALLING

Item Excluding site overheads and profit	Unit	Total rate £
BRICK/BLOCK WALLING WITH PIERS		
Half brick thick wall; 102.5 mm thick; with one brick piers at 2.00 m centres; excavate foundation trench 500 mm deep; remove spoil off site; lay site mixed concrete foundations 1:3:6 350 × 150 mm thick; laid in cement: lime: sand (1:1:6) mortar with flush joints; fair face one side; DPC two courses underground; engineering brick in cement: sand (1:3) mortar; coping of headers on end		
Wall 900 mm high above DPC		
engineering brick (class B) PC £239.00/1000	m	155.00
brick PC £300.00/1000	m	97.00
brick PC £800.00/1000	m	130.00
One brick thick wall; 225 mm thick; with one and a half brick piers at 3.0 m centres; excavation 400 mm deep; remove arisings off site; lay GEN 1 concrete foundations 450 mm wide × 250 mm thick; all in English Garden Wall bond; laid in cement: lime: sand (1:1:6) mortar with flush joints; fair face one side; DPC two courses engineering brick in cement: sand (1:3) mortar; precast concrete coping 152 × 75 mm		
Wall 900 mm high above DPC		
engineering brick (class B) PC £239.00/1000	m	260.00
brick PC £300.00/1000	m	175.00
brick PC £800.00/1000	m	400.00
Wall 1200 mm high above DPC		
engineering brick (class B) PC £239.00/1000	m	290.00
brick PC £300.00/1000	m	205.00
brick PC £800.00/1000	m	395.00
Wall 1800 mm high above DPC		
engineering brick (class B) PC £239.00/1000	m	420.00
brick PC £300.00/1000	m	300.00
brick PC £800.00/1000	m	580.00
BRICK PIERS		
Brick piers on concrete footings 500 × 500 × 250 mm thick; inclusive of excavations and disposal off site; coping of engineering brick on edge; 2 courses underground		
Pier 215 × 215 mm; one brick thick; brick PC £300/1000		
500 mm high	nr	41.00
750 mm high	nr	51.00
1.00 m high	nr	62.00
1.25 m high	nr	72.00
1.50 m high	nr	82.00
1.80 m high	nr	94.00
Pier 215 × 215 mm; one brick thick; brick PC £800/1000		
500 mm high	nr	51.00
750 mm high	nr	65.00
1.00 m high	nr	78.00
1.25 m high	nr	92.00
1.50 m high	nr	105.00
1.80 m high	nr	120.00

BRICK/BLOCK WALLING

Item Excluding site overheads and profit	Unit	Total rate £
BRICK PIERS – cont		
Brick piers on concrete footings 500 × 500 × 250 mm thick – cont		
Pier 337.5 × 337.5 mm; one and a half brick thick; brick PC £300/1000		
500 mm high	nr	71.00
750 mm high	nr	93.00
1.00 m high	nr	110.00
1.25 m high	nr	135.00
1.50 m high	nr	155.00
1.80 m high	nr	180.00
Pier 337.5 × 337.5 mm; one and a half brick thick; brick PC £800/1000		
500 mm high	nr	230.00
750 mm high	nr	310.00
1.00 m high	nr	395.00
1.25 m high	nr	475.00
1.50 m high	nr	560.00
1.80 m high	nr	660.00
Brick piers on concrete footings 800 × 800 × 600 mm thick; inclusive of excavations and disposal off site; coping of Haddonstone S150C classically moulded corbelled piercap PC £114.00; 120 mm thick overall		
Pier 450 × 450 mm; two bricks thick; brick PC £300/1000; height of brickwork above ground		
500 mm high	nr	265.00
750 mm high	nr	300.00
1.00 m high	nr	330.00
1.25 m high	nr	365.00
1.50 m high	nr	400.00
1.80 m high	nr	435.00
2.00 m high	nr	460.00
Pier 450 × 450 mm; two bricks thick; brick PC £800/1000; height of brickwork above ground		
500 mm high	nr	300.00
750 mm high	nr	350.00
1.00 m high	nr	395.00
1.25 m high	nr	440.00
1.50 m high	nr	490.00
1.80 m high	nr	550.00
2.00 m high	nr	580.00
Brick piers on concrete footings 1000 × 1000 × 600 mm thick; inclusive of excavations and disposal off site; coping Haddonstone S215C classically moulded corbelled pyramidal piercap PC £214.00		
Pier 600 × 600 mm; three bricks thick; brick PC £800/1000; height of brickwork above ground		
1.50 m high	nr	970.00
1.80 m high	nr	1075.00
2.00 m high	nr	1175.00

BRICK/BLOCK WALLING

Item Excluding site overheads and profit	Unit	Total rate £
BLOCK PIERS		
Block piers; blocks 440 × 215 × 100 mm thick laid on flat piers on concrete footings 600 × 600 × 250 mm thick; inclusive of excavations and disposal off site; coping of engineering brick on edge; 2 courses underground		
Pier 450 × 450 mm; to receive cladding treatment priced separately	nr	56.00
500 mm high	nr	51.00
750 mm high	nr	62.00
1.00 m high	nr	72.00
1.25 m high	nr	82.00
1.50 m high	nr	94.00
1.80 m high		
FACED BLOCK WALLING		
Concrete block wall with brick face 112.5 mm thick to bond pattern using snap headers; including excavation of foundation trench 450 mm deep; remove spoil off site; lay GEN 1 concrete foundations 600 × 300 mm thick; place stainless steel ties at 4 nr/m² of wall face; lay coping of brick on edge in engineering brick		
Solid blocks 7 N/mm² with snap header brick face; bricks PC £800.00/1000	m²	110.00
500 mm high	m²	140.00
750 mm high	m²	160.00
1.00 m high	m²	220.00
1.50 m high	m²	270.00
1.80 m high		

ROADS AND PAVINGS

Item Excluding site overheads and profit	Unit	Total rate £
BASES FOR PAVING		
Excavate ground and reduce levels to receive 38 mm thick slab and 25 mm mortar bed; dispose of excavated material off site		
Lay granular fill Type 1; limestone aggregate 150 mm thick laid to falls and compacted		
all by machine	m²	10.60
all by hand except disposal by grab	m²	44.00
Lay granular fill Type 1; recycled material; 150 mm thick laid to falls and compacted		
all by machine	m²	8.65
all by hand except disposal by grab	m²	42.00
Lay 1:2:4 concrete base 150 mm thick laid to falls		
all by machine	m²	22.00
all by hand except disposal by grab	m²	56.00
150 mm hardcore base with concrete base 150 mm deep		
concrete 1:2:4 site mixed	m²	29.00
concrete PAV 1 35 ready mixed	m²	27.00
Hardcore base 150 mm deep; concrete reinforced with A142 mesh		
site mixed concrete 1:2:4; 150 mm deep	m²	43.00
site mixed concrete 1:2:4; 250 mm deep	m²	59.00
concrete PAV 1 35 N/mm ready mixed; 150 mm deep	m²	33.50
concrete PAV 1 35 N/mm ready mixed; 250 mm deep	m²	48.00
KERBS AND EDGINGS		
Note: excavation is by machine unless otherwise mentioned		
Excavate trench and construct concrete foundation 300 mm wide × 150 mm deep; lay precast concrete kerb units bedded in semi-dry concrete; slump 35 mm maximum; haunching one side; disposal of arisings off site		
Kerbs laid straight		
125 mm high × 125 mm thick; bullnosed; type BN	m	25.50
125 × 255 mm; ref HB2; SP	m	27.50
150 × 305 mm; ref HB1	m	33.00
Excavate and construct concrete foundation 450 mm wide × 150 mm deep; lay precast concrete kerb units bedded in semi-dry concrete; slump 35 mm maximum; haunching one side; lay channel units bedded in 1:3 sand mortar; jointed in cement: sand mortar 1:3		
Kerbs; 125 mm high × 125 mm thick; bullnosed; type BN		
dished channel; 125 × 225 mm; Ref CS	m	46.50
square channel; 125 × 150 mm; Ref CS2	m	44.00
bullnosed channel; 305 × 150 mm; Ref CBN	m	54.00
Granite kerbs		
straight; 125 × 250 mm	m	50.00
curved; 125 × 250 mm	m	55.00
Brick channel; class B engineering bricks		
Excavate and construct concrete foundation 600 mm wide × 200 mm deep; lay channel to depths and falls bricks to be laid as headers along the channel; bedded in 1:3 lime: sand mortar bricks close jointed in cement: sand mortar 1:3		
3 courses wide	m	60.00

ROADS AND PAVINGS

Item Excluding site overheads and profit	Unit	Total rate £
Precast concrete edging on concrete foundation 100 × 150 mm deep and haunching one side 1:2:4 including all necessary excavation disposal and formwork		
Rectangular, chamfered or bullnosed; to one side of straight path		
50 × 150 mm	m	22.00
50 × 200 mm	m	27.00
50 × 250 mm	m	27.50
Rectangular, chamfered or bullnosed as above but to both sides of straight paths		
50 × 150 mm	m	47.00
50 × 200 mm	m	54.00
50 × 250 mm	m	55.00
Timber edgings; softwood		
Straight		
150 × 38 mm	m	6.50
150 × 50 mm	m	7.20
Curved; over 5 m radius		
150 × 38 mm	m	8.10
150 × 50 mm	m	9.50
Curved; 4–5 m radius		
150 × 38 mm	m	8.70
150 × 50 mm	m	10.10
Curved; 3–4 m radius		
150 × 38 mm	m	9.70
150 × 50 mm	m	11.40
Curved; 1–3 m radius		
150 × 38 mm	m	11.30
150 × 50 mm	m	13.10
Timber edgings; hardwood		
150 × 38 mm	m	10.50
150 × 50 mm	m	13.40
Brick or concrete block edge restraint; excavate for foundation; lay concrete 1:2:4 200 mm wide × 150 mm deep; on 50 mm thick sharp sand bed; lay blocks or bricks inclusive of haunching one side		
Blocks 200 × 100 × 60 mm; PC £7.63/m²; butt jointed		
header course	m	14.60
stretcher course	m	11.50
Bricks 215 × 112.5 × 50 mm; PC £300.00/1000; with mortar joints		
header course	m	16.20
stretcher course	m	15.30
Sawn yorkstone edgings; excavate for foundations; lay concrete 1:2:4 150 mm deep × 33.3% wider than the edging; on 35 mm thick mortar bed; inclusive of haunching one side		
Yorkstone 50 mm thick		
100 mm wide × random lengths	m	22.00
100 × 100 mm	m	26.00
100 mm wide × 200 mm long	m	33.50
250 mm wide × random lengths	m	35.00
500 mm wide × random lengths	m	57.00

ROADS AND PAVINGS

Item Excluding site overheads and profit	Unit	Total rate £
KERBS AND EDGINGS – cont		
Granite edgings; excavate for footings; lay concrete 1:2:4 150 mm deep × 33.3% wider than the edging; on 35 mm thick mortar bed; inclusive of haunching one side		
Granite 50 mm thick		
100 mm wide × random lengths	m	22.00
100 × 100 mm	m	26.00
100 mm wide × 200 mm long	m	33.50
250 mm wide × random lengths	m	35.00
500 mm wide × random lengths	m	57.00
ROADS		
Excavate weak points of excavation by hand and fill with well rammed Type 1 granular material		
100 mm thick	m²	9.00
200 mm thick	m²	11.80
Reinforced concrete roadbed 150 mm thick		
Excavate road bed and dispose excavated material off site; bring to grade; lay reinforcing mesh A142 lapped and joined; lay 150 mm thick in situ reinforced concrete roadbed 21 N/mm²; with 15 impregnated fibreboard expansion joints and polysulphide based sealant at 50 m centres; on 150 mm hardcore blinded with ash or sand; kerbs 155 × 255 mm to both sides; including foundations haunched one side in 11.5 N/mm² concrete with all necessary formwork; falls, crossfalls and cambers not exceeding 15°		
4.90 m wide	m	220.00
6.10 m wide	m	290.00
7.32 m wide	m	310.00
Macadam roadway over 1000 m²		
Excavate 350 mm for pathways or roadbed; bring to grade; lay 100 mm well-rolled hardcore; lay 150 mm Type 1 granular material; lay precast edgings kerbs 50 × 150 mm on both sides; including foundations haunched one side in 11.5 N/mm² concrete with all necessary formwork; machine lay surface of 90 mm macadam 60 mm base course and 30 mm wearing course; all disposal off site		
1.50 m wide	m	93.00
2.00 m wide	m	98.00
3.00 m wide	m	135.00
4.00 m wide	m	165.00
5.00 m wide	m	195.00
6.00 m wide	m	225.00
7.00 m wide	m	255.00
Macadam roadway 400–1000 m²		
Excavate 350 mm for pathways or roadbed; bring to grade; lay 100 mm well-rolled hardcore; lay 150 mm Type 1 granular material; lay precast edgings kerbs 50 × 150 mm on both sides; including foundations haunched one side in 11.5 N/mm² concrete with all necessary formwork; machine lay surface of 90 mm macadam 60 mm base course and 30 mm wearing course; all disposal off site		
1.50 m wide	m	98.00
2.00 m wide	m	110.00
3.00 m wide	m	145.00

ROADS AND PAVINGS

Item Excluding site overheads and profit	Unit	Total rate £
4.00 m wide	m	180.00
5.00 m wide	m	210.00
6.00 m wide	m	245.00
7.00 m wide	m	280.00
CAR PARKS		
Excavate 350 mm for pathways or roadbed to receive surface 100 mm thick priced separately; bring to grade; lay 100 mm well-rolled hardcore; lay 150 mm Type 1 granular material; lay kerbs 125 × 255 mm on both sides; including foundations haunched one side in 11.5 N/mm² concrete with all necessary formwork		
Work to falls, crossfalls and cambers not exceeding 15°		
1.50 m wide	m	75.00
2.00 m wide	m	84.00
3.00 m wide	m	99.00
4.00 m wide	m	110.00
5.00 m wide	m	130.00
6.00 m wide	m	140.00
7.00 m wide	m	160.00
As above but excavation 450 mm deep and base of Type 1 at 250 mm thick		
1.50 m wide	m	135.00
2.00 m wide	m	160.00
3.00 m wide	m	220.00
4.00 m wide	m	270.00
5.00 m wide	m	325.00
6.00 m wide	m	380.00
7.00 m wide	m	435.00
To excavated and prepared base above, lay roadbase of 40 mm size dense bitumen macadam 70 mm thick; lay wearing course of 10 mm size dense bitumen macadam 30 mm thick; mark out car parking bays 5.0 × 2.4 m with thermoplastic road paint; surfaces all mechanically laid		
Per bay; 5.0 × 2.4 m	each	210.00
Gangway	m²	14.40
As above but stainless steel road studs 100 × 100 mm; two per bay in lieu of thermoplastic paint		
per bay; 5.0 × 2.4 m	each	215.00
Car park as above but with interlocking concrete blocks 200 × 100 × 80 mm; grey		
per bay; 5.0 × 2.4 m	each	465.00
gangway	m²	38.50
Car park as above but with interlocking concrete blocks 200 × 100 × 80 mm; colours		
per bay; 5.0 × 2.4 m	each	480.00
gangway	m²	40.00
Car park as above but with interlocking concrete blocks 200 × 100 × 60 mm; grey		
per bay; 5.0 × 2.4 m	each	450.00
gangway	m²	37.50
Car park as above but with interlocking concrete blocks 200 × 100 × 60 mm; colours		
per bay; 5.0 × 2.4 m	each	460.00
gangway	m²	38.50

ROADS AND PAVINGS

Item Excluding site overheads and profit	Unit	Total rate £
CAR PARKS – cont		
Car park as above but with laying of Grass Concrete Grasscrete in situ continuously reinforced cellular surfacing; including expansion joints at 10 m centres; fill with topsoil and peat (5:1) and fertilizer at 35 g/m²; seed with dwarf rye grass at 35 g/m²		
GC1; 100 mm thick for cars and light traffic		
per bay; 5.0 × 2.4 m	each	385.00
gangway	m²	32.00
GC2; 150 mm thick for HGV traffic including dust carts		
per bay; 5.0 × 2.4 m	each	480.00
gangway	m²	40.00
CAR PARKING FOR DISABLED PEOPLE		
To excavated and prepared base above, lay roadbase of 20 mm size dense bitumen macadam 80 mm thick; lay wearing course of 10 mm size dense bitumen macadam 30 mm thick; mark out car parking bays with thermoplastic road paint		
per bay; 5.80 × 3.25 ambulant	each	270.00
per bay; 6.55 × 3.80 wheelchair	each	360.00
gangway	m²	14.40
Car park as above but with interlocking concrete blocks 200 × 100 × 60 mm; grey; but mark out bays		
per bay; 5.80 × 3.25 ambulant	each	3600.00
per bay; 6.55 × 3.80 wheelchair	each	4200.00
gangway	m²	37.50
PLAY AREA FOR BALL GAMES		
Excavate for playground; bring to grade; lay 225 mm consolidated hardcore; blind with 100 mm type 1; lay macadam base of 40 mm size dense bitumen macadam to 75 mm thick; lay wearing course 30 mm thick		
Over 1000 m²	100 m²	3500.00
400–1000 m²	100 m²	3600.00
Excavate for playground; bring to grade; lay 100 mm consolidated hardcore; blind with 100 mm type 1; lay macadam base of dense bitumen macadam to 50 mm thick; lay wearing course 20 mm thick		
Over 1000 m²	100 m²	2650.00
400–1000 m²	100 m²	3200.00
Excavate for playground; bring to grade; lay 100 mm consolidated hardcore; blind with 100 mm type 1; lay macadam base of dense bitumen macadam to 50 mm thick; lay wearing course of Addagrip resin coated aggregate 3 mm diameter × 6 mm thick		
Over 1000 m²	100 m²	4500.00
400–1000 m²	100 m²	4950.00
Safety surfaced play area		
Excavate for playground; bring to grade; lay with 150 mm type 1; lay macadam base of dense bitumen macadam to 50 mm thick		
lay safety surface 35 mm; coloured	100 m²	9100.00

ROADS AND PAVINGS

Item Excluding site overheads and profit	Unit	Total rate £
INTERLOCKING BLOCK PAVING		
Edge restraint to block paving; excavate for groundbeam; lay concrete 1:2:4 200 mm wide × 150 mm deep; on 50 mm thick sharp sand bed; inclusive of haunching one side		
Blocks 200 × 100 × 60 mm; PC £7.63/m²; butt jointed		
header course	m	14.60
stretcher course	m	11.50
Bricks 215 × 112.5 × 50 mm; PC £300.00/1000; with mortar joints		
header course	m	16.20
stretcher course	m	15.30
Excavate ground; supply and lay granular fill Type 1 150 mm thick laid to falls and compacted; supply and lay block pavers; laid on 50 mm compacted sharp sand; vibrated; joints filled with loose sand excluding edgings or kerbs measured separately		
Concrete blocks		
200 × 100 × 60 mm	m²	49.00
200 × 100 × 80 mm	m²	51.00
Reduce levels; lay 150 mm granular material Type 1; lay vehicular block paving to 90° herringbone pattern; on 50 mm compacted sand bed; vibrated; jointed in sand and vibrated; excavate and lay precast concrete edging 50 × 150 mm; on concrete foundation 1:2:4		
Blocks 200 × 100 × 60 mm		
1.0 m wide clear width between edgings	m	93.00
1.5 m wide clear width between edgings	m	120.00
2.0 m wide clear width between edgings	m	140.00
Blocks 200 × 100 × 60 mm but blocks laid 45° herringbone pattern including cutting edging blocks		
1.0 m wide clear width between edgings	m	99.00
1.5 m wide clear width between edgings	m	125.00
2.0 m wide clear width between edgings	m	150.00
Blocks 200 × 100 × 60 mm but all excavation by hand disposal off site by grab		
1.0 m wide clear width between edgings	m	200.00
1.5 m wide clear width between edgings	m	235.00
2.0 m wide clear width between edgings	m	270.00
BRICK PAVING		
WORKS BY MACHINE		
Excavate and lay base Type 1 150 mm thick remove arisings; all by machine; lay clay brick paving		
200 × 100 × 50 mm thick; butt jointed on 50 mm sharp sand bed		
PC £300.00/1000	m²	60.00
PC £600.00/1000	m²	75.00
200 × 100 × 50 mm thick; 10 mm mortar joints on 35 mm mortar bed		
PC £300.00/1000	m²	74.00
PC £600.00/1000	m²	87.00
Excavate and lay base Type 1 250 mm thick all by machine; remove arisings; lay clay brick paving		
200 × 100 × 50 mm thick; 10 mm mortar joints on 35 mm mortar bed		
PC £300.00/1000	m²	80.00

ROADS AND PAVINGS

Item Excluding site overheads and profit	Unit	Total rate £
BRICK PAVING – cont		
Excavate and lay 150 mm readymix concrete base reinforced with A393 mesh; all by machine; remove arisings; lay clay brick paving		
200 × 100 × 50 mm thick; 10 mm mortar joints on 35 mm mortar bed; running or stretcher bond		
PC £300.00/1000	m²	93.00
PC £600.00/1000	m²	105.00
200 × 100 × 50 mm thick; 10 mm mortar joints on 35 mm mortar bed; butt jointed; herringbone bond		
PC £300.00/1000	m²	80.00
PC £600.00/1000	m²	95.00
Excavate and lay base readymix concrete base 150 mm thick reinforced with A393 mesh; all by machine; remove arisings; lay clay brick paving		
215 × 102.5 × 50 mm thick; 10 mm mortar joints on 35 mm mortar bed		
PC £300.00/1000; herringbone	m²	95.00
PC £600.00/1000; herringbone	m²	110.00
WORKS BY HAND		
Excavate and lay base Type 1 150 mm thick by hand; arisings barrowed to spoil heap maximum distance 25 m and removal off site by grab; lay clay brick paving		
200 × 100 × 50 mm thick; butt jointed on 50 mm sharp sand bed		
PC £300.00/1000	m²	94.00
PC £600.00/1000	m²	120.00
Excavate and lay 150 mm concrete base; 1:3:6: site mixed concrete reinforced with A393 mesh; remove arisings to stockpile and then off site by grab; lay clay brick paving		
215 × 102.5 × 50 mm thick; 10 mm mortar joints on 35 mm mortar bed		
PC £300.00/1000	m²	94.00
PC £600.00/1000	m²	120.00
FLAG PAVING TO PEDESTRIAN AREAS Prices are inclusive of all mechanical excavation and disposal.		
Supply and lay precast concrete flags; excavate ground and reduce levels; treat substrate with total herbicide; supply and lay granular fill Type 1 150 mm thick laid to falls and compacted		
Standard precast concrete flags bedded and jointed in lime: sand mortar (1:3)		
450 × 450 × 70 mm; chamfered	m²	39.00
450 × 450 × 50 mm; chamfered	m²	35.50
600 × 300 × 50 mm	m²	34.00
600 × 450 × 50 mm	m²	33.00
600 × 600 × 50 mm	m²	30.50
750 × 600 × 50 mm	m²	30.00
900 × 600 × 50 mm	m²	29.00

ROADS AND PAVINGS

Item Excluding site overheads and profit	Unit	Total rate £
Coloured flags bedded and jointed in lime: sand mortar (1:3)		
600 × 600 × 50 mm	m²	32.50
450 × 450 × 70 mm; chamfered	m²	44.00
400 × 400 × 65 mm	m²	44.50
750 × 600 × 50 mm	m²	32.00
900 × 600 × 50 mm	m²	31.00
Marshalls Saxon; textured concrete flags; reconstituted yorkstone in colours; butt jointed bedded in lime: sand mortar (1:3)		
300 × 300 × 35 mm	m²	66.00
450 × 450 × 50 mm	m²	55.00
600 × 300 × 35 mm	m²	51.00
600 × 600 × 50 mm	m²	47.50
Tactile flags; Marshalls blister tactile pavings; red or buff; for blind pedestrian guidance laid to designed pattern		
450 × 450 mm	m²	46.50
400 × 400 mm	m²	50.00
PEDESTRIAN DETERRENT PAVING		
Excavate ground and bring to levels; treat substrate with total herbicide; supply and lay granular fill Type 1 150 mm thick laid to falls and compacted; supply and lay precast deterrent paving units bedded in lime: sand mortar (1:3) and jointed in lime: sand mortar (1:3)		
Marshalls Mono		
Lambeth pyramidal paving; 600 × 600 × 75 mm	m²	38.00
Townscape		
Abbey square cobble pattern pavings; reinforced; 600 × 600 × 60 mm	m²	66.00
Geoset chamfered studs; 600 × 600 × 60 mm	m²	62.00
IMITATION YORKSTONE PAVING		
Excavate ground and bring to levels; treat substrate with total herbicide; supply and lay granular fill Type 1 150 mm thick laid to falls and compacted; supply and lay imitation yorkstone paving laid to coursed patterns bedded in lime: sand mortar (1:3) and jointed in lime: sand mortar (1:3)		
Marshalls Heritage; square or rectangular		
300 × 300 × 38 mm	m²	69.00
600 × 300 × 30 mm	m²	56.00
600 × 450 × 38 mm	m²	53.00
450 × 450 × 38 mm	m²	51.00
600 × 600 × 38 mm	m²	47.00
As above but laid to random rectangular patterns		
various sizes selected from the above	m²	59.00
Imitation yorkstone laid random rectangular as above but on concrete base 150 mm thick		
by machine	m²	70.00
by hand	m²	100.00

ROADS AND PAVINGS

Item Excluding site overheads and profit	Unit	Total rate £
NATURAL STONE SLAB PAVING		
WORKS BY MACHINE (For works by hand please see Minor Works)		
Excavate ground by machine and reduce levels; to receive 65 mm thick slab and 35 mm mortar bed; dispose of excavated material off site; treat substrate with total herbicide; lay granular fill Type 1 150 mm thick laid to falls and compacted; lay to random rectangular pattern on 35 mm mortar bed		
New riven slabs		
laid random rectangular	m²	125.00
New riven slabs; but to 150 mm plain concrete base		
laid random rectangular	m²	135.00
New riven slabs; disposal by grab		
laid random rectangular	m²	130.00
Reclaimed Cathedral grade riven slabs		
laid random rectangular	m²	160.00
Reclaimed Cathedral grade riven slabs; but to 150 mm plain concrete base		
laid random rectangular	m²	170.00
Reclaimed Cathedral grade riven slabs; disposal by grab		
laid random rectangular	m²	165.00
New slabs sawn 6 sides		
laid random rectangular	m²	100.00
three sizes; laid to coursed pattern	m²	110.00
New slabs sawn 6 sides; but to 150 mm plain concrete base		
laid random rectangular	m²	115.00
three sizes; laid to coursed pattern	m²	120.00
New slabs sawn 6 sides; disposal by grab		
laid random rectangular	m²	105.00
3 sizes, laid to coursed pattern	m²	110.00
WORKS BY HAND		
Excavate ground by hand and reduce levels, to receive 65 mm thick slab and 35 mm mortar bed; barrow all materials and arisings 25 m; dispose of excavated material off site by grab; treat substrate with total herbicide; lay granular fill Type 1 150 thick laid to falls and compacted; lay to random rectangular pattern on 35 mm mortar bed		
New riven slabs		
laid random rectangular	m²	160.00
New riven slabs laid random rectangular; but to 150 mm plain concrete base		
laid random rectangular	m²	170.00
New riven slabs; but disposal to skip		
laid random rectangular	m²	175.00
Reclaimed Cathedral grade riven slabs		
laid random rectangular	m²	200.00
Reclaimed Cathedral grade riven slabs; but to 150 mm plain concrete base		
laid random rectangular	m²	210.00
Reclaimed Cathedral grade riven slabs; but disposal to skip		
laid random rectangular	m²	215.00
New slabs sawn 6 sides		
laid random rectangular	m²	135.00
three sizes; sawn 6 sides laid to coursed pattern	m²	140.00

ROADS AND PAVINGS

Item Excluding site overheads and profit	Unit	Total rate £
New slabs sawn 6 sides; but to 150 mm plain concrete base		
laid random rectangular	m²	150.00
three sizes; sawn 6 sides laid to coursed pattern	m²	155.00
GRANITE SETT PAVING – PEDESTRIAN		
Excavate ground and bring to levels; lay 100 mm hardcore to falls; compacted with 5 tonne roller; blind with compacted Type 1 50 mm thick; lay 100 mm concrete 1:2:4; supply and lay granite setts 100 × 100 × 100 mm bedded in cement: sand mortar (1:3) 25 mm thick minimum; close butted and jointed in fine sand; all excluding edgings or kerbs measured separately		
Setts laid to bonded pattern and jointed		
new setts 100 × 100 × 100 mm	m²	100.00
second-hand cleaned 100 × 100 × 100 mm	m²	110.00
Setts laid in curved pattern		
new setts 100 × 100 × 100 mm	m²	105.00
second-hand cleaned 100 × 100 × 100 mm	m²	115.00
GRANITE SETT PAVING – TRAFFICKED AREAS		
Excavate ground and bring to levels; lay 100 mm hardcore to falls; compacted with 5 tonne roller; blind with compacted Type 1 50 mm thick; lay 150 mm site mixed concrete 1:2:4 reinforced with steel fabric; supply and lay granite setts 100 × 100 × 100 mm bedded in cement: sand mortar (1:3) 25 mm thick minimum; close butted and jointed in fine sand; all excluding edgings or kerbs measured separately		
Site mixed concrete		
new setts 100 × 100 × 100 mm	m²	110.00
second-hand cleaned 100 × 100 × 100 mm	m²	120.00
Ready mixed concrete		
new setts 100 × 100 × 100 mm	m²	105.00
second-hand cleaned 100 × 100 × 100 mm	m²	115.00
CONCRETE PAVING		
Pedestrian areas		
Excavate to reduce levels; lay 100 mm Type 1; lay PAV 1 air entrained concrete; joints at maximum width of 6.0 m cut out and sealed with sealant; inclusive of all formwork and stripping		
100 mm thick	m²	63.00
Trafficked areas		
Excavate to reduce levels; lay 150 mm Type 1 lay PAV 2 40 N/mm² air entrained concrete; reinforced with steel mesh 200 × 200 mm square at 2.22 kg/m²; joints at max width of 6.0 m cut out and sealed with sealant		
150 mm thick	m²	81.00

ROADS AND PAVINGS

Item Excluding site overheads and profit	Unit	Total rate £
BEACH COBBLE PAVING		
Excavate ground and bring to levels and fill with compacted Type 1 fill 100 mm thick; lay GEN 1 concrete base 100 mm thick; supply and lay cobbles individually laid by hand bedded in cement: sand mortar (1:3) 25 mm thick minimum; dry grout with 1:3 cement: sand grout; brush off surplus grout and water in; sponge off cobbles as work proceeds; all excluding formwork edgings or kerbs measured separately		
Scottish beach cobbles 50–75 mm	m²	110.00
Scottish beach cobbles 100–200 mm	m²	115.00
Kidney flint cobbles 75–100 mm	m²	110.00
CONCRETE SETT PAVING		
Excavate ground and bring to levels; supply and lay 150 mm granular fill Type 1 laid to falls and compacted; supply and lay setts bedded in 50 mm sand and vibrated; joints filled with dry sand and vibrated; all excluding edgings or kerbs measured separately		
Marshalls Mono; Tegula precast concrete setts		
random sizes 60 mm thick	m²	45.00
single size 60 mm thick	m²	43.00
random size 80 mm thick	m²	49.50
single size 80 mm thick	m²	61.00
GRASS CONCRETE PAVING		
Reduce levels; lay 150 mm Type 1 granular fill compacted; supply and lay precast grass concrete blocks on 20 sand and level by hand; fill blocks with 3 mm sifted topsoil and pre-seeding fertilizer at 50 g/m²; sow with perennial ryegrass/chewings fescue seed at 35 g/m²		
Grass concrete		
GB103 406 × 406 × 103 mm	m²	37.00
GB83 406 × 406 × 83 mm	m²	34.00
extra for geotextile fabric underlayer	m²	0.75
Firepaths		
Excavate to reduce levels; lay 300 mm well rammed hardcore; blinded with 100 mm type 1; supply and lay Marshalls Mono Grassguard 180 precast grass concrete blocks on 50 mm sand and level by hand; fill blocks with sifted topsoil and pre-seeding fertilizer at 50 g/m²		
firepath 3.8 m wide	m	230.00
firepath 4.4 m wide	m	265.00
firepath 5.0 m wide	m	300.00
turning areas	m²	61.00
Charcon Hard Landscaping Grassgrid; 366 × 274 × 100 mm	m²	31.50

ROADS AND PAVINGS

Item Excluding site overheads and profit	Unit	Total rate £
SLAB/BRICK PATHS		
Stepping stone path inclusive of hand excavation and mechanical disposal, 100 mm Type 1 and sand blinding; slabs laid 100 mm apart to existing turf		
600 mm wide; 600 × 600 × 50 mm slabs		
natural finish	m	45.50
coloured	m	46.00
exposed aggregate	m	55.00
900 mm wide; 600 × 900 × 50 mm slabs		
natural finish	m	50.00
coloured	m	52.00
exposed aggregate	m	66.00
Pathway inclusive of mechanical excavation and disposal; 100 mm Type 1 and sand blinding; slabs close butted		
Straight butted path 900 mm wide; 600 × 900 × 50 mm slabs		
natural finish	m	18.70
coloured	m	19.70
exposed aggregate	m	29.00
Straight butted path 1200 mm wide; double row; 600 × 900 × 50 mm slabs; laid stretcher bond		
natural finish	m	33.50
coloured	m	36.00
exposed aggregate	m	57.00
Straight butted path 1200 mm wide; one row of 600 × 600 × 50 mm; one row 600 × 900 × 50 mm slabs		
natural finish	m	32.50
coloured	m	34.50
exposed aggregate	m	63.00
Straight butted path 1500 mm wide; slabs of 600 × 900 × 50 mm and 600 × 600 × 50 mm; laid to bond		
natural finish	m	40.50
coloured	m	42.00
exposed aggregate	m	63.00
Straight butted path 1800 wide; two rows of 600 × 900 × 50 slabs; laid bonded		
natural finish	m	44.00
coloured	m	54.00
exposed aggregate	m	85.00
Straight butted path 1800 mm wide; three rows of 600 × 900 × 50 mm slabs; laid stretcher bond		
natural finish	m	49.00
coloured	m	58.00
exposed aggregate	m	90.00
Brick paved paths; bricks 215 × 112.5 × 65 mm with mortar joints; prices are inclusive of excavation, disposal off site, 100 mm hardcore, 100 mm 1:2:4 concrete bed; edgings of brick 215 mm wide, haunched; all jointed in cement: lime: sand mortar (1:1:6)		
Path 1015 mm wide laid stretcher bond; edging course of headers		
rough stocks PC £450.00/1000	m	82.00
engineering brick PC £239.00/1000	m	77.00

ROADS AND PAVINGS

Item Excluding site overheads and profit	Unit	Total rate £
SLAB/BRICK PATHS – cont		
Brick paved paths – cont		
Path 1015 mm wide laid stack bond		
rough stocks PC £450.00/1000	m	86.00
engineering brick PC £239.00/1000	m	82.00
Path 1115 mm wide laid header bond		
rough stocks PC £450.00/1000	m	90.00
engineering brick PC £239.00/1000	m	85.00
Path 1330 mm wide laid basketweave bond		
rough stocks PC £450.00/1000	m	100.00
engineering brick PC £239.00/1000	m	97.00
Path 1790 mm wide laid basketweave bond		
rough stocks PC £450.00/1000	m	135.00
engineering brick PC £239.00/1000	m	99.00
Brick paved paths; brick paviors 200 × 100 × 50 mm chamfered edge with butt joints;		
prices are inclusive of excavation, 100 mm Type 1 and 50 mm sharp sand; jointing in		
kiln dried sand brushed in; exclusive of edge restraints		
Path 1000 mm wide laid stretcher bond; edging course of headers		
rough stocks PC £450.00/1000	m	75.00
engineering brick PC £239.00/1000	m	78.00
Path 1330 mm wide laid basketweave bond		
rough stocks PC £450.00/1000	m	105.00
engineering brick PC £239.00/1000	m	99.00
Path 1790 mm wide laid basketweave bond		
rough stocks PC £450.00/1000	m	140.00
engineering brick PC £239.00/1000	m	100.00
GRAVEL PATHS		
Reduce levels and remove spoil to dump on site; lay 150 mm hardcore well rolled; lay		
geofabric; fix timber edge 150 × 38 mm to both sides of straight paths; lay 50 mm		
Cedec gravel watered and rolled		
Lay Cedec gravel 50 mm thick; watered and rolled		
1.0 m wide	m	26.50
1.5 m wide	m	35.00
2.0 m wide	m	48.00
Lay Breedon gravel 50 mm thick; watered and rolled		
1.0 m wide	m	26.00
1.5 m wide	m	35.00
2.0 m wide	m	43.50

ROADS AND PAVINGS

Item Excluding site overheads and profit	Unit	Total rate £
BARK PAVING		
Excavate to reduce levels; remove all topsoil to dump on site; treat area with herbicide; lay 150 mm clean hardcore; blind with sand; lay 0.7 mm geotextile filter fabric water flow 50 l/m²/sec; supply and fix treated softwood edging boards 50 × 150 mm to bark area on hardcore base extended 150 mm beyond the bark area; boards fixed with galvanized nails to treated softwood posts 750 × 50 × 75 mm driven into firm ground at 1.0 m centres; edging boards to finish 25 mm above finished bark surface; tops of posts to be flush with edging boards and once weathered		
Supply and lay 100 mm Melcourt conifer walk chips 10-40 mm size		
1.00 m wide	m	21.00
2.00 m wide	m	33.00
3.00 m wide	m	45.00
4.00 m wide	m	55.00
Less for wood fibre	m²	−1.36
Less for hardwood chips	m²	−1.36
FOOTPATHS		
Excavate footpath to reduce level; remove arisings to tip on site maximum distance 25 m; lay Type 1 granular fill 100 mm thick; lay base course of 28 mm size dense bitumen macadam 50 mm thick; wearing course of 10 mm size dense bitumen macadam 20 mm thick; timber edge 150 × 38 mm		
Areas over 1000 m²		
1.0 m wide	m	30.00
1.5 m wide	m	41.00
2.0 m wide	m	52.00
Areas 400–1000 m²		
1.0 m wide	m	33.00
1.5 m wide	m	45.00
2.0 m wide	m	71.00

SPECIAL SURFACES FOR SPORT/PLAYGROUNDS

Item Excluding site overheads and profit	Unit	Total rate £
BARK PLAY AREA		
Excavate playground area to 450 mm depth; lay 150 mm broken stone or clean **hardcore; lay filter membrane; lay bark surface**		
Melcourt Industries		
Playbark 10/50; 300 mm thick	100 m²	3150.00
Playbark 8/25; 300 mm thick	100 m²	3700.00
BOWLING GREEN CONSTRUCTION		
Bowling green; complete		
Excavate 300 mm deep and grade to level; excavate and install 100 mm land drain to		
perimeter and backfill with shingle; install 60 mm land drain to surface at 4.5 m centres		
backfilled with shingle; install 50 m non perforated pipe 50 m long; install 100 mm		
compacted and levelled drainage stone, blind with grit and sand; spread 150 mm imported		
70:30 sand: soil accurately compacted and levelled; lay bowling green turf and top dress		
twice luted into surface; exclusive of perimeter ditches and bowls protection		
6 rink green 38.4 × 38.4 m	each	56000.00
install Toro automatic bowling green irrigation system with pump, tank, controller,		
electrics, pipework and 8 nr Toro 780 sprinklers; excluding pumphouse (optional)	each	8600.00
Supply and install to perimeter of green; Sportsmark preformed bowling green ditch channels		
Ultimate Design 99 steel reinforced concrete channel; 600 mm long section	each	8000.00
Supply and fit bowls protection material to rear hitting face of Ultimate channels 1 and 2 above		
Curl Grass artificial grass; 0.45 m wide	each	2450.00
Astroturf artificial grass; 0.45 m wide	each	2050.00
bowls protection ditch liner laid loose; 300 mm wide	each	1250.00
JOGGING TRACK		
Excavate track 250 mm deep; treat substrate with Casoron G4; lay filter membrane; lay		
100 mm depth gravel waste or similar; lay 100 mm compacted gravel	100 m²	5700.00
Extra for treated softwood edging 50 × 150 mm on 50 × 50 mm posts		
both sides	m	8.10
PLAYGROUNDS		
Excavate playground area and dispose of arisings to tip; lay Type 1 granular fill; lay **macadam surface two coat work 80 mm thick; base course of 28 mm size dense** **bitumen macadam 50 mm thick; wearing course of 10 mm size dense bitumen** **macadam 30 mm thick**		
Areas over 1000 m²		
excavation 180 mm base 100 mm thick	m²	23.00
excavation 225 mm base 150 mm thick	m²	26.00
Areas 400–1000 m²		
excavation 180 mm base 100 mm thick	m²	26.00
excavation 225 mm base 150 mm thick	m²	29.00
Excavate playground area to given levels and falls; remove soil off site and backfill **with compacted Type 1 granular fill; lay ready mixed concrete to fall 2% in all** **direction**		
Base 150 mm thick; surface 100 mm thick	m²	21.50
Base 150 mm thick; surface 150 mm thick	m²	27.50

SPECIAL SURFACES FOR SPORT/PLAYGROUNDS

Item Excluding site overheads and profit	Unit	Total rate £
SAFETY SURFACING		
Excavate ground and reduce levels to receive safety surface; dispose of excavated material off site; treat substrate with total herbicide; lay granular fill Type 1 150 mm thick laid to falls and compacted; lay macadam base 40 mm thick; supply and lay Rubaflex wet pour safety system to thicknesses as specified		
All by machine except macadam by hand		
black		
15 mm thick	100 m²	**4850.00**
35 mm thick	100 m²	**6300.00**
60 mm thick	100 m²	**7700.00**
coloured		
15 mm thick	100 m²	**8600.00**
35 mm thick	100 m²	**9200.00**
60 mm thick	100 m²	**10000.00**
All by hand except disposal by grab		
black		
15 mm thick	100 m²	**8000.00**
35 mm thick	100 m²	**9600.00**
60 mm thick	100 m²	**11000.00**
coloured		
15 mm thick	100 m²	**12000.00**
35 mm thick	100 m²	**12000.00**
60 mm thick	100 m²	**13000.00**
SPORTSGROUND CONSTRUCTION		
Plain sports pitches; site clearance, grading and drainage not included		
Cultivate ground and grade to levels; apply pre-seeding fertilizer at 900 kg/ha; apply pre-seeding selective weedkiller; seed in two operations with sports pitch type grass seed at 350 kg/ha; harrow and roll lightly; including initial cut; size		
association football; senior 114 × 72 m	each	**3500.00**
association football; junior 106 × 58 m	each	**2850.00**
rugby union pitch; 156 × 81 m	each	**5200.00**
rugby league pitch; 134 × 60 m	each	**3400.00**
hockey pitch; 95 × 60 m	each	**2600.00**
shinty pitch; 186 × 96 m	each	**7300.00**
men's lacrosse pitch; 100 × 55 m	each	**2500.00**
women's lacrosse pitch; 110 × 73 m	each	**3650.00**
target archery ground; 150 × 50 m	each	**2950.00**
cricket outfield; 160 × 142 m	each	**9300.00**
cycle track outfield; 160 × 80 m	each	**5700.00**
polo ground; 330 × 220 m	each	**27000.00**
Cricket square; excavate to depth of 150 mm; pass topsoil through 6 mm screen; return and mix evenly with imported marl or clay loam; bring to accurate levels; apply pre-seeding fertilizer at 50 g/m²; apply selective weedkiller; seed with cricket square type g grass seed at 50 g/m²; rake in and roll lightly; erect and remove temporary protective chestnut fencing; allow for initial cut and watering three times		
22.8 × 22.8 m	each	**15000.00**

PREPARATION FOR PLANTING/TURFING

Item Excluding site overheads and profit	Unit	Total rate £
SURFACE PREPARATION BY MACHINE		
Treat area with systemic non selective herbicide one month before starting cultivation operations; rip up subsoil using subsoiling machine to a depth of 250 mm below topsoil at 1.20 m centres in light to medium soils; rotavate to 200 mm deep in two passes; cultivate with chain harrow; roll lightly; clear stones over 50 mm		
By tractor	100 m²	8.85
As above but carrying out operations in clay or compacted gravel	100 m²	9.50
As above but ripping by tractor rotavation by pedestrian operated rotavator; clearance and raking by hand; herbicide application by knapsack sprayer	100 m²	41.00
As above but carrying out operations in clay or compacted gravel	100 m²	55.00
Spread and lightly consolidate topsoil brought from spoil heap not exceeding 100 m; in layers not exceeding 150 mm; grade to specified levels; remove stones over 25 mm; all by machine		
100 mm thick	100 m²	66.00
150 mm thick	100 m²	99.00
300 mm thick	100 m²	200.00
450 mm thick	100 m²	300.00
Extra to above for imported topsoil PC £22.00 m³ allowing for 20% settlement		
100 mm thick	100 m²	260.00
150 mm thick	100 m²	395.00
300 mm thick	100 m²	790.00
450 mm thick	100 m²	1175.00
500 mm thick	100 m²	1325.00
600 mm thick	100 m²	1575.00
750 mm thick	100 m²	1975.00
1.00 m thick	100 m²	2600.00
Extra for incorporating mushroom compost at 50 mm/m² into the top 150 mm of topsoil (compost delivered in 20 m³) loads		
manually spread; mechanically rotavated	100 m²	160.00
mechanically spread and rotavated	100 m²	110.00
Extra for incorporating manure at 50 mm/m² into the top 150 mm of topsoil loads		
manually spread; mechanically rotavated; 20 m³ loads	100 m²	280.00
mechanically spread and rotavated; 60 m³ loads	100 m²	170.00
SURFACE PREPARATION BY HAND		
Spread only and lightly consolidate topsoil brought from spoil heap in layers not exceeding 150 mm; grade to specified levels; remove stones over 25 mm; all by hand		
100 mm thick	100 m²	370.00
150 mm thick	100 m²	560.00
300 mm thick	100 m²	1100.00
450 mm thick	100 m²	1675.00
As above but inclusive of loading to location by barrow maximum distance 25 m; finished topsoil depth		
100 mm thick	100 m²	810.00
150 mm thick	100 m²	1225.00
300 mm thick	100 m²	2450.00

PREPARATION FOR PLANTING/TURFING

Item Excluding site overheads and profit	Unit	Total rate £
450 mm thick	100 m²	3650.00
500 mm thick	100 m²	4050.00
600 mm thick	100 m²	4900.00
As above but loading to location by barrow maximum distance 100 m; finished topsoil depth		
100 mm thick	100 m²	830.00
150 mm thick	100 m²	1250.00
300 mm thick	100 m²	2475.00
450 mm thick	100 m²	3750.00
500 mm thick	100 m²	4150.00
600 mm thick	100 m²	5000.00
Extra to the above for incorporating mushroom compost at 50 mm/m² into the top 150 mm of topsoil; by hand	100 m²	190.00
Extra to above for imported topsoil PC £22.00 m³ allowing for 20% settlement		
100 mm thick	100 m²	260.00
150 mm thick	100 m²	395.00
300 mm thick	100 m²	790.00
450 mm thick	100 m²	1175.00
500 mm thick	100 m²	1325.00
600 mm thick	100 m²	1575.00
750 mm thick	100 m²	1975.00
1.00 m thick	100 m²	2600.00

SEEDING AND TURFING

Item Excluding site overheads and profit	Unit	Total rate £
SEEDING AND TURFING		
Bring top 200 mm of topsoil to a fine tilth using tractor drawn implements; remove stones over 25 mm by mechanical stone rake and bring to final tilth by harrow; apply pre-seeding fertilizer at 50 g/m² and work into top 50 mm during final cultivation; seed with certified grass seed in two operations; roll seedbed lightly after sowing		
General amenity grass at 35 g/m²; BSH A3	100 m²	39.50
General amenity grass at 25 g/m²; DLF Trifolium J Court	100 m²	28.00
Shaded areas; BSH A6 at 50 g/m²	100 m²	44.00
Motorway and road verges; DLF Trifolium Promaster 120 at 25–35 g/m²	100 m²	32.00
Bring top 200 mm of topsoil to a fine tilth using pedestrian operated rotavator; remove stones over 25 mm; apply pre-seeding fertilizer at 50 g/m² and work into top 50 mm during final hand cultivation; seed with certified grass seed in two operations; rake and roll seedbed lightly after sowing		
General amenity grass at 35 g/m²; BSH A3	100 m²	110.00
General amenity grass at 25 g/m²; DLF Trifolium J Court	100 m²	97.00
Shaded areas; BSH A6 at 50 g/m²	100 m²	110.00
Motorway and road verges; DLF Trifolium Promaster 120 at 25–35 g/m²	100 m²	100.00
Extra to above for using imported topsoil spread by machine		
100 mm minimum depth	100 m²	290.00
150 mm minimum depth	100 m²	440.00
extra for slopes over 30°	50%	–
Extra to above for using imported topsoil spread by hand; maximum distance for transporting soil 100 m		
100 mm minimum depth	m²	630.00
150 mm minimum depth	m²	950.00
extra for slopes over 30°	50%	–
Extra for mechanically screening top 25 mm of topsoil through 6 mm screen and spreading on seedbed; debris carted to dump on site not exceeding 100 m	m³	6.45
Bring top 200 mm of topsoil to a fine tilth; remove stones over 50 mm; apply pre-seeding fertilizer at 50 g/m² and work into top 50 mm of topsoil during final cultivation; seed with certified grass seed in two operations; harrow and roll seedbed lightly after sowing		
Areas inclusive of fine levelling by specialist machinery		
outfield grass at 350 kg/ha; DLF Trifolium Promaster 40	ha	4100.00
outfield grass at 350 kg/ha; DLF Trifolium Promaster 70	ha	3800.00
sportsfield grass at 300 kg/ha; DLF Trifolium J Pitch	ha	3700.00
Seeded areas prepared by chain harrow		
low maintenance grass at 350 kg/ha	ha	4650.00
verge mixture grass at 150 kg/ha	ha	4000.00
Extra for wild flora mixture at 30 kg/ha		
BSH WSF 75 kg/ha	ha	3250.00
extra for slopes over 30°	50%	–
Cut existing turf to 1.0 × 1.0 × 0.5 m turves; roll up and move to stack not exceeding 100 m		
by pedestrian operated machine; roll up and stack by hand	100 m²	70.00
all works by hand	100 m²	200.00
Extra for boxing and cutting turves	100 m²	3.60

SEEDING AND TURFING

Item Excluding site overheads and profit	Unit	Total rate £
Bring top 200 mm of topsoil to a fine tilth in two passes; remove stones over 25 mm and bring to final tilth; apply pre-seeding fertilizer at 50 g/m² and work into top 50 mm during final cultivation; roll turf bed lightly		
Using tractor drawn implements and mechanical stone rake	m²	0.20
Cultivation by pedestrian rotavator; all other operations by hand	m²	0.60
As above but bring turf from stack not exceeding 100 m; lay turves to stretcher bond using plank barrow runs; firm turves using wooden turf beater		
using tractor drawn implements and mechanical stone rake	m²	1.80
cultivation by pedestrian rotavator; all other operations by hand	m²	2.05
As above but including imported turf; Rolawn Medallion		
using tractor drawn implements and mechanical stone rake	m²	3.55
cultivation by pedestrian rotavator; all other operations by hand	m²	3.80
Extra over to all of the above for watering on two occasions and carrying out initial cut		
by ride on triple mower	100 m²	1.10
by pedestrian mower	100 m²	4.75
by pedestrian mower; box cutting	100 m²	5.30
Extra for using imported topsoil spread and graded by machine		
25 mm minimum depth	m²	0.80
75 mm minimum depth	m²	2.20
100 mm minimum depth	m²	2.90
150 mm minimum depth	m²	4.40
Extra for using imported topsoil spread and graded by hand; distance of barrow run 25 m		
25 mm minimum depth	m²	1.65
75 mm minimum depth	m²	4.25
100 mm minimum depth	m²	5.40
150 mm minimum depth	m²	8.10
Extra for work on slopes over 30° including pegging with 200 galvanized wire pins	m²	1.70
Inturf Big Roll		
Supply, deliver in one consignment, fully prepare the area and install in Big Roll format Inturf 553, a turfgrass comprising dwarf perennial ryegrass, smooth stalked rneadowgrass and fescues; installation by tracked machine		
Preparation by tractor drawn rotavator	m²	3.10
Cultivation by pedestrian rotavator; all other operations by hand	m²	3.55
Erosion control		
On ground previously cultivated, bring area to level, treat with herbicide; lay 20 mm thick open texture erosion control mat with 100 mm laps, fixed with 8 × 400 mm steel pegs at 1.0 m centres; sow with low maintenance grass suitable for erosion control on slopes at 35 g/m²; spread imported topsoil 25 mm thick and fertilizer at 35 g/m²; water lightly using sprinklers on two occasions	m²	6.50
as above but hand-watering by hose pipe maximum distance from mains supply 50 m	m²	6.60

PLANTING

Item Excluding site overheads and profit	Unit	Total rate £
HEDGE PLANTING		
Works by machine; excavate trench for hedge 300 mm wide × 450 mm deep; deposit spoil alongside and plant hedging plants in single row at 200 mm centres; backfill with excavated material incorporating organic manure at 1 m³ per 5 m³; carry out initial cut; including delivery of plants from nursery		
Bare root hedging plants		
PC £0.30	100 m	1225.00
PC £0.60	100 m	1525.00
PC £1.00	100 m	1925.00
PC £1.20	100 m	2125.00
PC £1.50	100 m	2425.00
As above but two rows of hedging plants at 300 mm centres staggered rows		
PC £0.30	100 m	800.00
PC £0.60	100 m	1000.00
PC £1.00	100 m	1275.00
PC £1.50	100 m	1600.00
Works by hand; excavate trench for hedge 300 mm wide × 450 mm deep; deposit spoil alongside and plant hedging plants in single row at 200 mm centres; backfill with excavated material incorporating organic manure at 1 m³ per 5 m³; carry out initial cut; including delivery of plants from nursery		
Two rows of hedging plants at 300 mm centres staggered rows		
PC £0.30	100 m	1000.00
PC £0.60	100 m	1200.00
PC £1.00	100 m	1475.00
PC £1.50	100 m	1800.00
TREE PLANTING		
Excavate tree pit by hand; fork over bottom of pit; plant tree with roots well spread out; backfill with excavated material incorporating treeplanting compost at 1 m³ per 3 m³ of soil; one tree stake and two ties; tree pits square in sizes shown		
Light standard bare root tree in pit; PC £9.75		
600 × 600 mm deep	each	33.50
900 × 600 mm deep	each	48.00
Standard bare root tree in pit; PC £16.00		
600 × 600 mm deep	each	45.00
900 × 600 mm deep	each	53.00
Standard root balled tree in pit; PC £23.50		
600 × 600 mm deep	each	54.00
900 × 600 mm deep	each	62.00
1.00 × 1.00 m deep	each	84.00
Selected standard bare root tree in pit; PC £22.50		
900 × 900 mm deep	each	58.00
1.00 × 1.00 m deep	each	85.00
Selected standard root ball tree in pit; PC £32.50		
900 × 900 mm deep	each	84.00
1.00 × 1.00 m deep	each	125.00

PLANTING

Item Excluding site overheads and profit	Unit	Total rate £
Heavy standard bare root tree in pit; PC £39.25		
900 × 900 mm deep	each	96.00
1.00 × 1.00 m deep	each	105.00
1.50 m × 750 mm deep	each	140.00
Heavy standard root ball tree in pit; PC £54.25		
900 × 900 mm deep	each	110.00
1.00 × 1.00 m deep	each	125.00
1.50 m × 750 mm deep	each	155.00
Extra heavy standard bare root tree in pit; PC £63.00		
1.00 × 1.00 m deep	each	140.00
1.20 × 1.00 m deep	each	135.00
1.50 m × 750 mm deep	each	180.00
Extra heavy standard root ball tree in pit; PC £62.50		
1.00 × 1.00 m deep	each	140.00
1.50 m × 750 mm deep	each	170.00
1.50 × 1.00 m deep	each	205.00
Excavate tree pit by machine; fork over bottom of pit; plant tree with roots well spread out; backfill with excavated material, incorporating organic manure at 1 m³ per 3 m³ of soil; one tree stake and two ties; tree pits square in sizes shown		
Light standard bare root tree in pit; PC £9.75		
600 × 600 mm deep	each	28.50
900 × 900 mm deep	each	41.00
Standard bare root tree in pit; PC £16.00		
600 × 600 mm deep	each	40.00
900 × 600 mm deep	each	42.00
900 × 900 mm deep	each	48.00
Standard root balled tree in pit; PC £23.50		
600 × 600 mm deep	each	49.00
900 × 600 mm deep	each	51.00
900 × 900 mm deep	each	57.00
Selected standard bare root tree in pit; PC £22.50		
900 × 900 mm deep	each	58.00
1.00 × 1.00 m deep	each	69.00
Selected standard root ball tree in pit; PC £32.50		
900 × 900 mm deep	each	84.00
1.00 × 1.00 m deep	each	110.00
Heavy standard bare root tree in pit; PC £39.25		
900 × 900 mm deep	each	96.00
1.00 × 1.00 m deep	each	88.00
1.20 × 1.00 m deep	each	110.00
Heavy standard root ball tree in pit; PC £54.25		
900 × 900 mm deep	each	110.00
1.00 × 1.00 m deep	each	110.00
1.20 × 1.00 m deep	each	120.00
Extra heavy standard bare root tree in pit; PC £63.00		
1.00 × 1.00 m deep	each	120.00
1.20 × 1.00 m deep	each	135.00
1.50 × 1.00 m deep	each	165.00

PLANTING

Item Excluding site overheads and profit	Unit	Total rate £
TREE PLANTING – cont		
Excavate tree pit by machine – cont		
Extra heavy standard root ball tree in pit; PC £62.50		
1.00 × 1.00 m deep	each	125.00
1.20 × 1.00 m deep	each	140.00
1.50 × 1.00 m deep	each	170.00
SEMI MATURE TREE PLANTING		
Excavate tree pit deep by machine; fork over bottom of pit; plant rootballed tree Acer platanoides 'Emerald Queen' using telehandler where necessary; backfill with excavated material, incorporating Melcourt Topgrow bark/manure mixture at 1 m³ per 3 m³ of soil; Platipus underground guying system; tree pits 1500 × 1500 × 1500 mm deep inclusive of Platimats; excavated material not backfilled to treepit spread to surrounding area		
16–18 cm girth; PC £85.00	each	250.00
18–20 cm girth; PC £100.00	each	275.00
20–25 cm girth; PC £125.00	each	365.00
25–30 cm girth; PC £165.00	each	460.00
30–35 cm girth; PC £300.00	each	700.00
As above but treepits 2.00 × 2.00 × 1.5 m deep		
40–45 cm girth; PC £500.00	each	1125.00
45–50 cm girth; PC £700.00	each	1400.00
55–60 cm girth; PC £1,200.00	each	1800.00
67–70 cm girth; PC £2,200.00	each	3000.00
75–80 cm girth; PC £4,500.00	each	5600.00
Extra to the above for imported topsoil moved 25 m from tipping area and disposal off site of excavated material		
Tree pits 1500 × 1500 × 1500 mm deep		
16–18 cm girth	each	170.00
18–20 cm girth	each	160.00
20–25 cm girth	each	150.00
25–30 cm girth	each	140.00
30–35 cm girth	each	135.00
TREE PLANTING WITH MOBILE CRANES		
Excavate treepit 1.50 × 1.50 × 1.00 m deep; supply and plant semi mature trees delivered in full loads; trees lifted by crane; inclusive of backfilling tree pit with imported topsoil, compost, fertilizers and underground guying using Platipus anchors		
Self managed lift; local authority applications, health and safety, traffic management or road closures not included; tree size and distance of lift; 35 tonne crane		
25–30 cm; maximum 25 m distance	each	630.00
30–35 cm; maximum 25 m distance	each	730.00
35–40 cm; maximum 25 m distance	each	1075.00
55–60 cm; maximum 15 m distance	each	1775.00
80–90 cm; maximum 10 m distance	each	6800.00

PLANTING

Item Excluding site overheads and profit	Unit	Total rate £
Managed lift; inclusive of all local authority applications, health and safety, traffic management or road closures all by crane hire company; tree size and distance of lift; 35 tonne crane		
25–30 cm; maximum 25 m distance	each	640.00
30–35 cm; maximum 25 m distance	each	750.00
35–40 cm; maximum 25 m distance	each	1100.00
55–60 cm; maximum 15 m distance	each	1825.00
80–90 cm; maximum 10 m distance	each	7000.00
Self managed lift; local authority applications, health and safety, traffic management or road closures not included; tree size and distance of lift; 80 tonne crane		
25–30 cm; maximum 40 m distance	each	630.00
30–35 cm; maximum 40 m distance	each	740.00
35–40 cm; maximum 40 m distance	each	1075.00
55–60 cm; maximum 33 m distance	each	1800.00
80–90 cm; maximum 23 m distance	each	6900.00
Managed lift; inclusive of all local authority applications, health and safety, traffic management or road closures all by crane hire company; tree size and distance of lift; 35 tonne crane		
25–30 cm; maximum 40 m distance	each	640.00
30–35 cm; maximum 40 m distance	each	750.00
35–40 cm; maximum 40 m distance	each	1125.00
55–60 cm; maximum 33 m distance	each	1850.00
80–90 cm; maximum 23 m distance	each	7000.00
SHRUBS, GROUND COVERS AND BULBS		
Excavate planting holes 250 × 250 × 300 mm deep to area previously ripped and rotavated; excavated material left alongside planting hole		
By mechanical auger		
250 mm centres (16 plants per m^2)	m^2	10.60
300 mm centres (11.11 plants per m^2)	m^2	7.30
400 mm centres (6.26 plants per m^2)	m^2	4.10
450 mm centres (4.93 plants per m^2)	m^2	3.25
500 mm centres (4 plants per m^2)	m^2	2.65
600 mm centres (2.77 plants per m^2)	m^2	1.80
750 mm centres (1.77 plants per m^2)	m^2	1.15
900 mm centres (1.23 plants per m^2)	m^2	0.80
1.00 m centres (1 plant per m^2)	m^2	0.65
1.50 m centres (0.44 plants per m^2)	m^2	0.30
As above but excavation by hand		
250 mm centres (16 plants per m^2)	m^2	10.60
300 mm centres (11.11 plants per m^2)	m^2	7.30
400 mm centres (6.26 plants per m^2)	m^2	4.10
450 mm centres (4.93 plants per m^2)	m^2	3.25
500 mm centres (4 plants per m^2)	m^2	2.65
600 mm centres (2.77 plants per m^2)	m^2	1.80
750 mm centres (1.77 plants per m^2)	m^2	1.15
900 mm centres (1.23 plants per m^2)	m^2	0.80
1.00 m centres (1 plant per m^2)	m^2	0.65
1.50 m centres (0.44 plants per m^2)	m^2	0.30

PLANTING

Item Excluding site overheads and profit	Unit	Total rate £
SHRUBS, GROUND COVERS AND BULBS – cont		
Clear light vegetation from planting area and remove to dump on site; dig planting holes; plant whips with roots well spread out; backfill with excavated topsoil; including one 38 × 38 mm treated softwood stake, two tree ties and mesh guard 1.20 m high; planting matrix 1.5 × 1.5 m; allow for beating up once at 10% of original planting, cleaning and weeding round whips once, applying fertilizer once at 35 g/m²; using the following mix of whips, bare rooted		
Plant bare root plants average PC £0.27 each to a required matrix		
plant mix as above	100 m²	240.00
Plant bare root plants average PC £0.75 each to a required matrix		
plant mix as above	100 m²	220.00
Cultivate and grade shrub bed; bring top 300 mm of topsoil to a fine tilth, incorporating mushroom compost at 50 mm and Enmag slow release fertilizer; rake and bring to given levels; remove all stones and debris over 50 mm; dig planting holes average 300 × 300 × 300 mm deep; supply and plant specified shrubs in quantities as shown below; backfill with excavated material as above; water to field capacity and mulch 50 mm bark chips 20–40 mm size; water and weed regularly for 12 months and replace failed plants		
Shrubs 3 l PC £2.80; ground covers 9 cm PC £1.50		
100% shrub area		
300 mm centres	100 m²	5800.00
400 mm centres	100 m²	3500.00
500 mm centres	100 m²	2400.00
600 mm centres	100 m²	1825.00
100% groundcovers		
200 mm centres	100 m²	7300.00
300 mm centres	100 m²	3500.00
400 mm centres	100 m²	2175.00
500 mm centres	100 m²	1575.00
groundcover 30%/shrubs 70% at the distances shown below		
200/300 mm	100 m²	6200.00
300/400 mm	100 m²	3450.00
300/500 mm	100 m²	2700.00
400/500 mm	100 m²	2325.00
groundcover 50%/shrubs 50% at the distances shown below		
200/300 mm	100 m²	6500.00
300/400 mm	100 m²	3500.00
300/500 mm	100 m²	2950.00
400/500 mm	100 m²	2275.00
Cultivate ground by machine and rake to level; plant bulbs as shown; bulbs PC £25.00/100		
15 bulbs per m²	100 m²	660.00
25 bulbs per m²	100 m²	1075.00
50 bulbs per m²	100 m²	2125.00

PLANTING

Item Excluding site overheads and profit	Unit	Total rate £
Cultivate ground by machine and rake to level; plant bulbs as shown; bulbs PC £13.00/100		
15 bulbs per m²	100 m²	455.00
25 bulbs per m²	100 m²	740.00
50 bulbs per m²	100 m²	1450.00
Form holes in grass areas and plant bulbs using bulb planter, backfill with organic manure and turf plug; bulbs PC £13.00/100		
15 bulbs per m²	100 m²	660.00
25 bulbs per m²	100 m²	1075.00
50 bulbs per m²	100 m²	2175.00
BEDDING		
Spray surface with glyphosate; lift and dispose of turf when herbicide action is complete; cultivate new area for bedding plants to 400 mm deep; spread compost 100 mm deep and chemical fertilizer Enmag and rake to fine tilth to receive new bedding plants; remove all arisings		
Existing turf area		
disposal to skip	m²	4.90
disposal to compost area on site; distance 25 m	m²	5.30
Plant bedding to existing planting area; bedding planting PC £0.25 each		
Clear existing bedding; cultivate soil to 230 mm deep; incorporate compost 75 mm and rake to fine tilth; collect bedding from nursery and plant at 100 mm centres; irrigate on completion; maintain weekly for 12 weeks		
mass planted; 100 mm centres	m²	33.00
to patterns; 100 mm centres	m²	36.00
mass planted; 150 mm centres	m²	18.90
to patterns; 150 mm centres	m²	22.00
mass planted; 200 mm centres	m²	18.90
to patterns; 200 mm centres	m²	22.00
Extra for watering by hand held hose pipe		
Flow rate 25 litres/minute		
10 litres/m²	100 m²	0.10
15 litres/m²	100 m²	0.20
20 litres/m²	100 m²	0.20
25 litres/m²	100 m²	0.30
Flow rate 40 litres/minute		
10 litres/m²	100 m²	0.10
15 litres/m²	100 m²	0.10
20 litres/m²	100 m²	0.10
25 litres/m²	100 m²	0.20
PLANTING PLANTERS		
To brick planter; coat insides with 2 coats RIW liquid asphaltic composition; fill with 50 mm shingle and cover with geofabric; fill with screened topsoil incorporating 25% Topgrow compost and Enmag		

PLANTING

Item Excluding site overheads and profit	Unit	Total rate £
PLANTING PLANTERS – cont		
Planters 1.00 m deep		
1.00 × 1.00 m	each	130.00
1.00 × 2.00 m	each	230.00
1.00 × 3.00 m	each	335.00
Planters 1.50 m deep		
1.00 × 1.00 m	each	195.00
1.00 × 2.00 m	each	320.00
1.00 × 3.00 m	each	510.00
Container planting; fill with 50 mm shingle and cover with geofabric; fill with screened topsoil incorporating 25% Topgrow compost and Enmag		
Planters 1.00 m deep		
400 × 400 × 400 mm deep	each	10.50
400 × 400 × 600 mm deep	each	14.50
1.00 m × 400 mm wide × 400 mm deep	each	24.50
1.00 m × 600 mm wide × 600 mm deep	each	43.00
1.00 m × 100 mm wide × 400 mm deep	each	46.50
1.00 m × 100 mm wide × 600 mm deep	each	68.00
1.00 m × 100 mm wide × 1.00 m deep	each	110.00
1.00 m diameter × 400 mm deep	each	36.50
1.00 m diameter × 1.00 m deep	each	89.00
2.00 m diameter × 1.00 m deep	each	350.00

LANDSCAPE MAINTENANCE

Item Excluding site overheads and profit	Unit	Total rate £
MAINTENANCE OF GRASSED AREAS		
Maintenance executed as part of a landscape construction contract		
Grass cutting		
Grass cutting; fine turf; using pedestrian guided machinery; arisings boxed and disposed of off site		
per occasion	100 m²	5.50
per annum 26 cuts	m²	1.45
per annum 18 cuts	m²	1.00
Grass cutting; standard turf; using self propelled three gang machinery		
per occasion	100 m²	0.40
per annum 26 cuts	m²	0.10
per annum 18 cuts	m²	0.10
Maintenance for one year; recreation areas, parks, amenity grass areas; using tractor drawn machinery		
per occasion	ha	29.00
per annum 26 cuts	ha	760.00
per annum 18 cuts	ha	530.00
Aeration of turfed areas		
Aerate ground with spiked aerator; apply spring/summer fertilizer once; apply autumn/winter fertilizer once; cut grass, 16 cuts; sweep up leaves twice		
as part of a landscape contract, defects liability	ha	1775.00
as part of a long term maintenance contract	ha	1150.00
MAINTENANCE OF PLANTED AREAS		
Vegetation control; native planting; roadside railway or forestry planted areas		
Post planting maintenance; control of weeds and grass; herbicide spray applications; maintain weed free circles 1.00 m diameter to planting less than 5 years old in roadside, rail or forestry planting environments and the like; strim grass to 50–75 mm; prices per occasion (three applications of each operation normally required)		
Knapsack spray application; glyphosate; planting at		
1.50 mm centres	ha	1000.00
1.75 mm centres	ha	590.00
2.00 mm centres	ha	640.00
Maintain planted areas; control of weeds and grass; maintain weed free circles 1.00 m diameter to planting less than 5 years old in roadside, rail or forestry planting environments and the like; strim surrounding grass to 50–75 mm; prices per occasion (three applications of each operation normally required)		
Herbicide spray applications; CDA (controlled droplet application) glyphosate and strimming; plants planted at the following centres		
1.50 mm centres	ha	520.00
1.75 mm centres	ha	590.00
2.00 mm centres	ha	640.00

LANDSCAPE MAINTENANCE

Item Excluding site overheads and profit	Unit	Total rate £
MAINTENANCE OF PLANTED AREAS – cont		
Post planting maintenance; control of weeds and grass; herbicide spray applications; CDA (controlled droplet application); Xanadu glyphosate/diuron; maintain weed free circles 1.00 m diameter to planting less than 5 years old in roadside, rail or forestry planting environments and the like; strim grass to 50-75 mm; prices per occasion (1.5 applications of herbicide and three strim operations normally required)		
Plants planted at the following centres		
1.50 mm centres	ha	**550.00**
1.75 mm centres	ha	**620.00**
2.00 mm centres	ha	**660.00**
Ornamental shrub beds		
Hand weed ornamental shrub bed during the growing season; planting less than 2 years old		
Mulched beds; weekly visits; planting centres		
600 mm centres	100 m²	**9.25**
400 mm centres	100 m²	**11.10**
300 mm centres	100 m²	**14.80**
ground covers	100 m²	**18.50**
Mulched beds; monthly visits; planting centres		
600 mm centres	100 m²	**13.90**
400 mm centres	100 m²	**18.50**
300 mm centres	100 m²	**22.00**
ground covers	100 m²	**28.00**
Non-mulched beds; weekly visits; planting centres		
600 mm centres	100 m²	**13.90**
400 mm centres	100 m²	**14.80**
300 mm centres	100 m²	**28.00**
ground covers	100 m²	**37.00**
Non-mulched beds; monthly visits; planting centres		
600 mm centres	100 m²	**18.50**
400 mm centres	100 m²	**22.00**
300 mm centres	100 m²	**28.00**
ground covers	100 m²	**46.00**
Remulch planting bed at the start of the planting season; top up mulch 25 mm thick; Melcourt Ltd		
Larger areas maximum distance 25 m; 80 m³ loads		
Ornamental bark mulch	100 m²	**140.00**
Melcourt Bark Nuggets	100 m²	**130.00**
Amenity Bark	100 m²	**97.00**
Forest biomulch	100 m²	**87.00**
Smaller areas; maximum distance 25 m; 25 m³ loads		
Ornamental bark mulch	100 m²	**170.00**
Melcourt Bark Nuggets	100 m²	**165.00**
Amenity Bark	100 m²	**130.00**
Forest biomulch	100 m²	**120.00**

FENCING

Item Excluding site overheads and profit	Unit	Total rate £
TEMPORARY FENCING		
Site fencing; supply and erect temporary protective fencing and remove at completion of works		
Cleft chestnut paling		
1.20 m high 75 mm larch posts at 3 m centres	100 m	800.00
1.50 m high 75 mm larch posts at 3 m centres	100 m	960.00
Heras fencing		
2 week hire period	100 m	870.00
4 week hire period	100 m	1275.00
8 week hire period	100 m	2125.00
12 week hire period	100 m	3000.00
16 week hire period	100 m	3800.00
24 week hire period	100 m	5500.00
CHAIN LINK AND WIRE FENCING		
Chain link fencing; supply and erect chain link fencing; form post holes and erect concrete posts and straining posts with struts at 50 m centres all set in 1:3:6 concrete; fix line wires		
3 mm galvanized wire 50 mm chainlink fencing		
900 mm high	m	22.50
1200 mm high	m	25.00
1800 mm high	m	34.50
Plastic coated 3.15 gauge galvanized wire mesh		
900 mm high	m	21.50
1200 mm high	m	23.50
1800 mm high	m	34.00
Extra for additional concrete straining posts with 1 strut set in concrete		
900 mm high	each	63.00
1200 mm high	each	66.00
1800 mm high	each	80.00
Extra for additional concrete straining posts with 2 struts set in concrete		
900 mm high	each	91.00
1200 mm high	each	93.00
1800 mm high	each	120.00
Extra for additional angle iron straining posts with 2 struts set in concrete		
900 mm high	each	68.00
1200 mm high	each	78.00
1400 mm high	each	92.00
1800 mm high	each	95.00
2400 mm high	each	105.00

FENCING

Item Excluding site overheads and profit	Unit	Total rate £
TIMBER FENCING		
Clear fenceline of existing evergreen shrubs 3 m high average; grub out roots by machine chip on site and remove off site; erect close boarded timber fence in treated softwood, pales 100 × 22 mm lapped, 150 × 22 mm gravel boards		
Concrete posts 100 × 100 mm at 3.0 m centres set into ground in 1:3:6 concrete		
900 mm high	m	63.00
1500 mm high	m	68.00
1800 mm high	m	73.00
As above but with softwood posts; three arris rails		
1350 mm high	m	65.00
1650 mm high	m	73.00
1800 mm high	m	74.00
Erect chestnut pale fencing; cleft chestnut pales; two lines galvanized wire, galvanized tying wire, treated softwood posts at 3.0 m centres and straining posts and struts at 50 m centres driven into firm ground		
900 mm high; posts 75 mm diameter × 1200 mm long	m	5.60
1200 mm high; posts 75 mm diameter × 1500 mm long	m	7.75
Construct timber rail; horizontal hit and miss type; rails 150 × 25 mm; posts 100 × 100 mm at 1.8 m centres; twice stained with coloured wood preservative; including excavation for posts and concreting into ground (C7P)		
In treated softwood		
1800 mm high	m	55.00
Construct cleft oak rail fence with rails 300 mm minimum girth tennoned both ends; 125 × 100 mm treated softwood posts double mortised for rails; corner posts 125 × 125 mm, driven into firm ground at 2.5 m centres		
3 rails	m	17.00
4 rails	m	20.50
DEER STOCK RABBIT FENCING		
Construct rabbit-stop fencing; erect galvanized wire netting; mesh 31 mm; 900 mm above ground, 150 mm below ground turned out and buried; on 75 mm diameter treated timber posts 1.8 m long driven 700 mm into firm ground at 4.0 m centres; netting clipped to top and bottom straining wires 2.63 mm diameter; straining post 150 mm diameter × 2.3 m long driven into firm ground at 50 m intervals		
turned in 150 mm	100 m	920.00
buried 150 mm	100 m	1000.00
Deer fence		
Construct deer-stop fencing; erect five 4 mm diameter plain galvanized wires and five 2 ply galvanized barbed wires at 150 mm spacing on 45 × 45 × 5 mm angle iron posts 2.4 m long driven into firm ground at 3.0 m centres, driven into firm ground at 3.0 m centres, with timber droppers 25 × 38 mm × 1.0 m long at 1.5 m centres	100 m	1550.00

FENCING

Item Excluding site overheads and profit	Unit	Total rate £
Forestry fencing Supply and erect forestry fencing of three lines of 3 mm plain galvanized wire tied to 1700 × 65 mm diameter angle iron posts at 2750 m centres with 1850 × 100 mm diameter straining posts and 1600 × 80 mm diameter struts at 50.0 m centres driven into firm ground		
1800 mm high; three wires	100 m	880.00
1800 mm high; three wires including cattle fencing	100 m	1150.00
CONCRETE FENCING		
Supply and erect precast concrete post and panel fence in 2 m bays; panels to be shiplap profile, aggregate faced one side; posts set 600 mm in ground in concrete		
1500 mm high	m	28.00
1800 mm high	m	33.00
2100 mm high	m	37.00
SECURITY FENCING		
Supply and erect chainlink fence; 51 × 3 mm mesh; with line wires and stretcher bars bolted to concrete posts at 3.0 m centres and straining posts at 10 m centres; posts set in concrete 450 × 450 mm × 33% of height of post deep; fit straight extension arms of 45 × 45 × 5 mm steel angle with three lines of barbed wire and droppers; all metalwork to be factory hot-dip galvanized for painting on site		
Galvanized 3 mm mesh		
900 mm high	m	26.50
1200 mm high	m	30.50
1800 mm high	m	41.00
As above but with PVC coated 3.15 mm mesh (diameter of wire 2.5 mm)		
900 mm high	m	25.50
1200 mm high	m	29.00
1800 mm high	m	40.00
Add to fences above for base of fence to be fixed with hairpin staples cast into concrete ground beam 1:3:6 site mixed concrete; mechanical excavation disposal to on site spoil heaps		
125 × 225 mm deep	m	5.10
Add to fences above for straight extension arms of 45 × 45 × 5 mm steel angle with three lines of barbed wire and droppers	m	4.65
Supply and erect palisade security fence Jacksons Barbican 2500 mm high with rectangular hollow section steel pales at 150 mm centres on three 50 × 50 × 6 mm rails; rails bolted to 80 × 60 mm posts set in concrete 450 × 450 × 750 mm deep at 2750 mm centres; tops of pales to points and set at 45° angle; all metalwork to be hot-dip factory galvanized for painting on site	m	110.00
Supply and erect single gate to match above complete with welded hinges and lock		
1.0 m wide	each	1250.00
4.0 m wide	each	1375.00
8.0 m wide	pair	2175.00

FENCING

Item Excluding site overheads and profit	Unit	Total rate £
SECURITY FENCING – cont		
Supply and erect Orsogril proprietary welded steel mesh panel fencing on steel posts set 750 mm deep in concrete foundations 600 × 600 mm; supply and erect proprietary single gate 2.0 m wide to match fencing		
930 mm high	100 m	**12000.00**
1326 mm high	100 m	**15000.00**
1722 mm high	100 m	**20000.00**
RAILINGS		
Conservation of historic railings; Eura Conservation Ltd		
Remove railings to workshop off site; shotblast and repair mildly damaged railings; remove rust and paint with three coats; transport back to site and re-erect		
railings with finials 1.80 m high	m	**540.00**
railings ornate cast or wrought iron	m	**840.00**
Supply and erect mild steel bar railing of 19 mm balusters at 115 mm centres welded to mild steel top and bottom rails 40 × 10 mm; bays 2.0 m long bolted to 51 × 51 mm ms hollow section posts set in C15P concrete; all metal work galvanized after fabrication		
900 mm high	m	**75.00**
1200 mm high	m	**98.00**
1500 mm high	m	**110.00**
Supply and erect mild steel pedestrian guard rail Class A; rails to be rectangular hollow sealed section 50 × 30 × 2.5 mm, vertical support 25 × 19 mm central between intermediate and top rail; posts to be set 300 mm into paving base; all components factory welded and factory primed for painting on site		
Panels 1000 mm high × 2000 mm wide with 150 mm toe space and 200 mm visibility gap at top	m	**89.00**
BALLSTOP FENCING		
Supply and erect plastic coated 30 × 30 mm netting fixed to 60.3 mm diameter 12 mm solid bar lattice galvanized dual posts; top, middle and bottom rails with 3 horizontal rails on 60.3 mm diameter nylon coated tubular steel posts at 3.0 m centres and 60.3 mm diameter straining posts with struts at 50 m centres set 750 mm into FND2 concrete footings 300 × 300 × 600 mm deep; include framed chain link gate 900 × 1800 mm high to match complete with hinges and locking latch		
4500 mm high	100 m	**10000.00**
5000 mm high	100 m	**13000.00**
6000 mm high	100 m	**14500.00**
STOCK GATE		
Erect stock gate, ms tubular field gate, diamond braced 1.8 m high hung on tubular steel posts set in concrete (C7P); complete with ironmongery; all galvanized		
width 3.00 m	each	**300.00**
width 4.20 m	each	**325.00**

FENCING

Item	Unit	Total
Excluding site overheads and profit		rate £

TRIP RAILS

Steel trip rail

Erect trip rail of galvanized steel tube 38 mm internal diameter with sleeved joint fixed to 38 mm diameter steel posts as above 700 mm long; set in C7P concrete at 1.20 m centres; metalwork primed and painted two coats metal preservative paint	m	140.00

Birdsmouth fencing; timber

600 mm high	m	19.10
900 mm high	m	20.00

STREET FURNITURE

Item Excluding site overheads and profit	Unit	Total rate £
BENCHES, SEATS AND BOLLARDS		
Bollards and access restriction		
Supply and install powder coated steel parking posts and bases to specified colour; fold down top locking type, complete with keys and instructions; posts and bases set in concrete footing 200 × 200 × 300 mm deep	each	8.30
Supply and install 10 nr cast iron Doric bollards 920 mm high above ground × 170 mm diameter bedded in concrete base 400 mm diameter × 400 mm deep	10 nr	1375.00
Benches and seating		
In grassed area excavate for base 2500 × 1575 mm and lay 100 mm hardcore, 100 mm concrete, brick pavers in stack bond bedded in 25 mm cement: lime: sand mortar; supply and fix where shown on drawing proprietary seat, hardwood slats on black powder coated steel frame, bolted down with 4 nr 24 × 90 mm recessed hex-head stainless steel anchor bolts set into concrete	set	1050.00
Cycle stand		
Supply and fix cycle stand 1250 m × 550 mm of 60.3 mm black powder coated hollow steel sections, one-piece with rounded top corners; set 250 mm into paving	each	380.00
Street planters		
Supply and locate in position precast concrete planters; fill with topsoil placed over 50 mm shingle and terram; plant with assorted 5 litre and 3 litre shrubs to provide instant effect		
970 mm diameter × 470 mm high; white exposed aggregate finish	each	1.85

DRAINAGE

Item Excluding site overheads and profit	Unit	Total rate £
SURFACE WATER DRAINAGE		
Clay gully Excavate hole; supply and set in concrete (C10P) vitrified clay trapped mud (dirt) gully with rodding eye; complete with galvanized bucket and cast iron hinged locking grate and frame, flexible joint to pipe; connect to drainage system with flexible joints	each	150.00
Concrete road gully Excavate hole and lay 100 mm concrete base (1:3:6) 150 × 150 mm to suit given invert level of drain; supply and connect trapped precast concrete road gully 450 mm diameter × 1.07 m deep with 160 mm outlet; set in concrete surround; connect to vitrified clay drainage system with flexible joints; supply and fix straight bar dished top cast iron grating and frame; bedded in cement: sand mortar (1:3)	each	320.00
Gullies PVC-u Excavate hole and lay 100 mm concrete (C20P) base 150 × 150 mm to suit given invert level of drain; connect to drainage system; backfill with DoT Type 1 granular fill; install gully; complete with cast iron grate and frame		
trapped PVC-u gully	each	67.00
bottle gully 228 × 228 × 317 mm deep	each	68.00
bottle gully 228 × 228 × 642 mm deep	each	82.00
yard gully 300 mm diameter × 600 mm deep	each	230.00
Inspection chambers; brick manhole; excavate pit for inspection chamber including earthwork support and disposal of spoil to dump on site not exceeding 100 m; lay concrete (1:2:4) base 1500 mm diameter × 200 mm thick; 110 mm vitrified clay channels; benching in concrete (1:3:6) allowing one outlet and two inlets for 110 mm diameter pipe; construct inspection chamber 1 brick thick walls of engineering brick Class B; backfill with excavated material; complete with 2 nr cast iron step irons 1200 × 1200 × 1200 mm		
cover slab of precast concrete	each	880.00
access cover; Group 2; 600 × 450 mm	each	920.00
1200 × 1200 × 1500 mm		
access cover; Group 2; 600 × 450 mm	each	1075.00
recessed cover 5 tonne load; 600 × 450 mm; filled with block paviors	each	940.00
Pipe laying Excavate trench by excavator 600 mm deep; lay Type 2 bedding; backfill to 150 mm above pipe with gravel rejects; lay non woven geofabric and fill with topsoil to ground level		
160 mm PVC-u drainpipe	100 m	2075.00
110 mm PVC-u drainpipe	100 m	1250.00
150 mm vitrified clay	100 m	1975.00
100 mm vitrified clay	100 m	1350.00
Linear drainage to design sensitive areas Excavate trench by machine; lay Aco Brickslot channel drain on concrete base and surround to falls; all to manufacturers specifications		
paving surround to both sides of channel	m	160.00

DRAINAGE

Item Excluding site overheads and profit	Unit	Total rate £
SURFACE WATER DRAINAGE – cont		
Linear drainage to pedestrian area		
Excavate trench by machine; lay Aco MultiDrain MD polymer concrete channel drain on concrete base and surround to falls; all to manufacturers specifications; paving surround to channel with brick paving PC £300.00/1000		
Brickslot galvanized grating; paving to both sides	m	160.00
Brickslot stainless steel grating; paving to both sides	m	280.00
slotted galvanized steel grating; paving surround to one side of channel	m	120.00
Excavate trench by machine; lay Aco MultiDrain PPD recycled polypropylene channel drain on concrete base and surround to falls; all to manufacturers specifications; paving surround to channel with brick paving PC £300.00/1000		
Heelguard composite black; paving surround to both sides of channel	m	120.00
Linear drainage to light vehicular area		
Excavate trench by machine; lay Aco MultiDrain PPD recycled polypropylene channel drain on concrete base and surround to falls; all to manufacturers specifications; paving surround to channel with brick paving PC £300.00/1000		
Heelguard composite black; paving surround to both sides of channel	m	140.00
ductile iron; paving surround to both sides of channel	m	190.00
Accessories for channel drain		
Sump unit with sediment bucket	nr	150.00
End cap; inlet/outlet	nr	17.20
AGRICULTURAL DRAINAGE		
Excavate and form ditch and bank with 45° sides in light to medium soils; all widths taken at bottom of ditch		
300 mm wide × 600 mm deep	100 m	100.00
600 mm wide × 900 mm deep	100 m	120.00
1.20 m wide × 900 mm deep	100 m	750.00
1.50 m wide × 1.20 m deep	100 m	1250.00
Clear and bottom existing ditch average 1.50 m deep, trim back vegetation and remove debris to licensed tip not exceeding 13 km, lay jointed concrete pipes; including bedding, haunching and topping with 150 mm concrete; 11.50 N/mm² – 40 mm aggregate; backfill with approved spoil from site		
Pipes 300 mm diameter	100 m	5900.00
Pipes 450 mm diameter	100 m	7500.00
Pipes 600 mm diameter	100 m	9900.00
Clay land drain		
Excavate trench by excavator to 450 mm deep; lay 100 mm vitrified clay drain with butt joints, bedding Class B; backfill with excavated material screened to remove stones over 40 mm; backfill to be laid in layers not exceeding 150 mm; top with 150 mm topsoil remove surplus material to approved dump on site not exceeding 100 m; final level of fill to allow for settlement	100 m	1575.00

DRAINAGE

Item	Unit	Total
Excluding site overheads and profit		rate £
SUBSOIL DRAINAGE – BY MACHINE		
Main drain; remove 150 mm topsoil and deposit alongside trench, excavate drain		
trench by machine and lay flexible perforated drain; lay bed of gravel rejects 100 mm;		
backfill with gravel rejects or similar to within 150 mm of finished ground level;		
complete fill with topsoil; remove surplus spoil to approved dump on site		
Main drain 160 mm supplied in 35 m lengths		
450 mm deep	100 m	550.00
600 mm deep	100 m	630.00
900 mm deep	100 m	1000.00
Extra for couplings	each	2.00
As above but with 100 mm main drain supplied in 100 m lengths		
450 mm deep	100 m	850.00
600 mm deep	100 m	1125.00
900 mm deep	100 m	1825.00
Extra for couplings	each	1.60
Laterals to mains above; herringbone pattern; excavation and backfilling as above;		
inclusive of connecting lateral to main drain		
160 mm pipe to 450 mm deep trench		
laterals at 1.0 m centres	100 m²	550.00
laterals at 2.0 m centres	100 m²	275.00
laterals at 3.0 m centres	100 m²	180.00
laterals at 5.0 m centres	100 m²	110.00
laterals at 10.0 m centres	100 m²	55.00
160 mm pipe to 600 mm deep trench		
laterals at 1.0 m centres	100 m²	630.00
laterals at 2.0 m centres	100 m²	310.00
laterals at 3.0 m centres	100 m²	210.00
laterals at 5.0 m centres	100 m²	125.00
laterals at 10.0 m centres	100 m²	63.00
160 mm pipe to 900 mm deep trench		
laterals at 1.0 m centres	100 m²	1000.00
laterals at 2.0 m centres	100 m²	500.00
laterals at 3.0 m centres	100 m²	330.00
laterals at 5.0 m centres	100 m²	200.00
laterals at 10.0 m centres	100 m²	100.00
Extra for 160/160 mm couplings connecting laterals to main drain		
laterals at 1.0 m centres	10 m	78.00
laterals at 2.0 m centres	10 m	39.00
laterals at 3.0 m centres	10 m	26.00
laterals at 5.0 m centres	10 m	15.60
laterals at 10.0 m centres	10 m	7.80
100 mm pipe to 450 mm deep trench		
laterals at 1.0 m centres	100 m²	850.00
laterals at 2.0 m centres	100 m²	430.00
laterals at 3.0 m centres	100 m²	280.00
laterals at 5.0 m centres	100 m²	170.00
laterals at 10.0 m centres	100 m²	85.00

DRAINAGE

Item Excluding site overheads and profit	Unit	Total rate £
SUBSOIL DRAINAGE – BY MACHINE – cont		
Laterals to mains above – cont		
100 mm pipe to 600 mm deep trench		
laterals at 1.0 m centres	100 m²	1125.00
laterals at 2.0 m centres	100 m²	560.00
laterals at 3.0 m centres	100 m²	370.00
laterals at 5.0 m centres	100 m²	220.00
laterals at 10.0 m centres	100 m²	110.00
100 mm pipe to 900 mm deep trench		
laterals at 1.0 m centres	100 m²	1825.00
laterals at 2.0 m centres	100 m²	920.00
laterals at 3.0 m centres	100 m²	600.00
laterals at 5.0 m centres	100 m²	365.00
laterals at 10.0 m centres	100 m²	180.00
80 mm pipe to 450 mm deep trench		
laterals at 1.0 m centres	100 m²	820.00
laterals at 2.0 m centres	100 m²	410.00
laterals at 3.0 m centres	100 m²	270.00
laterals at 5.0 m centres	100 m²	160.00
laterals at 10.0 m centres	100 m²	82.00
80 mm pipe to 600 mm deep trench		
laterals at 1.0 m centres	100 m²	1075.00
laterals at 2.0 m centres	100 m²	540.00
laterals at 3.0 m centres	100 m²	360.00
laterals at 5.0 m centres	100 m²	215.00
laterals at 10.0 m centres	100 m²	110.00
80 mm pipe to 900 mm deep trench		
laterals at 1.0 m centres	100 m²	1775.00
laterals at 2.0 m centres	100 m²	900.00
laterals at 3.0 m centres	100 m²	590.00
laterals at 5.0 m centres	100 m²	360.00
laterals at 10.0 m centres	100 m²	180.00
Extra for 100/80 mm junctions connecting laterals to main drain		
laterals at 1.0 m centres	10 m	37.00
laterals at 2.0 m centres	10 m	18.40
laterals at 3.0 m centres	10 m	12.30
laterals at 5.0 m centres	10m	7.35
laterals at 10.0 m centres	10m	3.70
SUBSOIL DRAINAGE – BY HAND		
Main drain; remove 150 mm topsoil and deposit alongside trench; excavate drain trench by machine and lay flexible perforated drain; lay bed of gravel rejects 100 mm; backfill with gravel rejects or similar to within 150 mm of finished ground level; complete fill with topsoil; remove surplus spoil to approved dump on site		
Main drain 160 mm in supplied in 35 m lengths		
450 mm deep	100 m	1750.00
600 mm deep	100 m	2175.00
900 mm deep	100 m	2900.00
Extra for couplings	each	2.00

DRAINAGE

Item Excluding site overheads and profit	Unit	Total rate £
As above but with 100 mm main drain supplied in 100 m lengths		
450 mm deep	100 m	1150.00
600 mm deep	100 m	1450.00
900 mm deep	100 m	2025.00
Extra for couplings	each	1.60
Laterals to mains above; herringbone pattern; excavation and backfilling as above; inclusive of connecting lateral to main drain		
160 mm pipe to 450 mm deep trench		
laterals at 1.0 m centres	100 m²	1750.00
laterals at 2.0 m centres	100 m²	870.00
laterals at 3.0 m centres	100 m²	570.00
laterals at 5.0 m centres	100 m²	350.00
laterals at 10.0 m centres	100 m²	175.00
160 mm pipe to 600 mm deep trench		
laterals at 1.0 m centres	100 m²	2175.00
laterals at 2.0 m centres	100 m²	1075.00
laterals at 3.0 m centres	100 m²	720.00
laterals at 5.0 m centres	100 m²	435.00
laterals at 10.0 m centres	100 m²	220.00
160 mm pipe to 900 mm deep trench		
laterals at 1.0 m centres	100 m²	2900.00
laterals at 2.0 m centres	100 m²	1450.00
laterals at 3.0 m centres	100 m²	960.00
laterals at 5.0 m centres	100 m²	580.00
laterals at 10.0 m centres	100 m²	290.00
Extra for 160/160 mm couplings connecting laterals to main drain		
laterals at 1.0 m centres	10 m	78.00
laterals at 2.0 m centres	10 m	39.00
laterals at 3.0 m centres	10 m	26.00
laterals at 5.0 m centres	10 m	15.60
laterals at 10.0 m centres	10 m	7.80
100 mm pipe to 450 mm deep trench		
laterals at 1.0 m centres	100 m²	1150.00
laterals at 2.0 m centres	100 m²	580.00
laterals at 3.0 m centres	100 m²	380.00
laterals at 5.0 m centres	100 m²	230.00
laterals at 10 m centres	100 m²	115.00
100 mm pipe to 600 mm deep trench		
laterals at 1.0 m centres	100 m²	1450.00
laterals at 2.0 m centres	100 m²	730.00
laterals at 3.0 m centres	100 m²	480.00
laterals at 5.0 m centres	100 m²	290.00
laterals at 10.0 m centres	100 m²	145.00
100 mm pipe to 900 mm deep trench		
laterals at 1.0 m centres	100 m²	2025.00
laterals at 2.0 m centres	100 m²	1000.00
laterals at 3.0 m centres	100 m²	670.00
laterals at 5.0 m centres	100 m²	405.00
laterals at 10.0 m centres	100 m²	200.00

DRAINAGE

Item Excluding site overheads and profit	Unit	Total rate £
SUBSOIL DRAINAGE – BY HAND – cont		
Laterals to mains above – cont		
80 mm pipe to 450 mm deep trench		
laterals at 1.0 m centres	100 m²	1125.00
laterals at 2.0 m centres	100 m²	560.00
laterals at 3.0 m centres	100 m²	370.00
laterals at 5.0 m centres	100 m²	220.00
laterals at 10.0 m centres	100 m²	110.00
80 mm pipe to 600 mm deep trench		
laterals at 1.0 m centres	100 m²	1425.00
laterals at 2.0 m centres	100 m²	710.00
laterals at 3.0 m centres	100 m²	470.00
laterals at 5.0 m centres	100 m²	280.00
laterals at 10.0 m centres	100 m²	140.00
80 mm pipe to 900 mm deep trench		
laterals at 1.0 m centres	100 m²	1975.00
laterals at 2.0 m centres	100 m²	990.00
laterals at 3.0 m centres	100 m²	660.00
laterals at 5.0 m centres	100 m²	400.00
laterals at 10.0 m centres	100 m²	200.00
Extra for 100/80 mm couplings connecting laterals to main drain		
laterals at 1.0 m centres	10 m	37.00
laterals at 2.0 m centres	10 m	18.40
laterals at 3.0 m centres	10 m	12.30
laterals at 5.0 m centres	10 m	7.35
laterals at 10.0 m centres	10 m	3.70
SOAKAWAYS		
Construct soakaway from perforated concrete rings; excavation, casting in situ concrete ring beam base; filling with gravel 250 mm deep; placing perforated concrete rings; surrounding with geofabric and backfilling with 250 mm granular surround and excavated material; step irons and cover slab; inclusive of all earthwork retention and disposal off site of surplus material		
900 mm diameter		
1.00 m deep	nr	540.00
2.00 m deep	nr	890.00
1200 mm diameter		
1.00 m deep	nr	720.00
2.00 m deep	nr	1175.00
2400 mm diameter		
1.00 m deep	nr	2000.00
2.00 m deep	nr	3300.00

IRRIGATION

Item Excluding site overheads and profit	Unit	Total rate £
LANDSCAPE IRRIGATION		
Automatic irrigation; KAR UK Ltd		
Large garden consisting of 24 stations; 7000 m² irrigated area		
turf only	nr	11000.00
70/30% turf/shrub beds	nr	14000.00
Medium sized garden consisting of 12 stations; 3500 m² irrigated area		
turf only	nr	7700.00
70/30% turf/shrub beds	nr	11000.00
Medium sized garden consisting of 6 stations of irrigated area; 1000 m²		
turf only	nr	5200.00
70/30% turf/shrub beds	nr	5800.00
50/50% turf/shrub beds	nr	6300.00
Leaky pipe irrigation; Leaky Pipe Ltd		
Works by machine; main supply and connection to laterals; excavate trench for main or ring main 450 mm deep; supply and lay pipe; backfill and lightly compact trench		
20 mm LDPE	100 m	270.00
16 mm LDPE	100 m	230.00
Works by hand; main supply and connection to laterals; excavate trench for main or ring main 450 mm deep; supply and lay pipe; backfill and lightly compact trench		
20 mm LDPE	100 m	1350.00
16 mm LDPE	100 m	1300.00
Turf area irrigation; laterals to mains; to cultivated soil; excavate trench 150 mm deep using hoe or mattock; lay moisture leaking pipe laid 150 mm below ground at centres of 350 mm		
low leak	100 m²	620.00
high leak	100 m²	560.00
Landscape area irrigation; laterals to mains; moisture leaking pipe laid to the surface of irrigated areas at 600 mm centres		
low leak	100 m²	330.00
high leak	100 m²	290.00
Landscape area irrigation; laterals to mains; moisture leaking pipe laid to the surface of irrigated areas at 900 mm centres		
low leak	100 m²	220.00
high leak	100 m²	195.00
multistation controller	each	415.00
solenoid valves connected to automatic controller	each	530.00

WATER FEATURES

Item Excluding site overheads and profit	Unit	Total rate £
LAKES AND PONDS		
FAIRWATER LTD		
Excavate for small pond or lake maximum depth 1.00 m; remove arisings off site; grade and trim to shape; lay 75 mm sharp sand; line with 0.75 mm butyl liner 75 mm sharp sand and geofabric; cover over with sifted topsoil; anchor liner to anchor trench; install balancing tank and automatic top-up system		
Pond or lake of organic shape		
100 m²; perimeter 50 m	each	5200.00
250 m²; perimeter 90 m	each	7800.00
500 m²; perimeter 130 m	each	17500.00
1000 m²; perimeter 175 m	each	35000.00
Excavate for lake average depth 1.0 m, allow for bringing to specified levels; reserve topsoil; remove spoil to approved dump on site; remove all stones and debris over 75 mm; lay polythene sheet including welding all joints and seams by specialist; screen and replace topsoil 200 mm thick		
Prices are for lakes of regular shape		
500 micron sheet	1000 m	15000.00
1000 micron sheet	1000 m	18500.00
Extra for removing spoil to tip	m³	17.90
Extra for 25 mm sand blinding to lake bed	100 m²	105.00
Extra for screening topsoil	m²	1.10
Extra for spreading imported topsoil	100 m²	620.00
Plant aquatic plants in lake topsoil		
Aponogeton distachyum PC £312.00/100	100	74.00
Acorus calamus PC £198.00/100	100	74.00
Butomus umbellatus PC £198.00/100	100	74.00
Typha latifolia PC £198.00/100	100	74.00
Nymphaea PC £816.00/100	100	74.00
Formal water features		
Excavate and construct water feature of regular shape; lay 100 mm hardcore base and 150 mm concrete 1:2:4 site mixed; line base and vertical face with butyl liner 0.75 micron and construct vertical sides of reinforced blockwork; rendering two coats; anchor the liner behind blockwork; install pumps balancing tanks and all connections to mains supply		
1.00 × 1.00 × 1.00 m deep	nr	2425.00
2.00 × 1.00 × 1.00 m deep	nr	3050.00

TIMBER DECKING

Item Excluding site overheads and profit	Unit	Total rate £
AVS FENCING		
Timber decking; support structure of timber joists for decking laid on blinded base **(measured separately)**		
joists 50 × 150 mm	10 m²	360.00
joists 50 × 200 mm	10 m²	350.00
joists 50 × 250 mm	10 m²	300.00
As above but inclusive of timber decking boards in yellow cedar 142 mm wide × 42 mm thick		
joists 50 × 150 mm	10 m²	840.00
joists 50 × 200 mm	10 m²	830.00
joists 50 × 250 mm	10 m²	780.00
As above but boards 141 × 26 mm thick		
joists 50 × 200 mm	10 m²	350.00
joists 50 × 250 mm	10 m²	300.00
As above but decking boards in red cedar 131 mm wide × 42 mm thick		
joists 50 × 150 mm	10 m²	730.00
joists 50 × 200 mm	10 m²	730.00
joists 50 × 250 mm	10 m²	680.00
Add to all of the above for handrails fixed to posts 100 × 100 × 1370 mm high		
square balusters at 100 mm centres	m	72.00
square balusters at 300 mm centres	m	46.00
turned balusters at 100 mm centres	m	92.00
turned balusters at 300 mm centres	m	53.00

SPON'S PRICE BOOKS 2011

Spon's Architects' and Builders' Price Book 2011

DAVIS LANGDON

The most detailed, professionally relevant source of UK construction price information currently available anywhere.

New Measured Works items include bio diverse roofs; Clayboard void formers; fire resisting glass blocks; more UPVC window options; glazing; insulating panels; more internal door options; blister tactile paving; Metsec SFS framing; Ecosil paint. Approximate Estimating items include lift pits; Corium brick tiles; solar hot water; photovoltaic cells; and polished plaster. New elemental building cost models are added on land remediation; school refurbishment; and office refurbishment.

Hbk & electronic package*:
1000pp approx.: 978-0-415-58845-4

electronic package only:
978-0-203-84612-4
(inc. sales tax where appropriate)

Spon's Mechanical and Electrical Services Price Book 2011

DAVIS LANGDON ENGINEERING SERVICES

Still the only comprehensive and up to date annual services engineering price book available for the UK. This year the information on energy has been reworked, along with additional engineering design details and more design schematics.

Hbk & electronic package*:
640pp approx.: 978-0-415-58851-5

electronic package only:
978-0-203-84609-4
(inc. sales tax where appropriate)

Spon's External Works and Landscape Price Book 2011

DAVIS LANGDON

The only comprehensive source of information for detailed external works and landscape costs.

This year provides a number of new Major Works items: a priced preliminary section A, as NBC layout, Klargester petrol interceptors, Terravent soil decompaction, bespoke steel edging and green walls. And Minor Works items: garden lighting infrastructure and low voltage lighting, as well as green walls and Terravent soil decompaction …and lots of new approximate estimate items.

Hbk & electronic package*:
424pp approx.: 978-0-415-58849-2

electronic package only:
978-0-203-84610-0
(inc. sales tax where appropriate)

Spon's Civil Engineering and Highway Works Price Book 2011

DAVIS LANGDON

Materials prices are still rising quickly, and tender prices are falling… Assumptions on overheads and profits and preliminaries have been kept low. Several items, for example bridge bearing prices, have been significantly revised.

Hbk & electronic package*:
800pp approx.: 978-0-415-58847-8

electronic package only:
978-0-203-84611-7
(inc. sales tax where appropriate)

---

*Receive our eBook free when you order any hard copy Spon 2011 Price Book, with free estimating software to help you produce tender documents, customise data, perform word searches and simple calculations. Or buy just the ebook, with free estimating software. Visit **www.pricebooks.co.uk**

To Order: Tel: +44 (0) 1235 400524 **Fax:** +44 (0) 1235 400525

or Post: Taylor and Francis Customer Services,
Bookpoint Ltd, Unit T1, 200 Milton Park, Abingdon, Oxon, OX14 4SB

Email: book.orders@tandf.co.uk

A complete listing of all our books is on:
www.sponpress.com

Spon Press
an imprint of Taylor & Francis

PART 3

Prices for Measured Works – Major Works

INTRODUCTION

Typical Project Profile

Contract value	£100,000.00–£300,000.00
Labour rate (see page 6–8)	£18.50 per hour
Labour rate for maintenance contracts	£15.50 per hour
Number of site staff	30
Project area	6000 m^2
Project location	Outer London
Project components	50% hard landscape 50% soft landscape and planting
Access to works areas	Very good
Contract	Main contract
Delivery of materials	Full loads
Profit and site overheads	Excluded

NEW ITEMS FOR THIS EDITION

Item Excluding site overheads and profit	PC £	Labour hours	Labour £	Plant £	Material £	Unit	Total rate £
A PRELIMINARY COSTS							
Note: The preliminary requirements are different for all projects. Many are time related many are set costs. Others vary on the project requirements. We have attempted to provide a guide to the cost related items in Section A of the National Building Specification. These costs are a guide only. Users of the book should evaluate each project separately. There are other preliminary cost items which may be added in to the preliminary costs of any project which may not be featured here.							
A11 TENDER AND CONTRACT DOCUMENTS							
Tender and contract documents							
Drawing and plan printing and distribution costs; plans issued by employer on CD; project value							
£30,000	–	–	–	–	19.00	nr	**19.00**
£30,000 to £80,000	–	–	–	–	47.50	nr	**47.50**
£80,000 to £150,000	–	–	–	–	95.00	nr	**95.00**
£150,000 to £300,000	–	–	–	–	133.00	nr	**133.00**
£300,000 to £1,000,000	–	–	–	–	228.00	nr	**228.00**
Contract drawings for project management and distribution to suppliers subcontractors and site staff							
£30,000	–	–	–	–	38.00	nr	**38.00**
£30,000 to £80,000	–	–	–	–	95.00	nr	**95.00**
£80,000 to £150,000	–	–	–	–	190.00	nr	**190.00**
£150,000 to £300,000	–	–	–	–	266.00	nr	**266.00**
£300,000 to £1,000,000	–	–	–	–	456.00	nr	**456.00**
A12 THE SITE/EXISTING BUILDINGS							
A12/140 – Existing mains and services; mark positions of existing mains and services; locate and mark; site area							
less than 500 m²	–	1.50	27.75	–	–	nr	**27.75**
up to 1000 m²	–	3.00	55.50	–	–	nr	**55.50**
up to 2000 m²	–	8.00	148.00	–	–	nr	**148.00**
up to 4000 m²	–	16.00	296.00	–	–	nr	**296.00**
A12/200 – Access to the site							
For pedestrians							
security kiosk	–	–	–	65.00	–	week	**65.00**
security guard	–	40.00	400.00	–	–	week	**400.00**
protected walkways; Heras fencing on 2 sides	–	–	–	4.21	–	m	**4.21**
A12/210 – Parking							
Parking expenses where vehicles do not park on the site area; per vehicle							
metropolitan area; city centre	–	–	–	–	–	week	**200.00**
metropolitan area; outer areas	–	–	–	–	–	week	**160.00**
suburban restricted parking areas	–	–	–	–	–	week	**40.00**

NEW ITEMS FOR THIS EDITION

Item Excluding site overheads and profit	PC £	Labour hours	Labour £	Plant £	Material £	Unit	Total rate £
A12/210 – Congestion charging							
London only	–	–	–	–	–	week	40.00
A12/250 – Site visit; pre-tender for purposes of understanding site and tender requirements; prices below are based on a single senior manager attending site							
City location; average distance of travel 20 miles inclusive of travel and parking costs							
less than 500 m²	–	4.00	140.00	–	36.00	nr	176.00
up to 1000 m²	–	5.00	175.00	–	38.00	nr	213.00
up to 2000 m²	–	6.00	210.00	–	44.00	nr	254.00
up to 4000 m²	–	8.00	280.00	–	52.00	nr	332.00
Town or rural location; average distance of travel 20 miles inclusive of travel and parking costs							
less than 500 m²	–	3.00	105.00	–	22.00	nr	127.00
up to 1000 m²	–	4.00	140.00	–	23.00	nr	163.00
up to 2000 m²	–	5.00	175.00	–	24.00	nr	199.00
up to 4000 m²	–	6.00	210.00	–	24.00	nr	234.00
A20 THE CONTRACT/SUBCONTRACT							
Contract/subcontract evaluation							
Due diligence on evaluation or examination of clauses in contract documents; evaluation and report by a suitably qualified quantity surveyor or legal advisor where necessary							
minor works contract	–	1.00	35.00	–	–	nr	35.00
JCLI	–	–	–	–	70.00	nr	70.00
JCT intermediate contract or subcontract	–	1.00	35.00	–	140.00	nr	175.00
A30 EMPLOYER'S REQUIREMENTS: TENDERING/SUBLETTING/SUPPLY							
A30/480 – Programmes; tender stage programmes							
Allow for production of an outline works programmes for submission with the tenders; project value							
£30,000	–	1.50	52.50	–	–	nr	52.50
£50,000	–	2.00	70.00	–	–	nr	70.00
£75,000	–	2.50	87.50	–	–	nr	87.50
£100,000	–	3.00	105.00	–	–	nr	105.00
£200,000	–	4.00	140.00	–	–	nr	140.00
£500,000	–	4.50	157.50	–	–	nr	157.50
£1,000,000	–	6.00	210.00	–	–	nr	210.00
Method statements for private or commercial projects where award is not points based; provide detailed method statements on all aspects of the works; project value							
£30,000	–	2.50	87.50	–	–	nr	87.50
£50,000	–	3.50	122.50	–	–	nr	122.50
£75,000	–	4.00	140.00	–	–	nr	140.00
£100,000	–	5.00	175.00	–	–	nr	175.00
£200,000	–	6.00	210.00	–	–	nr	210.00
£500,000	–	7.00	245.00	–	–	nr	245.00
£1,000,000	–	8.00	280.00	–	–	nr	280.00

NEW ITEMS FOR THIS EDITION

Item Excluding site overheads and profit	PC £	Labour hours	Labour £	Plant £	Material £	Unit	Total rate £
A30 EMPLOYER'S REQUIREMENTS: TENDERING/SUBLETTING/SUPPLY – cont							
A30/500 – Method statements for local authority type projects where award is points rated; provide detailed method statements on all aspects of the works; project value							
£30,000	–	5.00	175.00	–	–	nr	175.00
£50,000	–	5.00	175.00	–	–	nr	175.00
£75,000	–	6.00	210.00	–	–	nr	210.00
£100,000	–	7.00	245.00	–	–	nr	245.00
£200,000	–	8.00	280.00	–	–	nr	280.00
£500,000	–	9.00	315.00	–	–	nr	315.00
£1,000,000	–	16.00	560.00	–	–	nr	560.00
A30/550-570 – Health and safety Produce health and safety file including preliminary meeting and subsequent progress meetings with external planning officer in connection with health and safety; project value							
£35,000	–	8.00	280.00	–	–	nr	280.00
£75,000	–	12.00	420.00	–	–	nr	420.00
£100,000	–	16.00	560.00	–	–	nr	560.00
£200,000 to £500,000	–	40.00	1400.00	–	–	nr	1400.00
Maintain health and safety file for project duration; project value							
£35,000	–	4.00	140.00	–	–	week	140.00
£75,000	–	4.00	140.00	–	–	week	140.00
£100,000	–	8.00	280.00	–	–	week	280.00
£200,000 to £500,000	–	8.00	280.00	–	–	week	280.00
Produce written risk assessments on all areas of operations within the scope of works of the contract; project value							
£35,000	–	2.00	70.00	–	–	nr	70.00
£75,000	–	3.00	105.00	–	–	nr	105.00
£100,000	–	5.00	175.00	–	–	nr	175.00
£200,000 to £500,000	–	8.00	280.00	–	–	nr	280.00
Produce COSHH assessments on all substances to be used in connection with the contract; project value							
£35,000	–	1.50	52.50	–	–	nr	52.50
£75,000	–	2.00	70.00	–	–	nr	70.00
£100,000	–	3.00	105.00	–	–	nr	105.00
£200,000 to £500,000	–	3.00	105.00	–	–	nr	105.00
A31 EMPLOYER'S REQUIREMENTS: PROVISION, CONTENT AND USE OF DOCUMENTS							
Provision, content and use of documents Supply as built drawings for elements of the project that may have varied from the original design drawings							
£35,000	–	2.00	70.00	–	–	nr	70.00
£75,000	–	2.00	70.00	–	–	nr	70.00
£100,000	–	4.00	140.00	–	–	nr	140.00
£200,000 to £500,000	–	5.00	175.00	–	–	nr	175.00

NEW ITEMS FOR THIS EDITION

Item Excluding site overheads and profit	PC £	Labour hours	Labour £	Plant £	Material £	Unit	Total rate £
A31/155 – Fees for Considerate **Contractors Scheme (CCS)**							
Project value							
up to £100,000	–	–	–	–	100.00	nr	100.00
£100,000–£500000	–	–	–	–	200.00	nr	200.00
£500,000–£5,000,000	–	–	–	–	400.00	nr	400.00
over £5,000,000	–	–	–	–	600.00	nr	600.00
Climatic conditions – keep records of							
temperature and rainfall; delays due to							
weather including descriptions of weather							
daily cost	–	0.08	2.92	–	–	day	2.92
weekly cost	–	0.42	14.58	–	–	day	14.58
A31/211 – Programmes; master programme **for the contract works**							
Allow for production of works programmes							
prior to the start of the works; project value							
£30,000	–	3.00	105.00	–	–	nr	105.00
£50,000	–	6.00	210.00	–	–	nr	210.00
£75,000	–	8.00	280.00	–	–	nr	280.00
£100,000	–	10.00	350.00	–	–	nr	350.00
£200,000	–	14.00	490.00	–	–	nr	490.00
£500,000	–	15.00	525.00	–	–	nr	525.00
£1,000,000	–	18.00	630.00	–	–	nr	630.00
A31/212 – Indicative staff resource chart							
Project value							
up to £100000	–	1.00	35.00	–	–	nr	35.00
up to £500000	–	1.50	52.50	–	–	nr	52.50
up to £1,000,000	–	2.00	70.00	–	–	nr	70.00
A31/213 – Curriculum vitae of staff							
Prepare and submit curriculum vitae of							
pre-construction and construction phase staff							
per staff member	–	0.75	26.25	–	–	nr	26.25
A33 SUBMISSION OF SAMPLES							
Set up sample panels for the works **inclusive of arranging delivery of sample** **materials; paving samples on 100 mm base**							
Costs of labours only							
brick paving panel (pointed)	–	6.00	111.00	–	–	m²	111.00
brick paving panel (butt jointed)	–	4.00	74.00	–	–	m²	74.00
block paving panel	–	4.00	74.00	–	–	m²	74.00
stone slab paving panel	–	6.00	111.00	–	–	m²	111.00
cladding panel	–	7.00	129.50	–	–	m²	129.50
brick wall panel	–	5.00	92.50	–	–	m²	92.50
render panel	–	3.50	64.75	–	–	m²	64.75
paint panel	–	3.00	55.50	–	–	m²	55.50

NEW ITEMS FOR THIS EDITION

Item Excluding site overheads and profit	PC £	Labour hours	Labour £	Plant £	Material £	Unit	Total rate £
F10 PIERS							
Isolated brick piers in English bond							
Engineering brick PC £239.00/1000 (not SMM)							
one brick thick (225 mm)	–	1.63	30.23	–	8.63	m	**38.86**
one and a half bricks thick (337.5 mm)	–	3.00	55.50	–	25.57	m	**81.07**
two bricks thick (450 mm)	–	4.50	83.25	–	40.87	m	**124.12**
Brick PC £300.00/1000 (not SMM)							
one brick thick (225 mm)	–	1.63	30.23	–	10.43	m	**40.66**
one and a half bricks thick (337.5 mm)	–	3.00	55.50	–	29.46	m	**84.96**
two bricks thick (450 mm)	–	4.50	83.25	–	47.79	m	**131.04**
three bricks thick (675 mm)	–	8.37	154.84	–	99.08	m	**253.92**
Brick PC £800.00/1000 (not SMM)							
one brick thick (225 mm)	–	1.63	30.23	–	25.13	m	**55.36**
one and a half bricks thick (337.5 mm)	–	3.00	55.50	–	61.36	m	**116.86**
two bricks thick (450 mm)	–	4.50	83.25	–	104.49	m	**187.74**
three bricks thick (675 mm)	–	8.37	154.84	–	226.65	m	**381.49**
Q31 PLATIPUS TREE ANCHORS							
Platipus anchors; anchoring kits for **semi-mature trees**							
Deadman kits; anchoring to concrete kerbs							
placed in base of tree pit							
trees 12–25 cm girth; 2.5–4.0 m high	26.62	0.75	13.88	–	41.24	nr	**55.12**
trees 25–45 cm girth; 4.0–7.5 m high	45.82	1.50	27.75	–	60.44	nr	**88.19**
trees 45–75 cm girth; 7.5–12.0 m high	150.86	2.00	37.00	–	165.48	nr	**202.48**
Q31 PLATIPUS TREE WATERING **SYSTEMS**							
Tree irrigation kits; direct water delivery to **the rootball area of trees**							
Piddler tree irrigation system; permeable							
membrane delivering targeted irrigation to							
rootball surround							
root balls up to 550 mm diameter; Irrigation							
Kit PID0	7.00	0.25	4.63	–	7.00	nr	**11.63**
root balls up to 900 mm diameter; Irrigation							
Kit PID1	12.50	0.30	5.55	–	12.50	nr	**18.05**
root balls up to 1.55 m diameter; Irrigation							
Kit PID2	16.50	0.40	7.40	–	16.50	nr	**23.90**
root balls up to 2.40 m diameter; Irrigation							
Kit PID3	20.50	0.45	8.32	–	20.50	nr	**28.82**
root balls up to 3.10 m diameter; Irrigation							
Kit PID4	25.50	0.50	9.25	–	25.50	nr	**34.75**

NEW ITEMS FOR THIS EDITION

Item Excluding site overheads and profit	PC £	Labour hours	Labour £	Plant £	Material £	Unit	Total rate £
Q31 LORENZ VON EHREN – PLEACHED/ PRUNED TREES							
Pleached trees; Carpinus betulus; specimen trees; Lorenz Von Ehren; supply and planting only; excavation, support and treepit additives not included							
Box shaped trees to provide 'floating hedge' or screen effect; planting distance 1 tree per m run; trunk and box alignment; wire root balls							
4 × transplanted; 20–25 cm; 1.00 m centres	300.00	5.00	92.50	16.62	300.00	m	**409.12**
5 × transplanted; 25–30 cm; 1.00 m centres	430.00	6.50	120.25	20.68	430.00	m	**570.93**
5 × transplanted; 30–35 cm; 1.20 m centres	720.00	11.00	203.50	27.68	720.00	m	**951.18**
Pleached trees; Tilia europaea 'Pallida'; specimen trees; Lorenz Von Ehren; supply and planting only; excavation, support and treepit additives not included							
Box pleached trees to provide 'floating hedge' or screen effect; wire root balls							
4 × transplanted; 20–25 cm; 1.00 m centres	300.00	5.00	92.50	16.62	300.00	m	**409.12**
4 × transplanted; 25–30 cm; 1.00 m centres	370.00	5.00	92.50	16.62	370.00	m	**479.12**
5 × transplanted; 30–35 cm; 1.20 m centres	358.32	5.00	92.50	16.62	358.32	m	**467.44**
5 × transplanted; 35–40 cm; 1.50 m centres	380.02	5.00	92.50	16.62	380.02	m	**489.14**
6 × transplanted; 40–45 cm; 1.80 m centres	444.48	5.00	92.50	16.62	444.48	m	**553.60**
6 × transplanted; 45–50 cm; 1.80 m centres	550.04	5.00	92.50	16.62	550.04	m	**659.16**
6 × transplanted; 50–60 cm; 2.00 m centres	700.00	5.00	92.50	16.62	700.00	m	**809.12**
Espalier pleached trees; specimen trees; Lorenz Von Ehren; supply and planting only; excavation, support and treepit additives not included							
Frame pleached trees; to provide floating screen effects; wire root balls							
Carpinus betulus; 20–25 cm; 1.00 m centres	420.00	5.00	92.50	16.62	420.00	m	**529.12**
Carpinus betulus; 25–30 cm; 1.00 m centres	550.00	5.00	92.50	16.62	550.00	m	**659.12**
Tilia europaea 'Pallida'; 20–25 cm; 1.00 m centres	420.00	5.00	92.50	16.62	420.00	m	**529.12**
Tilia europaea 'Pallida'; 25–30 cm; 1.00 m centres	550.00	5.00	92.50	16.62	550.00	m	**659.12**
Tilia europaea 'Pallida'; 30–35 cm; 1.20 m centres	658.31	5.00	92.50	16.62	658.31	m	**767.43**
Umbrella shaped pleaches							
Umbrella or roof pleached trees to provide umbrella effect; wire root balls							
Tilia euchlora; specimen; 4 × transplanted; 20–25 cm	390.00	5.00	92.50	16.62	390.00	ea	**499.12**
Tilia euchlora; specimen; 5 × transplanted; 25–30 cm	480.00	5.00	92.50	16.62	480.00	ea	**589.12**
Platanus acerifolia; clear stem; 4 × transplanted; 20–25 cm	390.00	5.00	92.50	16.62	390.00	ea	**499.12**
Platanus acerifolia; specimen; 5 × transplanted; 30–35 cm	700.00	5.00	92.50	16.62	700.00	ea	**809.12**
Tilia europaea; specimen; 5 × transplanted; 30–35 cm	700.00	5.00	92.50	16.62	700.00	ea	**809.12**
Platanus acerifolia; specimen; 5 × transplanted; 25–30 cm	480.00	5.00	92.50	16.62	480.00	ea	**589.12**

NEW ITEMS FOR THIS EDITION

Item Excluding site overheads and profit	PC £	Labour hours	Labour £	Plant £	Material £	Unit	Total rate £
Q31 GREEN SCREEN; MOBILANE LTD							
Green Screen; Mobilane Ltd Fully installed vertical green screen of Hedera hibernica; 65 plants per linear m; all on 5 mm galvanized steel weldmesh; attached to existing wall fence or hoarding							
screen 1.80 high	–	–	–	–	170.00	m	**170.00**
screen 2.20 m high	–	–	–	–	230.00	m	**230.00**
screen 4.00 m high	–	–	–	–	500.00	m	**500.00**
Living wall; Mobilane Ltd LivePanel; fully installed living wall adjusted to fit modules with developed substrate; panels are stacked onto building or wall face of the building and secured by an aluminium frame	–	–	–	–	600.00	m²	**600.00**
Q50 STREET FURNITURE							
Anti-skateboard devices Stainless steel; Furnitubes International Ltd							
for fitting to seats	18.00	0.20	3.70	–	18.00	ea	**21.70**
for fitting to benches	36.00	0.20	3.70	–	36.00	ea	**39.70**
Bollards Recycled plastic bollards; Furnitubes International Ltd							
Aberdeen Circular ABR150; 1000 × 150 mm diameter	56.00	2.00	37.00	–	61.21	ea	**98.21**
Aberdeen Square ABR140; 1000 × 140 × 140 mm	35.00	2.00	37.00	–	40.21	ea	**77.21**
Service bollards; Furnitubes International Ltd; stainless steel; for housing services connection points to electricity, water etc. (services connections not included)							
Kenton KEN717; 900 × 250 mm diameter	595.00	2.00	37.00	–	600.21	ea	**637.21**
Zenith ZEN707; 900 × 250 mm diameter	550.00	2.00	37.00	–	555.21	ea	**592.21**
Cigarette bins Furnitubes International Ltd							
ZEN 275 Zenith cigarette bin; stainless steel; wall or post mounted; 1.7 L	29.99	0.75	13.88	–	35.71	ea	**49.59**
LVR250 PC Liverpool cigarette bin; steel; wall mounted; 1.9 L	16.75	0.50	9.25	–	16.75	ea	**26.00**
SMK500F Smoke King cigarette bin; cast aluminium; bolt down; 3.5 L	199.99	0.75	13.88	–	205.71	ea	**219.59**
Outdoor seats; recycled plastic Furnitubes International Ltd							
Aberdeen seat, brown recycled plastic; 1 slat; 2 m long	296.00	2.00	37.00	–	315.57	ea	**352.57**
Aberdeen seat, brown recycled plastic; 2 slats; 2 m long	401.00	2.00	37.00	–	420.57	ea	**457.57**
Picnic table; recycled plastic Furnitubes International Ltd							
Dundee picnic table; brown recycled plastic; 1.8 m long	643.00	2.00	37.00	–	662.57	ea	**699.57**

NEW ITEMS FOR THIS EDITION

Item Excluding site overheads and profit	PC £	Labour hours	Labour £	Plant £	Material £	Unit	Total rate £
Cycle shelters Furnitubes International Ltd; Academy freestanding bolt down shelter, galvanized steel frame and roof; 3450 mm overall width; 2150 mm maximum height; 2670 mm depth							
corrugated roof	1873.00	6.00	111.00	–	1884.45	ea	**1995.45**
clear polycarbonate roof	2024.00	6.00	111.00	–	2035.45	ea	**2146.45**
Smoking shelters Furnitubes International Ltd; Ashby freestanding bolt down; aluminium frame; PET glazing; 2050 mm length × 2330 mm high							
3290 mm depth; maximum 3 people	2067.00	4.00	74.00	–	2078.45	ea	**2152.45**
6050 mm depth; maximum 6 people	2597.00	5.00	92.50	–	2608.45	ea	**2700.95**
Monolith signage boards Furnitubes International Ltd; Fulham monolith signage board; stainless steel frame; vinyl graphics; 2600 mm above ground; 600 mm below ground; 350 mm width × 120 mm depth							
without baseplate	3550.00	2.00	37.00	–	3554.83	ea	**3591.83**
with baseplate	3700.00	2.00	37.00	–	3704.83	ea	**3741.83**
V ELECTRICAL – ELECTRICAL CABLES							
Cable to trenches (trenching operations not included); laying only cable Twin core steel wire armoured cable; 50 m drums							
1.5 mm core	0.80	–	–	–	0.80	m	**0.80**
2.5 mm core	0.96	–	–	–	0.96	m	**0.96**
Twin core steel wire armoured cable; lengths less than 50 m runs							
1.5 mm core	0.86	–	–	–	0.86	m	**0.86**
2.5 mm core	1.04	–	–	–	1.04	m	**1.04**
Three core steel wire armoured cable; 50 m drums							
1.5 mm core	0.90	–	–	–	0.90	m	**0.90**
2.5 mm core	1.23	–	–	–	1.23	m	**1.23**
4.0 mm core	1.48	–	–	–	1.48	m	**1.48**
6.0 mm core	1.92	–	–	–	1.92	m	**1.92**
10.0 mm core	3.39	–	–	–	3.39	m	**3.39**
Three core steel wire armoured cable; lengths less than 50 m runs							
1.5 mm core	0.93	–	–	–	0.93	m	**0.93**
2.5 mm core	1.33	–	–	–	1.33	m	**1.33**
Ducts to trenches; twin wall flexible cable ducts with drawstrings; laid on 150 mm clean sharp sand Twin wall duct; laying to trenches							
63 mm × 50 m coils	1.00	0.02	0.37	–	1.00	m	**1.37**
110 mm × 50 m coils	1.50	0.02	0.37	–	1.50	m	**1.87**
Cable drawing through ducts							
straight runs up to 50 m lengths	–	1.00	18.50	–	–	nr	**18.50**

NEW ITEMS FOR THIS EDITION

Item Excluding site overheads and profit	PC £	Labour hours	Labour £	Plant £	Material £	Unit	Total rate £
V ELECTRICAL – ELECTRICAL CABLES – cont							
Terminations to armoured cables; cutting coiling and taping length of cable to receive connection to light fitting, transformer and the like; fixing to temporary post							
Armoured cable	–	0.33	6.16	–	–	m	6.16
Plain ducted cable	–	0.13	2.31	–	–	m	2.31
Junction boxes; fixing to cables to receive connections to light fittings or transformers; to IP68							
Underground jointing box							
2 way	29.00	0.33	6.83	–	29.00	m	35.83
3 way	34.00	0.57	11.71	–	34.00	m	45.71
Underground jointing box							
2 way	3.50	0.33	6.83	–	3.50	m	10.33
3 way	4.90	0.40	8.20	–	4.90	m	13.10
Electrical connections							
Connections to existing distribution boards per switching circuit located within 1 m of the distribution board; chasing of cables to walls not included	–	–	–	–	–	nr	41.67
Connections to switches; external power cable circuits; connections to internal wall switches; drilling through walls and making good; chasing of cables to walls not included							
price for the first circuit	4.00	1.00	18.50	–	45.67	nr	64.17
additional circuits	4.00	–	–	–	14.00	nr	14.00
V ELECTRICAL – TRENCHING FOR ELECTRICAL SERVICES							
Trenching for electrical services							
3 tonne excavator (bucket volume 0.13 m³); arisings laid alongside							
600 mm deep	–	–	–	1.84	–	m	1.84
800 mm deep	–	–	–	2.15	–	m	2.15
1.00 m deep	–	–	–	2.57	–	m	2.57
Extra over any types of excavating irrespective of depth for breaking out existing materials; heavy duty 110 volt breaker tool							
hard rock	–	5.00	92.50	19.10	–	m³	111.60
concrete	–	3.00	55.50	11.46	–	m³	66.96
reinforced concrete	–	4.00	74.00	19.50	–	m³	93.50
brickwork, blockwork or stonework	–	1.50	27.75	5.73	–	m³	33.48
By hand							
600 mm deep	–	0.24	4.45	–	–	m	4.45
800 mm deep	–	0.43	7.91	–	–	m	7.91
1.00 m deep	–	0.67	12.33	–	–	m	12.33

NEW ITEMS FOR THIS EDITION

Item Excluding site overheads and profit	PC £	Labour hours	Labour £	Plant £	Material £	Unit	Total rate £
Backfilling of trenches; including laying of electrical marker tape; compacting lightly as work proceeds							
To trenches containing armoured cable							
by machine	0.13	–	–	2.28	0.13	m	**2.41**
by hand	0.13	0.20	3.70	–	0.13	m	**3.83**
To trenches containing ducted cable; including 150 mm sharp sand over the duct							
by machine	1.97	–	–	2.85	1.97	m	**4.82**
by hand	1.97	0.25	4.63	–	1.97	m	**6.60**

A PRELIMINARIES

Item Excluding site overheads and profit	PC £	Labour hours	Labour £	Plant £	Material £	Unit	Total rate £
A11 TENDER AND CONTRACT DOCUMENTS							
Method statements							
Provide detailed method statements on all aspects of the works; project value							
£30,000	–	2.50	87.50	–	–	nr	87.50
£50,000	–	3.50	122.50	–	–	nr	122.50
£75,000	–	4.00	140.00	–	–	nr	140.00
£100,000	–	5.00	175.00	–	–	nr	175.00
£200,000	–	6.00	210.00	–	–	nr	210.00
Tender and contract documents							
Drawing and plan printing and distribution costs; plans issued by employer on CD; project value							
£30,000	–	–	–	–	19.00	nr	19.00
£30,000 to £80,000	–	–	–	–	47.50	nr	47.50
£80,000 to £150,000	–	–	–	–	95.00	nr	95.00
£150,000 to £300,000	–	–	–	–	133.00	nr	133.00
£300,000 to £1,000,000	–	–	–	–	228.00	nr	228.00
Contract drawings for project management and distribution to suppliers, subcontractors and site staff							
£30,000	–	–	–	–	38.00	nr	38.00
£30,000 to £80,000	–	–	–	–	95.00	nr	95.00
£80,000 to £150,000	–	–	–	–	190.00	nr	190.00
£150,000 to £300,000	–	–	–	–	266.00	nr	266.00
£300,000 to £1,000,000	–	–	–	–	456.00	nr	456.00
A12 THE SITE/EXISTING BUILDINGS							
A12/140 – Existing mains and services; mark positions of existing mains and services; locate and mark; site area							
less than 500 m²	–	1.50	27.75	–	–	nr	27.75
up to 1000 m²	–	3.00	55.50	–	–	nr	55.50
up to 2000 m²	–	8.00	148.00	–	–	nr	148.00
up to 4000 m²	–	16.00	296.00	–	–	nr	296.00
A12/200 – Access to the site							
For pedestrians							
security kiosk	–	–	–	65.00	–	week	65.00
security guard	–	40.00	400.00	–	–	week	400.00
protected walkways; Heras fencing on 2 sides	–	–	–	4.21	–	m	4.21
A12/210 – Parking							
Parking expenses where vehicles do not park on the site area; per vehicle							
metropolitan area; city centre	–	–	–	–	–	week	200.00
metropolitan area; outer areas	–	–	–	–	–	week	160.00
suburban restricted parking areas	–	–	–	–	–	week	40.00
A12/210 – Congestion charging							
London only	–	–	–	–	–	week	40.00

A PRELIMINARIES

Item Excluding site overheads and profit	PC £	Labour hours	Labour £	Plant £	Material £	Unit	Total rate £
A12/250 – Site visit; pre-tender for purposes of understanding site and tender requirements; prices below are based on a single senior manager attending site City location; average distance of travel 20 miles inclusive of travel and parking costs							
less than 500 m²	–	4.00	140.00	–	36.00	nr	**176.00**
up to 1000 m²	–	5.00	175.00	–	38.00	nr	**213.00**
up to 2000 m²	–	6.00	210.00	–	44.00	nr	**254.00**
up to 4000 m²	–	8.00	280.00	–	52.00	nr	**332.00**
Town or rural location; average distance of travel 20 miles inclusive of travel and parking costs							
less than 500 m²	–	3.00	105.00	–	22.00	nr	**127.00**
up to 1000 m²	–	4.00	140.00	–	23.00	nr	**163.00**
up to 2000 m²	–	5.00	175.00	–	24.00	nr	**199.00**
up to 4000 m²	–	6.00	210.00	–	24.00	nr	**234.00**
A20 THE CONTRACT/SUBCONTRACT							
Contract/subcontract evaluation Due diligence on evaluation or examination of clauses in contract documents; evaluation and report by a suitably qualified quantity surveyor or legal advisor where necessary							
minor works contract	–	1.00	35.00	–	–	nr	**35.00**
JCLI	–	–	–	–	70.00	nr	**70.00**
JCT intermediate contract or subcontract	–	1.00	35.00	–	140.00	nr	**175.00**
A30 EMPLOYER'S REQUIREMENTS: TENDERING/SUBLETTING/SUPPLY							
Tendering costs for an employed estimator for the acquisition of a landscape main or subcontract; inclusive of measurement, sourcing of materials and suppliers, cost and area calculations, pre- and post-tender meetings, bid compliance method statements and submission documents Remeasureable contract from prepared bills; contract value							
£25,000	–	–	–	–	–	nr	**180.00**
£50,000	–	–	–	–	–	nr	**300.00**
£100,000	–	–	–	–	–	nr	**720.00**
£200,000	–	–	–	–	–	nr	**1200.00**
£500,000	–	–	–	–	–	nr	**1560.00**
£1,000,000	–	–	–	–	–	nr	**2400.00**
Lump sum contract from specifications and drawings only; contract value							
£25,000	–	–	–	–	–	nr	**480.00**
£50,000	–	–	–	–	–	nr	**600.00**
£100,000	–	–	–	–	–	nr	**1200.00**
£200,000	–	–	–	–	–	nr	**2160.00**
£500,000	–	–	–	–	–	nr	**3000.00**
£1,000,000	–	–	–	–	–	nr	**4800.00**

A PRELIMINARIES

Item Excluding site overheads and profit	PC £	Labour hours	Labour £	Plant £	Material £	Unit	Total rate £
A30 EMPLOYER'S REQUIREMENTS: **TENDERING/SUBLETTING/SUPPLY – cont**							
Material costs for tendering Drawing and plan printing and distribution costs; plans issued by employer on CD; project value							
£30,000	–	–	–	–	19.00	nr	**19.00**
£30,000 to £80,000	–	–	–	–	47.50	nr	**47.50**
£80,000 to £150,000	–	–	–	–	95.00	nr	**95.00**
£150,000 to £300,000	–	–	–	–	133.00	nr	**133.00**
£300,000 to £1,000,000	–	–	–	–	228.00	nr	**228.00**
A31 EMPLOYER'S REQUIREMENTS: **PROVISION, CONTENT AND USE OF** **DOCUMENTS**							
Provision, content and use of documents Supply as built drawings for elements of the project that may have varied from the original design drawings							
£35,000	–	2.00	70.00	–	–	nr	**70.00**
£75,000	–	2.00	70.00	–	–	nr	**70.00**
£100,000	–	4.00	140.00	–	–	nr	**140.00**
£200,000 to £500,000	–	5.00	175.00	–	–	nr	**175.00**
A31/155 – Fees for Considerate **Contractors Scheme (CCS)** Project value							
up to £100,000	–	–	–	–	100.00	nr	**100.00**
£100,000–£500000	–	–	–	–	200.00	nr	**200.00**
£500,000–£5,000,000	–	–	–	–	400.00	nr	**400.00**
over £5,000,000	–	–	–	–	600.00	nr	**600.00**
Climatic conditions – keep records of temperature and rainfall; delays due to weather including descriptions of weather							
daily cost	–	0.08	2.92	–	–	day	**2.92**
weekly cost	–	0.42	14.58	–	–	day	**14.58**
A31/211 – Programmes; master programme **for the contract works** Allow for production of works programmes prior to the start of the works; project value							
£30,000	–	3.00	105.00	–	–	nr	**105.00**
£50,000	–	6.00	210.00	–	–	nr	**210.00**
£75,000	–	8.00	280.00	–	–	nr	**280.00**
£100,000	–	10.00	350.00	–	–	nr	**350.00**
£200,000	–	14.00	490.00	–	–	nr	**490.00**
£500,000	–	15.00	525.00	–	–	nr	**525.00**
£1,000,000	–	18.00	630.00	–	–	nr	**630.00**
A31/212 – Indicative staff resource chart Project value							
up to £100,000	–	1.00	35.00	–	–	nr	**35.00**
up to £500,000	–	1.50	52.50	–	–	nr	**52.50**
up to £1,000,000	–	2.00	70.00	–	–	nr	**70.00**

A PRELIMINARIES

Item Excluding site overheads and profit	PC £	Labour hours	Labour £	Plant £	Material £	Unit	Total rate £
A31/213 – Curriculum vitae of staff Prepare and submit curriculum vitae of pre-construction and construction phase staff							
per staff member	–	0.75	26.25	–	–	nr	26.25
A32 EMPLOYER'S REQUIREMENTS: **MANAGEMENT OF THE WORKS**							
Allow for updating the works programme during the course of the works; project value							
£30,000	–	1.00	35.00	–	–	nr	35.00
£50,000	–	1.50	52.50	–	–	nr	52.50
£75,000	–	2.00	70.00	–	–	nr	70.00
£100,000	–	3.00	105.00	–	–	nr	105.00
£200,000	–	5.00	175.00	–	–	nr	175.00
Setting out Setting out for external works operations comprising hard and soft works elements; placing of pegs and string lines to Landscape Architect's drawings; surveying levels and placing level pegs; obtaining approval from the Landscape Architect to commence works; areas of entire site							
1,000 m²	–	5.00	92.50	–	6.20	nr	98.70
2,500 m²	–	8.00	148.00	–	12.40	nr	160.40
5,000 m²	–	8.00	148.00	–	12.40	nr	160.40
10,000 m²	–	32.00	592.00	–	31.00	nr	623.00
A33 SUBMISSION OF SAMPLES							
Set up sample panels for the works **inclusive of arranging delivery of sample** **materials; paving samples on 100 mm base** Costs of labours only							
brick paving panel (pointed)	–	6.00	111.00	–	–	m²	111.00
brick paving panel (butt jointed)	–	4.00	74.00	–	–	m²	74.00
block paving panel	–	4.00	74.00	–	–	m²	74.00
stone slab paving panel	–	6.00	111.00	–	–	m²	111.00
cladding panel	–	7.00	129.50	–	–	m²	129.50
brick wall panel	–	5.00	92.50	–	–	m²	92.50
render panel	–	3.50	64.75	–	–	m²	64.75
paint panel	–	3.00	55.50	–	–	m²	55.50
A34 EMPLOYER'S REQUIREMENTS: **SECURITY/SAFETY/PROTECTION**							
Temporary security fence; HSS Hire; mesh **framed unclimbable fencing; including** **precast concrete supports and couplings** Weekly hire; 2.85 × 2.00 m high							
weekly hire rate	–	–	–	2.11	–	m	2.11
erection of fencing; labour only	–	0.10	1.85	–	–	m	1.85
removal of fencing loading to collection vehicle	–	0.07	1.23	–	–	m	1.23
delivery charge	–	–	–	0.80	–	m	0.80
return haulage charge	–	–	–	0.60	–	m	0.60

Prices for Measured Works – Major Works

A PRELIMINARIES

Item Excluding site overheads and profit	PC £	Labour hours	Labour £	Plant £	Material £	Unit	Total rate £
A41 CONTRACTOR'S GENERAL COST ITEMS: SITE ACCOMODATION							
General							
The following items are instances of the commonly found preliminary costs associated with external works contracts. The assumption is made that the external works contractor is subcontracted to a main contractor.							
Elliot Hire; erect temporary accommodation and storage on concrete base measured separately							
Prefabricated office hire; jackleg; open plan							
3.6 × 2.4 m	–	–	–	36.75	–	week	**36.75**
4.8 × 2.4 m	–	–	–	36.75	–	week	**36.75**
Armoured store; erect temporary secure storage container for tools and equipment							
3.0 × 2.4 m	–	–	–	12.00	–	week	**12.00**
3.6 × 2.4 m	–	–	–	12.00	–	week	**12.00**
Delivery and collection charges on site offices							
delivery charge	–	–	–	126.00	–	load	**126.00**
collection charge	–	–	–	126.00	–	load	**126.00**
Toilet facilities							
John Anderson Ltd; serviced self-contained toilet delivered to and collected from site; maintained by toilet supply company							
single chemical toilet including wash hand basin and water	–	–	–	27.00	–	week	**27.00**
delivery and collection; each way	–	–	–	25.00	–	nr	**25.00**
A43 CONTRACTOR'S GENERAL COST ITEMS: MECHANICAL PLANT							
Southern Conveyors; moving only of granular material by belt conveyor; conveyors fitted with troughed belts, receiving hopper, front and rear undercarriages and driven by electrical and air motors; support work for conveyor installation (scaffolding), delivery, collection and installation all excluded							
Conveyor belt width 400 mm; mechanically loaded and removed at offload point; conveyor length							
up to 5 m	–	1.50	27.75	6.64	–	m³	**34.39**
10 m	–	1.50	27.75	7.40	–	m³	**35.15**
12.5 m	–	1.50	27.75	7.76	–	m³	**35.51**
15 m	–	1.50	27.75	8.18	–	m³	**35.93**
20 m	–	1.50	27.75	8.96	–	m³	**36.71**
25 m	–	1.50	27.75	9.65	–	m³	**37.40**
30 m	–	1.50	27.75	10.41	–	m³	**38.16**

A PRELIMINARIES

Item Excluding site overheads and profit	PC £	Labour hours	Labour £	Plant £	Material £	Unit	Total rate £
Conveyor belt width 600 mm; mechanically loaded and removed at offload point; conveyor length							
up to 5 m	–	1.00	18.50	4.83	–	m³	**23.33**
10 m	–	1.00	18.50	5.44	–	m³	**23.94**
12.5 m	–	1.00	18.50	5.80	–	m³	**24.30**
15 m	–	1.00	18.50	6.13	–	m³	**24.63**
20 m	–	1.00	18.50	6.84	–	m³	**25.34**
25 m	–	1.00	18.50	7.50	–	m³	**26.00**
30 m	–	1.00	18.50	8.16	–	m³	**26.66**
Conveyor belt width 400 mm; mechanically loaded and removed by hand at offload point; conveyor length							
up to 5 m	–	3.19	59.02	5.13	–	m³	**64.15**
10 m	–	3.19	59.02	6.15	–	m³	**65.17**
12.5 m	–	3.19	59.02	6.63	–	m³	**65.65**
15 m	–	3.19	59.02	7.19	–	m³	**66.21**
20 m	–	3.19	59.02	8.23	–	m³	**67.25**
25 m	–	3.19	59.02	9.15	–	m³	**68.17**
30 m	–	3.19	59.02	10.16	–	m³	**69.18**
Terranova Cranes Ltd; crane hire; materials handling and lifting; telescopic cranes supply and management; exclusive of roadway management or planning applications; prices below illustrate lifts of 1 tonne at maximum crane reach; lift cycle of 0.25 hours; mechanical filling of material skip if appropriate							
35 tonne mobile crane; lifting capacity of 1 tonne at 26 m; C.P.A hire (all management by hirer)							
granular materials including concrete	–	0.75	13.88	16.04	–	tonne	**29.92**
palletized or packed materials	–	0.50	9.25	13.26	–	tonne	**22.51**
35 tonne mobile crane; lifting capacity of 1 tonne at 26 m; contract lift							
granular materials or concrete	–	0.75	13.88	43.82	–	tonne	**57.70**
palletized or packed materials	–	0.50	9.25	41.03	–	tonne	**50.28**
50 tonne mobile crane; lifting capacity of 1 tonne at 34 m; C.P.A hire (all management by hirer)							
granular materials including concrete	–	0.75	13.88	20.15	–	tonne	**34.03**
palletized or packed materials	–	1.00	18.50	17.36	–	tonne	**35.86**
50 tonne mobile crane; lifting capacity of 1 tonne at 34 m; contract lift							
granular materials or concrete	–	0.75	13.88	50.07	–	tonne	**63.95**
palletized or packed materials	–	0.50	9.25	47.28	–	tonne	**56.53**
80 tonne mobile crane; lifting capacity of 1 tonne at 44 m; C.P.A. hire (all management by hirer)							
granular materials including concrete	–	0.75	13.88	28.04	–	tonne	**41.92**
palletized or packed materials	–	0.50	9.25	25.25	–	tonne	**34.50**
80 tonne mobile crane; lifting capacity of 1 tonne at 44 m; contract lift							
granular materials or concrete	–	0.75	13.88	56.45	–	tonne	**70.33**
palletized or packed materials	–	0.50	9.25	53.66	–	tonne	**62.91**

A PRELIMINARIES

Item Excluding site overheads and profit	PC £	Labour hours	Labour £	Plant £	Material £	Unit	Total rate £
A44 CONTRACTOR'S GENERAL COST ITEMS: TEMPORARY WORKS							
Eve Trakway; portable roadway systems							
Temporary roadway system laid directly onto existing surface or onto PVC matting to protect existing surface; most systems are based on a weekly hire charge with transportation, installation and recovery charges included							
Heavy Duty Trakpanel; per panel (3.05 × 2.59 m) per week	–	–	–	–	–	m²	5.06
Outrigger Mats for use in conjunction with Heavy Duty Trakpanels; per set of 4 mats per week	–	–	–	–	–	set	85.00
Medium Duty Trakpanel; per panel (2.44 × 3.00 m) per week	–	–	–	–	–	m²	5.46
LD20 Eveolution; Light Duty Trakway; roll out system minimum delivery 50 m	–	–	–	–	–	m²	5.00
Terraplas Walkways; turf protection system; per section (1 × 1 m); per week	–	–	–	–	–	m²	–

B COMPLETE BUILDINGS/STRUCTURES/UNITS

Item Excluding site overheads and profit	PC £	Labour hours	Labour £	Plant £	Material £	Unit	Total rate £
B10 PREFABRICATED BUILDINGS/ STRUCTURES							
Cast stone buildings; Haddonstone Ltd; ornamental garden buildings in Portland Bath or Terracotta finished cast stone; prices for stonework and facades only; excavations, foundations, reinforcement, concrete infill, roofing and floors all priced separately							
Pavilion Venetian Folly L9400; Tuscan columns, pedimented arch, quoins and optional balustrading							
4184 mm high × 4728 mm wide × 3147 mm deep	11282.00	175.00	3237.50	231.00	11407.91	nr	**14876.41**
Pavilion L9300; Tuscan columns							
3496 mm high × 3634 mm wide	6805.00	144.00	2664.00	198.00	6930.91	nr	**9792.91**
Small Classical Temple L9250; 6 column with fibreglass lead effect finish dome roof							
overall height 3610 mm; diameter 2540 mm	5879.00	130.00	2405.00	165.00	6004.91	nr	**8574.91**
Large Classical Temple L9100; 8 column with fibreglass lead effect finish dome roof							
overall height 4664 mm; diameter 3190 mm	11025.00	165.00	3052.50	165.00	11150.91	nr	**14368.41**
Stepped floors to temples							
single step; Small Classical Temple	1080.00	24.00	444.00	–	1130.06	nr	**1574.06**
single step; Large Classical Temple	1525.00	26.00	481.00	–	1585.07	nr	**2066.07**
Stone structures; Architectural Heritage Ltd; hand carved from solid natural limestone with wrought iron domed roof, decorated frieze and base and integral seats; supply and erect only; excavations and concrete bases priced separately							
The Park Temple; 5 columns; 3500 mm high × 1650 mm diameter	9800.00	96.00	1776.00	61.20	9800.00	nr	**11637.20**
The Estate Temple; 6 columns; 4000 mm high × 2700 mm diameter	17600.00	120.00	2220.00	122.40	17600.00	nr	**19942.40**
Stone structures; Architectural Heritage Ltd; hand carved from solid natural limestone with solid oak trelliage; prices for stonework and facades only; supply and erect only; excavations and concrete bases priced separately							
The Pergola; 2240 mm high × 2640 mm wide × 6990 mm long	14000.00	96.00	1776.00	244.80	14029.49	nr	**16050.29**
Ornamental timber buildings; Architectural Heritage Ltd; hand built in English oak with fire retardant wheat straw thatch; hand painted, fired lead glass windows							
The Thatched Edwardian Summer House; 3710 mm high × 3540 mm wide × 2910 mm deep overall; 2540 × 1910 mm internal	22000.00	32.00	592.00	–	22000.00	nr	**22592.00**

B COMPLETE BUILDINGS/STRUCTURES/UNITS

Item Excluding site overheads and profit	PC £	Labour hours	Labour £	Plant £	Material £	Unit	Total rate £
B10 PREFABRICATED BUILDINGS/ STRUCTURES – cont							
Ornamental stone structures; Architectural Heritage Ltd; setting to bases or plinths (not included)							
The Obelisk; classic natural stone obelisk; tapering square form on panelled square base; 1860 mm high × 360 mm square	1200.00	2.00	37.00	–	1200.00	nr	**1237.00**
The Narcissus Column; natural limestone column on pedestal surmounted by bronze sculpture; overall height 3440 mm with 440 mm square base	3200.00	2.00	37.00	–	3200.00	nr	**3237.00**
The Armillary Sundial; artificial stone baluster pedestal base surmounted by verdigris copper armillary sphere calibrated sundial	2200.00	1.00	18.50	–	2200.00	nr	**2218.50**

C EXISTING SITE/BUILDINGS/SERVICES

Item Excluding site overheads and profit	PC £	Labour hours	Labour £	Plant £	Material £	Unit	Total rate £
C12 UNDERGROUND SERVICES SURVEY							
Ground penetrating radar surveys for artefacts, services and cable avoidance							
Areas not exceeding 1000 m²	–	–	–	–	–	nr	4800.00
C13 BUILDING FABRIC SURVEY							
Surveys							
Asbestos surveys							
type 3 asbestos survey (destructive testing) to single storey free standing buildings such as park buildings, storage sheds and the like; total floor area less than 100 m²	–	–	–	–	–	nr	950.00
extra over for the removal only of asbestos off site to tip (tipping charges only)	–	–	–	–	–	tonne	80.00
C20 DEMOLITION							
Demolish existing structures; disposal off site; mechanical demolition; with 3 tonne excavator and dumper							
Brick wall							
112.5 mm thick	–	0.13	2.47	0.70	3.41	m²	6.58
225 mm thick	–	0.17	3.08	0.88	6.82	m²	10.78
337.5 mm thick	–	–	–	8.03	11.45	m²	19.48
450 mm thick	–	–	–	10.70	13.65	m²	24.35
Demolish existing structures; disposal off site mechanically loaded; all other works by hand							
Brick wall							
112.5 mm thick	–	0.33	6.17	4.16	3.41	m²	13.74
225 mm thick	–	0.50	9.25	–	6.82	m²	16.07
337.5 mm thick	–	0.67	12.33	–	11.45	m²	23.78
450 mm thick	–	1.00	18.50	–	13.65	m²	32.15
Demolish existing structures; disposal off site mechanically loaded; by diesel or electric breaker; all other works by hand							
Brick wall							
112.5 mm thick	–	0.17	3.08	0.43	3.41	m²	6.92
225 mm thick	–	0.20	3.70	0.51	6.82	m²	11.03
337.5 mm thick	–	0.25	4.63	0.64	10.12	m²	15.39
450 mm thick	–	0.33	6.17	0.85	13.65	m²	20.67
Break out concrete footings associated with freestanding walls; inclusive of all excavation and backfilling with excavated material; disposal off site mechanically loaded							
By mechanical breaker; diesel or electric							
plain concrete	–	1.50	27.75	9.11	17.12	m³	53.98
reinforced concrete	–	2.50	46.25	12.93	30.33	m³	89.51

C EXISTING SITE/BUILDINGS/SERVICES

Item Excluding site overheads and profit	PC £	Labour hours	Labour £	Plant £	Material £	Unit	Total rate £
C20 DEMOLITION – cont							
Remove existing freestanding buildings;							
demolition by hand							
Timber building with suspended timber floor;							
hardstanding or concrete base not included;							
disposal off site							
shed 6.0 m²	–	2.00	37.00	–	31.59	nr	68.59
shed 10.0 m²	–	3.00	55.51	–	56.87	nr	112.38
shed 15.0 m²	–	3.50	64.75	–	88.47	nr	153.22
Timber building; insulated; with timber or							
concrete posts set in concrete, felt covered							
timber or tiled roof; internal walls cladding with							
timber or plasterboard; load arisings to skip							
timber structure 6.0 m²	–	2.50	46.25	–	136.49	nr	182.74
timber structure 12.0 m²	–	3.50	64.75	–	190.83	nr	255.58
timber structure 20.0 m²	–	8.00	148.00	377.81	202.21	nr	728.02
Demolition of freestanding brick buildings							
with tiled or sheet roof; concrete							
foundations measured separately;							
mechanical demolition; maximum distance							
to stockpile 25 m; inclusive for all access							
scaffolding and the like; maximum height							
of roof 4.0 m; inclusive of all doors,							
windows, guttering and down pipes;							
including disposal by grab							
Half brick thick							
10 m²	–	8.00	148.00	145.72	147.39	nr	441.11
20 m²	–	16.00	296.00	258.39	189.56	nr	743.95
1 brick thick							
10 m²	–	10.00	185.00	225.38	294.78	nr	705.16
20 m²	–	16.00	296.00	338.16	379.12	nr	1013.28
Cavity wall with blockwork inner skin and brick							
outer skin; insulated							
10 m²	–	12.00	222.09	305.15	409.19	nr	936.43
20 m²	–	20.00	370.00	417.97	527.17	nr	1315.14
Extra over to the above for disconnection							
of services							
Electrical							
disconnection	–	2.00	41.00	–	–	nr	41.00
grub out cables and dispose; backfilling; by							
machine	–	–	–	0.48	0.51	m	0.99
grub out cables and dispose; backfilling; by							
hand	–	0.50	9.25	–	0.51	m	9.76
Water supply, foul or surface water drainage							
disconnection; capping off	–	1.00	18.50	–	41.00	nr	59.50
grub out pipes and dispose; backfilling; by							
machine	–	–	–	0.48	0.51	m	0.99
grub out pipes and dispose; backfilling; by							
hand	–	0.50	9.25	–	0.51	m	9.76

C EXISTING SITE/BUILDINGS/SERVICES

Item Excluding site overheads and profit	PC £	Labour hours	Labour £	Plant £	Material £	Unit	Total rate £
C50 REPAIRING/RENOVATING/ CONSERVING METAL							
Eura Conservation Ltd; conservation of metal railings; works to heritage conservation standards							
Taking down and transporting existing metalwork to an off site workshop for conservation							
pair of gates; maximum overall width 4.00 m	–	–	–	–	–	pair	542.00
side screens to gates	–	–	–	–	–	pair	440.00
railings; 1.20 m high; plain	–	–	–	–	–	m	47.00
railings; 1.80 m high; with finials	–	–	–	–	–	m	89.00
ornate cast or wrought iron railings	–	–	–	–	–	m	98.00
Conserving metalwork off site; inclusive of repairs to metalwork; rust removal, rubbing down and preparing for re-erection on site							
gates; 2.50 m high; width not exceeding 4.00 m	–	–	–	–	–	m	1076.00
railings; 1.20 m high; plain	–	–	–	–	–	m	142.00
railings; 1.80 m high; with finials	–	–	–	–	–	m	215.00
ornate cast or wrought iron railings	–	–	–	–	–	m	425.00
Transporting from store and re-erection of existing gates into locations recorded with position orientation and features of the original installation							
gates; 2.50 m high; width not exceeding 4.00 m	–	–	–	–	–	nr	925.00
Refixing existing railings previously taken down and repaired on site							
railings; 1.20 m high; plain	–	–	–	–	–	m	94.00
railings; 1.80 m high; with finials	–	–	–	–	–	m	182.00
ornate cast or wrought iron railings	–	–	–	–	–	m	226.00
Eura Conservation Ltd; supply of conservation grade railings to match existing materials on site; inclusive of all surveys, measurements and analysis of materials, installation methods and the like							
Wavy bar railings							
1.20 m high; plain	–	–	–	–	–	m	92.00
1.70 m high; straight	–	–	–	–	–	m	295.00
1.70 m high; curved	–	–	–	–	–	m	355.00
1.80 m high; with finials	–	–	–	–	–	m	287.00
ornate cast or wrought iron railings	–	–	–	–	–	m	787.00
Painting of railings: see section M60							

D GROUNDWORK

Item Excluding site overheads and profit	PC £	Labour hours	Labour £	Plant £	Material £	Unit	Total rate £
D11 SOIL STABILIZATION							
Soil stabilization – General							
Preamble: Earth-retaining and stabilizing							
materials are often specified as part of the							
earth-forming work in landscape contracts,							
and therefore this section lists a number of							
products specially designed for large-scale							
earth control. There are two types: rigid units							
for structural retention of earth on steep							
slopes; and flexible meshes and sheets for							
control of soil erosion where structural							
strength is not required. Prices for these items							
depend on quantity, difficulty of access to the							
site and availability of suitable filling material;							
estimates should be obtained from the							
manufacturer when the site conditions have							
been determined.							
Retaining walls							
Preamble: Earth retaining and stabilization can							
take numerous forms depending on the height,							
soil type and loadings. The main types include							
gravity walls, reinforced soil and soil nailing.							
Prices for these items depend on quantity,							
access for the installation and the strength of							
the soils. Estimates should be obtained from							
the manufacturer when site conditions and							
layout have been determined.							
Crib walls; Phi Group							
Permacrib (timber) or Andacrib (concrete) crib							
walling system; machine filled infill with							
crushed rock; inclusive of concrete footing and							
rear wall drain; excluding excavation and							
backfill material							
retaining walls up to 2.0 m	–	–	–	–	–	m²	150.00
retaining walls up to 4.0 m	–	–	–	–	–	m²	180.00
retaining walls up to 6.0 m	–	–	–	–	–	m²	200.00
Green faced reinforced soil; Phi Group							
Textomur reinforced soil system embankments;							
to reinforced soil slopes 55–70° to the							
horizontal; including facing and geogrid;							
excluding backfill material							
retaining walls up to 3.0 m	–	–	–	–	–	m²	125.00
retaining walls up to 6.0 m	–	–	–	–	–	m²	145.00
Split face concrete blocks reinforced soil;							
Phi Group							
Modular blocks reinforced soil system							
embankments; to reinforced soil slopes near							
vertical; including geogrid; excluding backfill							
material							
retaining walls up to 3.0 m	–	–	–	–	–	m²	185.00
retaining walls up to 6.0 m	–	–	–	–	–	m²	210.00

D GROUNDWORK

Item Excluding site overheads and profit	PC £	Labour hours	Labour £	Plant £	Material £	Unit	Total rate £
Retaining walls; RCC; Tarmac Precast Concrete Retaining walls of units with plain concrete finish; prices based on 24 tonne loads but other quantities available (excavation, temporary shoring, foundations and backfilling not included)							
1000 mm wide × 1000 mm high	88.00	1.50	27.75	6.66	120.87	m	**155.28**
1000 mm wide × 1250 mm high	127.00	1.27	23.50	6.66	167.37	m	**197.53**
1000 mm wide × 1750 mm high	145.00	1.27	23.50	6.66	201.03	m	**231.19**
1000 mm wide × 2400 mm high	220.00	1.27	23.50	6.66	297.24	m	**327.40**
1000 mm wide × 2690 mm high	265.00	1.27	23.50	44.18	352.32	m	**420.00**
1000 mm wide × 3000 mm high	275.00	1.47	27.20	55.22	373.51	m	**455.93**
1000 mm wide × 3750 mm high	393.00	1.57	29.14	66.27	519.53	m	**614.94**
Retaining walls; Maccaferri Ltd Wire mesh gabions; galvanized mesh 80 × 100 mm; filling with broken stones 125–200 mm size; wire down securely to manufacturer's instructions; filling front face by hand							
2 × 1 × 0.50 m	21.05	2.00	37.00	4.44	88.25	nr	**129.69**
2 × 1 × 1.0 m	29.49	4.00	74.00	8.88	163.89	nr	**246.77**
PVC coated gabions							
2 × 1 × 0.5 m	27.24	2.00	37.00	4.44	94.44	nr	**135.88**
2 × 1 × 1.0 m	38.34	4.00	74.00	8.88	172.74	nr	**255.62**
Reno mattress gabions							
6 × 2 × 0.17 m	80.06	3.00	55.50	6.66	217.15	nr	**279.31**
6 × 2 × 0.23 m	86.66	4.50	83.25	8.88	272.13	nr	**364.26**
6 × 2 × 0.30 m	101.86	6.00	111.00	9.99	343.78	nr	**464.77**
Retaining walls; Tensar International Tensartech TW1 retaining wall system; modular dry laid concrete blocks; 220 × 400 mm long × 150 mm high connected to Tensar RE geogrid with proprietary connectors; geogrid laid horizontally within the fill at 300 mm centres; on 150 × 450 mm concrete foundation; filling with imported granular material							
1.00 m high	76.88	3.00	55.50	8.41	83.22	m²	**147.13**
2.00 m high	76.88	4.00	74.00	8.41	83.22	m²	**165.63**
3.00 m high	76.88	4.50	83.25	8.41	83.22	m²	**174.88**
Retaining walls; Grass Concrete Ltd Betoflor precast concrete landscape retaining walls including soil filling to pockets (excavation, concrete foundations, backfilling stones to rear of walls and planting not included)							
Betoflor interlocking units; 250 mm long × 250 × 200 mm modular deep; in walls 250 mm wide	–	–	–	–	–	m²	**85.50**
extra over Betoflor interlocking units for colours	–	–	–	–	–	m²	**7.99**
Betoatlas earth retaining walls; 250 mm long × 500 mm wide × 200 mm modular deep; in walls 500 mm wide	–	–	–	–	–	m²	**137.28**

D GROUNDWORK

Item Excluding site overheads and profit	PC £	Labour hours	Labour £	Plant £	Material £	Unit	Total rate £
D11 SOIL STABILIZATION – cont							
Retaining walls – cont							
Betoflor precast concrete landscape retaining walls including soil filling to pockets (excavation, concrete foundations, backfilling stones to rear of walls and planting not included) – cont							
extra over Betoatlas interlocking units for colours	–	–	–	–	–	m²	15.38
ABG Ltd; Webwall; honeycomb structured HDPE polymer strips with interlocking joints; laid to prepared terraces and backfilled with excavated material to form raked wall for planting							
500 mm high units laid and backfilled horizontally within the face of the slope to produce stepped gradient; measured as bank face area; wall height							
up to 1.00 m high	–	–	–	–	–	m²	80.00
up to 2.00 m high	–	–	–	–	–	m²	100.00
over 2.00 m high	–	–	–	–	–	m²	110.00
250 mm high units laid and backfilled horizontally within the face of the slope to produce stepped gradient; measured as bank face area; wall height							
up to 1.00 m high	–	–	–	–	–	m²	90.00
up to 2.00 m high	–	–	–	–	–	m²	110.00
over 2.00 m high	–	–	–	–	–	m²	120.00
Setting out; grading and levelling; compacting bottoms of excavations	–	0.13	2.47	0.44	–	m	2.91
Extra over Betoflor retaining walls for C7P concrete foundations							
700 × 300 mm deep	14.53	0.27	4.94	–	14.53	m	19.47
Retaining walls; Forticrete Ltd							
Keystone precast concrete block retaining wall; geogrid included for walls over 1.00 m high; excavation, concrete foundation, stone backfill to rear of wall all measured separately							
1.0 m high	88.15	2.40	44.40	–	88.15	m²	132.55
2.0 m high	88.15	2.40	44.40	8.25	99.95	m²	152.60
3.0 m high	88.15	2.40	44.40	11.00	108.15	m²	163.55
4.0 m high	88.15	2.40	44.40	13.20	114.50	m²	172.10
Retaining walls; Forticrete Ltd							
Stepoc Blocks; interlocking blocks; 10 mm reinforcing laid loose horizontally to preformed notches and vertical reinforcing nominal size 10 mm fixed to starter bars; infilling with concrete; foundations and starter bars measured separately							
type 325; 400 × 225 × 325 mm	47.08	1.40	25.90	–	65.78	m²	91.68
type 256; 400 × 225 × 256 mm	44.61	1.20	22.20	–	61.85	m²	84.05
type 190; 400 × 225 × 200 mm	34.21	1.00	18.50	–	48.51	m²	67.01

D GROUNDWORK

Item Excluding site overheads and profit	PC £	Labour hours	Labour £	Plant £	Material £	Unit	Total rate £
Embankments; Tensar International							
Embankments; reinforced with Tensar Geogrid RE520; geogrid laid horizontally within fill to 100% of vertical height of slope at 1.00 m centres; filling with excavated material							
slopes less than 45°; Tensartech Natural Green System	2.10	0.19	3.43	1.76	2.10	m³	7.29
slopes exceeding 45°; Tensartech Greenslope System	4.51	0.19	3.43	2.60	4.51	m³	10.54
extra over to slopes exceeding 45° for temporary shuttering; shuttering measured per m² of vertical elevation	–	–	–	–	10.00	m²	10.00
Embankments; reinforced with Tensar Geogrid RE540; geogrid laid horizontally within fill to 100% of vertical height of slope at 1.00 m centres; filling with excavated material							
slopes less than 45°; Tensartech Natural Green System	2.63	0.19	3.43	1.76	2.63	m³	7.82
slopes exceeding 45°; Tensartech Greenslope System	5.64	0.19	3.43	2.60	5.64	m³	11.67
extra over to slopes exceeding 45° for temporary shuttering; shuttering measured per m² of vertical elevation	–	–	–	–	10.00	m²	10.00
Extra for Tensar Mat 400; erosion control mats; to faces of slopes of 45° or less; filling with 20 mm fine topsoil; seeding	3.59	0.02	0.37	0.17	4.72	m²	5.26
Embankments; reinforced with Tensar Geogrid RE540; geogrid laid horizontally within fill to 50% of horizontal length of slopes at specified centres; filling with imported fill PC £12.00/m³							
300 mm centres; slopes up to 45°; Tensartech Natural Green System	4.36	0.23	4.32	2.00	18.76	m³	25.08
600 mm centres; slopes up to 45°; Tensartech Natural Green System	2.18	0.19	3.43	2.60	16.58	m³	22.61
extra over to slopes exceeding 45° for temporary shuttering; shuttering measured per m² of vertical elevation	–	–	–	–	10.00	m²	10.00
Extra for Tensar Mat 400; erosion control mats; to faces of slopes of 45° or less; filling with 20 mm fine topsoil; seeding	3.59	0.02	0.37	0.17	4.72	m²	5.26
Embankments; reinforced with Tensar Geogrid RE540; geogrid laid horizontally within fill; filling with excavated material; wrapping around at faces							
300 mm centres; slopes exceeding 45°; Tensartech Natural Green System	8.74	0.19	3.43	1.76	8.74	m³	13.93
600 mm centres; slopes exceeding 45°; Tensartech Greenslope System	4.36	0.19	3.43	1.76	4.36	m³	9.55
extra over to slopes exceeding 45° for temporary shuttering; shuttering measured per m² of vertical elevation	–	–	–	–	10.00	m²	10.00
Extra over embankments for bagwork face supports for slopes exceeding 45°	–	0.50	9.25	–	13.00	m²	22.25
Extra over embankments for seeding of bags	–	0.10	1.85	–	0.21	m²	2.06
Extra over embankments for Bodkin joints	–	–	–	–	1.00	m	1.00

D GROUNDWORK

Item Excluding site overheads and profit	PC £	Labour hours	Labour £	Plant £	Material £	Unit	Total rate £
D11 SOIL STABILIZATION – cont							
Anchoring systems; surface stabilization; Platipus Anchors Ltd; centres and depths shown should be verified with a design engineer; they may vary either way dependent on circumstances							
Concrete revetments; hard anodized stealth anchors with stainless steel accessories installed to a depth of 1 m							
SO4 anchors to a depth of 1.00–1.50 m at 1.00 m centres	25.36	0.33	6.17	1.58	25.36	m²	33.11
Loadlocking and crimping tool; for stressing and crimping the anchors							
hire rate per week	69.46	–	–	–	69.46	nr	69.46
Surface erosion; geotextile anchoring on slopes; aluminium alloy Stealth anchors							
SO4/SO6 to a depth of 1.0–2.0 m at 2.00 m centres	8.66	0.20	3.70	1.58	8.66	m²	13.94
Brick block or in situ concrete distressed retaining walls; anchors installed in two rows at varying tensions between 18–50 kN along the length of the retaining wall; core drilling of wall not included							
SO8/BO6/BO8 bronze Stealth and Bat anchors installed in combination with stainless steel accessories to a depth of 6 m; average cost base on 1.50 m centres; long term solution	363.83	0.50	9.25	1.58	363.83	m²	374.66
SO8/BO6/BO8 cast SG iron Stealth and Bat anchors installed in combination with galvanized steel accessories to a depth of 6 m; average cost base on 1.50 m centres; short term solution	155.45	0.50	9.25	1.58	155.45	m²	166.28
Timber retaining wall support; anchors installed 19 kN along the length of the retaining wall							
SO6/SO8 bronze Stealth anchors including load plate and wedge grip; installed in combination with stainless steel accessories to a depth of 3.0 m; 1.00 centres; average cost	176.00	0.50	9.25	1.27	176.00	m²	186.52
Gabions; anchors for support and stability to moving, overturning or rotating gabion retaining walls							
SO8/BO6 anchors including base plate at 1.00 m centres to a depth of 4.00 m	275.63	0.33	6.17	1.58	275.63	m²	283.38
Soil nailing of geofabrics; anchors fixed through surface of fabric or erosion control surface treatment (not included)							
SO8/BO6 anchors including base plate at 1.50 m centres to a depth of 3.00 m	110.25	0.17	3.08	1.06	110.25	m²	114.39

D GROUNDWORK

Item Excluding site overheads and profit	PC £	Labour hours	Labour £	Plant £	Material £	Unit	Total rate £
Embankments; Grass Concrete Ltd							
Grasscrete; in situ reinforced concrete surfacing; to 20 mm thick sand blinding layer (not included); including soiling and seeding							
GC1; 100 mm thick	–	–	–	–	–	m²	**32.12**
GC2; 150 mm thick	–	–	–	–	–	m²	**39.86**
Grassblock 103; solid matrix precast concrete blocks; to 20 mm thick sand blinding layer; excluding edge restraint; including soiling and seeding							
406 × 406 × 103 mm; fully interlocking	–	–	–	–	–	m²	**33.95**
Grass reinforcement; Farmura Environmental Ltd							
Matrix grass paver; recycled polyethylene and polypropylene mixed interlocking erosion control and grass reinforcement system laid to rootzone prepared separately and filled with screened topsoil and seeded with grass seed; green							
640 × 330 × 38 mm	13.50	0.13	2.31	–	18.17	m²	**20.48**
Embankments; Cooper Clarke Civils and Lintels							
Ecoblock polyethylene; 925 × 310 × 50 mm; heavy duty for car parking and fire paths							
to firm subsoil (not included)	18.38	0.04	0.74	–	18.91	m²	**19.65**
to 100 mm granular fill and Geotextile	18.38	0.08	1.48	0.13	23.20	m²	**24.81**
to 250 mm granular fill and Geotextile	18.38	0.20	3.70	0.22	28.55	m²	**32.47**
Extra for filling Ecoblock with topsoil; seeding with rye grass at 50 g/m²	1.34	0.02	0.37	0.18	1.34	m²	**1.89**
Neoweb polyethylene soil stabilizing panels; to soil surfaces brought to grade (not included); filling with excavated material							
panels; 2.5 × 8.0 m × 100 mm deep	7.86	0.10	1.85	2.51	7.86	m²	**12.22**
panels; 2.5 × 8.0 m × 200 mm deep	15.70	0.13	2.47	3.00	15.70	m²	**21.17**
Neoweb polyethylene soil stabilizing panels; to soil surfaces brought to grade (not included); filling with ballast							
panels; 2.5 × 8.0 m × 100 mm deep	7.86	0.11	2.06	2.51	9.71	m²	**14.28**
panels; 2.5 × 8.0 m × 200 mm deep	15.70	0.15	2.85	3.00	19.40	m²	**25.25**
Neoweb polyethylene soil stabilizing panels; to soil surfaces brought to grade (not included); filling with ST2 10 N/mm² concrete							
panels; 2.5 × 8.0 m × 100 mm deep	7.86	0.16	2.96	–	14.27	m²	**17.23**
panels; 2.5 × 8.0 m × 200 mm deep	15.70	0.20	3.70	–	28.52	m²	**32.22**
Extra over for filling Neoweb with imported topsoil; seeding with rye grass at 50 g/m²	2.79	0.07	1.23	0.29	2.79	m²	**4.31**

D GROUNDWORK

Item Excluding site overheads and profit	PC £	Labour hours	Labour £	Plant £	Material £	Unit	Total rate £
D11 SOIL STABILIZATION – cont							
Timber log retaining walls; AVS Fencing Supplies Ltd; all timber posts are kiln dried redwood with 15 year guarantee; posts which have been cut would not be subject to the supplier's guarantee							
Machine rounded softwood logs to trenches priced separately; disposal of excavated material priced separately; inclusive of 75 mm hardcore blinding to trench and backfilling trench with site mixed concrete 1:3:6; geofabric pinned to rear of logs; heights of logs above ground							
500 mm (constructed from 1.80 m lengths)	28.05	1.50	27.75	–	36.74	m	64.49
1.20 m (constructed from 1.80 m lengths)	56.10	1.30	24.05	–	74.90	m	98.95
1.60 m (constructed from 2.40 m lengths)	74.90	2.50	46.25	–	98.83	m	145.08
2.00 m (constructed from 3.00 m lengths)	93.60	3.50	64.75	–	122.65	m	187.40
As above but with 150 mm machine rounded timbers							
500 mm	38.44	2.50	46.25	–	47.13	m	93.38
1.20 m	128.01	1.75	32.38	–	146.81	m	179.19
1.60 m	128.13	3.00	55.50	–	152.06	m	207.56
As above but with 200 mm machine rounded timbers							
1.80 m (constructed from 2.40 m lengths)	170.80	4.00	74.00	–	199.85	m	273.85
Railway sleeper walls; AVS Fencing Supplies Ltd; retaining wall from railway sleepers; fixed with steel galvanized pins 12 mm driven into the ground; sleepers laid flat							
Grade 1 softwood; 2590 × 250 × 125 mm							
150 mm; 1 sleeper high	6.81	0.50	9.25	–	7.29	m	16.54
300 mm; 2 sleepers high	13.62	1.00	18.50	–	14.58	m	33.08
450 mm; 3 sleepers high	20.27	1.50	27.75	–	21.53	m	49.28
600 mm; 4 sleepers high	27.30	2.00	37.00	–	28.56	m	65.56
Grade 1 softwood as above but with 2 nr galvanized angle iron stakes set into concrete internally and screwed to the inside face of the sleepers							
750 mm; 5 sleepers high	33.77	2.50	46.25	–	59.70	m²	105.95
900 mm; 6 sleepers high	40.42	2.75	50.88	–	69.31	m²	120.19
Grade 1 hardwood; 2590 × 250 × 150 mm							
150 mm; 1 sleeper high	7.00	0.50	9.25	–	7.48	m	16.73
300 mm; 2 sleepers high	14.01	1.00	18.50	–	14.97	m	33.47
450 mm; 3 sleepers high	20.84	1.50	27.75	–	22.10	m	49.85
600 mm; 4 sleepers high	28.08	2.00	37.00	–	29.34	m	66.34
Grade 1 hardwood as above but with 2 nr galvanized angle iron stakes set into concrete internally and screwed to the inside face of the sleepers							
750 mm; 5 sleepers high	34.74	2.50	46.25	–	60.67	m²	106.92
900 mm; 6 sleepers high	41.58	2.75	50.88	–	70.46	m²	121.34

D GROUNDWORK

Item Excluding site overheads and profit	PC £	Labour hours	Labour £	Plant £	Material £	Unit	Total rate £
New pine softwood; 2400 × 250 × 125 mm							
120 mm; 1 sleeper high	7.40	0.50	9.25	–	7.88	m	17.13
240 mm; 2 sleepers high	14.80	1.00	18.50	–	15.76	m	34.26
360 mm; 3 sleepers high	22.20	1.50	27.75	–	23.46	m	51.21
480 mm; 4 sleepers high	29.60	2.00	37.00	–	30.86	m	67.86
New pine softwood as above but with 2 nr galvanized angle iron stakes set into concrete internally and screwed to the inside face of the sleepers							
600 mm; 5 sleepers high	37.00	2.50	46.25	–	62.93	m²	109.18
720 mm; 6 sleepers high	44.40	2.75	50.88	–	73.28	m²	124.16
New oak hardwood 2600 × 220 × 130 mm							
130 mm; 1 sleeper high	9.80	0.50	9.25	–	10.28	m	19.53
260 mm; 2 sleepers high	19.60	1.00	18.50	–	20.56	m	39.06
390 mm; 3 sleepers high	29.40	1.50	27.75	–	30.66	m	58.41
520 mm; 4 sleepers high	39.20	2.00	37.00	–	40.46	m	77.46
New oak hardwood as above but with 2 nr galvanized angle iron stakes set into concrete internally and screwed to the inside face of the sleepers							
640 mm; 5 sleepers high	49.00	2.50	46.25	–	74.93	m²	121.18
760 mm; 6 sleepers high	58.80	2.75	50.88	–	87.68	m²	138.56
Excavate foundation trench; set railway sleepers vertically on end in concrete 1:3:6 continuous foundation to 33.3% of their length to form retaining wall							
Grade 1 hardwood; finished height above ground level							
300 mm	18.63	3.00	55.50	1.83	22.83	m	80.16
500 mm	27.00	3.00	55.50	1.83	31.20	m	88.53
600 mm	37.53	3.00	55.50	1.83	45.57	m	102.90
750 mm	40.50	3.50	64.75	1.83	48.54	m	115.12
1.00 m	46.13	3.75	69.38	1.83	54.94	m	126.15
Excavate and place vertical steel universal beams 165 mm wide in concrete base at 2.59 m centres; fix railway sleepers set horizontally between beams to form horizontal fence or retaining wall							
Grade 1 hardwood; bay length 2.59 m							
500 mm high; 2 sleepers	16.50	2.50	46.25	–	55.29	bay	101.54
750 mm high; 3 sleepers	24.25	2.60	48.10	–	82.13	bay	130.23
1.00 m high; 4 sleepers	32.50	3.00	55.50	–	106.60	bay	162.10
1.25 m high; 5 sleepers	41.50	3.50	64.75	–	136.93	bay	201.68
1.50 m high; 6 sleepers	48.75	3.00	55.50	–	162.96	bay	218.46
1.75 m high; 7 sleepers	56.75	3.00	55.50	–	170.96	bay	226.46
Willow walling to riverbanks; LDC Ltd							
Woven willow walling as retention to riverbanks; driving or concreting posts in at 2 m centres; intermediate posts at 500 mm centres							
1.20 m high	–	–	–	–	–	m	126.35
1.50 m high	–	–	–	–	–	m	161.08

D GROUNDWORK

Item Excluding site overheads and profit	PC £	Labour hours	Labour £	Plant £	Material £	Unit	Total rate £
D11 SOIL STABILIZATION – cont							
Flexible sheet materials; Tensar **International**							
Tensar Mat 400 erosion mats; 3.0–4.5 m wide; securing with Tensar pegs; lap rolls 100 mm; anchors at top and bottom of slopes; in trenches	4.01	0.02	0.31	–	4.01	m²	**4.32**
Topsoil filling to Tensar Mat; including brushing and raking	0.69	0.02	0.31	0.10	0.69	m²	**1.10**
Tensar Bi-axial Geogrid; to graded compacted base; filling with 200 mm granular fill; compacting (turf or paving to surfaces not included); 400 mm laps							
TriAx 150; 39 × 39 mm mesh	1.30	0.01	0.25	0.13	4.18	m²	**4.56**
TriAx 160; 39 × 39 mm mesh	1.75	0.01	0.25	0.13	4.63	m²	**5.01**
TriAx 170; 39 × 39 mm mesh	2.10	0.01	0.25	0.13	4.98	m²	**5.36**
Flexible sheet materials; Terram Ltd							
Terram synthetic fibre filter fabric; to graded base (not included)							
Terram 1000; 0.70 mm thick; mean water flow 50 l/m²/s	0.43	0.02	0.31	–	0.43	m²	**0.74**
Terram 2000; 1.00 mm thick; mean water flow 33 l/m²/s	1.09	0.02	0.31	–	1.09	m²	**1.40**
Terram Minipack	0.55	–	0.06	–	0.55	m²	**0.61**
Flexible sheet materials; Greenfix Ltd							
Greenfix; erosion control mats; 10–15 mm thick; fixing with 4 nr crimped pins in accordance with manufacturer's instructions; to graded surface (not included)							
unseeded Eromat 1; 2.40 m wide	103.47	2.00	37.00	–	163.47	100 m²	**200.47**
unseeded Eromat 2; 2.40 m wide	130.69	2.00	37.00	–	190.69	100 m²	**227.69**
unseeded Eromat 3; 2.40 m wide	147.03	2.00	37.00	–	207.03	100 m²	**244.03**
seeded Covamat 1; 2.40 m wide	166.25	2.00	37.00	–	226.25	100 m²	**263.25**
seeded Covamat 2; 2.40 m wide	210.00	2.00	37.00	–	270.00	100 m²	**307.00**
seeded Covamat 3; 2.40 m wide	227.50	2.00	37.00	–	287.50	100 m²	**324.50**
Bioroll; 300 mm diameter; to river banks and revetments	19.00	0.13	2.31	–	20.76	m	**23.07**
Extra over Greenfix erosion control mats for fertilizer applied at 70 g/m²	7.42	0.20	3.70	–	7.42	100 m²	**11.12**
Extra over Greenfix erosion control mats for Geojute; fixing with steel pins	1.28	0.02	0.41	–	1.28	m²	**1.69**
Extra over Greenfix erosion control mats for laying to slopes exceeding 30°	–	–	–	–	–	25%	**–**

D GROUNDWORK

Item Excluding site overheads and profit	PC £	Labour hours	Labour £	Plant £	Material £	Unit	Total rate £
Extra for the following operations							
Spreading 25 mm approved topsoil							
by machine	0.66	0.01	0.11	0.10	0.66	m^2	0.87
by hand	0.66	0.04	0.65	–	0.66	m^2	1.31
Grass seed; PC £4.50/kg; spreading in two operations; by hand							
35 g/m^2	15.75	0.17	3.08	–	15.75	100 m^2	18.83
50 g/m^2	22.50	0.17	3.08	–	22.50	100 m^2	25.58
70 g/m^2	31.50	0.17	3.08	–	31.50	100 m^2	34.58
100 g/m^2	45.00	0.20	3.70	–	45.00	100 m^2	48.70
125 g/m^2	56.25	0.20	3.70	–	56.25	100 m^2	59.95
Extra over seeding by hand for slopes over 30° (allowing for the actual area but measured in plan)							
35 g/m^2	2.34	–	0.07	–	2.34	100 m^2	2.41
50 g/m^2	3.38	–	0.07	–	3.38	100 m^2	3.45
70 g/m^2	4.72	–	0.07	–	4.72	100 m^2	4.79
100 g/m^2	6.75	–	0.08	–	6.75	100 m^2	6.83
125 g/m^2	8.41	–	0.08	–	8.41	100 m^2	8.49
Grass seed; PC £4.50/kg; spreading in two operations; by machine							
35 g/m^2	15.75	–	–	0.51	15.75	100 m^2	16.26
50 g/m^2	22.50	–	–	0.51	22.50	100 m^2	23.01
70 g/m^2	31.50	–	–	0.51	31.50	100 m^2	32.01
100 g/m^2	45.00	–	–	0.51	45.00	100 m^2	45.51
125 kg/ha	562.50	–	–	50.95	562.50	ha	613.45
150 kg/ha	675.00	–	–	50.95	675.00	ha	725.95
200 kg/ha	900.00	–	–	50.95	900.00	ha	950.95
250 kg/ha	1125.00	–	–	50.95	1125.00	ha	1175.95
300 kg/ha	1350.00	–	–	50.95	1350.00	ha	1400.95
350 kg/ha	1575.00	–	–	50.95	1575.00	ha	1625.95
400 kg/ha	1800.00	–	–	50.95	1800.00	ha	1850.95
500 kg/ha	2250.00	–	–	50.95	2250.00	ha	2300.95
700 kg/ha	3150.00	–	–	50.95	3150.00	ha	3200.95
1400 kg/ha	6300.00	–	–	50.95	6300.00	ha	6350.95
Extra over seeding by machine for slopes over 30° (allowing for the actual area but measured in plan)							
35 g/m^2	2.36	–	–	0.08	2.36	100 m^2	2.44
50 g/m^2	3.38	–	–	0.08	3.38	100 m^2	3.46
70 g/m^2	4.72	–	–	0.08	4.72	100 m^2	4.80
100 g/m^2	6.75	–	–	0.08	6.75	100 m^2	6.83
125 kg/ha	84.38	–	–	7.64	84.38	ha	92.02
150 kg/ha	101.25	–	–	7.64	101.25	ha	108.89
200 kg/ha	135.00	–	–	7.64	135.00	ha	142.64
250 kg/ha	168.75	–	–	7.64	168.75	ha	176.39
300 kg/ha	202.50	–	–	7.64	202.50	ha	210.14
350 kg/ha	236.25	–	–	7.64	236.25	ha	243.89
400 kg/ha	270.00	–	–	7.64	270.00	ha	277.64
500 kg/ha	337.50	–	–	7.64	337.50	ha	345.14
700 kg/ha	472.50	–	–	7.64	472.50	ha	480.14
1400 kg/ha	945.00	–	–	7.64	945.00	ha	952.64

D GROUNDWORK

Item Excluding site overheads and profit	PC £	Labour hours	Labour £	Plant £	Material £	Unit	Total rate £
D20 EXCAVATING AND FILLING							
MACHINE SELECTION TABLE							
Road Equipment Ltd; machine volumes for excavating/filling only and placing excavated material alongside or to a dumper; no bulkages are allowed for in the material volumes; these rates should be increased by user-preferred percentages to suit prevailing site conditions; the figures in the next section for 'Excavation mechanical' and filling allow for the use of banksmen within the rates shown below							
1.5 tonne excavators: digging volume							
1 cycle/minute; 0.04 m³	–	0.42	7.71	2.45	–	m³	**10.16**
2 cycles/minute; 0.08 m³	–	0.21	3.85	1.55	–	m³	**5.40**
3 cycles/minute; 0.12 m³	–	0.14	2.57	1.26	–	m³	**3.83**
3 tonne excavators; digging volume							
1 cycle/minute; 0.13 m³	–	0.13	2.37	1.40	–	m³	**3.77**
2 cycles/minute; 0.26 m³	–	0.06	1.19	2.20	–	m³	**3.39**
3 cycles/minute; 0.39 m³	–	0.04	0.79	0.96	–	m³	**1.75**
5 tonne excavators; digging volume							
1 cycle/minute; 0.28 m³	–	0.06	1.10	1.94	–	m³	**3.04**
2 cycles/minute; 0.56 m³	–	0.03	0.55	1.62	–	m³	**2.17**
3 cycles/minute; 0.84 m³	–	0.02	0.37	1.68	–	m³	**2.05**
7 tonne excavators; supplied with operator; digging volume							
1 cycle/minute; 0.28 m³	–	0.06	1.10	2.97	–	m³	**4.07**
2 cycles/minute; 0.56 m³	–	0.03	0.55	1.45	–	m³	**2.00**
3 cycles/minute; 0.84 m³	–	0.02	0.37	1.69	–	m³	**2.06**
21 tonne excavators; supplied with operator; digging volume							
1 cycle/minute; 1.21 m³	–	–	–	0.70	–	m³	**0.70**
2 cycles/minute; 2.42 m³	–	–	–	0.29	–	m³	**0.29**
3 cycles/minute; 3.63 m³	–	–	–	0.19	–	m³	**0.19**
Backhoe loader; excavating; JCB 3 CX rear bucket capacity 0.28 m³							
1 cycle/minute; 0.28 m³	–	–	–	1.82	–	m³	**1.82**
2 cycles/minute; 0.56 m³	–	–	–	0.91	–	m³	**0.91**
3 cycles/minute; 0.84 m³	–	–	–	0.61	–	m³	**0.61**
Backhoe loader; loading from stockpile; JCB 3 CX front bucket capacity 1.00 m³							
1 cycle/minute; 1.00 m³	–	–	–	0.51	–	m³	**0.51**
2 cycles/minute; 2.00 m³	–	–	–	0.25	–	m³	**0.25**
Note: All volumes below are based on excavated 'earth' moist at 1,997 kg/m³ solid or 1,598 kg/m³ loose; a 25% bulkage factor has been used; the weight capacities below exceed the volume capacities of the machine in most cases; see the memorandum section at the back of this book for further weights of materials.							

D GROUNDWORK

Item Excluding site overheads and profit	PC £	Labour hours	Labour £	Plant £	Material £	Unit	Total rate £
Dumpers; Road Equipment Ltd							
1 tonne high tip skip loader; volume 0.485 m³							
(775 kg)							
5 loads per hour	–	0.41	7.63	1.31	–	m³	8.94
7 loads per hour	–	0.29	5.45	0.98	–	m³	6.43
10 loads per hour	–	0.21	3.81	0.74	–	m³	4.55
2.5 tonne dumper; excavated material volume							
1.3 m³							
4 loads per hour	–	–	–	4.46	–	m³	4.46
5 loads per hour	–	–	–	3.57	–	m³	3.57
7 loads per hour	–	–	–	3.17	–	m³	3.17
10 loads per hour	–	–	–	1.74	–	m³	1.74
6 tonne dumper; maximum volume 3.40 m³							
(5.4 t); available volume 3.77 m³							
4 loads per hour	–	0.07	1.23	0.30	–	m³	1.53
5 loads per hour	–	0.05	0.98	0.25	–	m³	1.23
7 loads per hour	–	0.04	0.70	0.19	–	m³	0.89
10 loads per hour	–	0.03	0.49	0.15	–	m³	0.64
Market prices of topsoil; prices shown							
include for 20% settlement							
Recycled topsoil	–	–	–	–	21.60	m³	21.60
Multiple source screened topsoil	–	–	–	–	26.40	m³	26.40
Single source topsoil; British Sugar PLC	–	–	–	–	39.60	m³	39.60
High grade topsoil for planting	–	–	–	–	60.00	m³	60.00
Site preparation							
Felling and removing trees off site							
girth 600 mm–1.50 m (95–240 mm trunk							
diameter)	–	5.00	92.50	19.04	–	nr	111.54
girth 1.50–3.00 m (240–475 mm trunk							
diameter)	–	17.00	314.50	76.15	–	nr	390.65
girth 3.00–4.00 m (475–630 mm trunk							
diameter)	–	48.53	897.77	121.83	–	nr	1019.60
Removing tree stumps							
girth 600 mm–1.50 m	–	2.00	37.00	67.32	–	nr	104.32
girth 1.50–3.00 m	–	7.00	129.50	93.27	–	nr	222.77
girth over 3.00 m	–	12.00	222.00	159.88	–	nr	381.88
Stump grinding; disposing to spoil heaps							
girth 600 mm–1.50 m	–	2.00	37.00	32.20	–	nr	69.20
girth 1.50–3.00 m	–	2.50	46.25	64.41	–	nr	110.66
girth over 3.00 m	–	4.00	74.00	56.35	–	nr	130.35
Clearing site vegetation							
mechanical clearance	–	0.25	4.63	11.55	–	100 m²	16.18
hand clearance	–	2.00	37.00	–	–	100 m²	37.00
Lifting turf for preservation							
sod cutter machine lift and stack	–	0.75	13.88	9.96	–	100 m²	23.84
hand lift and stack	–	8.33	154.16	–	–	100 m²	154.16
Lifting turf for disposal (disposal not included)							
by excavator; 5 tonne	–	–	–	35.35	–	100 m²	35.35
by excavator; 7 tonne	–	–	–	26.43	–	100 m²	26.43
by excavator; 7 tonne; working with two							
dumpers	–	–	–	24.81	–	100 m²	24.81
by excavator; 7 tonne; working with two							
dumpers	–	–	–	23.63	–	100 m²	23.63
hand lift and stack	–	8.33	154.16	–	–	100 m²	154.16

D GROUNDWORK

Item Excluding site overheads and profit	PC £	Labour hours	Labour £	Plant £	Material £	Unit	Total rate £
D20 EXCAVATING AND FILLING – cont							
Site clearance; by machine; clear site of							
mature shrubs from existing cultivated							
beds; dig out roots by machine							
Mixed shrubs in beds; planting centres							
500 mm average							
height less than 1 m	–	0.03	0.46	0.72	–	m²	1.18
1.00–1.50 m	–	0.04	0.74	1.15	–	m²	1.89
1.50–2.00 m; pruning to ground level by							
hand	–	0.10	1.85	2.88	–	m²	4.73
2.00–3.00 m; pruning to ground level by							
hand	–	0.10	1.85	5.77	–	m²	7.62
3.00–4.00 m; pruning to ground level by							
hand	–	0.20	3.70	9.61	–	m²	13.31
Site clearance; by hand; clear site of							
mature shrubs from existing cultivated							
beds; dig out roots							
Mixed shrubs in beds; planting centres							
500 mm average							
height less than 1 m	–	0.33	6.17	–	–	m²	6.17
1.00–1.50 m	–	0.50	9.25	–	–	m²	9.25
1.50–2.00 m	–	1.00	18.50	–	–	m²	18.50
2.00–3.00 m	–	2.00	37.00	–	–	m²	37.00
3.00–4.00 m	–	3.00	55.51	–	–	m²	55.51
Disposal of material from site clearance							
operations							
Shrubs and groundcovers less than 1.00 m							
height; disposal by 15 m³ self loaded truck							
deciduous shrubs; not chipped; winter	–	0.05	0.93	0.51	8.00	m²	9.44
deciduous shrubs; chipped; winter	–	0.05	0.93	1.30	1.20	m²	3.43
evergreen or deciduous shrubs; chipped;							
summer	–	0.08	1.39	0.80	4.80	m²	6.99
Shrubs 1.00–2.00 m height; disposal by 15 m³							
self loaded truck							
deciduous shrubs; not chipped; winter	–	0.17	3.08	0.51	24.00	m²	27.59
deciduous shrubs; chipped; winter	–	0.25	4.63	0.80	4.80	m²	10.23
evergreen or deciduous shrubs; chipped;							
summer	–	0.30	5.55	0.89	9.60	m²	16.04
Shrubs or hedges 2.00–3.00 m height;							
disposal by 15 m³ self loaded truck							
deciduous plants non woody growth; not							
chipped; winter	–	0.25	4.63	0.51	24.00	m²	29.14
deciduous shrubs; chipped; winter	–	0.50	9.25	0.80	12.00	m²	22.05
evergreen or deciduous shrubs non woody							
growth; chipped; summer	–	0.25	4.63	0.80	4.80	m²	10.23
evergreen or deciduous shrubs woody							
growth; chipped; summer	–	0.67	12.33	1.08	14.40	m²	27.81
Shrubs and groundcovers less than 1.00 m							
height; disposal to spoil heap							
deciduous shrubs; not chipped; winter	–	0.05	0.93	–	–	m²	0.93
deciduous shrubs; chipped; winter	–	0.05	0.93	0.29	–	m²	1.22
evergreen or deciduous shrubs; chipped;							
summer	–	0.08	1.39	0.29	–	m²	1.68

D GROUNDWORK

Item Excluding site overheads and profit	PC £	Labour hours	Labour £	Plant £	Material £	Unit	Total rate £
Shrubs 1.00–2.00 m height; disposal to spoil heap							
deciduous shrubs; not chipped; winter	–	0.17	3.08	–	–	m²	**3.08**
deciduous shrubs; chipped; winter	–	0.25	4.63	0.29	–	m²	**4.92**
evergreen or deciduous shrubs; chipped; summer	–	0.30	5.55	0.38	–	m²	**5.93**
Shrubs or hedges 2.00–3.00 m height; disposal to spoil heaps							
deciduous plants non woody growth; not chipped; winter	–	0.25	4.63	–	–	m²	**4.63**
deciduous shrubs; chipped; winter	–	0.50	9.25	0.29	–	m²	**9.54**
evergreen or deciduous shrubs non woody growth; chipped; summer	–	0.67	12.33	0.57	–	m²	**12.90**
evergreen or deciduous shrubs woody growth; chipped; summer	–	1.50	27.75	1.14	–	m²	**28.89**
Note: The figures in this section relate to the machine capacities shown earlier in this section. The figures below however allow for dig efficiency based on depth. The figures below also allow for a banksman. The figures below allow for bulkages of 25% on loamy soils; adjustments should be made for different soil types.							
Excavating; mechanical; topsoil for preservation							
3 tonne tracked excavator (bucket volume 0.13 m³)							
average depth 100 mm	–	0.02	0.31	0.95	–	m²	**1.26**
average depth 150 mm	–	0.02	0.36	1.07	–	m²	**1.43**
average depth 200 mm	–	0.03	0.46	1.43	–	m²	**1.89**
average depth 250 mm	–	0.03	0.51	1.59	–	m²	**2.10**
average depth 300 mm	–	0.03	0.62	1.90	–	m²	**2.52**
JCB Sitemaster 3CX (bucket volume 0.28 m³)							
average depth 100 mm	–	1.00	18.50	33.66	–	100 m²	**52.16**
average depth 150 mm	–	1.25	23.13	42.08	–	100 m²	**65.21**
average depth 200 mm	–	1.78	32.93	59.91	–	100 m²	**92.84**
average depth 250 mm	–	1.90	35.15	63.95	–	100 m²	**99.10**
average depth 300 mm	–	2.00	37.00	67.32	–	100 m²	**104.32**
Excavating; mechanical; to reduce levels							
5 tonne excavator (bucket volume 0.28 m³)							
maximum depth not exceeding 0.25 m	–	0.14	2.59	0.78	–	m³	**3.37**
maximum depth not exceeding 1.00 m	–	0.10	1.76	0.53	–	m³	**2.29**
JCB Sitemaster 3CX (bucket volume 0.28 m³)							
maximum depth not exceeding 0.25 m	–	0.07	1.29	2.36	–	m³	**3.65**
maximum depth not exceeding 1.00 m	–	0.06	1.10	2.00	–	m³	**3.10**
21 tonne 360 tracked excavator (bucket volume 1.21 m³)							
maximum depth not exceeding 1.00 m	–	0.01	0.20	0.46	–	m³	**0.66**
maximum depth not exceeding 2.00 m	–	0.02	0.31	0.69	–	m³	**1.00**
Pits; 3 tonne tracked excavator							
maximum depth not exceeding 0.25 m	–	0.33	6.17	2.44	–	m³	**8.61**
maximum depth not exceeding 1.00 m	–	0.50	9.25	3.67	–	m³	**12.92**
maximum depth not exceeding 2.00 m	–	0.60	11.10	4.40	–	m³	**15.50**

Prices for Measured Works – Major Works

D GROUNDWORK

Item Excluding site overheads and profit	PC £	Labour hours	Labour £	Plant £	Material £	Unit	Total rate £
D20 EXCAVATING AND FILLING – cont							
Excavating – cont							
Trenches; width not exceeding 0.30 m; 3 tonne excavator							
maximum depth not exceeding 0.25 m	–	1.33	24.67	5.87	–	m³	30.54
maximum depth not exceeding 1.00 m	–	0.69	12.67	3.01	–	m³	15.68
maximum depth not exceeding 2.00 m	–	0.60	11.10	2.64	–	m³	13.74
Trenches; width exceeding 0.30 m; 3 tonne excavator							
maximum depth not exceeding 0.25 m	–	0.60	11.10	2.64	–	m³	13.74
maximum depth not exceeding 1.00 m	–	0.50	9.25	2.20	–	m³	11.45
maximum depth not exceeding 2.00 m	–	0.38	7.12	1.69	–	m³	8.81
Extra over any types of excavating irrespective of depth for breaking out existing materials; JCB with breaker attachment							
hard rock	–	0.50	9.25	63.75	–	m³	73.00
concrete	–	0.50	9.25	23.91	–	m³	33.16
reinforced concrete	–	1.00	18.50	40.31	–	m³	58.81
brickwork, blockwork or stonework	–	0.25	4.63	23.91	–	m³	28.54
Extra over any types of excavating irrespective of depth for breaking out existing hard pavings; JCB with breaker attachment							
concrete; 100 mm thick	–	–	–	1.75	–	m²	1.75
concrete; 150 mm thick	–	–	–	2.92	–	m²	2.92
concrete; 200 mm thick	–	–	–	3.51	–	m²	3.51
concrete; 300 mm thick	–	–	–	5.26	–	m²	5.26
reinforced concrete; 100 mm thick	–	0.08	1.54	2.12	–	m²	3.66
reinforced concrete; 150 mm thick	–	0.08	1.39	2.87	–	m²	4.26
reinforced concrete; 200 mm thick	–	0.10	1.85	3.82	–	m²	5.67
reinforced concrete; 300 mm thick	–	0.15	2.77	5.73	–	m²	8.50
tarmacadam; 75 mm thick	–	–	–	1.75	–	m²	1.75
tarmacadam and hardcore; 150 mm thick	–	–	–	2.81	–	m²	2.81
Extra over any types of excavating irrespective of depth for taking up							
precast concrete paving slabs	–	0.07	1.23	0.63	–	m²	1.86
natural stone paving	–	0.10	1.85	0.95	–	m²	2.80
cobbles	–	0.13	2.31	1.19	–	m²	3.50
brick paviors	–	0.13	2.31	1.19	–	m²	3.50
Excavating; hand							
Topsoil for preservation; loading to barrows							
average depth 100 mm	–	0.24	4.44	–	–	m²	4.44
average depth 150 mm	–	0.36	6.66	–	–	m²	6.66
average depth 200 mm	–	0.58	10.66	–	–	m²	10.66
average depth 250 mm	–	0.72	13.32	–	–	m²	13.32
average depth 300 mm	–	0.86	15.98	–	–	m²	15.98
Excavating; hand							
Topsoil to reduce levels							
maximum depth not exceeding 0.25 m	–	2.40	44.40	–	–	m³	44.40
maximum depth not exceeding 1.00 m	–	3.12	57.72	–	–	m³	57.72
Pits							
maximum depth not exceeding 0.25 m	–	2.67	49.33	–	–	m³	49.33
maximum depth not exceeding 1.00 m	–	3.47	64.13	–	–	m³	64.13
maximum depth not exceeding 2.00 m (includes earthwork support)	–	6.93	128.27	–	–	m³	182.19

D GROUNDWORK

Item Excluding site overheads and profit	PC £	Labour hours	Labour £	Plant £	Material £	Unit	Total rate £
Trenches; width not exceeding 0.30 m							
maximum depth not exceeding 0.25 m	–	2.86	52.86	–	–	m³	**52.86**
maximum depth not exceeding 1.00 m	–	3.72	68.84	–	–	m³	**68.84**
maximum depth not exceeding 2.00 m							
(includes earthwork support)	–	3.72	68.84	–	–	m³	**95.80**
Trenches; width exceeding 0.30 m wide							
maximum depth not exceeding 0.25 m	–	2.86	52.86	–	–	m³	**52.86**
maximum depth not exceeding 1.00 m	–	4.00	74.00	–	–	m³	**74.00**
maximum depth not exceeding 2.00 m							
(includes earthwork support)	–	6.00	111.00	–	–	m³	**164.92**
Extra over any types of excavating irrespective of depth for breaking out existing materials; hand held pneumatic breaker							
rock	–	5.00	92.50	31.70	–	m³	**124.20**
concrete	–	2.50	46.25	15.85	–	m³	**62.10**
reinforced concrete	–	4.00	74.00	29.58	–	m³	**103.58**
brickwork, blockwork or stonework	–	1.50	27.75	9.51	–	m³	**37.26**
Filling to make up levels; mechanical (JCB rear bucket)							
Arising from the excavations							
average thickness not exceeding 0.25 m	–	0.07	1.36	2.25	–	m³	**3.61**
average thickness less than 500 mm	–	0.06	1.16	1.91	–	m³	**3.07**
average thickness 1.00 m	–	0.05	0.86	1.42	–	m³	**2.28**
Obtained from on site spoil heaps; average 25 m distance; multiple handling							
average thickness less than 250 mm	–	0.07	1.32	2.70	–	m³	**4.02**
average thickness less than 500 mm	–	0.06	1.09	2.22	–	m³	**3.31**
average thickness 1.00 m	–	0.05	0.93	1.89	–	m³	**2.82**
Obtained off site; planting quality topsoil PC £22.00/m³							
average thickness less than 250 mm	–	0.07	1.32	2.70	26.40	m³	**30.42**
average thickness less than 500 mm	–	0.06	1.09	2.22	26.40	m³	**29.71**
average thickness 1.00 m	–	0.05	0.93	1.89	26.40	m³	**29.22**
Obtained off site; hardcore; PC £12.20/m³							
average thickness less than 250 mm	–	0.08	1.54	3.15	14.64	m³	**19.33**
average thickness less than 500 mm	–	0.06	1.16	2.36	14.64	m³	**18.16**
average thickness 1.00 m	–	0.05	0.97	1.99	14.64	m³	**17.60**
Filling to make up levels; hand							
Arising from the excavations							
average thickness exceeding 0.25 m; depositing in layers 150 mm maximum thickness	–	0.60	11.10	–	–	m³	**11.10**
Obtained from on site spoil heaps; average 25 m distance; multiple handling							
average thickness exceeding 0.25 m thick; depositing in layers 150 mm maximum thickness	–	1.00	18.50	–	–	m³	**18.50**

D GROUNDWORK

Item Excluding site overheads and profit	PC £	Labour hours	Labour £	Plant £	Material £	Unit	Total rate £
D20 EXCAVATING AND FILLING – cont							
Disposal							
Note: Most commercial site disposal is carried							
out by 20 tonne, 8 wheeled vehicles. It has							
been customary to calculate disposal from							
construction sites in terms of full 15 m³ loads.							
Spon's research has found that based on							
weights of common materials such as clean							
hardcore and topsoil, vehicles could not load							
more than 12 m³ at a time. Most hauliers do							
not make it apparent that their loads are							
calculated by weight and not by volume. The							
rates below reflect these lesser volumes which							
are limited by the 20 tonne limit.							
The volumes shown below are based on							
volumes 'in the solid'. Weight and bulking							
factors have been applied. For further							
information please see the weights of typical							
materials in the Earthworks section of the							
Memoranda at the back of this book.							
Disposal; mechanical							
Light soils and loams (bulking factor of 1.25);							
40 tonnes (2 loads per hour)							
Excavated material; off site; to tip;							
mechanically loaded (JCB)							
inert	–	0.04	0.77	1.28	15.84	m³	17.89
hardcore	–	0.04	0.77	1.28	14.40	m³	16.45
macadam	–	0.04	0.77	1.28	18.00	m³	20.05
clean concrete	–	0.04	0.77	1.28	10.80	m³	12.85
soil (sandy and loam) dry	–	0.04	0.77	1.28	15.84	m³	17.89
soil (sandy and loam) wet	–	0.04	0.77	1.28	16.80	m³	18.85
broken out compacted materials such as							
road bases and the like	–	0.04	0.77	1.28	18.00	m³	20.05
soil (clay) dry	–	0.04	0.77	1.28	20.56	m³	22.61
soil (clay) wet	–	0.04	0.77	1.28	21.36	m³	23.41
rubbish (mixed loads)	–	0.04	0.77	1.28	24.00	m³	26.05
green waste	–	0.04	0.77	1.28	24.00	m³	26.05
As above but allowing for 3 loads (36 m³, 60							
tonne) per hour removed							
inert material	–	0.03	0.51	0.85	15.84	m³	17.20
Excavated material; off site to tip;							
mechanically loaded by grab; capacity of load							
7.5 m³ (12 tonne)							
inert material	–	–	–	–	–	m³	28.60
hardcore	–	–	–	–	–	m³	38.57
macadam	–	–	–	–	–	m³	23.20
clean concrete	–	–	–	–	–	m³	26.00
soil (sandy and loam) dry	–	–	–	–	–	m³	28.60
soil (sandy and loam) wet	–	–	–	–	–	m³	30.33
broken out compacted materials such as							
road bases and the like	–	–	–	–	–	m³	31.78
soil (clay) dry	–	–	–	–	–	m³	37.12
soil (clay) wet	–	–	–	–	–	m³	38.57
rubbish (mixed loads)	–	–	–	–	–	m³	36.67
green waste	–	–	–	–	–	m³	33.33

D GROUNDWORK

Item Excluding site overheads and profit	PC £	Labour hours	Labour £	Plant £	Material £	Unit	Total rate £
Disposal by skip; 6 yd³ (4.6 m³)							
Excavated material loaded to skip							
by machine	–	–	–	48.40	–	m³	48.40
by hand	–	3.00	55.50	46.19	–	m³	101.69
Excavated material; on site							
In spoil heaps							
average 25 m distance	–	0.02	0.42	0.87	–	m³	1.29
average 50 m distance	–	0.04	0.72	1.51	–	m³	2.23
average 100 m distance	–	0.06	1.15	2.40	–	m³	3.55
average 200 m distance	–	0.12	2.22	4.64	–	m³	6.86
Spreading on site							
average 25 m distance	–	0.02	0.42	4.12	–	m³	4.54
average 50 m distance	–	0.04	0.72	4.62	–	m³	5.34
average 100 m distance	–	0.06	1.15	5.65	–	m³	6.80
average 200 m distance	–	0.12	2.22	5.31	–	m³	7.53
Disposal; hand							
Excavated material; onsite; in spoil heaps							
average 25 m distance	–	2.40	44.40	–	–	m³	44.40
average 50 m distance	–	2.64	48.84	–	–	m³	48.84
average 100 m distance	–	3.00	55.50	–	–	m³	55.50
average 200 m distance	–	3.60	66.60	–	–	m³	66.60
Excavated material; spreading on site							
average 25 m distance	–	2.64	48.84	–	–	m³	48.84
average 50 m distance	–	3.00	55.50	–	–	m³	55.50
average 100 m distance	–	3.60	66.60	–	–	m³	66.60
average 200 m distance	–	4.20	77.70	–	–	m³	77.70
Grading operations; surface previously excavated to reduce levels to prepare to receive subsequent treatments; grading to accurate levels and falls 20 mm tolerances							
Clay or heavy soils or hardcore							
5 tonne excavator	–	0.04	0.74	0.22	–	m²	0.96
JCB Sitemaster 3CX	–	0.02	0.37	0.67	–	m²	1.04
21 tonne 360 tracked excavator	–	0.01	0.19	0.42	–	m²	0.61
by hand	–	0.10	1.85	–	–	m²	1.85
Loamy topsoils							
5 tonne excavator	–	0.03	0.49	0.15	–	m²	0.64
JCB Sitemaster 3CX	–	0.01	0.25	0.45	–	m²	0.70
21 tonne 360 tracked excavator	–	0.01	0.14	0.31	–	m²	0.45
by hand	–	0.05	0.93	–	–	m²	0.93
Sand or graded granular materials							
5 tonne excavator	–	0.02	0.37	0.11	–	m²	0.48
JCB Sitemaster 3CX	–	0.01	0.19	0.34	–	m²	0.53
21 tonne 360 tracked excavator	–	0.01	0.11	0.24	–	m²	0.35
by hand	–	0.03	0.62	–	–	m²	0.62
Grading operations; surface recently filled to raise levels to prepare to receive subsequent treatments; grading to accurate levels and falls 20 mm tolerances							
Clay or heavy soils or hardcore							
5 tonne excavator	–	0.03	0.56	0.17	–	m²	0.73
JCB Sitemaster 3CX	–	0.02	0.28	0.50	–	m²	0.78
21 tonne 360 tracked excavator	–	0.01	0.12	0.28	–	m²	0.40
by hand	–	0.08	1.54	–	–	m²	1.54

D GROUNDWORK

Item Excluding site overheads and profit	PC £	Labour hours	Labour £	Plant £	Material £	Unit	Total rate £
D20 EXCAVATING AND FILLING – cont							
Grading operations – cont							
Loamy topsoils							
5 tonne excavator	–	0.02	0.38	0.11	–	m²	0.49
JCB Sitemaster 3CX	–	0.01	0.19	0.35	–	m²	0.54
21 tonne 360 tracked excavator	–	0.01	0.12	0.28	–	m²	0.40
by hand	–	0.04	0.74	–	–	m²	0.74
Sand or graded granular materials							
5 tonne excavator	–	0.02	0.28	0.08	–	m²	0.36
JCB Sitemaster 3CX	–	0.01	0.14	0.25	–	m²	0.39
21 tonne 360 tracked excavator	–	0.01	0.09	0.21	–	m²	0.30
by hand	–	0.03	0.53	–	–	m²	0.53
Surface grading; Agripower Ltd; using							
laser controlled equipment							
Maximum variation of existing ground level							
75 mm; laser box grader and tractor	–	–	–	–	–	ha	1200.00
100 mm; laser controlled low ground							
pressure bulldozer	–	–	–	–	–	ha	1900.00
Fraise mowing; scarification of existing turf							
layer to rootzone level; Agripower Ltd							
Cart arisings to tip on site							
average thickness 50–75 mm	–	–	–	–	–	ha	2963.00
Cultivating							
Ripping up subsoil; using approved subsoiling							
machine; minimum depth 250 mm below							
topsoil; at 1.20 m centres; in							
gravel or sandy clay	–	–	–	2.40	–	100 m²	2.40
soil compacted by machines	–	–	–	2.80	–	100 m²	2.80
clay	–	–	–	3.00	–	100 m²	3.00
chalk or other soft rock	–	–	–	6.01	–	100 m²	6.01
Extra for subsoiling at 1 m centres	–	–	–	0.60	–	100 m²	0.60
Breaking up existing ground; using pedestrian							
operated tine cultivator or rotavator; loam or							
sandy soil							
100 mm deep	–	0.22	4.07	2.12	–	100 m²	6.19
150 mm deep	–	0.28	5.09	2.65	–	100 m²	7.74
200 mm deep	–	0.37	6.78	3.54	–	100 m²	10.32
As above but in heavy clay or wet soils							
100 mm deep	–	0.44	8.14	4.25	–	100 m²	12.39
150 mm deep	–	0.66	12.21	6.37	–	100 m²	18.58
200 mm deep	–	0.82	15.26	7.96	–	100 m²	23.22
Breaking up existing ground; using tractor							
drawn tine cultivator or rotavator							
100 mm deep	–	–	–	0.28	–	100 m²	0.28
150 mm deep	–	–	–	0.36	–	100 m²	0.36
200 mm deep	–	–	–	0.47	–	100 m²	0.47
400 mm deep	–	–	–	1.42	–	100 m²	1.42
Cultivating ploughed ground; using disc, drag,							
or chain harrow							
4 passes	–	–	–	1.71	–	100 m²	1.71
Rolling cultivated ground lightly; using							
self-propelled agricultural roller	–	0.06	1.03	0.60	–	100 m²	1.63

D GROUNDWORK

Item Excluding site overheads and profit	PC £	Labour hours	Labour £	Plant £	Material £	Unit	Total rate £
Importing and storing selected and approved topsoil; 20 tonne load = 11.88 m³ average							
20 tonne loads	26.40	–	–	–	26.40	m³	**26.40**
Surface treatments							
Compacting							
bottoms of excavations	–	0.01	0.09	0.03	–	m²	**0.12**
Surface preparation							
Trimming surfaces of cultivated ground to final levels; removing roots, stones and debris exceeding 50 mm in any direction to tip off site; slopes less than 15°							
clean ground with minimal stone content	–	0.25	4.63	–	–	100 m²	**4.63**
slightly stony; 0.5 kg stones per m²	–	0.33	6.16	–	–	100 m²	**6.16**
very stony; 1.0–3.0 kg stones per m²	–	0.50	9.25	–	0.01	100 m²	**9.26**
clearing mixed, slightly contaminated rubble; inclusive of roots and vegetation	–	0.50	9.25	–	0.06	100 m²	**9.31**
clearing brick-bats, stones and clean rubble	–	0.60	11.10	–	0.02	100 m²	**11.12**

E IN SITU CONCRETE/LARGE PRECAST CONCRETE

Item Excluding site overheads and profit	PC £	Labour hours	Labour £	Plant £	Material £	Unit	Total rate £
E10 MIXING/CASTING/CURING IN SITU CONCRETE							
General The concrete mixes used here are referred to as 'Designed', 'Standard' and 'Designated' mixes.							
Designed mix User specifies the performance of the concrete. Producer is responsible for selecting the appropriate mix. Strength testing is essential.							
Prescribed mix User specifies the mix constituents and is responsible for ensuring that the concrete meets performance requirements. Accurate mix proportions are essential.							
Standard mix Made with a restricted range of materials. Specification to include the proposed use of the material as well as the standard mix reference, type of cement, type and size of aggregate, and slump (workability). Quality assurance is required.							
Designated mix Producer to hold current product conformity certification and quality approval. Quality assurance is essential. The mix must not be modified. Note: The mixes below are all mixed in 113 litre mixers. Please see the Minor Works section of this book for mixes containing aggregates delivered in bulk bags.							
Concrete mixes; mixed on site; costs for producing concrete; prices for commonly used mixes for various types of work; based on bulk load 20 tonne rates for aggregates Roughest type mass concrete such as footings and road haunchings; 300 mm thick							
1:3:6	77.60	–	–	–	77.60	m³	**77.60**
1:3:6 sulphate-resisting	91.19	–	–	–	91.19	m³	**91.19**
As above but aggregates delivered in 10 tonne loads							
1:3:6	87.07	–	–	–	87.07	m³	**87.07**
1:3:6 sulphate-resisting	96.29	–	–	–	96.29	m³	**96.29**
As above but aggregates delivered in 850 kg bulk bags							
1:3:6	114.45	–	–	–	114.45	m³	**114.45**
1:3:6 sulphate-resisting	120.29	–	–	–	120.29	m³	**120.29**

E IN SITU CONCRETE/LARGE PRECAST CONCRETE

Item Excluding site overheads and profit	PC £	Labour hours	Labour £	Plant £	Material £	Unit	Total rate £
Most ordinary use of concrete such as mass walls above ground, road slabs, etc. and general reinforced concrete work							
1:2:4	89.42	–	–	–	89.42	m³	**89.42**
1:2:4 sulphate-resisting	108.63	–	–	–	108.63	m³	**108.63**
As above but aggregates delivered in 10 tonne loads							
1:2:4	94.52	–	–	–	94.52	m³	**94.52**
1:2:4 sulphate-resisting	113.73	–	–	–	113.73	m³	**113.73**
As above but aggregates delivered in 850 kg bulk bags							
1:2:4	121.90	–	–	–	121.90	m³	**121.90**
1:2:4 sulphate-resisting	141.12	–	–	–	141.12	m³	**141.12**
Watertight floors, pavements, walls, tanks, pits, steps, paths, surface of two course roads; reinforced concrete where extra strength is required							
1:1.5:3	84.06	1.20	22.14	–	84.06	m³	**106.20**
As above but aggregates delivered in 10 tonne loads							
1:1.5:3	89.16	1.20	22.14	–	89.16	m³	**111.30**
As above but aggregates delivered in 850 kg bulk bags							
1:1.5:3	116.54	1.20	22.14	–	116.54	m³	**138.68**
Plain in situ concrete; site mixed; **10 N/mm² – 40 mm aggregate (1:3:6)** **(aggregate delivery indicated)**							
Foundations							
ordinary portland cement; 20 tonne ballast loads	79.54	1.00	18.50	–	79.54	m³	**98.04**
ordinary portland cement; 10 tonne ballast loads	89.25	1.00	18.50	–	89.25	m³	**107.75**
ordinary portland cement; 850 kg bulk bags	117.31	1.00	18.50	–	117.31	m³	**135.81**
sulphate-resistant cement; 10 tonne ballast loads	96.88	1.00	18.50	–	96.88	m³	**115.38**
sulphate-resistant cement; 850 kg bulk bags	123.30	1.00	18.50	–	123.30	m³	**141.80**
Foundations; poured on or against earth or unblinded hardcore							
ordinary portland cement; 20 tonne ballast loads	81.48	1.00	18.50	–	81.48	m³	**99.98**
ordinary portland cement; 10 tonne ballast loads	91.42	1.00	18.50	–	91.42	m³	**109.92**
ordinary portland cement; 850 kg bulk bags	120.18	1.00	18.50	–	120.18	m³	**138.68**
sulphate-resistant cement; 20 tonne ballast loads	114.07	1.00	18.50	–	114.07	m³	**132.57**
sulphate-resistant cement; 850 kg bulk bags	126.30	1.00	18.50	–	126.30	m³	**144.80**
Isolated foundations							
ordinary portland cement; 20 tonne ballast loads	79.54	1.10	20.35	–	79.54	m³	**99.89**
ordinary portland cement; 850 kg bulk bags	117.31	1.10	20.35	–	117.31	m³	**137.66**
sulphate-resistant cement; 20 tonne ballast loads	108.63	1.10	20.35	–	108.63	m³	**128.98**
sulphate-resistant cement; 850 kg bulk bags	123.30	1.10	20.35	–	123.30	m³	**143.65**

E IN SITU CONCRETE/LARGE PRECAST CONCRETE

Item Excluding site overheads and profit	PC £	Labour hours	Labour £	Plant £	Material £	Unit	Total rate £
E10 MIXING/CASTING/CURING IN SITU CONCRETE – cont							
Plain in situ concrete; site mixed; **21 N/mm² – 20 mm aggregate (1:2:4)**							
Foundations							
ordinary portland cement; 20 tonne ballast loads	91.66	1.00	18.50	–	91.66	m³	**110.16**
ordinary portland cement; 850 kg bulk bags	124.95	1.00	18.50	–	124.95	m³	**143.45**
sulphate-resistant cement; 20 tonne ballast loads	111.35	1.00	18.50	–	111.35	m³	**129.85**
sulphate-resistant cement; 850 kg bulk bags	123.30	1.00	18.50	–	123.30	m³	**141.80**
Foundations; poured on or against earth or unblinded hardcore							
ordinary portland cement; 20 tonne ballast loads	93.89	1.00	18.50	–	93.89	m³	**112.39**
ordinary portland cement; 850 kg bulk bags	128.00	1.00	18.50	–	128.00	m³	**146.50**
sulphate-resistant cement; 20 tonne ballast loads	114.07	1.00	18.50	–	114.07	m³	**132.57**
sulphate-resistant cement; 850 kg bulk bags	126.30	1.00	18.50	–	126.30	m³	**144.80**
Isolated foundations							
ordinary portland cement; 20 tonne ballast loads	91.66	1.10	20.35	–	91.66	m³	**112.01**
ordinary portland cement; 850 kg bulk bags	124.95	1.10	20.35	–	124.95	m³	**145.30**
sulphate-resistant cement; 20 tonne ballast loads	111.35	1.10	20.35	–	111.35	m³	**131.70**
sulphate-resistant cement; 850 kg bulk bags	123.30	1.10	20.35	–	123.30	m³	**143.65**
Ready mix concrete; concrete mixed on site; Euromix							
Ready mix concrete mixed on site; placed by barrow not more than 25 m distance (16 barrows/m³)							
concrete 1:3:6	95.00	1.00	18.50	–	95.00	m³	**113.50**
concrete 1:2:4	95.00	1.00	18.50	–	95.00	m³	**113.50**
concrete C30	105.00	1.00	18.50	–	105.00	m³	**123.50**
Reinforced in situ concrete; site mixed; **21 N/mm² – 20 mm aggregate (1:2:4);** **aggregates delivered in 10 tonne loads**							
Foundations							
ordinary portland cement	91.66	2.20	40.70	–	91.66	m³	**132.36**
sulphate-resistant cement	111.35	2.20	40.70	–	111.35	m³	**152.05**
Foundations; poured on or against earth or unblinded hardcore							
ordinary portland cement	91.66	2.20	40.70	–	91.66	m³	**132.36**
sulphate-resistant cement	99.25	2.20	40.70	–	99.25	m³	**139.95**
Isolated foundations							
ordinary portland cement	91.66	2.75	50.88	–	91.66	m³	**142.54**
sulphate-resistant cement	99.25	2.75	50.88	–	99.25	m³	**150.13**

E IN SITU CONCRETE/LARGE PRECAST CONCRETE

Item Excluding site overheads and profit	PC £	Labour hours	Labour £	Plant £	Material £	Unit	Total rate £
Plain in situ concrete; ready mixed; Tarmac Southern; 10 N/mm mixes; suitable for mass concrete fill and blinding							
Foundations							
GEN 1; Designated mix	62.80	1.50	27.75	–	62.80	m³	**90.55**
ST 2; Standard mix	67.31	1.50	27.75	–	67.31	m³	**95.06**
Foundations; poured on or against earth or unblinded hardcore							
GEN 1; Designated mix	62.80	1.57	29.14	–	62.80	m³	**91.94**
ST 2; Standard mix	68.91	1.57	29.14	–	68.91	m³	**98.05**
Isolated foundations							
GEN 1; Designated mix	62.80	2.00	37.00	–	62.80	m³	**99.80**
ST 2; Standard mix	67.31	2.00	37.00	–	67.31	m³	**104.31**
Plain in situ concrete; ready mixed; Tarmac Southern; 15 N/mm mixes; suitable for oversite below suspended slabs and strip footings in non aggressive soils							
Foundations							
GEN 2; Designated mix	67.62	1.50	27.75	–	67.62	m³	**95.37**
ST 3; Standard mix	70.03	1.50	27.75	–	70.03	m³	**97.78**
Foundations; poured on or against earth or unblinded hardcore							
GEN 2; Designated mix	64.40	1.57	29.14	–	64.40	m³	**93.54**
ST 3; Standard mix	66.70	1.57	29.14	–	66.70	m³	**95.84**
Isolated foundations							
GEN 2; Designated mix	64.40	2.00	37.00	–	64.40	m³	**101.40**
ST 3; Standard mix	66.70	2.00	37.00	–	66.70	m³	**103.70**
Plain in situ concrete; ready mixed; Tarmac Southern; air entrained mixes suitable for paving							
Beds or slabs; house drives, parking and external paving							
PAV 1; 35 N/mm²; Designated mix	71.92	1.50	27.75	–	71.92	m³	**99.67**
Beds or slabs; heavy duty external paving							
PAV 2; 40 N/mm²; Designated mix	73.29	1.50	27.75	–	73.29	m³	**101.04**
Reinforced in situ concrete; ready mixed; Tarmac Southern; 35 N/mm² mix; suitable for foundations in class 2 sulphate conditions							
Foundations							
RC 35; Designated mix	73.29	2.00	37.00	–	73.29	m³	**110.29**
Foundations; poured on or against earth or unblinded hardcore							
RC 35; Designated mix	75.03	2.10	38.85	–	75.03	m³	**113.88**
Isolated foundations							
RC 35; Designated mix	69.80	2.00	37.00	–	69.80	m³	**106.80**

E IN SITU CONCRETE/LARGE PRECAST CONCRETE

Item Excluding site overheads and profit	PC £	Labour hours	Labour £	Plant £	Material £	Unit	Total rate £
E20 FORMWORK FOR IN SITU CONCRETE							
Plain vertical formwork; basic finish							
Sides of foundations							
height not exceeding 250 mm	–	1.00	18.50	–	1.25	m	**19.75**
height 250–500 mm	–	1.00	18.50	–	2.14	m	**20.64**
height 500 mm–1.00 m	–	1.50	27.75	–	4.29	m	**32.04**
height exceeding 1.00 m	–	2.00	37.00	–	4.29	m²	**41.29**
Sides of foundations; left in							
height not exceeding 250 mm	–	1.00	18.50	–	4.48	m	**22.98**
height 250–500 mm	–	1.00	18.50	–	8.61	m	**27.11**
height 500 mm–1.00 m	–	1.50	27.75	–	17.22	m	**44.97**
height over 1.00 m	–	2.00	37.00	–	12.37	m²	**49.37**
E30 REINFORCEMENT FOR IN SITU CONCRETE							
The rates for reinforcement shown for steel bars below are based on prices which would be supplied on a typical landscape contract. The steel prices shown have been priced on a selection of steel delivered to site where the total order quantity is in the region of 2 tonnes. The assumption is that should larger quantities be required, the work would fall outside the scope of the typical landscape contract defined in the front of this book. Keener rates can be obtained for larger orders.							
Reinforcement bars; hot rolled plain round mild steel; straight or bent							
Bars							
8 mm nominal size	600.00	27.00	499.50	–	600.00	tonne	**1099.50**
10 mm nominal size	600.00	26.00	481.00	–	600.00	tonne	**1081.00**
12 mm nominal size	600.00	25.00	462.50	–	600.00	tonne	**1062.50**
16 mm nominal size	600.00	24.00	444.00	–	600.00	tonne	**1044.00**
20 mm nominal size	600.00	23.00	425.50	–	600.00	tonne	**1025.50**
Reinforcement bar to concrete formwork							
8 mm bar							
100 ccs	2.34	0.33	6.11	–	2.34	m²	**8.45**
200 ccs	1.14	0.25	4.63	–	1.14	m²	**5.77**
300 ccs	0.78	0.15	2.77	–	0.78	m²	**3.55**
10 mm bar							
100 ccs	3.66	0.33	6.11	–	3.66	m²	**9.77**
200 ccs	1.80	0.25	4.63	–	1.80	m²	**6.43**
300 ccs	1.20	0.15	2.77	–	1.20	m²	**3.97**
12 mm bar							
100 ccs	5.28	0.33	6.11	–	5.28	m²	**11.39**
200 ccs	2.64	0.25	4.63	–	2.64	m²	**7.27**
300 ccs	1.74	0.15	2.77	–	1.74	m²	**4.51**
16 mm bar							
100 ccs	9.42	0.40	7.40	–	9.42	m²	**16.82**
200 ccs	4.74	0.33	6.11	–	4.74	m²	**10.85**
300 ccs	3.18	0.25	4.63	–	3.18	m²	**7.81**

E IN SITU CONCRETE/LARGE PRECAST CONCRETE

Item Excluding site overheads and profit	PC £	Labour hours	Labour £	Plant £	Material £	Unit	Total rate £
25 mm bar							
100 ccs	23.10	0.40	7.40	–	23.10	m²	**30.50**
200 ccs	11.52	0.33	6.11	–	11.52	m²	**17.63**
300 ccs	7.68	0.25	4.63	–	7.68	m²	**12.31**
32 mm bar							
100 ccs	37.86	0.40	7.40	–	37.86	m²	**45.26**
200 ccs	18.90	0.33	6.11	–	18.90	m²	**25.01**
300 ccs	12.60	0.25	4.63	–	12.60	m²	**17.23**
Reinforcement fabric; lapped; in beds or suspended slabs							
Fabric							
A98 (1.54 kg/m²)	1.78	0.22	4.07	–	1.78	m²	**5.85**
A142 (2.22 kg/m²)	2.56	0.22	4.07	–	2.56	m²	**6.63**
A193 (3.02 kg/m²)	3.50	0.22	4.07	–	3.50	m²	**7.57**
A252 (3.95 kg/m²)	4.14	0.24	4.44	–	4.14	m²	**8.58**
A393 (6.16 kg/m²)	6.45	0.28	5.18	–	6.45	m²	**11.63**

F MASONRY

Item Excluding site overheads and profit	PC £	Labour hours	Labour £	Plant £	Material £	Unit	Total rate £
MARKET PRICES OF MATERIALS							
Cement; Builder Centre							
Portland cement	–	–	–	–	3.32	25 kg	3.32
Sulphate-resistant cement	–	–	–	–	4.90	25 kg	4.90
White cement	–	–	–	–	7.42	25 kg	7.42
Sand; Builder Centre							
Building sand							
loose	–	–	–	–	42.00	m^3	42.00
850 kg bulk bags	–	–	–	–	59.96	nr	59.96
Sharp sand							
loose	–	–	–	–	30.60	m^3	30.60
850 kg bulk bags	–	–	–	–	59.96	nr	59.96
Sand; Yeoman Aggregates Ltd							
Sharp sand	–	–	–	–	17.00	tonne	17.00
F10 BRICK/BLOCK WALLING DATA							
Note: Batching quantities for these mortar mixes may be found in the Tables and Memoranda section of this book.							
Mortar mixes; common mixes for various types of work; mortar mixed on site; prices based on builders merchant rates for cement							
Aggregates delivered in 850 kg bulk bags; mechanically mixed							
1:3	–	0.75	13.88	–	139.36	m^3	153.24
1:4	–	0.75	13.88	–	111.11	m^3	124.99
1:1:6	–	0.75	13.88	–	135.49	m^3	149.37
1:1:6 sulphate-resisting	–	0.75	13.88	–	152.56	m^3	166.44
Mortar mixes; common mixes for various types of work; mortar mixed on site; prices based on builders merchant rates for cement							
Aggregates delivered in 10 tonne loads; mechanically mixed							
1:3	–	0.75	13.88	–	111.31	m^3	125.19
1:4	–	0.75	13.88	–	86.24	m^3	100.12
1:1:6	–	0.75	13.88	–	112.03	m^3	125.91
1:1:6 sulphate-resisting	–	0.75	13.88	–	129.10	m^3	142.98
Variation in brick prices (area × price difference × value shown below)							
Add or subtract the following amounts for every £1.00/1000 difference in the PC price of the measured items below							
half brick thick	–	–	–	–	0.06	m^2	0.06
one brick thick	–	–	–	–	0.13	m^2	0.13
one and a half brick thick	–	–	–	–	0.19	m^2	0.19
two brick thick	–	–	–	–	0.25	m^2	0.25

F MASONRY

Item Excluding site overheads and profit	PC £	Labour hours	Labour £	Plant £	Material £	Unit	Total rate £
Mortar (1:3) required per m² of brickwork; brick size 215 × 102.5 × 65 mm							
Half brick wall (103 mm)							
no frog	–	–	–	–	2.89	m²	**2.89**
single frog	–	–	–	–	3.56	m²	**3.56**
double frog	–	–	–	–	4.21	m²	**4.21**
2 × half brick cavity wall (270 mm)							
no frog	–	–	–	–	5.00	m²	**5.00**
single frog	–	–	–	–	5.92	m²	**5.92**
double frog	–	–	–	–	7.24	m²	**7.24**
One brick wall (215 mm)							
no frog	–	–	–	–	6.05	m²	**6.05**
single frog	–	–	–	–	7.24	m²	**7.24**
double frog	–	–	–	–	8.41	m²	**8.41**
One and a half brick wall (328 mm)							
no frog	–	–	–	–	8.29	m²	**8.29**
single frog	–	–	–	–	9.73	m²	**9.73**
double frog	–	–	–	–	11.57	m²	**11.57**
Mortar (1:3) required per m² of blockwork; blocks 440 × 215 mm							
Block thickness							
100 mm	–	–	–	–	0.93	m²	**0.93**
140 mm	–	–	–	–	1.19	m²	**1.19**
hollow blocks 440 × 215 mm	–	–	–	–	0.30	m²	**0.30**
Movement of materials							
Loading to wheelbarrows and transporting to location; per 215 mm thick walls; maximum distance 25 m	–	0.42	7.71	–	–	m²	**7.71**
F10 BRICK WALLING							
Class B engineering bricks; PC £239.00/ 1000; double Flemish bond in cement mortar (1:3)							
Offloading mechanically; loading to wheelbarrows; transporting to location maximum 25 m distance per 215 mm thick walls	–	0.42	7.71	–	–	m²	**7.71**
Construct walls							
half brick thick	–	1.80	33.30	–	16.93	m²	**50.23**
one brick thick	–	3.60	66.59	–	33.87	m²	**100.46**
one and a half brick thick	–	5.40	99.89	–	63.72	m²	**163.61**
two brick thick	–	7.20	133.19	–	67.74	m²	**200.93**
Construct walls; curved; mean radius 6 m							
half brick thick	–	2.70	50.00	–	17.53	m²	**67.53**
one brick thick	–	5.41	100.00	–	33.87	m²	**133.87**
Construct walls; curved; mean radius 1.50 m							
half brick thick	–	3.60	66.59	–	17.56	m²	**84.15**
one brick thick	–	7.20	133.19	–	33.87	m²	**167.06**
Construct walls; tapering; one face battering; average							
one and a half brick thick	–	6.65	123.00	–	50.80	m²	**173.80**
two brick thick	–	8.87	164.01	–	67.74	m²	**231.75**

F MASONRY

Item Excluding site overheads and profit	PC £	Labour hours	Labour £	Plant £	Material £	Unit	Total rate £
F10 BRICK WALLING – cont							
Class B engineering bricks – cont							
Construct walls; battering (retaining)							
one and a half brick thick	–	6.65	123.00	–	50.80	m²	**173.80**
two brick thick	–	8.87	164.01	–	67.74	m²	**231.75**
Projections; vertical							
one brick × half brick	–	0.70	12.95	–	3.78	m	**16.73**
one brick × one brick	–	1.40	25.90	–	7.57	m	**33.47**
one and a half brick × one brick	–	2.10	38.85	–	11.35	m	**50.20**
two brick by one brick	–	2.30	42.55	–	15.14	m	**57.69**
Walls; half brick thick							
in honeycomb bond	–	1.80	33.30	–	12.30	m²	**45.60**
in quarter bond	–	1.67	30.83	–	16.58	m²	**47.41**
Facing bricks; PC £300.00/1000; English **garden wall bond; in gauged mortar (1:1:6);** **facework one side**							
Mechanically offloading; maximum 25 m distance; loading to wheelbarrows and transporting to location; walls	–	0.42	7.71	–	–	m²	**7.71**
Construct walls							
half brick thick	–	1.80	33.30	–	20.79	m²	**54.09**
half brick thick (using site cut snap headers to form bond)	–	2.41	44.68	–	20.79	m²	**65.47**
one brick thick	–	3.60	66.59	–	41.58	m²	**108.17**
one and a half brick thick	–	5.40	99.89	–	62.37	m²	**162.26**
two brick thick	–	7.20	133.19	–	83.15	m²	**216.34**
Walls; curved; mean radius 6 m							
half brick thick	–	2.70	50.00	–	22.32	m²	**72.32**
one brick thick	–	5.41	100.00	–	43.38	m²	**143.38**
Walls; curved; mean radius 1.50 m							
half brick thick	–	3.60	66.59	–	21.87	m²	**88.46**
one brick thick	–	7.20	133.19	–	42.48	m²	**175.67**
Walls; tapering; one face battering; average							
one and a half brick thick	–	6.65	123.00	–	65.07	m²	**188.07**
two brick thick	–	8.87	164.01	–	86.75	m²	**250.76**
Walls; battering (retaining)							
one and a half brick thick	–	5.98	110.70	–	65.07	m²	**175.77**
two brick thick	–	7.98	147.61	–	86.75	m²	**234.36**
Projections; vertical							
one brick × half brick	–	0.70	12.95	–	4.64	m	**17.59**
one brick × one brick	–	1.40	25.90	–	9.28	m	**35.18**
one and a half brick × one brick	–	2.10	38.85	–	13.92	m	**52.77**
two brick × one brick	–	2.30	42.55	–	18.56	m	**61.11**
Brickwork fair faced both sides; facing bricks in gauged mortar (1:1:6)							
extra for fair face both sides; flush, struck, weathered, or bucket-handle pointing	–	0.67	12.33	–	–	m²	**12.33**
Less for cement mortar (1:3) in lieu of gauged mortar							
half brick thick	–	–	–	–	–0.01	m²	**–0.01**
one brick thick	–	–	–	–	–0.02	m²	**–0.02**
one and a half brick thick	–	–	–	–	–0.03	m²	**–0.03**
two brick thick	–	–	–	–	–0.04	m²	**–0.04**

F MASONRY

Item Excluding site overheads and profit	PC £	Labour hours	Labour £	Plant £	Material £	Unit	Total rate £
Bricks; PC £800.00/1000; English garden wall bond; in gauged mortar (1:1:6)							
Walls							
half brick thick (stretcher bond)	–	1.80	33.30	–	52.29	m²	85.59
half brick thick (using site cut snap headers to form bond)	–	2.41	44.68	–	52.29	m²	96.97
one brick thick	–	3.60	66.59	–	102.66	m²	169.25
one and a half brick thick	–	5.40	99.89	–	153.99	m²	253.88
two brick thick	–	7.20	133.19	–	205.31	m²	338.50
Walls; curved; mean radius 6 m							
half brick thick	–	2.70	50.00	–	55.32	m²	105.32
one brick thick	–	7.20	133.19	–	109.38	m²	242.57
Walls; curved; mean radius 1.50 m							
half brick thick	–	1.60	29.60	–	57.72	m²	87.32
one brick thick	–	3.20	59.20	–	114.18	m²	173.38
Less for cement mortar (1:3) in lieu of gauged mortar							
half brick thick	–	–	–	–	–0.01	m²	–0.01
one brick thick	–	–	–	–	–0.02	m²	–0.02
one and a half brick thick	–	–	–	–	–0.03	m²	–0.03
two brick thick	–	–	–	–	–0.04	m²	–0.04
Walls; stretcher bond; wall ties at 450 mm centres vertically and horizontally							
one brick thick	1.39	3.27	60.56	–	104.05	m²	164.61
two brick thick	4.18	4.91	90.82	–	209.50	m²	300.32
Brickwork fair faced both sides; facing bricks in gauged mortar (1:1:6)							
extra for fair face both sides; flush, struck, weathered or bucket-handle pointing	–	0.67	12.33	–	–	m²	12.33
Brick copings							
Copings; all brick headers-on-edge; two angles rounded 53 mm radius; flush pointing top and both sides as work proceeds; one brick wide; horizontal							
machine-made specials	48.05	0.49	9.11	–	48.81	m	57.92
handmade specials	61.80	0.49	9.11	–	62.55	m	71.66
Extra over copings for two courses machine-made tile creasings; projecting 25 mm each side; 260 mm wide copings; horizontal	5.94	0.50	9.25	–	6.32	m	15.57
Copings; all brick headers-on-edge; flush pointing top and both sides as work proceeds; one brick wide; horizontal							
facing bricks PC £300.00/1000	4.30	0.49	9.11	–	4.30	m	13.41
engineering bricks PC £239.00/1000	3.35	0.49	9.11	–	3.35	m	12.46
F10 BLOCK WALLING							
Dense aggregate concrete blocks; Tarmac Topblock or other equal and approved; in gauged mortar (1:2:9); one course underground							
Walls							
Solid blocks 7 N/mm²							
440 × 215 × 100 mm thick	7.30	1.20	22.20	–	8.30	m²	30.50
440 × 215 × 140 mm thick	11.75	1.33	24.61	–	12.88	m²	37.49

F MASONRY

Item Excluding site overheads and profit	PC £	Labour hours	Labour £	Plant £	Material £	Unit	Total rate £
F10 BLOCK WALLING – cont							
Dense aggregate concrete blocks – cont							
Solid blocks 7 N/mm² laid flat							
440 × 100 × 215 mm thick	14.89	3.22	59.57	–	17.58	m²	77.15
Hollow concrete blocks							
440 × 215 × 215 mm thick	19.00	1.30	24.05	–	20.13	m²	44.18
Filling of hollow concrete blocks with concrete							
as work proceeds; tamping and compacting							
440 × 215 × 215 mm thick	16.35	0.20	3.70	–	16.35	m²	20.05
F10 BRICK PIERS							
Isolated brick piers in English bond							
Engineering brick PC £239.00/1000							
one brick thick (225 mm)	–	7.00	129.50	–	37.63	m²	167.13
one and a half brick thick (337.5 mm)	–	9.00	166.50	–	56.44	m²	222.94
two brick thick (450 mm)	–	10.00	185.00	–	76.50	m²	261.50
three brick thick (675 mm)	–	12.40	229.40	–	112.88	m²	342.28
Engineering brick PC £239.00/1000 (not SMM)							
one brick (225 mm)	–	1.63	30.23	–	8.63	m	38.86
one and a half brick thick (337.5 mm)	–	3.00	55.50	–	25.57	m	81.07
two brick thick (450 mm)	–	4.50	83.25	–	40.87	m	124.12
three brick thick (675 mm)	–	8.37	154.84	–	83.52	m	238.36
Brick PC £300.00/1000							
one brick thick (225 mm)	–	7.00	129.50	–	45.31	m²	174.81
one and a half brick thick (337.5 mm)	–	9.00	166.50	–	67.97	m²	234.47
two brick thick (450 mm)	–	10.00	185.00	–	91.87	m²	276.87
three brick thick (675 mm)	–	12.40	229.40	–	135.93	m²	365.33
Brick PC £300.00/1000 (not SMM)							
one brick (225 mm)	–	1.63	30.23	–	10.43	m	40.66
one and a half brick thick (337.5 mm)	–	3.00	55.50	–	29.46	m	84.96
two brick thick (450 mm)	–	4.50	83.25	–	47.79	m	131.04
three brick thick (675 mm)	–	8.37	154.84	–	99.08	m	253.92
Brick PC £800.00/1000							
one brick thick (225 mm)	–	7.00	129.50	–	108.31	m²	237.81
one and a half brick thick (337.5 mm)	–	9.00	166.50	–	162.47	m²	328.97
two brick thick (450 mm)	–	10.00	185.00	–	217.87	m²	402.87
three brick thick (675 mm)	–	12.40	229.40	–	324.93	m²	554.33
Brick PC £800.00/1000 (not SMM)							
one brick (225 mm)	–	1.63	30.23	–	25.13	m	55.36
one and a half brick thick (337.5 mm)	–	3.00	55.50	–	61.36	m	116.86
two brick thick (450 mm)	–	4.50	83.25	–	104.49	m	187.74
three brick thick (675 mm)	–	8.37	154.84	–	226.65	m	381.49
F10 BLOCK PIERS							
Isolated blockwork piers; to receive facing							
treatments (not included)							
Solid blocks; 440 × 100 × 215 mm thick;							
7 N/mm² laid flat							
450 mm wide × 450 mm thick	64.77	8.80	162.80	–	122.35	m²	285.15
440 × 100 × 215 mm thick (not SMM)	29.20	3.22	59.57	–	31.89	m	91.46

F MASONRY

Item Excluding site overheads and profit	PC £	Labour hours	Labour £	Plant £	Material £	Unit	Total rate £
F20 NATURAL STONE RUBBLE WALLING							
Granite walls							
Granite random rubble walls; laid dry							
200 mm thick; single faced	48.09	6.66	123.21	–	48.09	m²	**171.30**
Granite walls; one face battering to 50°;							
pointing faces							
450 mm (average) thick	116.32	5.00	92.50	–	117.11	m²	**209.61**
Dry stone walling – General							
Preamble: In rural areas where natural stone							
is a traditional material, it may be possible to							
use dry stone walling or dyking as an							
alternative to fences or brick walls. Many local							
authorities are willing to meet the extra cost of							
stone walling in areas of high landscape							
value, and they may hold lists of available							
craftsmen. DSWA Office, Westmorland County							
Showground, Lane Farm, Crooklands,							
Milnthorpe, Cumbria, LA7 7NH; Tel: 01539							
567953; E-mail: information@dswa.org.uk							
Note: Traditional walls are not built on							
concrete foundations.							
Dry stone wall; wall on concrete foundation							
(not included); dry stone coursed wall inclusive							
of locking stones and filling to wall with broken							
stone or rubble; walls up to 1.20 m high;							
battered; 2 sides fair faced							
Yorkstone	–	6.50	120.25	–	66.73	m²	**186.98**
Cotswold stone	–	6.50	120.25	–	81.57	m²	**201.82**
Purbeck	–	6.50	120.25	–	88.98	m²	**209.23**
F22 CAST STONE ASHLAR WALLING/							
DRESSINGS							
Haddonstone Ltd; cast stone piers;							
ornamental masonry in Portland Bath or							
Terracotta finished cast stone							
Gate Pier S120; to foundations and							
underground work measured separately;							
concrete infill							
S120G base unit to pier; 699 × 699 ×							
172 mm	165.68	0.75	13.88	–	170.04	nr	**183.92**
S120F/F shaft base unit; 533 × 533 ×							
280 mm	131.60	1.00	18.50	–	136.11	nr	**154.61**
S120E/E main shaft unit; 533 × 533 ×							
280 mm; nr of units required dependent on							
height of pier	131.60	1.00	18.50	–	136.11	nr	**154.61**
S120D/D top shaft unit; 33 × 533 × 280 mm	131.60	1.00	18.50	–	136.11	nr	**154.61**
S120C pier cap unit; 737 × 737 × 114 mm	156.28	0.50	9.25	–	160.79	nr	**170.04**
S120B pier block unit; base for finial; 533 ×							
533 × 64 mm	66.98	0.33	6.11	–	67.13	nr	**73.24**
Pier blocks; flat to receive gate finial							
S100B; 440 × 440 × 63 mm	43.48	0.33	6.17	–	43.83	nr	**50.00**
S120B; 546 × 546 × 64 mm	66.98	0.33	6.17	–	67.33	nr	**73.50**
S150B; 330 × 330 × 51 mm	23.50	0.33	6.17	–	23.85	nr	**30.02**

F MASONRY

Item Excluding site overheads and profit	PC £	Labour hours	Labour £	Plant £	Material £	Unit	Total rate £
F22 CAST STONE ASHLAR WALLING/ **DRESSINGS – cont**							
Haddonstone Ltd – cont							
Pier caps; part weathered							
S100C; 915 × 915 × 150 mm	372.48	0.50	9.25	–	373.19	nr	**382.44**
S120C; 737 × 737 × 114 mm	156.28	0.50	9.25	–	156.63	nr	**165.88**
S150C; 584 × 584 × 120 mm	113.98	0.50	9.25	–	114.33	nr	**123.58**
Pier caps; weathered							
S230C; 1029 × 1029 × 175 mm	488.80	0.50	9.25	–	489.15	nr	**498.40**
S215C; 687 × 687 × 175 mm	213.85	0.50	9.25	–	214.20	nr	**223.45**
S210C; 584 × 584 × 175 mm	133.95	0.50	9.25	–	134.30	nr	**143.55**
Pier strings							
S100S; 800 × 800 × 55 mm	126.90	0.50	9.25	–	127.11	nr	**136.36**
S120S; 555 × 555 × 44 mm	57.58	0.50	9.25	–	57.79	nr	**67.04**
S150S; 457 × 457 × 48 mm	35.25	0.50	9.25	–	35.46	nr	**44.71**
Balls and bases							
E150A Ball 535 mm and E150C collared base	354.00	0.50	9.25	–	354.21	nr	**363.46**
E120A Ball 330 mm and E120C collared base	133.00	0.50	9.25	–	133.21	nr	**142.46**
E110A Ball 230 mm and E110C collared base	93.00	0.50	9.25	–	93.21	nr	**102.46**
E100A Ball 170 mm and E100B plain base	58.00	0.50	9.25	–	58.21	nr	**67.46**
Haddonstone Ltd; cast stone copings; **ornamental masonry in Portland Bath or** **Terracotta finished cast stone**							
Copings for walls; bedded, jointed and pointed in approved coloured cement-lime mortar 1:2:9							
T100 weathered coping; 102 mm high × 178 mm wide × 914 mm	40.98	0.33	6.16	–	41.29	m	**47.45**
T140 weathered coping; 102 mm high × 337 mm wide × 914 mm	68.32	0.33	6.17	–	68.62	m	**74.79**
T200 weathered coping; 127 mm high × 508 mm wide × 750 mm	102.46	0.33	6.17	–	102.76	m	**108.93**
T170 weathered coping; 108 mm high × 483 mm wide × 914 mm	110.14	0.33	6.17	–	110.45	m	**116.62**
T340 raked coping; 75–100 mm high × 290 mm wide × 900 mm	63.06	0.33	6.17	–	63.36	m	**69.53**
T310 raked coping; 76–89 mm high × 381 mm wide × 914 mm	73.01	0.33	6.17	–	73.31	m	**79.48**
Bordeaux walling; Forticrete Ltd							
Dry stacked random sized units 150–400 mm long × 150 mm high cast stone wall mechanically interlocked with fibreglass pins; constructed to levelling pad of coarse compacted granular material back filled behind the elevation with 300 mm wide granular drainage material; walls to 5.00 m high (retaining walls over heights shown below require individual design)							
gravity wall; near vertical; 250 mm thick	81.00	1.00	18.50	–	81.00	m²	**99.50**
gravity wall; 9.5°; battered	81.00	2.00	37.00	–	81.00	m²	**118.00**
copings to Bordeaux wall; 70 mm thick; random lengths	13.00	0.25	4.63	–	13.00	m	**17.63**

F MASONRY

Item Excluding site overheads and profit	PC £	Labour hours	Labour £	Plant £	Material £	Unit	Total rate £
Retaining wall; as above but reinforced with Tensar Geogrid RE520 laid between every two courses horizontally into the face of the excavation (excavation not included)							
Near vertical; 250 mm thick; 1.50 m of geogrid length							
up to 1.20 m high; 2 layers of geogrid	81.00	1.50	27.75	–	87.00	m²	**114.75**
1.20–1.50 m high; 3 layers of geogrid	81.00	2.00	37.00	–	90.00	m²	**127.00**
1.50–1.80 m high; 4 layers of geogrid	81.00	2.25	41.63	–	93.00	m²	**134.63**
Battered; maximum 1:3 slope							
up to 1.20 m high; 3 layers of geogrid	81.00	2.50	46.25	–	90.00	m²	**136.25**
1.20–1.50 m high; 4 layers of geogrid	81.00	2.50	46.25	–	93.00	m²	**139.25**
1.50–1.80 m high; 5 layers of geogrid	81.00	3.00	55.50	–	95.50	m²	**151.00**
F30 ACCESSORIES/SUNDRY ITEMS FOR BRICK/BLOCK/STONE WALLING							
Damp-proof courses; pitch polymer; 150 mm laps							
Horizontal							
width not exceeding 225 mm	4.72	1.14	21.09	–	6.10	m²	**27.19**
width exceeding 225 mm	5.27	0.58	10.73	–	6.65	m²	**17.38**
Vertical							
width not exceeding 225 mm	4.72	1.72	31.82	–	6.10	m²	**37.92**
Two courses slates in cement mortar (1:3)							
Horizontal							
width exceeding 225 mm	9.04	3.46	64.01	–	12.79	m²	**76.80**
Vertical							
width exceeding 225 mm	9.04	5.18	95.83	–	12.79	m²	**108.62**
F31 PRECAST CONCRETE SILLS/LINTELS/COPINGS/FEATURES							
Precast concrete coping							
Copings; once weathered; twice grooved							
152 × 75 mm	5.36	0.40	7.40	–	5.58	m	**12.98**
178 × 65 mm	7.40	0.40	7.40	–	7.71	m	**15.11**
305 × 75 mm	10.50	0.50	9.25	–	11.00	m	**20.25**
Pier caps; four sides weathered							
305 × 305 mm	6.70	1.00	18.50	–	6.85	nr	**25.35**
381 × 381 mm	7.10	1.00	18.50	–	7.25	nr	**25.75**
533 × 533 mm	6.90	1.20	22.20	–	7.05	nr	**29.25**

G STRUCTURAL/CARCASSING METAL/TIMBER

Item Excluding site overheads and profit	PC £	Labour hours	Labour £	Plant £	Material £	Unit	Total rate £
G31 PREFABRICATED TIMBER UNIT DECKING							
Timber decking							
Supports for timber decking; softwood joists to receive decking boards; joists at 400 mm centres; Southern Yellow pine							
38 × 88 mm	15.62	1.00	18.50	–	17.33	m²	**35.83**
47 × 150 mm	14.69	1.00	18.50	–	16.39	m²	**34.89**
47 × 100 mm	9.79	1.00	18.50	–	11.50	m²	**30.00**
Hardwood decking; Yellow Balau; grooved or smooth; 6 mm joints							
deck boards; 90 mm wide × 19 mm thick	17.18	1.00	18.50	–	19.12	m²	**37.62**
deck boards; 145 mm wide × 21 mm thick	27.67	1.00	18.50	–	29.62	m²	**48.12**
deck boards; 145 mm wide × 28 mm thick	28.40	1.00	18.50	–	30.35	m²	**48.85**
Hardwood decking; Ipe; smooth; 6 mm joints							
deck boards; 90 mm wide × 19 mm thick	61.65	1.00	18.50	–	63.60	m²	**82.10**
deck boards; 145 mm wide × 19 mm thick	44.95	1.00	18.50	–	46.90	m²	**65.40**
AVS; Profile; Western Red Cedar; 6 mm joints smooth or reeded							
ex 125 wide × 38 mm thick	18.15	1.00	18.50	–	20.10	m²	**38.60**
ex 150 wide × 32 mm thick	16.39	1.00	18.50	–	18.34	m²	**36.84**
Handrails and base rail; fixed to posts at 2.00 m centres							
posts; 100 × 100 × 1370 mm high	9.29	1.00	18.50	–	9.74	m	**28.24**
posts; turned; 1220 mm high	15.40	1.00	18.50	–	15.85	m	**34.35**
Handrails; balusters							
square balusters at 100 mm centres	34.20	0.50	9.25	–	34.65	m	**43.90**
square balusters at 300 mm centres	11.39	0.33	6.16	–	11.75	m	**17.91**
turned balusters at 100 mm centres	54.00	0.50	9.25	–	54.45	m	**63.70**
turned balusters at 300 mm centres	17.98	0.33	6.11	–	18.34	m	**24.45**

H CLADDING/COVERING

Item Excluding site overheads and profit	PC £	Labour hours	Labour £	Plant £	Material £	Unit	Total rate £
H51 NATURAL STONE SLAB CLADDING/ FEATURES							
Sawn yorkstone cladding; Johnsons Wellfield Quarries Ltd Six sides sawn stone; rubbed face; sawn and jointed edges; fixed to blockwork (not included) with stainless steel fixings; Ancon Ltd; grade 304 stainless steel frame cramp and dowel 7 mm; cladding units drilled 4 × to receive dowels							
440 × 200 × 50 mm thick	84.50	1.87	34.59	–	87.90	m²	**122.49**
H52 CAST STONE SLAB CLADDING/ FEATURES							
Cast stone cladding; Haddonstone Ltd Reconstituted stone in Portland Bath or Terracotta; fixed to blockwork or concrete (not included) with stainless steel fixings; stainless steel frame cramp and dowel M6 mm; cladding units drilled 4 × to receive dowels							
1000 × 300 × 50 mm thick	42.00	1.87	34.59	–	45.40	m²	**79.99**

J WATERPROOFING

Item Excluding site overheads and profit	PC £	Labour hours	Labour £	Plant £	Material £	Unit	Total rate £
J10 SPECIALIST WATERPROOF RENDERING							
Sika waterproof rendering; steel trowelled							
Walls; 20 mm thick; three coats; to concrete base							
width exceeding 300 mm	–	–	–	–	–	m²	48.84
width not exceeding 300 mm	–	–	–	–	–	m²	76.53
Walls; 25 mm thick; three coats; to concrete base							
width exceeding 300 mm	–	–	–	–	–	m²	55.37
width not exceeding 300 mm	–	–	–	–	–	m²	87.93
J20 MASTIC ASPHALT TANKING/DAMP-PROOFING							
Tanking and damp-proofing; mastic asphalt; type T1097; Bituchem							
13 mm thick; one coat covering; to concrete base; flat; work subsequently covered							
width exceeding 300 mm	–	–	–	–	–	m²	11.24
20 mm thick; two coat coverings; to concrete base; flat; work subsequently covered							
width exceeding 300 mm	–	–	–	–	–	m²	14.09
30 mm thick; three coat coverings; to concrete base; flat; work subsequently covered							
width exceeding 300 mm	–	–	–	–	–	m²	19.13
13 mm thick; two coat coverings; to brickwork base; vertical; work subsequently covered							
width exceeding 300 mm	–	–	–	–	–	m²	37.34
20 mm thick; three coat coverings; to brickwork base; vertical; work subsequently covered							
width exceeding 300 mm	–	–	–	–	–	m²	50.98
Internal angle fillets; work subsequently covered	–	–	–	–	–	m	3.97
Turning asphalt nibs into grooves; 20 mm deep	–	–	–	–	–	m	2.52
J30 LIQUID APPLIED TANKING/DAMP-PROOFING							
Tanking and damp-proofing; Ruberoid Building Products; Synthaprufe cold applied bituminous emulsion waterproof coating							
Synthaprufe; to smooth finished concrete or screeded slabs; flat; blinding with sand							
two coats	2.38	0.22	4.11	–	2.51	m²	6.62
three coats	3.57	0.31	5.74	–	3.70	m²	9.44
Synthaprufe; to fair faced brickwork with flush joints, rendered brickwork or smooth finished concrete walls; vertical							
two coats	2.68	0.29	5.29	–	2.81	m²	8.10
three coats	3.93	0.40	7.40	–	4.06	m²	11.46

J WATERPROOFING

Item Excluding site overheads and profit	PC £	Labour hours	Labour £	Plant £	Material £	Unit	Total rate £
Tanking and damp-proofing; RIW Ltd Liquid asphaltic composition; to smooth finished concrete screeded slabs or screeded slabs; flat							
two coats	4.34	0.33	6.17	–	4.34	m²	**10.51**
Liquid asphaltic composition; fair faced brickwork with flush joints, rendered brickwork or smooth finished concrete walls; vertical							
two coats	4.34	0.50	9.25	–	4.34	m²	**13.59**
Heviseal; to smooth finished concrete or screeded slabs; to surfaces of ponds, tanks or planters; flat							
two coats	7.88	0.33	6.17	–	7.88	m²	**14.05**
Heviseal; to fair faced brickwork with flush joints, rendered brickwork or smooth finished concrete walls; to surfaces of retaining walls, ponds, tanks or planters; vertical							
two coats	7.88	0.50	9.25	–	7.88	m²	**17.13**
J40 FLEXIBLE SHEET TANKING/DAMP- **PROOFING**							
Tanking and damp-proofing; Grace **Construction Products** Bitu-thene 2000; 1.00 mm thick; overlapping and bonding; including sealing all edges							
to concrete slabs; flat	7.26	0.25	4.63	–	7.26	m²	**11.89**
to brick/concrete walls; vertical	7.26	0.40	7.40	–	7.59	m²	**14.99**

J50 GREEN ROOF SYSTEMS

Preamble: A variety of systems are available which address all the varied requirements for a successful Green Roof. For installation by approved contractors only. The prices shown are for budgeting purposes only, as each installation is site specific and may incorporate some or all of the resources shown. Specifiers should verify that the systems specified include for design liability and inspections by the suppliers. The systems below assume commercial insulation levels are required to the space below the proposed Green Roof. Extensive Green Roofs are those of generally lightweight construction with low maintenance planting and shallow soil designed for aesthetics only; Intensive Green Roofs are designed to allow use for recreation and trafficking. They require more maintenance and allow a greater variety of surfaces and plant types.

J WATERPROOFING

Item Excluding site overheads and profit	PC £	Labour hours	Labour £	Plant £	Material £	Unit	Total rate £
J50 GREEN ROOF SYSTEMS – cont							
Intensive Green Roof; Bauder Ltd; soil-based systems able to provide a variety of hard and soft landscaping; laid to the surface of an unprepared roof deck							
Vapour barrier laid to prevent intersticial condensation from spaces below the roof applied by torching to the roof deck							
VB4-Expal aluminium lined	–	–	–	–	–	m²	12.03
Insulation laid and hot bitumen bonded to vapour barrier							
PIR Insulation 100 mm	–	–	–	–	–	m²	25.52
Underlayer to receive root barrier partially bonded to insulation by torching							
G4E	–	–	–	–	–	m²	12.49
Root barrier							
Plant E; chemically treated root resistant capping sheet fully bonded to G4E underlayer by torching	–	–	–	–	–	m²	19.01
Slip layers to absorb differential movement							
PE Foil; 2 layers laid to root barriers	–	–	–	–	–	m²	3.34
Optional protection layer to prevent mechanical damage							
Protection Mat; 6 mm thick rubber matting; loose laid	–	–	–	–	–	m²	10.26
Drainage medium laid to root barrier							
Drainage Board; free draining EPS 50 mm thick	–	–	–	–	–	m²	12.48
Reservoir Board; up to 21.5 litre water storage capacity EPS; 50 mm thick	–	–	–	–	–	m²	17.09
Filtration to prevent soil migration to drainage system							
Filter Fleece; 3 mm thick polyester geotextile; loose laid over drainage/ reservoir layer	–	–	–	–	–	m²	3.46
For hard landscaped areas incorporate rigid drainage board laid to the protection mat							
PLT 60 drainage board	–	–	–	–	–	m²	27.42
Extensive Green Roof System; Bauder Ltd; low maintenance soil free system incorporating single layer growing and planting medium							
Vapour barrier laid to prevent intersticial condensation applied by torching to the roof deck							
VB4-Expal aluminium lined	–	–	–	–	–	m²	12.03
Insulation laid and hot bitumen bonded to vapour barrier							
PIR Insulation 100 mm	–	–	–	–	–	m²	25.52
Underlayer to receive rootbarrier partially bonded to insulation by torching							
G4E	–	–	–	–	–	m²	12.49
Root barrier							
Plant E; chemically treated root resistant capping sheet fully bonded to G4E underlayer by torching	–	–	–	–	–	m²	19.01

J WATERPROOFING

Item Excluding site overheads and profit	PC £	Labour hours	Labour £	Plant £	Material £	Unit	Total rate £
Landscape options to the above systems; **Bauder Ltd**							
Hydroplanting system; to Extensive Green Roof as detailed above							
SDF mat 20 mm thick drainage layer; loose laid	–	–	–	–	–	m²	9.83
Xeroflor vegetation blanket; Bauder Ltd; to Extensive Green Roof as detailed above							
Xeroflor Xf 301 pre-cultivated sedum blanket incorporating 800 gram recycled fibre water retention layer; laid loose	–	–	–	–	–	m²	40.08
Waterproofing to upstands; Bauder Ltd							
Bauder Vapour Barrier; Bauder G4E & Bauder Plant E							
up to 200 mm high	–	–	–	–	–	m	18.46
up to 400 mm high	–	–	–	–	–	m	24.88
up to 600 mm high	–	–	–	–	–	m	29.22
Reservoir and drainage boards; Bauder; **non-specialist installed products for laying** **under pavings or planting for collection or** **drainage**							
DSE20 Drainage/Reservoir Board; 20 mm thick drainage board	–	–	–	–	–	m²	9.03
DSE40 Drainage/Reservoir Board; for intensive green roof applications	–	–	–	–	–	m²	13.60
Versicell 20 Drainage Board; specialist installed drainage layer for intensive green roof applications; predominantly hard landscape	–	–	–	–	–	m²	21.88
Bauder Extensive Substrate; 80 mm deep; for use with sedum and hardy perennials and biodiverse vegetation	–	–	–	–	–	m²	17.58
Bauder Intensive Substrate; 200 mm deep; for use in traditional green roof applications e.g. grass and bedding plants	–	–	–	–	–	m²	43.12
Inverted waterproofing systems; Alumasc **Exterior Building Products Ltd; to roof** **surfaces to receive Green Roof systems**							
Hydrotech 6125; monolithic hot melt rubberized bitumen; applied in two 3 mm layers incorporating a polyester reinforcing sheet with 4 mm thick protection sheet and chemically impregnated root barrier; fully bonded into the Hydrotech; applied to plywood or suitably prepared wood float finish and primed concrete deck or screeds							
10 mm thick	–	–	–	–	–	m²	35.62
Alumasc Roofmate; extruded polystyrene insulation; optional system; thickness to suit required U value; calculated at design stage; indicative thicknesses; laid to Hydrotech 6125							
0.25 U value; average requirement 120 mm	–	–	–	–	–	m²	24.69

Prices for Measured Works – Major Works

J WATERPROOFING

Item Excluding site overheads and profit	PC £	Labour hours	Labour £	Plant £	Material £	Unit	Total rate £
J50 GREEN ROOF SYSTEMS – cont							
Warm Roof waterproofing systems; Alumasc							
Exterior Building Products Ltd (Euroroof); to							
roof surfaces to receive Green Roof systems							
Derbigum system							
Nilperm aluminium lined vapour barrier;							
2 mm thick; bonded in hot bitumen to the							
roof deck	–	–	–	–	–	m²	11.16
Korklite insulation bonded to the vapour							
barrier in hot bitumen; U value dependent;							
80 mm thick	–	–	–	–	–	m²	21.13
Hi-Ten Universal 2 mm thick underlayer;							
fully bonded to the insulation	–	–	–	–	–	m²	9.09
Derbigum Anti-Root cap sheet impregnated							
with root resisting chemical bonded to the							
underlayer	–	–	–	–	–	m²	18.70
Intensive Green Roof Systems; Alumasc							
Exterior Building Products Ltd;							
components laid to the insulation over the							
Hydrotech or Derbigum waterproofing							
above							
Optional inclusion; moisture retention layer							
SSM-45; moisture mat	–	–	–	–	–	m²	6.47
Drainage layer; Floradrain recycled							
polypropylene; providing water reservoir,							
multidirectional drainage and mechanical							
damage protection							
FD25; 25 mm deep; rolls inclusive of filter							
sheet SF	–	–	–	–	–	m²	21.97
FD40; 40 mm deep; rolls inclusive of filter							
sheet SF	–	–	–	–	–	m²	28.61
FD25; 25 mm deep; sheets excluding filter							
sheet	–	–	–	–	–	m²	17.81
FD40; 40 mm deep; sheets excluding filter							
sheet	–	–	–	–	–	m²	21.97
FD60; 60 mm deep; sheets excluding filter							
sheet	–	–	–	–	–	m²	34.67
Drainage layer; Elastodrain recycled rubber							
mat; providing multidirectional drainage and							
mechanical damage protection							
EL200; 20 mm deep	–	–	–	–	–	m²	28.50
Zincolit recycled crushed brick; optional							
drainage infill to Floradrain layers							
FD40; 17 litres per m²	–	0.03	0.63	–	7.12	m²	7.75
FD60; 27 litres per m²	–	0.05	1.00	–	8.43	m²	9.43
Filter sheet; rolled out onto drainage layer							
Filter sheet TG for Elastodrain range	–	–	–	–	–	m²	5.58
Filter sheet SF for Floradrain range	–	–	–	–	–	m²	4.75
Intensive substrate; lightweight growing							
medium laid to filter sheet	–	–	–	–	–	m²	24.93
Extensive substrate; sedum carpet laid to filter							
sheet	–	–	–	–	–	m²	17.68
Extensive substrate; rockery type plants laid to							
filter sheet	–	–	–	–	–	m²	23.15
Semi-intensive substrate; heather with							
lavender laid to filter sheet	–	–	–	–	–	m²	31.82

J WATERPROOFING

Item Excluding site overheads and profit	PC £	Labour hours	Labour £	Plant £	Material £	Unit	Total rate £
Extensive Green Roof Systems; Alumasc Exterior Building Products Ltd; components laid to the Hydrotech or insulation layers above							
Moisture retention layer							
SSM-45; moisture mat	–	–	–	–	–	m²	6.47
Drainage layer for flat roofs; Floradrain; recycled polypropylene; providing water reservoir, multidirectional drainage and mechanical damage protection; supplied in sheet or roll form							
FD25; 25 mm deep; rolls inclusive of filter sheet SF	–	–	–	–	–	m²	21.97
FD40; 40 mm deep; rolls inclusive of filter sheet SF	–	–	–	–	–	m²	28.61
FD25; 25 mm deep; sheets excluding filter sheet	–	–	–	–	–	m²	17.81
FD40; 40 mm deep; sheets excluding filter sheet	–	–	–	–	–	m²	21.97
FD60; 60 mm deep; sheets excluding filter sheet	–	–	–	–	–	m²	34.67
Drainage layer for pitched roofs; Floratec; recycled polystyrene; providing water reservoir and multidirectional drainage; laid to moisture mat or to the waterproofing							
FS50	–	–	–	–	–	m²	16.97
FS75	–	–	–	–	–	m²	19.94
Landscape options							
Sedum Mat vegetation layer; Alumasc Exterior Building Products Ltd; to Extensive Green Roof as detailed above	–	–	–	–	–	m²	41.55
Green Roof components; Alumasc Exterior Building Products Ltd							
Outlet inspection chambers							
KS15; 150 mm deep	–	–	–	–	–	nr	140.09
height extension piece 100 mm	–	–	–	–	–	nr	50.69
height extension piece 200 mm	–	–	–	–	–	nr	54.85
Outlet and irrigation control chambers							
B32; 300 × 300 × 300 mm high	–	–	–	–	–	nr	421.46
B52; 400 × 500 × 500 mm high	–	–	–	–	–	nr	534.24
Outlet damming piece for water retention	–	–	–	–	–	nr	111.60
Linear drainage channel; collects surface water from adjacent hard surfaces or down pipes for distribution to the drainage layer	–	–	–	–	–	m	110.41

M SURFACE FINISHES

Item Excluding site overheads and profit	PC £	Labour hours	Labour £	Plant £	Material £	Unit	Total rate £
M20 PLASTERED/RENDERED/ **ROUGHCAST COATINGS**							
Cement: lime: sand (1:1:6); 19 mm thick; **two coats; wood floated finish**							
Walls							
width exceeding 300 mm; to brickwork or blockwork base	–	–	–	–	–	m²	**41.80**
Extra over cement: sand: lime (1:1:6) coatings for decorative texture finish with water repellent cement							
combed or floated finish	–	–	–	–	–	m²	**3.80**
M40 STONE/CONCRETE/QUARRY/ **CERAMIC TILING/MOSAIC**							
Ceramic tiles; unglazed slip-resistant; **various colours and textures; jointing**							
Floors							
level or to falls only not exceeding 15° from horizontal; 150 × 150 × 8 mm thick	19.50	1.00	18.50	–	22.01	m²	**40.51**
level or to falls only not exceeding 15° from horizontal; 150 × 150 × 12 mm thick	23.76	1.00	18.50	–	26.27	m²	**44.77**
Clay tiles – General							
Preamble: Typical specification – Clay tiles shall be reasonably true to shape, flat, free from flaws, frost resistant and true to sample approved by the Landscape Architect prior to laying. Quarry tiles (or semi-vitrified tiles) shall be of external quality, either heather brown or blue, to size specified, laid on 1:2:4 concrete, with 20 mm maximum aggregate 100 mm thick, on 100 mm hardcore. The hardened concrete should be well wetted and the surplus water taken off. Clay tiles shall be thoroughly wetted immediately before laying and then drained and shall be bedded to 19 mm thick cement: sand (1:3) screed. Joints should be approximately 4 mm (or 3 mm for vitrified tiles) grouted in cement: sand (1:2) and cleaned off immediately.							
Quarry tiles; external quality; including **bedding; jointing**							
Floors							

level or to falls only not exceeding 15° from
   horizontal; 150 × 150 × 12.5 mm thick;
   heather brown
	24.38	0.80	14.80	–	26.88	m²	**41.68**

level or to falls only not exceeding 15° from
   horizontal; 225 × 225 × 29 mm thick;
   heather brown
	24.38	0.67	12.33	–	26.88	m²	**39.21**

level or to falls only not exceeding 15° from
   horizontal; 150 × 150 × 12.5 mm thick;
   blue/black
	15.48	1.00	18.50	–	17.98	m²	**36.48**

level or to falls only not exceeding 15° from
   horizontal; 194 × 194 × 12.5 mm thick;
   heather brown
	14.26	0.80	14.80	–	16.76	m²	**31.56**

M SURFACE FINISHES

Item Excluding site overheads and profit	PC £	Labour hours	Labour £	Plant £	Material £	Unit	Total rate £
M60 PAINTING/CLEAR FINISHING							
Eura Conservation Ltd; restoration of railings; works carried out off site; excludes removal of railings from site							
Shotblast railings; remove all traces of paint and corrosion; apply temporary protective primer; measured overall 2 sides							
to plain railings	–	–	–	–	–	m²	**40.00**
to decorative railings	–	–	–	–	–	m²	**79.00**
Paint new galvanized railings in situ with 3 coats system							
vertical bar railings or gates	–	–	–	–	–	m²	**11.00**
vertical bar railings or gates with dog mesh panels	–	–	–	–	–	m²	**12.00**
gate posts not exceeding 300 mm girth	–	–	–	–	–	m²	**6.00**
Paint previously painted railings; shotblasted off site							
vertical bar railings or gates	–	–	–	–	–	m²	**13.00**
gate posts not exceeding 300 mm girth	–	–	–	–	–	m²	**6.00**
Prepare; touch up primer; two undercoats and one finishing coat of gloss oil paint; on metal surfaces							
General surfaces							
girth exceeding 300 mm	1.38	0.67	12.33	–	1.38	m²	**13.71**
isolated surfaces; girth not exceeding 300 mm	4.18	0.40	7.40	–	4.18	m	**11.58**
isolated areas not exceeding 0.50 m² irrespective of girth	2.76	0.67	12.33	–	2.76	nr	**15.09**
Ornamental railings; each side measured separately							
girth exceeding 300 mm	1.38	0.57	10.57	–	1.38	m²	**11.95**
Prepare; one coat primer; two finishing coats of gloss paint; on wood surfaces							
New wood surfaces							
girth exceeding 300 mm	1.66	0.80	14.80	–	1.66	m²	**16.46**
isolated surfaces; girth not exceeding 300 mm	0.55	1.00	18.50	–	0.55	m	**19.05**
isolated areas not exceeding 0.50 m² irrespective of girth	2.48	0.50	9.25	–	2.48	nr	**11.73**
Previously painted wood surfaces							
girth exceeding 300 mm	0.56	0.50	9.25	–	0.56	m²	**9.81**
isolated surfaces; girth not exceeding 300 mm	0.18	0.67	12.33	–	0.18	m	**12.51**
isolated areas not exceeding 0.50 m² irrespective of girth	1.11	0.40	7.40	–	1.11	nr	**8.51**
Fences and sheds; prepare; two coats of Protek wood preserver on wood surfaces							
Planed surfaces							
girth exceeding 300 mm	0.54	0.07	1.23	–	0.54	m²	**1.77**
isolated surfaces; girth not exceeding 300 mm	0.18	0.17	3.08	–	0.18	m	**3.26**
isolated areas not exceeding 0.50 m² irrespective of girth	0.54	0.25	4.63	–	0.54	nr	**5.17**

M SURFACE FINISHES

Item Excluding site overheads and profit	PC £	Labour hours	Labour £	Plant £	Material £	Unit	Total rate £
M60 PAINTING/CLEAR FINISHING – cont							
Fences and sheds – cont							
Sawn surfaces							
girth exceeding 300 mm	0.20	0.10	1.85	–	0.20	m²	**2.05**
isolated surfaces; girth not exceeding							
300 mm	2.69	0.17	3.08	–	2.69	m	**5.77**
isolated areas not exceeding 0.50 m²							
irrespective of girth	0.81	0.25	4.63	–	0.81	nr	**5.44**
Fences and sheds; prepare; two coats of							
Sadolin wood preserver on wood surfaces							
Planed surfaces							
girth exceeding 300 mm	1.58	0.07	1.23	–	1.58	m²	**2.81**
isolated surfaces; girth not exceeding							
300 mm	0.53	0.17	3.08	–	0.53	m	**3.61**
isolated areas not exceeding 0.50 m²							
irrespective of girth	0.79	0.25	4.63	–	0.79	nr	**5.42**
Sawn surfaces							
girth exceeding 300 mm	1.29	0.10	1.85	–	1.29	m²	**3.14**
isolated surfaces; girth not exceeding							
300 mm	0.84	0.17	3.08	–	0.84	m	**3.92**
isolated areas not exceeding 0.50 m²							
irrespective of girth	1.27	0.25	4.63	–	1.27	nr	**5.90**
Prepare; proprietary solution primer; two							
coats of dark stain; on wood surfaces							
General surfaces							
girth exceeding 300 mm	1.33	0.17	3.08	–	1.33	m²	**4.41**
isolated surfaces; girth not exceeding							
300 mm	0.44	0.13	2.31	–	0.44	m	**2.75**
Two coats Weathershield; to clean, dry							
surfaces; in accordance with							
manufacturer's instructions							
Brick or block walls							
girth exceeding 300 mm	1.04	0.40	7.40	–	1.04	m²	**8.44**
Cement render or concrete walls							
girth exceeding 300 mm	1.04	0.33	6.17	–	1.04	m²	**7.21**

P BUILDING FABRIC SUNDRIES

Item Excluding site overheads and profit	PC £	Labour hours	Labour £	Plant £	Material £	Unit	Total rate £
P30 TRENCHES/PIPEWAYS/PITS FOR BURIED ENGINEERING SERVICES							
Excavating trenches; using 3 tonne tracked excavator; to receive pipes; grading bottoms; earthwork support; filling with excavated material to within 150 mm of finished surfaces and compacting; completing fill with topsoil; disposal of surplus soil							
Services not exceeding 200 mm nominal size							
average depth of run not exceeding 0.50 m	1.19	0.12	2.22	1.01	1.19	m	**4.42**
average depth of run not exceeding 0.75 m	1.19	0.16	3.01	1.39	1.19	m	**5.59**
average depth of run not exceeding 1.00 m	1.19	0.28	5.24	2.42	1.19	m	**8.85**
average depth of run not exceeding 1.25 m	1.19	0.38	7.09	3.27	1.19	m	**11.55**
Excavating trenches; using 3 tonne tracked excavator; to receive pipes; grading bottoms; earthwork support; filling with imported granular material and compacting; disposal of surplus soil							
Services not exceeding 200 mm nominal size							
average depth of run not exceeding 0.50 m	1.44	0.09	1.60	0.72	4.01	m	**6.33**
average depth of run not exceeding 0.75 m	2.16	0.11	2.00	0.90	6.01	m	**8.91**
average depth of run not exceeding 1.00 m	2.88	0.14	2.58	1.17	8.01	m	**11.76**
average depth of run not exceeding 1.25 m	3.60	0.23	4.23	1.95	10.02	m	**16.20**
Excavating trenches, using 3 tonne tracked excavator, to receive pipes; grading bottoms; earthwork support; filling with lean mix concrete; disposal of surplus soil							
Services not exceeding 200 mm nominal size							
average depth of run not exceeding 0.50 m	9.62	0.11	1.97	0.35	11.06	m	**13.38**
average depth of run not exceeding 0.75 m	14.42	0.13	2.41	0.44	16.58	m	**19.43**
average depth of run not exceeding 1.00 m	19.23	0.17	3.08	0.59	22.11	m	**25.78**
average depth of run not exceeding 1.25 m	24.04	0.23	4.16	0.88	27.64	m	**32.68**
Earthwork support; providing support to opposing faces of excavation; moving along as work proceeds							
Maximum depth not exceeding 2.00 m							
distance between opposing faces not exceeding 2.00 m	–	0.80	14.80	17.96	–	m	**32.76**

Q PAVING/PLANTING/FENCING/SITE FURNITURE

Item Excluding site overheads and profit	PC £	Labour hours	Labour £	Plant £	Material £	Unit	Total rate £
Q10 KERBS/EDGINGS/CHANNELS/PAVING ACCESSORIES							
Foundations to kerbs							
Excavating trenches; width 300 mm; 3 tonne excavator; disposal off site							
depth 300 mm	–	–	–	3.43	1.54	m	**4.97**
depth 400 mm	–	–	–	3.75	2.05	m	**5.80**
By hand; barrowing to spoil heap and disposal off site							
depth 300	–	0.18	3.33	–	17.12	m	**20.45**
depth 400	–	0.24	4.44	–	17.12	m	**21.56**
Excavating trenches; width 450 mm; 3 tonne excavator; disposal off site							
depth 300 mm	–	–	–	4.29	2.31	m	**6.60**
depth 400 mm	–	–	–	4.90	3.08	m	**7.98**
Foundations; to kerbs, edgings or channels; in situ concrete; 21 N/mm^2 – 20 mm aggregate ((1:2:4) site mixed); one side against earth face, other against formwork (not included); site mixed concrete							
Site mixed concrete							
150 mm wide × 100 mm deep	–	0.13	2.47	–	1.34	m	**3.81**
150 mm wide × 150 mm deep	–	0.17	3.08	–	2.01	m	**5.09**
200 mm wide × 150 mm deep	–	0.20	3.70	–	2.68	m	**6.38**
300 mm wide × 150 mm deep	–	0.23	4.32	–	4.02	m	**8.34**
600 mm wide × 200 mm deep	–	0.29	5.29	–	10.73	m	**16.02**
Ready mixed concrete							
150 mm wide × 100 mm deep	–	0.13	2.47	–	0.99	m	**3.46**
150 mm wide × 150 mm deep	–	0.17	3.08	–	1.48	m	**4.56**
200 mm wide × 150 mm deep	–	0.20	3.70	–	1.97	m	**5.67**
300 mm wide × 150 mm deep	–	0.23	4.32	–	2.96	m	**7.28**
600 mm wide × 200 mm deep	–	0.29	5.29	–	7.88	m	**13.17**
Formwork; sides of foundations (this will usually be required to one side of each kerb foundation adjacent to road subbases)							
100 mm deep	–	0.04	0.77	–	0.26	m	**1.03**
150 mm deep	–	0.04	0.77	–	0.39	m	**1.16**
Precast concrete kerbs, channels, edgings etc.; Marshalls Mono; bedding, jointing and pointing in cement mortar (1:3); including haunching with in situ concrete; 11.50 N/mm^2 – 40 mm aggregate one side							
Kerbs; straight							
150 × 305 mm; HB1	7.99	0.50	9.25	–	10.32	m	**19.57**
125 × 255 mm; HB2; SP	3.82	0.44	8.22	–	6.15	m	**14.37**
125 × 150 mm; BN	2.59	0.40	7.40	–	4.92	m	**12.32**
Dropper kerbs; left and right handed							
125 × 2 55–150 mm; DL1 or DR1	5.78	0.50	9.25	–	8.11	m	**17.36**
125 × 2 55–150 mm; DL2 or DR2	5.78	0.50	9.25	–	8.11	m	**17.36**
Quadrant kerbs							
305 mm radius	9.72	0.50	9.25	–	11.27	nr	**20.52**
455 mm radius	10.46	0.50	9.25	–	12.79	nr	**22.04**

Q PAVING/PLANTING/FENCING/SITE FURNITURE

Item Excluding site overheads and profit	PC £	Labour hours	Labour £	Plant £	Material £	Unit	Total rate £
Straight kerbs or channels; to radius; 125 × 255 mm							
0.90 m radius (2 units per quarter circle)	7.31	0.80	14.80	–	9.71	m	**24.51**
1.80 m radius (4 units per quarter circle)	7.31	0.73	13.45	–	9.71	m	**23.16**
2.40 m radius (5 units per quarter circle)	7.31	0.68	12.54	–	9.71	m	**22.25**
3.00 m radius (5 units per quarter circle)	7.31	0.67	12.33	–	9.71	m	**22.04**
4.50 m radius (2 units per quarter circle)	7.31	0.60	11.11	–	9.71	m	**20.82**
6.10 m radius (11 units per quarter circle)	7.31	0.58	10.73	–	9.71	m	**20.44**
7.60 m radius (14 units per quarter circle)	7.31	0.57	10.57	–	9.71	m	**20.28**
9.15 m radius (17 units per quarter circle)	7.31	0.56	10.28	–	9.71	m	**19.99**
10.70 m radius (20 units per quarter circle)	7.31	0.56	10.28	–	9.71	m	**19.99**
12.20 m radius (22 units per quarter circle)	7.31	0.53	9.74	–	9.71	m	**19.45**
Kerbs; Conservation Kerb units; to simulate natural granite kerbs							
255 × 150 × 914 mm; laid flat	20.13	0.57	10.57	–	24.31	m	**34.88**
150 × 255 × 914 mm; laid vertical	20.13	0.57	10.57	–	23.62	m	**34.19**
145 × 255 mm; radius internal 3.25 m	22.21	0.83	15.42	–	25.70	m	**41.12**
150 × 255 mm; radius external 3.40 m	21.21	0.83	15.42	–	24.71	m	**40.13**
145 × 255 mm; radius internal 6.50 m	19.85	0.67	12.33	–	23.34	m	**35.67**
145 × 255 mm; radius external 6.70 m	19.73	0.67	12.33	–	23.23	m	**35.56**
150 × 255 mm; radius internal 9.80 m	19.62	0.67	12.33	–	23.12	m	**35.45**
150 × 255 mm; radius external 10.00 m	18.60	0.67	12.33	–	22.10	m	**34.43**
305 × 305 × 255 mm; solid quadrants	30.61	0.57	10.57	–	34.10	nr	**44.67**
Channels; square							
125 × 225 × 915 mm long; CS1	3.65	0.40	7.40	–	11.78	m	**19.18**
125 × 150 × 915 mm long; CS2	2.95	0.40	7.40	–	9.67	m	**17.07**
Channels; dished							
305 × 150 × 915 mm long; CD	10.32	0.40	7.40	–	19.12	m	**26.52**
150 × 100 × 915 mm	6.05	0.40	7.40	–	11.05	m	**18.45**
Precast concrete edging units; including **haunching with in situ concrete;** **11.50 N/mm² – 40 mm aggregate both sides**							
Edgings; rectangular, bullnosed or chamfered							
50 × 150 mm	1.57	0.33	6.17	–	5.46	m	**11.63**
125 × 150 mm bullnosed	1.63	0.33	6.17	–	5.51	m	**11.68**
50 × 200 mm	2.37	0.33	6.17	–	6.25	m	**12.42**
50 × 250 mm	2.75	0.33	6.17	–	6.63	m	**12.80**
50 × 250 mm flat top	3.22	0.33	6.17	–	7.10	m	**13.27**
Marshalls Mono; small element precast **concrete kerb system; Keykerb Large (KL)** **upstand of 100–125 mm; on 150 mm** **concrete foundation including haunching** **with in situ concrete (1:3:6) 1 side**							
Bullnosed or half battered; 100 × 127 × 200 mm							
laid straight	11.64	0.80	14.80	–	12.45	m	**27.25**
radial blocks laid to curve; 8 blocks/¼ circle – 500 mm radius	8.96	1.00	18.50	–	9.77	m	**28.27**
radial blocks laid to curve; 8 radial blocks, alternating 8 standard blocks/¼ circle – 1000 mm radius	14.78	1.25	23.13	–	15.59	m	**38.72**

Q PAVING/PLANTING/FENCING/SITE FURNITURE

Item Excluding site overheads and profit	PC £	Labour hours	Labour £	Plant £	Material £	Unit	Total rate £
Q10 KERBS/EDGINGS/CHANNELS/PAVING ACCESSORIES – cont							
Marshalls Mono – cont							
Bullnosed or half battered – cont							
radial blocks, alternating 16 standard							
blocks/¼ circle – 1500 mm radius	13.59	1.50	27.75	–	14.41	m	**42.16**
internal angle 90°	4.92	0.20	3.70	–	4.92	nr	**8.62**
external angle	4.92	0.20	3.70	–	4.92	nr	**8.62**
drop crossing kerbs; KL half battered to KL							
Splay; LH and RH	20.56	1.00	18.50	–	21.37	pair	**39.87**
drop crossing kerbs; KL half battered to KS							
bullnosed	20.00	1.00	18.50	–	20.81	pair	**39.31**
Marshalls Mono; small element precast concrete kerb system; Keykerb Small (KS) upstand of 25–50 mm; on 150 mm concrete foundation including haunching with in situ concrete (1:3:6) 1 side							
Half battered							
laid straight	8.17	0.80	14.80	–	8.98	m	**23.78**
radial blocks laid to curve; 8 blocks/¼							
circle; 500 mm radius	6.71	1.00	18.50	–	7.52	m	**26.02**
radial blocks laid to curve; 8 radial blocks, alternating 8 standard blocks/¼ circle;							
1000 mm radius	10.79	1.25	23.13	–	11.61	m	**34.74**
radial blocks, alternating 16 standard							
blocks/¼ circle; 1500 mm radius	11.55	1.50	27.75	–	12.37	m	**40.12**
internal angle 90°	4.92	0.20	3.70	–	4.92	nr	**8.62**
external angle	4.92	0.20	3.70	–	4.92	nr	**8.62**
Dressed natural stone kerbs – General							
Preamble: The kerbs are to be good, sound and uniform in texture and free from defects; worked straight or to radius, square and out of wind, with the top front and back edges parallel or concentric to the dimensions specified. All drill and pick holes shall be removed from dressed faces. Standard dressings shall be designated as either fine picked, single axed or nidged or rough punched.							
Dressed natural stone kerbs; CED Ltd; on concrete foundations (not included); including haunching with in situ concrete; 11.50 N/mm² – 40 mm aggregate one side							
Granite kerbs; 125 × 250 mm							
special quality; straight; random lengths	17.93	0.80	14.80	–	20.73	m	**35.53**
Granite kerbs; 125 × 250 mm; curved to mean radius 3 m							
special quality; random lengths	20.02	0.91	16.84	–	22.82	m	**39.66**

Q PAVING/PLANTING/FENCING/SITE FURNITURE

Item Excluding site overheads and profit	PC £	Labour hours	Labour £	Plant £	Material £	Unit	Total rate £
New granite sett edgings; CED Ltd; 100 × **100 × 100 mm; bedding in cement mortar** **(1:4); including haunching with in situ** **concrete; 11.50 N/mm² – 40 mm aggregate** **one side; concrete foundations (not** **included)**							
Edgings 100 × 100 × 100 mm; ref 1R 'better quality' with 20 mm pointing gaps							
100 mm wide	2.74	1.33	24.61	–	5.06	m	**29.67**
2 rows; 220 mm wide	5.48	2.22	41.07	–	8.48	m	**49.55**
3 rows; 340 mm wide	8.21	3.00	55.50	–	10.58	m	**66.08**
Edgings 100 × 100 × 200 mm; ref 1R 'better quality' with 20 mm pointing gaps							
100 mm wide	2.79	0.75	13.88	–	5.12	m	**19.00**
2 rows; 220 mm wide	5.59	1.50	27.75	–	8.59	m	**36.34**
3 rows; 340 mm wide	8.38	2.25	41.63	–	10.74	m	**52.37**
Edgings 100 × 100 × 100 mm; ref 8R 'standard quality' with 20 mm pointing gaps							
100 mm wide	2.62	1.33	24.61	–	4.95	m	**29.56**
2 rows; 220 mm wide	5.25	2.22	41.07	–	8.25	m	**49.32**
3 rows; 340 mm wide	7.87	3.00	55.50	–	10.24	m	**65.74**
Edgings 100 × 100 × 200 mm; ref 8R 'standard quality' with 20 mm pointing gaps							
100 mm wide	3.68	0.75	13.88	–	6.00	m	**19.88**
2 rows; 220 mm wide	7.35	1.50	27.75	–	10.36	m	**38.11**
3 rows; 340 mm wide	11.03	2.25	41.63	–	13.39	m	**55.02**
Reclaimed granite setts edgings; CED Ltd; **100 × 100 mm; bedding in cement mortar** **(1:4); including haunching with in situ** **concrete; 11.50 N/mm² – 40 mm aggregate** **one side; on concrete foundations (not** **included)**							
Edgings; with 20 mm pointing gaps							
100 mm wide	3.58	1.33	24.61	–	5.91	m	**30.52**
2 rows; 220 mm wide	7.17	2.22	41.07	–	10.17	m	**51.24**
3 rows; 340 mm wide	10.75	3.00	55.50	–	13.11	m	**68.61**
Brick or block stretchers; bedding in **cement mortar (1:4); on 150 mm deep** **concrete foundations, including haunching** **with in situ concrete; 11.50 N/mm² – 40 mm** **aggregate one side**							
Single course							
concrete paving blocks; PC £8.45/m²; 200 × 100 × 60 mm	1.69	0.31	5.69	–	4.04	m	**9.73**
engineering bricks; PC £239.00/1000; 215 × 102.5 × 65 mm	1.12	0.40	7.40	–	3.46	m	**10.86**
paving bricks; PC £450.00/1000; 215 × 102.5 × 65 mm	2.00	0.40	7.40	–	4.35	m	**11.75**
Two courses							
concrete paving blocks; PC £8.45/m²; 200 × 100 × 60 mm	3.38	0.40	7.40	–	6.18	m	**13.58**
engineering bricks; PC £239.00/1000; 215 × 102.5 × 65 mm	2.23	0.57	10.57	–	5.03	m	**15.60**
paving bricks; PC £450.00/1000; 215 × 102.5 × 65 mm	4.00	0.57	10.57	–	6.80	m	**17.37**

Q PAVING/PLANTING/FENCING/SITE FURNITURE

Item Excluding site overheads and profit	PC £	Labour hours	Labour £	Plant £	Material £	Unit	Total rate £
Q10 KERBS/EDGINGS/CHANNELS/PAVING ACCESSORIES – cont							
Brick or block stretchers – cont							
Three courses							
concrete paving blocks; PC £8.45/m2;							
200 × 100 × 60 mm	5.07	0.44	8.22	–	7.87	m	**16.09**
engineering bricks; PC £239.00/1000;							
215 × 102.5 × 65 mm	3.35	0.67	12.33	–	6.15	m	**18.48**
paving bricks; PC £450.00/1000; 215 ×							
102.5 × 65 mm	6.00	0.67	12.33	–	8.80	m	**21.13**
Bricks on edge; bedding in cement mortar (1:4); on 150 mm deep concrete foundations; including haunching with in situ concrete; 11.50 N/mm^2 – 40 mm aggregate one side							
One brick wide							
engineering bricks; 215 × 102.5 × 65 mm	3.35	0.57	10.57	–	6.26	m	**16.83**
paving bricks; 215 × 102.5 × 65 mm	6.00	0.57	10.57	–	8.91	m	**19.48**
Two courses; stretchers laid on edge; 225 mm wide							
engineering bricks; 215 × 102.5 × 65 mm	6.69	1.20	22.20	–	10.60	m	**32.80**
paving bricks; 215 × 102.5 × 65 mm	12.00	1.20	22.20	–	15.91	m	**38.11**
Extra over bricks on edge for standard kerbs to one side; haunching in concrete							
125 × 255 mm; HB2; SP	3.82	0.44	8.22	–	6.15	m	**14.37**
Channels; bedding in cement mortar (1:3); joints pointed flush; on concrete foundations (not included)							
Three courses stretchers; 350 mm wide; quarter bond to form dished channels							
engineering bricks; PC £239.00/1000;							
215 × 102.5 × 65 mm	3.35	1.00	18.50	–	4.36	m	**22.86**
paving bricks; PC £450.00/1000; 215 ×							
102.5 × 65 mm	6.00	1.00	18.50	–	7.01	m	**25.51**
Three courses granite setts; 340 mm wide; to form dished channels							
340 mm wide	12.03	2.00	37.00	–	14.35	m	**51.35**
Timber edging boards; fixed with 50 × 50 × 750 mm timber pegs at 1000 mm centres (excavations and hardcore under edgings not included)							
Straight							
38 × 150 mm treated softwood edge boards	2.19	0.10	1.85	–	2.19	m	**4.04**
50 × 150 mm treated softwood edge boards	2.94	0.10	1.85	–	2.94	m	**4.79**
38 × 150 mm hardwood (iroko) edge boards	6.17	0.10	1.85	–	6.17	m	**8.02**
50 × 150 mm hardwood (iroko) edge boards	9.09	0.10	1.85	–	9.09	m	**10.94**
Curved							
38 × 150 mm treated softwood edge boards	2.19	0.20	3.70	–	2.19	m	**5.89**
50 × 150 mm treated softwood edge boards	2.94	0.25	4.63	–	2.94	m	**7.57**
38 × 150 mm hardwood (iroko) edge boards	6.17	0.20	3.70	–	6.17	m	**9.87**
50 × 150 mm hardwood (iroko) edge boards	9.09	0.25	4.63	–	9.09	m	**13.72**

Q PAVING/PLANTING/FENCING/SITE FURNITURE

Item Excluding site overheads and profit	PC £	Labour hours	Labour £	Plant £	Material £	Unit	**Total** **rate £**
Permaloc AshphaltEdge; RTS Ltd; **extruded aluminium alloy L shaped edging** **with 5.33 mm exposed upper lip; edging** **fixed to roadway base and edge profile** **with 250 mm steel fixing spike; laid to** **straight or curvilinear road edge;** **subsequently filled with macadam (not** **included)** Depth of macadam							
38 mm	7.52	0.02	0.31	–	7.52	m	**7.83**
51 mm	8.48	0.02	0.32	–	8.48	m	**8.80**
64 mm	9.37	0.02	0.34	–	9.37	m	**9.71**
76 mm	11.05	0.02	0.37	–	11.05	m	**11.42**
102 mm	13.18	0.02	0.38	–	13.18	m	**13.56**
Permaloc Cleanline; RTS Ltd; heavy duty **straight profile edging; for edgings to soft** **landscape beds or turf areas; 3.2 mm ×** **102 mm high; 3.2 mm thick with 4.75 mm** **exposed upper lip; fixed to form straight or** **curvilinear edge with 305 mm fixing spike** Milled aluminium							
100 mm deep	8.62	0.02	0.34	–	8.62	m	**8.96**
Black							
100 mm deep	8.62	0.02	0.34	–	8.62	m	**8.96**
Permaloc Permastrip; RTS Ltd; heavy duty **L shaped profile maintenance strip;** **3.2 mm × 89 mm high with 5.2 mm exposed** **top lip; for straight or gentle curves on** **paths or bed turf interfaces; fixed to form** **straight or curvilinear edge with standard** **305 mm stake; other stake lengths available** Milled aluminium							
89 mm deep	8.62	0.02	0.34	–	8.62	m	**8.96**
Black							
89 mm deep	8.62	0.02	0.34	–	8.62	m	**8.96**
Permaloc Proline; RTS Ltd; medium duty **straight profiled maintenance strip;** **3.2 mm × 102 mm high with 3.18 mm** **exposed top lip; for straight or gentle** **curves on paths or bed turf interfaces;** **fixed to form straight or curvilinear edge** **with standard 305 mm stake; other stake** **lengths available** Milled aluminium							
89 mm deep	7.50	0.02	0.34	–	7.50	m	**7.84**
Black							
89 mm deep	7.50	0.02	0.34	–	7.50	m	**7.84**

Q PAVING/PLANTING/FENCING/SITE FURNITURE

Item Excluding site overheads and profit	PC £	Labour hours	Labour £	Plant £	Material £	Unit	Total rate £
Q20 GRANULAR SUBBASES TO ROADS/ PAVINGS							
Hardcore bases; obtained off site; PC £12.00/m³							
By machine							
100 mm thick	1.20	0.05	0.93	1.01	1.20	m²	3.14
150 mm thick	1.80	0.07	1.23	1.29	1.80	m²	4.32
200 mm thick	2.40	0.08	1.48	1.39	2.40	m²	5.27
300 mm thick	3.60	0.07	1.23	1.43	3.60	m²	6.26
exceeding 300 mm thick	14.40	0.17	3.08	3.36	14.40	m³	20.84
By hand							
100 mm thick	1.20	0.20	3.70	0.17	1.20	m²	5.07
150 mm thick	2.16	0.30	5.56	0.17	2.16	m²	7.89
200 mm thick	2.88	0.40	7.40	0.27	2.88	m²	10.55
300 mm thick	4.32	0.60	11.10	0.17	4.32	m²	15.59
exceeding 300 mm thick	14.40	2.00	37.00	0.56	14.40	m³	51.96
Hardcore; difference for each £1.00 increase/ decrease in PC price per m³; price will vary with type and source of hardcore							
average 75 mm thick	–	–	–	–	0.08	m²	0.08
average 100 mm thick	–	–	–	–	0.10	m²	0.10
average 150 mm thick	–	–	–	–	0.15	m²	0.15
average 200 mm thick	–	–	–	–	0.20	m²	0.20
average 250 mm thick	–	–	–	–	0.25	m²	0.25
average 300 mm thick	–	–	–	–	0.30	m²	0.30
exceeding 300 mm thick	–	–	–	–	1.10	m³	1.10
Type 1 granular fill base; PC £15.95/tonne (£35.09/m³ compacted)							
By machine							
100 mm thick	3.51	0.03	0.52	0.33	3.51	m²	4.36
150 mm thick	5.26	0.03	0.46	0.51	5.26	m²	6.23
250 mm thick	8.77	0.02	0.38	0.84	8.77	m²	9.99
over 250 mm thick	35.09	0.20	3.70	2.94	35.09	m³	41.73
By hand (mechanical compaction)							
100 mm thick	3.51	0.17	3.08	0.06	3.51	m²	6.65
150 mm thick	5.26	0.25	4.63	0.08	5.26	m²	9.97
250 mm thick	8.77	0.42	7.71	0.14	8.77	m²	16.62
over 250 mm thick	35.09	0.47	8.64	0.14	35.09	m³	43.87
Type 1 (recycled material) granular fill base; PC £9.95/tonne (£21.89/m³ compacted)							
By machine							
100 mm thick	2.19	0.03	0.52	0.33	2.19	m²	3.04
150 mm thick	3.28	0.03	0.46	0.51	3.28	m²	4.25
250 mm thick	5.47	0.02	0.38	0.84	5.47	m²	6.69
over 250 mm thick	9.95	0.20	3.70	2.94	9.95	m³	16.59
By hand (mechanical compaction)							
100 mm thick	2.19	0.17	3.08	0.06	2.19	m²	5.33
150 mm thick	3.28	0.25	4.63	0.08	3.28	m²	7.99
250 mm thick	5.47	0.42	7.71	0.14	5.47	m²	13.32
over 250 mm thick	9.95	0.47	8.64	0.14	9.95	m³	18.73

Q PAVING/PLANTING/FENCING/SITE FURNITURE

Item Excluding site overheads and profit	PC £	Labour hours	Labour £	Plant £	Material £	Unit	Total rate £
Surface treatments							
Sand blinding; to hardcore base (not							
included); 25 mm thick	0.77	0.03	0.62	–	0.77	m²	**1.39**
Sand blinding; to hardcore base (not							
included); 50 mm thick	1.53	0.05	0.93	–	1.53	m²	**2.46**
Filter fabrics; to hardcore base (not included)	0.45	0.01	0.19	–	0.45	m²	**0.64**
Q21 IN SITU CONCRETE ROADS/PAVINGS							
Unreinforced concrete; on prepared							
subbase (not included)							
Roads; 21.00 N/mm² – 20 mm aggregate							
(1:2:4) mechanically mixed on site							
100 mm thick	9.39	0.13	2.31	–	9.39	m²	**11.70**
150 mm thick	14.08	0.17	3.08	–	14.08	m²	**17.16**
Reinforced in situ concrete; mechanically							
mixed on site; normal Portland cement; on							
hardcore base (not included);							
reinforcement (not included)							
Roads; 11.50 N/mm² – 40 mm aggregate							
(1:3:6)							
100 mm thick	7.95	0.40	7.40	–	7.95	m²	**15.35**
150 mm thick	11.94	0.60	11.10	–	11.94	m²	**23.04**
200 mm thick	16.30	0.80	14.80	–	16.30	m²	**31.10**
250 mm thick	19.88	1.00	18.50	0.22	19.88	m²	**38.60**
300 mm thick	23.86	1.20	22.20	0.22	23.86	m²	**46.28**
Roads; 21.00 N/mm² – 20 mm aggregate							
(1:2:4)							
100 mm thick	9.17	0.40	7.40	–	9.17	m²	**16.57**
150 mm thick	13.75	0.60	11.10	–	13.75	m²	**24.85**
200 mm thick	18.33	0.80	14.80	–	18.33	m²	**33.13**
250 mm thick	22.91	1.00	18.50	0.22	22.91	m²	**41.63**
300 mm thick	27.50	1.20	22.20	0.22	27.50	m²	**49.92**
Roads; 25.00 N/mm² – 20 mm aggregate GEN							
4 ready mixed							
100 mm thick	6.83	0.40	7.40	–	6.83	m²	**14.23**
150 mm thick	10.24	0.60	11.10	–	10.24	m²	**21.34**
200 mm thick	13.65	0.80	14.80	–	13.65	m²	**28.45**
250 mm thick	17.06	1.00	18.50	0.22	17.06	m²	**35.78**
300 mm thick	20.48	1.20	22.20	0.22	20.48	m²	**42.90**
Reinforced in situ concrete; ready mixed;							
discharged directly into location from							
supply lorry; normal Portland cement; on							
hardcore base (not included);							
reinforcement (not included)							
Roads; 11.50 N/mm² – 40 mm aggregate							
(1:3:6)							
100 mm thick	6.41	0.16	2.96	–	6.41	m²	**9.37**
150 mm thick	9.62	0.24	4.44	–	9.62	m²	**14.06**
200 mm thick	12.82	0.36	6.66	–	12.82	m²	**19.48**
250 mm thick	16.02	0.54	9.99	0.22	16.02	m²	**26.23**
300 mm thick	19.23	0.66	12.21	0.22	19.23	m²	**31.66**

Q PAVING/PLANTING/FENCING/SITE FURNITURE

Item Excluding site overheads and profit	PC £	Labour hours	Labour £	Plant £	Material £	Unit	Total rate £
Q21 IN SITU CONCRETE ROADS/ PAVINGS – cont							
Reinforced in situ concrete – cont							
Roads; 21.00 N/mm² – 20 mm aggregate (1:2:4)							
100 mm thick	6.92	0.16	2.96	–	6.92	m²	**9.88**
150 mm thick	10.38	0.24	4.44	–	10.38	m²	**14.82**
200 mm thick	13.84	0.36	6.66	–	13.84	m²	**20.50**
250 mm thick	17.30	0.54	9.99	0.22	17.30	m²	**27.51**
300 mm thick	20.76	0.66	12.21	0.22	20.76	m²	**33.19**
Roads; 26.00 N/mm² – 20 mm aggregate (1:1.5:3)							
100 mm thick	7.06	0.16	2.96	–	7.06	m²	**10.02**
150 mm thick	10.59	0.24	4.44	–	10.59	m²	**15.03**
200 mm thick	13.84	0.36	6.66	–	13.84	m²	**20.50**
250 mm thick	17.65	0.54	9.99	0.22	17.65	m²	**27.86**
300 mm thick	21.18	0.66	12.21	0.22	21.18	m²	**33.61**
Roads; PAV1 concrete – 35 N/mm²; Designated mix							
100 mm thick	7.02	0.16	2.96	–	7.02	m²	**9.98**
150 mm thick	10.54	0.24	4.44	–	10.54	m²	**14.98**
200 mm thick	14.04	0.36	6.66	–	14.04	m²	**20.70**
250 mm thick	17.55	0.54	9.99	0.22	17.55	m²	**27.76**
300 mm thick	21.06	0.66	12.21	0.22	21.06	m²	**33.49**
Concrete sundries							
Treating surfaces of unset concrete; grading to cambers; tamping with 75 mm thick steel shod tamper or similar	–	0.13	2.47	–	–	m²	**2.47**
Expansion joints							
13 mm thick joint filler; formwork							
width or depth not exceeding 150 mm	1.80	0.20	3.70	–	2.67	m	**6.37**
width or depth 150–300 mm	3.59	0.25	4.63	–	5.33	m	**9.96**
width or depth 300–450 mm	5.39	0.30	5.55	–	8.00	m	**13.55**
25 mm thick joint filler; formwork							
width or depth not exceeding 150 mm	3.80	0.20	3.70	–	4.67	m	**8.37**
width or depth 150–300 mm	7.60	0.25	4.63	–	9.34	m	**13.97**
width or depth 300–450 mm	11.40	0.30	5.55	–	14.01	m	**19.56**
Sealants; sealing top 25 mm of joint with rubberized bituminous compound	1.13	0.25	4.63	–	1.13	m	**5.76**
Formwork for in situ concrete							
Sides of foundations							
height not exceeding 250 mm	0.66	0.03	0.62	–	0.74	m	**1.36**
height 250–500 mm	0.88	0.04	0.74	–	1.07	m	**1.81**
height 500 mm–1.00 m	0.88	0.05	0.93	–	1.17	m	**2.10**
height exceeding 1.00 m	2.63	2.00	37.00	–	4.29	m²	**41.29**
Extra over formwork for curved work 6 m radius	–	0.25	4.63	–	–	m	**4.63**
Steel road forms; to edges of beds or faces of foundations							
150 mm wide	–	0.20	3.70	1.10	–	m	**4.80**

Q PAVING/PLANTING/FENCING/SITE FURNITURE

Item Excluding site overheads and profit	PC £	Labour hours	Labour £	Plant £	Material £	Unit	Total rate £
Reinforcement; fabric; side laps 150 mm; **head laps 300 mm; mesh 200 × 200 mm; in** **roads, footpaths or pavings**							
Fabric							
A142 (2.22 kg/m^2)	2.56	0.08	1.54	–	2.56	m^2	4.10
A193 (3.02 kg/m^2)	3.50	0.08	1.54	–	3.50	m^2	5.04
Q22 COATED MACADAM/ASPHALT **ROADS/PAVINGS**							
Coated macadam/asphalt roads/pavings – **General**							
Preamble: The prices for all in situ finishings to roads and footpaths include for work to falls, crossfalls or slopes not exceeding 15° from horizontal; for laying on prepared bases (not included) and for rolling with an appropriate roller.							
Users should note the new terminology for the surfaces described below which is to European standard descriptions. The now redundant descriptions for each course are shown in brackets.							
Macadam surfacing; Spadeoak **Construction Co Ltd; surface (wearing)** **course; 20 mm of 6 mm dense bitumen** **macadam**							
Machine lay; areas 1000 m^2 and over							
limestone aggregate	–	–	–	–	–	m^2	5.26
granite aggregate	–	–	–	–	–	m^2	5.32
red	–	–	–	–	–	m^2	9.45
Hand lay; areas 400 m^2 and over							
limestone aggregate	–	–	–	–	–	m^2	6.69
granite aggregate	–	–	–	–	–	m^2	6.75
red	–	–	–	–	–	m^2	11.12
Macadam surfacing; Spadeoak **Construction Co Ltd; surface (wearing)** **course; 30 mm of 10 mm dense bitumen** **macadam**							
Machine lay; areas 1000 m^2 and over							
limestone aggregate	–	–	–	–	–	m^2	6.35
granite aggregate	–	–	–	–	–	m^2	6.43
red	–	–	–	–	–	m^2	13.31
Hand lay; areas 400 m^2 and over							
limestone aggregate	–	–	–	–	–	m^2	7.84
granite aggregate	–	–	–	–	–	m^2	7.93
red	–	–	–	–	–	m^2	15.24

Q PAVING/PLANTING/FENCING/SITE FURNITURE

Item Excluding site overheads and profit	PC £	Labour hours	Labour £	Plant £	Material £	Unit	Total rate £
Q22 COATED MACADAM/ASPHALT **ROADS/PAVINGS – cont**							
Macadam surfacing; Spadeoak **Construction Co Ltd; surface (wearing)** **course; 40 mm of 10 mm dense bitumen** **macadam**							
Machine lay; areas 1000 m² and over							
limestone aggregate	–	–	–	–	–	m²	8.08
granite aggregate	–	–	–	–	–	m²	8.20
red	–	–	–	–	–	m²	15.09
Hand lay; areas 400 m² and over							
limestone aggregate	–	–	–	–	–	m²	9.69
granite aggregate	–	–	–	–	–	m²	9.81
red	–	–	–	–	–	m²	17.13
Macadam surfacing; Spadeoak **Construction Co Ltd; binder (base) course;** **50 mm of 20 mm dense bitumen macadam**							
Machine lay; areas 1000 m² and over							
limestone aggregate	–	–	–	–	–	m²	8.10
granite aggregate	–	–	–	–	–	m²	8.23
Hand lay; areas 400 m² and over							
limestone aggregate	–	–	–	–	–	m²	9.71
granite aggregate	–	–	–	–	–	m²	9.85
Macadam surfacing; Spadeoak **Construction Co Ltd; binder (base) course;** **60 mm of 20 mm dense bitumen macadam**							
Machine lay; areas 1000 m² and over							
limestone aggregate	–	–	–	–	–	m²	8.91
granite aggregate	–	–	–	–	–	m²	9.06
Hand lay; areas 400 m² and over							
limestone aggregate	–	–	–	–	–	m²	10.57
granite aggregate	–	–	–	–	–	m²	10.73
Macadam surfacing; Spadeoak **Construction Co Ltd; base (roadbase)** **course; 75 mm of 28 mm dense bitumen** **macadam**							
Machine lay; areas 1000 m² and over							
limestone aggregate	–	–	–	–	–	m²	10.93
granite aggregate	–	–	–	–	–	m²	11.12
Hand lay; areas 400 m² and over							
limestone aggregate	–	–	–	–	–	m²	12.73
granite aggregate	–	–	–	–	–	m²	12.94
Macadam surfacing; Spadeoak **Construction Co Ltd; base (roadbase)** **course; 100 mm of 28 mm dense bitumen** **macadam**							
Machine lay; areas 1000 m² and over							
limestone aggregate	–	–	–	–	–	m²	13.54
granite aggregate	–	–	–	–	–	m²	13.80
Hand lay; areas 400 m² and over							
limestone aggregate	–	–	–	–	–	m²	15.51
granite aggregate	–	–	–	–	–	m²	15.79

Q PAVING/PLANTING/FENCING/SITE FURNITURE

Item Excluding site overheads and profit	PC £	Labour hours	Labour £	Plant £	Material £	Unit	Total rate £
Macadam surfacing; Spadeoak Construction Co Ltd; base (roadbase) course; 150 mm of 28 mm dense bitumen macadam in two layers							
Machine lay; areas 1000 m² and over							
limestone aggregate	–	–	–	–	–	m²	21.35
granite aggregate	–	–	–	–	–	m²	21.74
Hand lay; areas 400 m² and over							
limestone aggregate	–	–	–	–	–	m²	24.92
granite aggregate	–	–	–	–	–	m²	25.33
Base (roadbase) course; 200 mm of 28 mm dense bitumen macadam in two layers							
Machine lay; areas 1000 m² and over							
limestone aggregate	–	–	–	–	–	m²	27.13
granite aggregate	–	–	–	–	–	m²	27.65
Hand lay; areas 400 m² and over							
limestone aggregate	–	–	–	–	–	m²	21.08
granite aggregate	–	–	–	–	–	m²	31.63
Resin bound macadam pavings; Bituchem; to pedestrian or vehicular hard landscape areas; laid to base course (not included)							
Natratex wearing course; clear resin bound macadam							
25 mm thick to pedestrian areas	16.60	0.13	2.31	–	16.60	m²	18.91
30 mm thick to vehicular areas	20.00	0.03	0.62	–	20.00	m²	20.62
Colourtex coloured resin bound macadam							
25 mm thick	16.60	0.13	2.31	–	16.60	m²	18.91
Marking car parks							
Car parking space division strips; laid hot at 115°C; on bitumen macadam surfacing							
minimum daily rate	–	–	–	–	–	item	500.00
Stainless metal road studs							
100 × 100 mm	5.61	0.25	4.63	–	5.61	nr	10.24
Bound aggregates; Addagrip Surface Treatments UK Ltd; natural decorative resin bonded surface dressing laid to concrete, macadam or to plywood panels priced separately							
Primer coat to macadam or concrete base	–	–	–	–	–	m²	4.00
Golden pea gravel; 1–3 mm							
buff adhesive	–	–	–	–	–	m²	20.00
red adhesive	–	–	–	–	–	m²	20.00
green adhesive	–	–	–	–	–	m²	20.00
Golden pea gravel; 2–5 mm							
buff adhesive	–	–	–	–	–	m²	24.00
Chinese bauxite; 1–3 mm							
buff adhesive	–	–	–	–	–	m²	20.00
Cobalt Blue Glass; 6 mm							
15 mm depth	–	–	–	–	–	m²	75.00

Q PAVING/PLANTING/FENCING/SITE FURNITURE

Item Excluding site overheads and profit	PC £	Labour hours	Labour £	Plant £	Material £	Unit	Total rate £
Q22 COATED MACADAM/ASPHALT ROADS/PAVINGS – cont							
Bound aggregates – cont							
Midnight Grey; 6 mm							
15 mm depth	–	–	–	–	–	m²	**45.00**
Chocolate; 6 mm							
15 mm depth	–	–	–	–	–	m²	**45.00**
18 mm depth	–	–	–	–	–	m²	**55.00**
Q23 GRAVEL/HOGGIN/WOODCHIP ROADS/ PAVINGS							
Excavation and path preparation							
Excavating; 300 mm deep; to width of path; depositing excavated material at sides of excavation							
width 1.00 m	–	–	–	2.25	–	m²	**2.25**
width 1.50 m	–	–	–	1.87	–	m²	**1.87**
width 2.00 m	–	–	–	1.60	–	m²	**1.60**
width 3.00 m	–	–	–	1.35	–	m²	**1.35**
Excavating trenches; in centre of pathways; 100 mm flexible drain pipes; filling with clean broken stone or gravel rejects							
300 × 450 mm deep	3.04	0.10	1.85	1.12	3.04	m	**6.01**
Hand trimming and compacting reduced surface of pathway; by machine							
width 1.00 m	–	0.05	0.93	0.17	–	m	**1.10**
width 1.50 m	–	0.04	0.82	0.15	–	m	**0.97**
width 2.00 m	–	0.04	0.74	0.13	–	m	**0.87**
width 3.00 m	–	0.04	0.74	0.13	–	m	**0.87**
Permeable membranes; to trimmed and compacted surface of pathway							
Terram 1000	0.45	0.02	0.37	–	0.45	m²	**0.82**
Permaloc AshphaltEdge; RTS Ltd; extruded aluminium alloy L shaped edging with 5.33 mm exposed upper lip; edging fixed to roadway base and edge profile with 250 mm steel fixing spike; laid to straight or curvilinear road edge; subsequently filled with macadam (not included)							
Depth of macadam							
38 mm	7.52	0.02	0.31	–	7.52	m	**7.83**
51 mm	8.48	0.02	0.32	–	8.48	m	**8.80**
64 mm	9.37	0.02	0.34	–	9.37	m	**9.71**
76 mm	11.05	0.02	0.37	–	11.05	m	**11.42**
102 mm	13.18	0.02	0.38	–	13.18	m	**13.56**

Q PAVING/PLANTING/FENCING/SITE FURNITURE

Item Excluding site overheads and profit	PC £	Labour hours	Labour £	Plant £	Material £	Unit	Total rate £
Permaloc Cleanline; RTS Ltd; heavy duty straight profile edging; for edgings to soft landscape beds or turf areas; 3.2 mm × 102 mm high; 3.2 mm thick with 4.75 mm exposed upper lip; fixed to form straight or curvilinear edge with 305 mm fixing spike							
Milled aluminium							
100 mm deep	8.62	0.02	0.34	–	8.62	m	**8.96**
Black							
100 mm deep	8.62	0.02	0.34	–	8.62	m	**8.96**
Permaloc Permastrip; RTS Ltd; heavy duty L shaped profile maintenance strip; 3.2 mm × 89 mm high with 5.2 mm exposed top lip; for straight or gentle curves on paths or bed turf interfaces; fixed to form straight or curvilinear edge with standard 305 mm stake; other stake lengths available							
Milled aluminium							
89 mm deep	8.62	0.02	0.34	–	8.62	m	**8.96**
Black							
89 mm deep	8.62	0.02	0.34	–	8.62	m	**8.96**
Permaloc Proline; RTS Ltd; medium duty straight profiled maintenance strip; 3.2 mm × 102 mm high with 3.18 mm exposed top lip; for straight or gentle curves on paths or bed turf interfaces; fixed to form straight or curvilinear edge with standard 305 mm stake; other stake lengths available							
Milled aluminium							
89 mm deep	7.50	0.02	0.34	–	7.50	m	**7.84**
Black							
89 mm deep	7.50	0.02	0.34	–	7.50	m	**7.84**
Filling to make up levels							
Obtained off site; hardcore; PC £12.00/m³							
150 mm thick	1.92	0.04	0.74	0.47	1.92	m²	**3.13**
Obtained off site; granular fill type 1; PC £15.95/tonne (£35.09/m³ compacted)							
100 mm thick	3.51	0.03	0.52	0.33	3.51	m²	**4.36**
150 mm thick	5.26	0.03	0.46	0.51	5.26	m²	**6.23**
Surface treatments							
Sand blinding; to hardcore (not included)							
50 mm thick	1.68	0.04	0.74	–	1.68	m²	**2.42**
Filter fabric; to hardcore (not included)	0.45	0.01	0.19	–	0.45	m²	**0.64**

Q PAVING/PLANTING/FENCING/SITE FURNITURE

Item Excluding site overheads and profit	PC £	Labour hours	Labour £	Plant £	Material £	Unit	Total rate £
Q23 GRAVEL/HOGGIN/WOODCHIP ROADS/ PAVINGS – cont							
Granular pavings							
Footpath gravels; porous self binding gravel							
CED Ltd; Cedec gravel; self-binding; laid to inert (non-limestone) base measured separately; compacting							
red, silver or gold; 50 mm thick	11.21	0.03	0.46	0.39	11.21	m²	**12.06**
Grundon Ltd; Coxwell self-binding path gravels laid and compacted to excavation or base measured separately							
50 mm thick	3.27	0.03	0.46	0.39	3.27	m²	**4.12**
Breedon Special Aggregates; Golden Gravel or equivalent; rolling wet; on hardcore base (not included); for pavements; to falls and crossfalls and to slopes not exceeding 15° from horizontal; over 300 mm wide							
50 mm thick	10.70	0.03	0.46	0.32	10.70	m²	**11.48**
75 mm thick	16.04	0.10	1.85	0.40	16.04	m²	**18.29**
Breedon Special Aggregates; Wayfarer specially formulated fine gravel for use on golf course pathways							
50 mm thick	10.05	0.03	0.46	0.32	10.05	m²	**10.83**
75 mm thick	15.06	0.05	0.93	0.57	15.06	m²	**16.56**
Hoggin (stabilized); PC £54.58/m³ on hardcore base (not included); to falls and crossfalls and to slopes not exceeding 15° from horizontal; over 300 mm wide							
100 mm thick	8.19	0.03	0.62	0.52	8.19	m²	**9.33**
150 mm thick	12.28	0.05	0.93	0.77	12.28	m²	**13.98**
Ballast; as dug; watering; rolling; on hardcore base (not included)							
100 mm thick	4.05	0.03	0.62	0.52	4.05	m²	**5.19**
150 mm thick	6.08	0.05	0.93	0.77	6.08	m²	**7.78**
Footpath gravels; porous loose gravels							
Breedon Special Aggregates; Breedon Buff decorative limestone chippings							
50 mm thick	4.97	0.01	0.19	0.09	4.97	m²	**5.25**
75 mm thick	14.05	0.01	0.23	0.11	14.05	m²	**14.39**
Breedon Special Aggregates; Brindle or Moorland Black chippings							
50 mm thick	6.54	0.01	0.19	0.09	6.54	m²	**6.82**
75 mm thick	9.82	0.01	0.23	0.11	9.82	m²	**10.16**
Breedon Special Aggregates; Slate Chippings; plum, blue or green							
50 mm thick	9.32	0.01	0.19	0.09	9.32	m²	**9.60**
75 mm thick	13.99	0.01	0.23	0.11	13.99	m²	**14.33**

Q PAVING/PLANTING/FENCING/SITE FURNITURE

Item Excluding site overheads and profit	PC £	Labour hours	Labour £	Plant £	Material £	Unit	Total rate £
Washed shingle; on prepared base (not included)							
25–50 mm size; 25 mm thick	0.83	0.02	0.33	0.08	0.83	m²	**1.24**
25–50 mm size; 75 mm thick	2.49	0.05	0.97	0.23	2.49	m²	**3.69**
50–75 mm size; 25 mm thick	0.83	0.02	0.37	0.09	0.83	m²	**1.29**
50–75 mm size; 75 mm thick	2.49	0.07	1.23	0.29	2.49	m²	**4.01**
Pea shingle; on prepared base (not included)							
10–15 mm size; 25 mm thick	0.95	0.02	0.33	0.08	0.95	m²	**1.36**
5–10 mm size; 75 mm thick	2.86	0.05	0.97	0.23	2.86	m²	**4.06**
Wood chip surfaces; Melcourt Industries Ltd (items labelled FSC are Forest Stewardship Council certified)							
Wood chips; to surface of pathways by machine; material delivered in 80 m³ loads; levelling and spreading by hand (excavation and preparation not included)							
Walk Chips; 100 mm thick; FSC (25 m³ loads)	3.85	0.03	0.62	0.28	3.85	m²	**4.75**
Walk Chips; 100 mm thick; FSC (80 m³ loads)	2.49	0.03	0.62	0.28	2.49	m²	**3.39**
Woodfibre; 100 mm thick; FSC (80 m³ loads)	2.49	0.03	0.62	0.28	2.49	m²	**3.39**
Q24 INTERLOCKING BRICK/BLOCK ROADS/PAVINGS							
Precast concrete block edgings; PC £7.63/ m2; 200 × 100 × 60 mm; on prepared base (not included); haunching one side							
Edgings; butt joints							
stretcher course	0.76	0.17	3.08	–	2.93	m	**6.01**
header course	1.53	0.27	4.93	–	3.85	m	**8.78**
Precast concrete vehicular paving blocks; Marshalls Plc; on prepared base (not included); on 50 mm compacted sharp sand bed; blocks laid in 7 mm loose sand and vibrated; joints filled with sharp sand and vibrated; level and to falls only							
Trafica paving blocks; 450 × 450 × 70 mm							
Perfecta finish; colour natural	28.50	0.50	9.25	0.10	30.18	m²	**39.53**
Perfecta finish; colour buff	33.00	0.50	9.25	0.10	34.68	m²	**44.03**
Saxon finish; colour natural	25.06	0.50	9.25	0.10	26.75	m²	**36.10**
Saxon finish; colour buff	28.80	0.50	9.25	0.10	30.49	m²	**39.84**
Precast concrete vehicular paving blocks; Keyblok Marshalls Plc; on prepared base (not included); on 50 mm compacted sharp sand bed; blocks laid in 7 mm loose sand and vibrated; joints filled with sharp sand and vibrated; level and to falls only							
Herringbone bond							
200 × 100 × 60 mm; natural grey	8.01	1.50	27.75	0.10	9.76	m²	**37.61**
200 × 100 × 60 mm; colours	8.87	1.50	27.75	0.10	10.62	m²	**38.47**
200 × 100 × 80 mm; natural grey	9.09	1.50	27.75	0.10	10.84	m²	**38.69**
200 × 100 × 80 mm; colours	10.27	1.50	27.75	0.10	12.01	m²	**39.86**

Q PAVING/PLANTING/FENCING/SITE FURNITURE

Item Excluding site overheads and profit	PC £	Labour hours	Labour £	Plant £	Material £	Unit	Total rate £
Q24 INTERLOCKING BRICK/BLOCK **ROADS/PAVINGS – cont**							
Precast concrete vehicular paving blocks – **cont**							
Basketweave bond							
200 × 100 × 60 mm; natural grey	8.01	1.20	22.20	0.10	9.76	m²	**32.06**
200 × 100 × 60 mm; colours	8.87	1.20	22.20	0.10	10.62	m²	**32.92**
200 × 100 × 80 mm; natural grey	9.09	1.20	22.20	0.10	10.84	m²	**33.14**
200 × 100 × 80 mm; colours	10.27	1.20	22.20	0.10	12.01	m²	**34.31**
Precast concrete vehicular paving blocks; **Charcon Hard Landscaping; on prepared** **base (not included); on 50 mm compacted** **sharp sand bed; blocks laid in 7 mm loose** **sand and vibrated; joints filled with sharp** **sand and vibrated; level and to falls only**							
Europa concrete blocks							
200 × 100 × 60 mm; natural grey	9.86	1.50	27.75	0.10	11.60	m²	**39.45**
200 × 100 × 60 mm; colours	10.75	1.50	27.75	0.10	12.50	m²	**40.35**
Parliament concrete blocks							
200 × 100 × 65 mm; natural grey	26.56	1.50	27.75	0.10	28.30	m²	**56.15**
200 × 100 × 65 mm; colours	26.56	1.50	27.75	0.10	28.30	m²	**56.15**
Recycled polyethylene grassblocks; **Boddingtons Grass Reinforcement** **Solutions; Boddingtons Ltd; interlocking** **units laid to prepared base or rootzone (not** **included)**							
BodPave85; load bearing <400 tonnes per m²; 500 × 500 × 50 mm deep; 35 mm ground spike							
1–50 m²	16.50	0.20	3.70	0.18	19.15	m²	**23.03**
51–500 m²	13.75	0.20	3.70	0.18	16.40	m²	**20.28**
501–1000 m²	13.20	0.20	3.70	0.18	15.85	m²	**19.73**
1001–1559 m²	12.50	0.20	3.70	0.18	15.15	m²	**19.03**
1560 m² or over	12.00	0.20	3.70	0.18	14.65	m²	**18.53**
GrassProtecta; extruded expanded polyethylene flexible mesh laid to existing grass surface or newly seeded areas to provide heavy surface protection from traffic and pedestrians							
Standard; 1.2 kg/m²; 20 × 2 m; up to 320 m²	7.50	0.01	0.15	–	7.50	m²	**7.65**
Standard; 1.2 kg/m²; 20 × 2 m; 321–3000 m²	6.50	0.01	0.15	–	6.50	m²	**6.65**
Standard; 1.2 kg/m²; 20 × 2 m; over 3000 m²	6.00	0.01	0.15	–	6.00	m²	**6.15**
Heavy; 2 kg/m²; 20 × 2 m; up to 320 m²	8.50	0.01	0.15	–	8.50	m²	**8.65**
Heavy; 2 kg/m²; 20 × 2 m; 321–3000 m²	8.00	0.01	0.15	–	8.00	m²	**8.15**
Heavy; 2 kg/m²; 20 × 2 m; over 3000 m²	7.50	0.01	0.15	–	7.50	m²	**7.65**
TurfProtecta; extruded polyethylene flexible mesh laid to existing grass surface or newly seeded areas to provide surface protection from traffic including vehicle or animal wear and tear							
Standard; 30 × 2 m; up to 300 m²	2.90	0.01	0.15	–	2.90	m²	**3.05**
Standard; 30 × 2 m; 301–600 m²	2.49	0.01	0.15	–	2.49	m²	**2.64**
Standard; 30 × 2 m; 601–1440 m²	2.41	0.01	0.15	–	2.41	m²	**2.56**

Q PAVING/PLANTING/FENCING/SITE FURNITURE

Item Excluding site overheads and profit	PC £	Labour hours	Labour £	Plant £	Material £	Unit	Total rate £
Standard; 30 × 2 m; over 1440 m²	2.24	0.01	0.15	–	2.24	m²	2.39
Premium; 30 × 2 m; up to 300 m²	2.99	0.01	0.15	–	2.99	m²	3.14
Premium; 30 × 2 m; 301–600 m²	2.66	0.01	0.15	–	2.66	m²	2.81
Premium; 30 × 2 m; 601–1440 m²	2.57	0.01	0.15	–	2.57	m²	2.72
Premium; 30 × 2 m; over 1440 m²	2.41	0.01	0.15	–	2.41	m²	2.56
Grassroad; Cooper Clarke Civils and Lintels; heavy duty for car parking and fire paths verge hardening and shallow embankments Honeycomb cellular polyproylene interconnecting paviors with integral downstead anti-shear cleats including topsoil but excluding edge restraints; to granular subbase (not included)							
635 × 330 × 42 mm overall laid to a module of 622 × 311 × 32 mm	–	–	–	–	–	m²	29.35
extra over for green colour	–	–	–	–	–	m²	1.20
Q25 SLAB/BRICK/SETT/COBBLE PAVINGS							
Bricks – General Preamble: Bricks shall be hard, well burnt, non-dusting, resistant to frost and sulphate attack and true to shape, size and sample.							
Movement of materials Mechanically offloading bricks; loading wheelbarrows; transporting maximum 25 m distance	–	0.20	3.70	–	–	m²	3.70
Edge restraints; to brick paving; on prepared base (not included); 65 mm thick bricks; PC £300.00/1000; haunching one side Header course							
200 × 100 mm; butt joints	3.00	0.27	4.93	–	5.33	m	10.26
210 × 105 mm; mortar joints	2.67	0.50	9.25	–	5.22	m	14.47
Stretcher course							
200 × 100 mm; butt joints	1.50	0.17	3.08	–	3.67	m	6.75
210 × 105 mm; mortar joints	1.36	0.33	6.17	–	3.64	m	9.81
Variation in brick prices; add or subtract the following amounts for every £1.00/1000 difference in the PC price Edgings							
100 mm wide	–	–	–	–	0.05	10 m	0.05
200 mm wide	–	–	–	–	0.10	10 m	0.10
102.5 mm wide	–	–	–	–	0.04	10 m	0.04
215 mm wide	–	–	–	–	0.09	10 m	0.09
Clay brick pavings; on prepared base (not included); bedding on 50 mm sharp sand; kiln dried sand joints Pavings; 200 × 100 × 65 mm wirecut chamfered paviors							
brick; PC £450.00/1000	23.06	1.44	26.71	0.22	24.98	m²	51.91

Q PAVING/PLANTING/FENCING/SITE FURNITURE

Item Excluding site overheads and profit	PC £	Labour hours	Labour £	Plant £	Material £	Unit	Total rate £
Q25 SLAB/BRICK/SETT/COBBLE PAVINGS – cont							
Clay brick pavings; 200 × 100 × 50 mm; laid to running stretcher or stack bond only; on prepared base (not included); bedding on cement: sand (1:4) pointing mortar as work proceeds							
PC £600.00/1000							
laid on edge	48.81	4.76	88.10	–	55.53	m²	**143.63**
laid on edge but pavior 65 mm thick	41.00	3.81	70.48	–	47.72	m²	**118.20**
laid flat	26.62	2.20	40.70	–	30.63	m²	**71.33**
PC £500.00/1000							
laid on edge	40.67	4.76	88.10	–	47.39	m²	**135.49**
laid on edge but pavior 65 mm thick	34.16	3.81	70.48	–	40.89	m²	**111.37**
laid flat	22.19	2.20	40.70	–	26.19	m²	**66.89**
PC £400.00/1000							
laid flat	17.75	2.20	40.70	–	21.75	m²	**62.45**
laid on edge	32.54	4.76	88.10	–	39.26	m²	**127.36**
laid on edge but pavior 65 mm thick	27.33	3.81	70.48	–	34.05	m²	**104.53**
PC £300.00/1000							
laid on edge	24.40	4.76	88.10	–	31.13	m²	**119.23**
laid on edge but pavior 65 mm thick	20.50	3.81	70.48	–	27.22	m²	**97.70**
laid flat	13.31	2.20	40.70	–	17.32	m²	**58.02**
Clay brick pavings; 200 × 100 × 50 mm; butt jointed laid herringbone or basketweave pattern only; on prepared base (not included); bedding on 50 mm sharp sand							
PC £600.00/1000							
laid flat	30.75	1.44	26.71	0.29	32.67	m²	**59.67**
PC £500.00/1000							
laid flat	25.63	1.44	26.71	0.29	27.54	m²	**54.54**
PC £400.00/1000							
laid flat	20.50	1.44	26.71	0.29	22.42	m²	**49.42**
PC £300.00/1000							
laid flat	15.38	1.44	26.71	0.29	17.29	m²	**44.29**
Clay brick pavings; 215 × 102.5 × 65 mm; on prepared base (not included); bedding on cement: sand (1:4) pointing mortar as work proceeds							
Paving bricks; PC £600.00/1000; herringbone bond							
laid on edge	36.44	3.55	65.77	–	41.78	m²	**107.55**
laid flat	23.70	2.37	43.85	–	29.04	m²	**72.89**
Paving bricks; PC £600.00/1000; basketweave bond							
laid on edge	36.44	2.37	43.85	–	41.78	m²	**85.63**
laid flat	23.70	1.58	29.24	–	29.04	m²	**58.28**
Paving bricks; PC £600.00/1000; running or stack bond							
laid on edge	36.44	1.90	35.09	–	41.78	m²	**76.87**
laid flat	23.70	1.26	23.39	–	29.04	m²	**52.43**

Q PAVING/PLANTING/FENCING/SITE FURNITURE

Item Excluding site overheads and profit	PC £	Labour hours	Labour £	Plant £	Material £	Unit	Total rate £
Paving bricks; PC £500.00/1000; herringbone bond							
laid on edge	30.37	3.55	65.77	–	33.43	m²	99.20
laid flat	19.75	2.37	43.85	–	25.09	m²	68.94
Paving bricks; PC £500.00/1000; basketweave bond							
laid on edge	30.37	2.37	43.85	–	33.43	m²	77.28
laid flat	19.75	1.58	29.24	–	25.09	m²	54.33
Paving bricks; PC £500.00/1000; running or stack bond							
laid on edge	30.37	1.90	35.09	–	33.43	m²	68.52
laid flat	19.75	1.26	23.39	–	25.09	m²	48.48
Paving bricks; PC £400.00/1000; herringbone bond							
laid on edge	24.29	3.55	65.77	–	27.36	m²	93.13
laid flat	15.80	2.37	43.85	–	21.04	m²	64.89
Paving bricks; PC £400.00/1000; basketweave bond							
laid on edge	24.29	2.37	43.85	–	27.36	m²	71.21
laid flat	15.80	1.58	29.24	–	21.04	m²	50.28
Paving bricks; PC £400.00/1000; running or stack bond							
laid on edge	24.29	1.90	35.09	–	27.36	m²	62.45
laid flat	15.80	1.26	23.39	–	21.04	m²	44.43
Paving bricks; PC £300.00/1000; herringbone bond							
laid on edge	17.77	3.55	65.77	–	20.84	m²	86.61
laid flat	11.85	2.37	43.85	–	17.19	m²	61.04
Paving bricks; PC £300.00/1000; basketweave bond							
laid on edge	17.77	2.37	43.85	–	20.84	m²	64.69
laid flat	11.85	1.58	29.24	–	17.19	m²	46.43
Paving bricks; PC £300.00/1000; running or stack bond							
laid on edge	17.77	1.90	35.08	–	20.84	m²	55.92
laid flat	11.85	1.26	23.39	–	17.19	m²	40.58
Cutting							
curved cutting	–	0.44	8.22	5.32	–	m	13.54
raking cutting	–	0.33	6.17	4.18	–	m	10.35
Add or subtract the following amounts for every £10.00/1000 difference in the prime cost of bricks							
Butt joints							
200 × 100 mm	–	–	–	–	0.50	m²	0.50
215 × 102.5 mm	–	–	–	–	0.45	m²	0.45
10 mm mortar joints							
200 × 100 mm	–	–	–	–	0.43	m²	0.43
215 × 102.5 mm	–	–	–	–	0.40	m²	0.40

Q PAVING/PLANTING/FENCING/SITE FURNITURE

Item Excluding site overheads and profit	PC £	Labour hours	Labour £	Plant £	Material £	Unit	Total rate £
Q25 SLAB/BRICK/SETT/COBBLE PAVINGS – cont							
SUDS paving; Concrete Block Permeable Paving (CBPP)							
Note: Sustainable Drainage Systems (SUDS) aim to reduce flood risk by controlling the rate and volume of surface water run off from developments. Permeable Paving is a SUDS technique. In addition to managing the quantity of surface water run off more effectively, one of the primary benefits of CBPP is the potential enhancement of water quality. Permeable pavements improve water quality by mirroring nature in providing filtration and allowing for natural biodegradation of hydrocarbons and the dilution of other contaminants as water passes through the system.							
Bases for Sustainable Urban Drainage System (SUDS) subbase for use below impervious paving; Aggregate Industries Ltd; Bardon DrainAgg 20							
4/20 mm open graded material laid loose on prepared subgrade; within edge restraints to paving area (not included); graded to levels and falls as per paving manufacturer's instructions							
150 mm thick	5.10	0.03	0.46	0.51	5.10	m²	6.07
200 mm thick	6.80	0.02	0.31	0.67	6.80	m²	7.78
250 mm thick	8.50	0.02	0.38	0.84	8.50	m²	9.72
350 mm thick	11.90	0.03	0.53	1.15	11.90	m²	13.58
450 mm thick	15.30	0.04	0.74	1.61	15.30	m²	17.65
Subbase replacement system for SUDS paving Charcon Permavoid; Aggregate Industries Ltd; geocellular, interlocking, high strength; high void capacity (92%); typically a 4:1 ratio when compared with alternative granular systems; allows for water storage to be confined at shallow depth within the subbase layer; inclusive of connection, ties, pins; does not include for encapsulation in geotextile/geomembranes							
708 × 354 × 150 mm							
150 mm thick	51.64	0.17	3.08	–	51.64	m²	54.72
300 mm thick	107.12	0.33	6.17	–	107.12	m²	113.29
Encapsulation of Permavoid; 150 mm thick systems; geomembrane geofabric or combinatiion of both for filtration or attenuation of water; please see the manufacturers design guide							
Charcon geomembrane, heat welded	12.70	0.05	0.93	–	12.70	m²	13.63
Charcon non reinforced geotextile	7.87	0.03	0.62	–	7.87	m²	8.49
Charcon reinforced geotextile	4.22	0.03	0.62	–	4.22	m²	4.84

Q PAVING/PLANTING/FENCING/SITE FURNITURE

Item Excluding site overheads and profit	PC £	Labour hours	Labour £	Plant £	Material £	Unit	Total rate £
Encapsulation of Permavoid; 300 mm thick systems; geomembrane geofabric or combinatiion of both for filtration or attenuation of water; please see the manufacturers design guide							
Charcon geomembrane, heat welded	14.35	0.06	1.03	–	14.35	m²	**15.38**
Charcon non reinforced geotextile	8.89	0.04	0.79	–	8.89	m²	**9.68**
Charcon reinforced geotextile	3.98	0.04	0.66	–	3.98	m²	**4.64**
Pervious surfacing materials; Aggregate Industries Ltd; Charcon Infilta CBPP Rectangular permeable block paving system; incorporating a 5 mm spacer design that provides a 5 mm void allowing ingress of water through to the sub base storage system (priced separately); various colours; bedded on 50 mm thick 2–6.3 mm clean, angular, free draining uncompacted aggregate; joints infilled with 3 mm clean grit							
Infilta; 200 × 100 × 80 mm thick; natural grey	14.25	0.44	8.22	0.10	16.92	m²	**25.24**
Infilta; 200 × 100 × 80 mm thick; coloured	14.75	0.44	8.22	0.10	17.42	m²	**25.74**
Slab paving; precast concrete pavings; Charcon Hard Landscaping; on prepared subbase (not included); bedding on 25 mm thick cement: sand mortar (1:4); butt joints; straight both ways; on 50 mm thick sharp sand base							
Pavings; natural grey							
450 × 450 × 70 mm chamfered	16.05	0.44	8.22	–	20.08	m²	**28.30**
450 × 450 × 50 mm chamfered	12.54	0.44	8.22	–	16.58	m²	**24.80**
600 × 300 × 50 mm	10.89	0.44	8.22	–	14.92	m²	**23.14**
400 × 400 × 65 mm chamfered	21.81	0.40	7.40	–	25.85	m²	**33.25**
450 × 600 × 50 mm	10.44	0.44	8.22	–	14.48	m²	**22.70**
600 × 600 × 50 mm	8.56	0.40	7.40	–	12.59	m²	**19.99**
750 × 600 × 50 mm	8.18	0.40	7.40	–	12.21	m²	**19.61**
900 × 600 × 50 mm	7.00	0.40	7.40	–	11.03	m²	**18.43**
Pavings; coloured							
450 × 450 × 70 mm chamfered	21.38	0.44	8.22	–	25.42	m²	**33.64**
450 × 600 × 50 mm	12.93	0.44	8.22	–	16.96	m²	**25.18**
400 × 400 × 65 mm chamfered	21.81	0.40	7.40	–	25.85	m²	**33.25**
600 × 600 × 50 mm	10.39	0.40	7.40	–	14.42	m²	**21.82**
750 × 600 × 50 mm	9.62	0.40	7.40	–	13.66	m²	**21.06**
900 × 600 × 50 mm	8.27	0.40	7.40	–	12.30	m²	**19.70**
Precast concrete pavings; Charcon Hard Landscaping; on prepared subbase (not included); bedding on 25 mm thick cement: sand mortar (1:4); butt joints; straight both ways; jointing in cement: sand (1:3) brushed in; on 50 mm thick sharp sand base							
Appalacian rough textured exposed aggregate pebble paving							
600 × 600 × 65 mm	27.37	0.50	9.25	–	29.87	m²	**39.12**

Q PAVING/PLANTING/FENCING/SITE FURNITURE

Item Excluding site overheads and profit	PC £	Labour hours	Labour £	Plant £	Material £	Unit	Total rate £
Q25 SLAB/BRICK/SETT/COBBLE PAVINGS – cont							
Pavings; Marshalls Plc; spot bedding on 5 nr pads of cement: sand mortar (1:4); on sharp sand							
Blister Tactile pavings; specially textured slabs for guidance of blind pedestrians; red or buff							
400 × 400 × 50 mm	27.87	0.50	9.25	–	29.45	m²	38.70
450 × 450 × 50 mm	24.55	0.50	9.25	–	26.13	m²	35.38
Metric Four Square pavings							
496 × 496 × 50 mm; exposed river gravel aggregate	95.41	0.50	9.25	–	96.99	m²	106.24
Metric Four Square cycle blocks							
496 × 496 × 50 mm; exposed aggregate	95.41	0.25	4.63	–	97.57	m²	102.20
Precast concrete pavings; Marshalls Plc; Heritage imitation riven yorkstone paving; on prepared subbase measured separately; bedding on 25 mm thick cement: sand mortar (1:4); pointed straight both ways cement: sand (1:3)							
Square and rectangular paving							
450 × 300 × 38 mm	36.61	1.00	18.50	–	40.16	m²	58.66
450 × 450 × 38 mm	22.54	0.75	13.88	–	26.10	m²	39.98
600 × 300 × 38 mm	27.03	0.80	14.80	–	30.59	m²	45.39
600 × 450 × 38 mm	24.90	0.75	13.88	–	28.46	m²	42.34
600 × 600 × 38 mm	23.58	0.50	9.25	–	27.19	m²	36.44
Extra labours for laying the a selection of the above sizes to random rectangular pattern	–	0.33	6.17	–	–	m²	6.17
Radial paving for circles							
circle with centre stone and first ring (8 slabs); 450 × 230/560 × 38 mm; diameter 1.54 m (total area 1.86 m²)	63.70	1.50	27.75	–	68.33	nr	96.08
circle with second ring (16 slabs); 450 × 300/460 × 38 mm; diameter 2.48 m (total area 4.83 m²)	63.70	4.00	74.00	–	176.06	nr	250.06
circle with third ring (16 slabs); 450 × 470/ 625 × 38 mm; diameter 3.42 m (total area 9.18 m²)	63.70	8.00	148.00	–	321.13	nr	469.13
Stepping stones							
380 mm diameter × 38 mm	4.60	0.20	3.70	–	9.61	nr	13.31
asymmetrical; 560 × 420 × 38 mm	6.15	0.20	3.70	–	11.16	nr	14.86
Precast concrete pavings; Marshalls Plc; Chancery imitation reclaimed riven yorkstone paving; on prepared subbase measured separately; bedding on 25 mm thick cement: sand mortar (1:4); pointed straight both ways cement: sand (1:3)							
Square and rectangular paving							
300 × 300 × 45 mm	25.89	1.00	18.50	–	29.44	m²	47.94
450 × 300 × 45 mm	24.41	0.90	16.65	–	27.97	m²	44.62
600 × 300 × 45 mm	23.68	0.80	14.80	–	27.24	m²	42.04
600 × 450 × 45 mm	23.79	0.75	13.88	–	27.34	m²	41.22
450 × 450 × 45 mm	21.46	0.75	13.88	–	25.02	m²	38.90
600 × 600 × 45 mm	23.74	0.50	9.25	–	27.30	m²	36.55
Extra labours for laying the a selection of the above sizes to random rectangular pattern	–	0.33	6.17	–	–	m²	6.17

Q PAVING/PLANTING/FENCING/SITE FURNITURE

Item Excluding site overheads and profit	PC £	Labour hours	Labour £	Plant £	Material £	Unit	Total rate £
Radial paving for circles							
circle with centre stone and first ring (8 slabs); 450 × 230/560 × 38 mm; diameter 1.54 m (total area 1.86 m^2)	78.35	1.50	27.75	–	82.98	nr	110.73
circle with second ring (16 slabs); 450 × 300/460 × 38 mm; diameter 2.48 m (total area 4.83 m^2)	202.22	4.00	74.00	–	214.11	nr	288.11
circle with third ring (16 slabs); 450 × 470/625 × 38 mm; diameter 3.42 m (total area 9.18 m^2)	368.12	8.00	148.00	–	390.50	nr	538.50
Squaring off set for 2 ring circle							
16 slabs; 2.72 m^2	215.30	1.00	18.50	–	215.30	nr	233.80
Marshalls Plc; La Linia Paving; fine textured exposed aggregate 80 mm thick pavings in various sizes to designed laying patterns; laid to 50 mm sharp sand bed on Type 1 base all priced separately							
Bonded laying patterns; 300 × 300 mm							
light granite/anthracite basalt	32.73	0.75	13.88	–	32.73	m^2	46.61
Indian granite/yellow	32.38	0.75	13.88	–	32.38	m^2	46.26
Random scatter pattern incorporating 100 × 200 mm, 200 × 200 mm and 300 × 200 mm units							
light granite/anthracite basalt	31.93	1.00	18.50	–	31.93	m^2	50.43
Indian granite/yellow	31.59	1.00	18.50	–	31.59	m^2	50.09
Marshalls Plc; La Linia Grande Paving; as above but 140 mm thick							
Bonded laying patterns; 300 × 300 mm							
all colours	61.48	0.75	13.88	–	61.48	m^2	75.36
Random scatter pattern incorporating 100 × 200 mm, 200 × 200 mm and 300 × 200 mm units							
light granite/anthracite basalt	61.48	1.00	18.50	–	61.48	m^2	79.98
Extra over to the above for incorporating inlay stones to patterns; blue, light granite or anthracite basalt							
to prescribed patterns	387.63	1.50	27.75	–	387.63	m^2	415.38
as individual units; triangular 200 × 200 × 282 mm	2.19	0.05	0.93	–	2.19	nr	3.12
Pedestrian deterrent pavings; Marshalls Plc; on prepared base (not included); bedding on 25 mm cement: sand (1:3); cement: sand (1:3) joints							
Lambeth pyramidal pavings							
600 × 600 × 75 mm	18.16	0.25	4.63	–	21.72	m^2	26.35
Thaxted pavings; granite sett appearance							
600 × 600 × 75 mm	21.42	0.33	6.17	–	24.98	m^2	31.15

Q PAVING/PLANTING/FENCING/SITE FURNITURE

Item Excluding site overheads and profit	PC £	Labour hours	Labour £	Plant £	Material £	Unit	Total rate £
Q25 SLAB/BRICK/SETT/COBBLE PAVINGS – cont							
Pedestrian deterrent pavings; Townscape Products Ltd; on prepared base (not included); bedding on 25 mm cement: sand (1:3); cement: sand (1:3) joints							
Strata striated textured slab pavings; giving bonded appearance; grey							
600 × 600 × 60 mm	38.17	0.50	9.25	–	41.73	m²	50.98
Geoset raised chamfered studs pavings; grey							
600 × 600 × 60 mm	38.17	0.50	9.25	–	41.73	m²	50.98
Abbey square cobble pattern pavings; reinforced							
600 × 600 × 65 mm	36.08	0.80	14.80	–	39.64	m²	54.44
Edge restraints; to block paving; on prepared base (not included); 200 × 100 × 80 mm; PC £8.66/m²; haunching one side							
Header course							
200 × 100 mm; butt joints	1.73	0.27	4.93	–	4.06	m	8.99
Stretcher course							
200 × 100 mm; butt joints	0.87	0.17	3.08	–	3.04	m	6.12
Concrete paviors; Marshalls Plc; on prepared base (not included); bedding on 50 mm sand; kiln dried sand joints swept in							
Keyblok paviors							
200 × 100 × 60 mm; grey	7.63	0.40	7.40	0.10	9.37	m²	16.87
200 × 100 × 60 mm; colours	8.45	0.40	7.40	0.10	10.19	m²	17.69
200 × 100 × 80 mm; grey	8.66	0.44	8.22	0.10	10.40	m²	18.72
200 × 100 × 80 mm; colours	9.78	0.44	8.22	0.10	11.52	m²	19.84
Concrete cobble paviors; Charcon Hard Landscaping; Concrete Products; on prepared base (not included); bedding on 50 mm sand; kiln dried sand joints swept in							
Paviors							
Woburn blocks; 100–201 × 134 × 80 mm; random sizes	27.91	0.67	12.33	0.10	29.65	m²	42.08
Woburn blocks; 100–201 × 134 × 80 mm; single size	27.91	0.50	9.25	0.10	29.65	m²	39.00
Woburn blocks; 100–201 × 134 × 60 mm; random sizes	23.00	0.67	12.33	0.10	24.75	m²	37.18
Woburn blocks; 100–201 × 134 × 60 mm; single size	22.98	0.50	9.25	0.10	24.72	m²	34.07
Concrete setts; on 25 mm sand; compacted; vibrated; joints filled with sand; natural or coloured; well rammed hardcore base (not included)							
Marshalls Plc; Tegula cobble paving							
60 mm thick; random sizes	21.56	0.57	10.57	0.10	23.45	m²	34.12
60 mm thick; single size	21.56	0.45	8.41	0.10	23.45	m²	31.96
80 mm thick; random sizes	25.33	0.57	10.57	0.10	27.22	m²	37.89
80 mm thick; single size	25.33	0.45	8.41	0.10	27.22	m²	35.73
cobbles; 80 × 80 × 60 mm thick; traditional	37.47	0.56	10.28	0.10	39.37	m²	49.75

Q PAVING/PLANTING/FENCING/SITE FURNITURE

Item Excluding site overheads and profit	PC £	Labour hours	Labour £	Plant £	Material £	Unit	Total rate £
Cobbles							
Charcon Hard Landscaping; Country setts							
100 mm thick; random sizes	37.19	1.00	18.50	0.10	39.09	m²	57.69
100 mm thick; single size	38.12	0.67	12.33	0.10	40.02	m²	52.45
Natural stone, slab or granite paving –							
General							
Preamble: Provide paving slabs of the							
specified thickness in random sizes but not							
less than 25 slabs per 10 m² of surface area,							
to be laid in parallel courses with joints							
alternately broken and laid to falls.							
Reconstituted yorkstone aggregate							
pavings; Marshalls Plc; Saxon on prepared							
subbase measured separately; bedding on							
25 mm thick cement: sand mortar (1:4); on							
50 mm thick sharp sand base							
Square and rectangular paving in buff; butt							
joints straight both ways							
300 × 300 × 35 mm	33.85	0.88	16.28	–	38.44	m²	54.72
600 × 300 × 35 mm	22.06	0.71	13.23	–	26.65	m²	39.88
450 × 450 × 50 mm	24.15	0.82	15.26	–	28.74	m²	44.00
600 × 600 × 35 mm	16.55	0.55	10.18	–	21.14	m²	31.32
600 × 600 × 50 mm	20.78	0.60	11.19	–	25.37	m²	36.56
Square and rectangular paving in natural; butt							
joints straight both ways							
300 × 300 × 35 mm	27.75	0.88	16.28	–	32.34	m²	48.62
450 × 450 × 50 mm	20.53	0.77	14.24	–	25.12	m²	39.36
600 × 300 × 35 mm	19.43	0.82	15.26	–	24.02	m²	39.28
600 × 600 × 35 mm	14.25	0.55	10.18	–	18.84	m²	29.02
600 × 600 × 50 mm	17.30	0.66	12.21	–	21.89	m²	34.10
Radial paving for circles; 20 mm joints							
circle with centre stone and first ring (8							
slabs); 450 × 230/560 × 35 mm; diameter							
1.54 m (total area 1.86 m²)	65.78	1.50	27.75	–	70.67	nr	98.42
circle with second ring (16 slabs); 450 ×							
300/460 × 35 mm; diameter 2.48 m (total							
area 4.83 m²)	160.34	4.00	74.00	–	172.96	nr	246.96
circle with third ring (24 slabs); 450 × 310/							
430 × 35 mm; diameter 3.42 m (total area							
9.18 m²)	302.18	8.00	148.00	–	326.31	nr	474.31
Granite setts; bedding on 25 mm cement:							
sand (1:3)							
Natural granite setts; 100 × 100 mm to 125 ×							
150 mm × 150–250 mm length; riven surface;							
silver grey							
new; standard grade	36.47	2.00	37.00	–	43.98	m²	80.98
new; high grade	27.71	2.00	37.00	–	35.22	m²	72.22
reclaimed; cleaned	36.26	2.00	37.00	–	43.77	m²	80.77

Q PAVING/PLANTING/FENCING/SITE FURNITURE

Item Excluding site overheads and profit	PC £	Labour hours	Labour £	Plant £	Material £	Unit	Total rate £
Q25 SLAB/BRICK/SETT/COBBLE **PAVINGS – cont**							
Natural stone, slate or granite flag pavings; **CED Ltd; on prepared base (not included);** **bedding on 25 mm cement: sand (1:3);** **cement: sand (1:3) joints**							
Yorkstone; riven laid random rectangular							
new slabs; 40–60 mm thick	78.97	1.71	31.71	–	82.26	m²	**113.97**
reclaimed slabs; Cathedral grade;							
50–75 mm thick	96.29	2.80	51.80	–	99.59	m²	**151.39**
Donegal quartzite slabs; standard tiles							
200 mm × random lengths × 15–25 mm	67.92	3.50	64.75	–	71.21	m²	**135.96**
250 mm × random lengths × 15–25 mm	67.92	3.30	61.05	–	71.21	m²	**132.26**
300 mm × random lengths × 15–25 mm	67.92	3.10	57.35	–	71.21	m²	**128.56**
350 mm × random lengths × 15–25 mm	67.92	2.80	51.80	–	71.21	m²	**123.01**
400 mm × random lengths × 15–25 mm	67.92	2.50	46.25	–	71.21	m²	**117.46**
450 mm × random lengths × 15–25 mm	67.92	2.33	43.17	–	71.21	m²	**114.38**
Natural yorkstone, pavings or edgings; **Johnsons Wellfield Quarries; sawn 6 sides;** **50 mm thick; on prepared base measured** **separately; bedding on 25 mm cement:** **sand (1:3); cement: sand (1:3) joints**							
Paving							
laid to random rectangular pattern	56.65	1.71	31.71	–	59.78	m²	**91.49**
laid to coursed laying pattern; 3 sizes	61.10	1.72	31.90	–	64.23	m²	**96.13**
Paving; single size							
600 × 600 mm	66.09	0.85	15.72	–	69.22	m²	**84.94**
600 × 400 mm	66.09	1.00	18.50	–	69.22	m²	**87.72**
300 × 200 mm	74.81	2.00	37.00	–	77.94	m²	**114.94**
215 × 102.5 mm	77.31	2.50	46.25	–	80.44	m²	**126.69**
Paving; cut to template off site; 600 × 600 mm; radius							
1.00 m	161.50	3.33	61.67	–	164.63	m²	**226.30**
2.50 m	161.50	2.00	37.00	–	164.63	m²	**201.63**
5.00 m	161.50	2.00	37.00	–	164.63	m²	**201.63**
Edgings							
100 mm wide × random lengths	6.98	0.50	9.25	–	7.30	m	**16.55**
100 × 100 mm	7.73	0.50	9.25	–	10.86	m	**20.11**
100 × 200 mm	15.46	0.50	9.25	–	18.59	m	**27.84**
250 mm wide × random lengths	15.27	0.40	7.40	–	18.40	m	**25.80**
500 mm wide × random lengths	30.55	0.33	6.17	–	33.68	m	**39.85**
Yorkstone edgings; 600 mm long × 250 mm wide; cut to radius							
1.00 m to 3.00	48.63	0.50	9.25	–	48.94	m	**58.19**
3.00 m to 5.00 m	48.63	0.44	8.22	–	48.94	m	**57.16**
exceeding 5.00 m	48.63	0.40	7.40	–	48.94	m	**56.34**
Natural yorkstone pavings or edgings; **Johnsons Wellfield Quarries; sawn 6 sides;** **75 mm thick; on prepared base measured** **separately; bedding on 25 mm cement:** **sand (1:3); cement: sand (1:3) joints**							
Paving							
laid to random rectangular pattern	67.98	0.95	17.57	–	71.11	m²	**88.68**
laid to coursed laying pattern; 3 sizes	72.96	0.95	17.57	–	76.09	m²	**93.66**

Q PAVING/PLANTING/FENCING/SITE FURNITURE

Item Excluding site overheads and profit	PC £	Labour hours	Labour £	Plant £	Material £	Unit	Total rate £
Paving; single size							
600 × 600 mm	77.95	0.95	17.57	–	81.08	m²	98.65
600 × 400 mm	77.95	0.95	17.57	–	81.08	m²	98.65
300 × 200 mm	87.78	0.75	13.88	–	90.91	m²	104.79
215 × 102.5 mm	89.78	2.50	46.25	–	92.90	m²	139.15
Paving; cut to template off site; 600 × 600 mm; radius							
1.00 m	190.00	4.00	74.00	–	193.13	m²	267.13
2.50 m	190.00	2.50	46.25	–	193.13	m²	239.38
5.00 m	190.00	2.50	46.25	–	193.13	m²	239.38
Edgings							
100 mm wide × random lengths	8.73	0.60	11.10	–	9.04	m	20.14
100 × 100 mm	0.90	0.60	11.10	–	4.03	m	15.13
100 × 200 mm	1.80	0.50	9.25	–	4.93	m	14.18
250 mm wide × random lengths	18.25	0.50	9.25	–	21.38	m	30.63
500 mm wide × random lengths	36.49	0.40	7.40	–	39.62	m	47.02
Edgings; 600 mm long × 250 mm wide; cut to radius							
1.00 m to 3.00	56.11	0.60	11.10	–	56.42	m	67.52
3.00 m to 5.00 m	56.11	0.50	9.25	–	56.42	m	65.67
exceeding 5.00 m	56.11	0.44	8.22	–	56.42	m	64.64
CED Ltd; Indian sandstone, riven pavings or edgings; 25–35 mm thick; on prepared base measured separately; bedding on 25 mm cement: sand (1:3); cement: sand (1:3) joints							
Paving							
laid to random rectangular pattern	21.50	2.40	44.41	–	24.63	m²	69.04
laid to coursed laying pattern; 3 sizes	21.50	2.00	37.00	–	24.63	m²	61.63
Paving; single size							
600 × 600 mm	21.50	1.00	18.50	–	24.63	m²	43.13
600 × 400 mm	21.50	1.25	23.13	–	24.63	m²	47.76
400 × 400 mm	21.50	1.67	30.83	–	24.63	m²	55.46
Natural stone, slate or granite flag pavings; CED Ltd; on prepared base (not included); bedding on 25 mm cement: sand (1:3); cement: sand (1:3)							
Granite paving; sawn 6 sides; textured top							
new slabs; silver grey; 50 mm thick	38.35	1.71	31.71	–	41.64	m²	73.35
new slabs; blue grey; 50 mm thick	44.96	1.71	31.71	–	48.25	m²	79.96
new slabs; yellow grey; 50 mm thick	47.61	1.71	31.71	–	50.90	m²	82.61
new slabs; black; 50 mm thick	70.09	1.71	31.71	–	73.38	m²	105.09
Edgings; silver grey							
100 mm wide × random lengths	6.77	1.00	18.50	–	7.08	m	25.58
100 × 100 mm	7.74	1.50	27.75	–	8.05	m	35.80
100 mm long × 200 mm wide	12.57	0.60	11.10	–	12.89	m	23.99
250 mm wide × random lengths	13.30	0.75	13.88	–	16.43	m	30.31
300 mm wide × random lengths	14.51	0.75	13.88	–	17.64	m	31.52

Q PAVING/PLANTING/FENCING/SITE FURNITURE

Item Excluding site overheads and profit	PC £	Labour hours	Labour £	Plant £	Material £	Unit	Total rate £
Q25 SLAB/BRICK/SETT/COBBLE PAVINGS – cont							
Cobble pavings – General Cobbles should be embedded by hand, tight-butted, endwise to a depth of 60% of their length. A dry grout of rapid-hardening cement: sand (1:2) shall be brushed over the cobbles until the interstices are filled to the level of the adjoining paving. Surplus grout shall then be brushed off and a light, fine spray of water applied over the area.							
Cobble pavings Cobbles; to present a uniform colour in panels; or varied in colour as required							
Scottish Beach Cobbles; 50–75 mm	21.50	3.33	61.67	–	27.22	m²	**88.89**
Scottish Beach Cobbles; 75–100 mm	36.92	2.50	46.25	–	42.65	m²	**88.90**
Scottish Beach Cobbles; 100–200 mm	52.20	2.00	37.00	–	57.92	m²	**94.92**
Concrete cycle blocks; Marshalls Plc; bedding in cement: sand (1:4) Metric 4 Square cycle stand blocks; smooth grey concrete							
496 × 496 × 100 mm	40.50	0.25	4.63	–	40.97	nr	**45.60**
Concrete cycle blocks; Townscape Products Ltd; on 100 mm concrete (1:2:4); on 150 mm hardcore; bedding in cement: sand (1:4) Cycle blocks							
Cycle Bloc; in white concrete	29.69	0.25	4.63	–	30.32	nr	**34.95**
Mountain Cycle Bloc; in white concrete	41.07	0.25	4.63	–	41.70	nr	**46.33**
Grass concrete – General Preamble: Grass seed should be a perennial ryegrass mixture, with the proportion depending on expected traffic. Hardwearing winter sportsground mixtures are suitable for public areas. Loose gravel, shingle or sand is liable to be kicked out of the blocks; rammed hoggin or other stabilized material should be specified.							
Grass concrete; Grass Concrete Ltd; on blinded granular Type 1 subbase (not included) Grasscrete in situ concrete continuously reinforced surfacing; including expansion joints at 10 m centres; soiling; seeding							
GC2; 150 mm thick; traffic up to 40.00 tonnes	–	–	–	–	–	m²	**39.86**
GC1; 100 mm thick; traffic up to 13.30 tonnes	–	–	–	–	–	m²	**32.12**
GC3; 76 mm thick; traffic up to 4.30 tonnes	–	–	–	–	–	m²	**28.43**

Q PAVING/PLANTING/FENCING/SITE FURNITURE

Item Excluding site overheads and profit	PC £	Labour hours	Labour £	Plant £	Material £	Unit	Total rate £
Grass concrete; Grass Concrete Ltd; 406 × **406 mm blocks; on 20 mm sharp sand; on** **blinded MOT type 1 subbase (not** **included); level and to falls only; including** **filling with topsoil; seeding with dwarf rye** **grass at PC £4.50/kg** Pavings							
GB103; 103 mm thick	15.75	0.40	7.40	0.18	17.94	m²	**25.52**
GB83; 83 mm thick	13.52	0.36	6.73	0.18	15.71	m²	**22.62**
Grass concrete; Charcon Hard **Landscaping; on 25 mm sharp sand;** **including filling with topsoil; seeding with** **dwarf rye grass at PC £4.50/kg** Grassgrid grass/concrete paving blocks							
366 × 274 × 100 mm thick	16.73	0.38	6.94	0.18	18.92	m²	**26.04**
Grass concrete; Marshalls Plc; on 25 mm **sharp sand; including filling with topsoil;** **seeding with dwarf rye grass at PC £4.50/kg** Concrete grass pavings							
Grassguard 130; for light duty applications (80 mm prepared base not included)	18.54	0.38	6.94	0.18	20.90	m²	**28.02**
Grassguard 160; for medium duty applications (80–150 mm prepared base not included)	21.86	0.46	8.54	0.18	24.22	m²	**32.94**
Grassguard 180; for heavy duty applications (150 mm prepared base not included)	24.48	0.60	11.10	0.18	26.84	m²	**38.12**
Full mortar bedding Extra over pavings for bedding on 25 mm cement: sand (1:4); in lieu of spot bedding on sharp sand	–	0.03	0.46	–	1.07	m²	**1.53**
STEPS							
In situ concrete steps; ready mixed **concrete; Gen 1; on 100 mm hardcore** **base; base concrete thickness 100 mm** **treads and risers cast simultaneously;** **inclusive of A142 mesh within the concrete;** **step riser 170 mm high; inclusive of all** **shuttering and concrete labours; floated** **finish; to graded ground (not included)** Treads 250 mm wide							
one riser	4.24	4.00	74.00	–	7.74	m	**81.74**
two risers	6.79	6.00	111.00	–	12.10	m	**123.10**
three risers	9.83	8.00	148.00	–	23.81	m	**171.81**
four risers	13.20	12.00	222.00	–	28.23	m	**250.23**
six risers	16.72	16.00	296.00	–	33.67	m	**329.67**

Q PAVING/PLANTING/FENCING/SITE FURNITURE

Item Excluding site overheads and profit	PC £	Labour hours	Labour £	Plant £	Material £	Unit	Total rate £
Q25 SLAB/BRICK/SETT/COBBLE PAVINGS – cont							
In situ concrete steps – cont							
Lay surface to concrete in situ steps above							
granite treads; pre-cut off site; 305 mm							
wide × 40 mm thick bullnose finish inclusive							
of 20 mm overhang; granite riser 20 mm							
thick all laid on 15 mm mortar bed	23.29	2.00	37.00	–	24.23	m	**61.23**
riven yorkstone tread and riser 50 mm							
thick; cut on site	33.02	2.50	46.25	–	33.96	m	**80.21**
riven yorkstone treads 225 mm wide ×							
50 mm thick cut on site; two courses brick							
risers laid on flat stretcher bond; bricks PC							
£500.00/1000	16.15	3.00	55.50	–	21.86	m	**77.36**
Marshalls Saxon paving slab 450 ×							
450 mm cut to 225 mm wide × 38 mm thick;							
cut on site; two courses brick risers laid on							
flat stretcher bond; bricks PC £500.00/1000	4.65	3.00	55.50	–	10.36	m	**65.86**
Treads 450 mm wide							
one riser	7.63	5.00	92.50	–	12.01	m	**104.51**
two risers	15.59	7.00	129.50	–	24.53	m	**154.03**
three risers	22.03	9.00	166.50	–	38.80	m	**205.30**
four risers	38.11	13.00	240.50	–	55.96	m	**296.46**
six risers	48.43	18.00	333.00	–	68.07	m	**401.07**
Lay surface to concrete in situ steps							
granite treads; pre-cut off site; 470 mm							
wide × 40 mm thick bullnose finish inclusive							
of 20 mm overhang; granite riser 20 mm							
thick all laid on 15 mm mortar bed	33.31	4.00	74.00	–	34.49	m	**108.49**
riven yorkstone tread and riser 50 mm							
thick; cut on site	44.87	4.50	83.25	–	46.05	m	**129.30**
riven yorkstone treads 460 mm wide 50 mm							
thick cut on site; two courses brick risers							
laid on flat stretcher bond; bricks PC							
£500.00/1000	33.02	5.00	92.50	–	38.92	m	**131.42**
Marshalls Saxon paving slab 450 × 450 ×							
38 mm thick cut on site; two courses brick							
risers laid on flat stretcher bond; bricks PC							
£500.00/1000	9.30	2.75	50.88	–	15.01	m	**65.89**
Specialized mortars; SteinTec Ltd							
Primer for specialized mortar; SteinTec							
Tuffbond priming mortar immediately prior to							
paving material being placed on bedding							
mortar							
maximum thickness 1.5 mm (1.5 kg/m^2)	–	0.02	0.37	–	72.00	m^2	**72.37**
Tuffbed; 2 pack hydraulic mortar for paving							
applications							
30 mm thick	7.20	0.17	3.08	–	7.20	m^2	**10.28**
40 mm thick	9.60	0.20	3.70	–	9.60	m^2	**13.30**
50 mm thick	12.00	0.25	4.63	–	12.00	m^2	**16.63**
SteinTec jointing mortar; joint size							
100 × 100 × 100 mm; 10 mm joints;							
31.24 kg/m^2	1249.60	0.50	9.25	–	1249.60	m^2	**1258.85**
100 × 100 × 100 mm; 20 mm joints;							
31.24 kg/m^2	2200.00	1.00	18.50	–	2200.00	m^2	**2218.50**

Q PAVING/PLANTING/FENCING/SITE FURNITURE

Item Excluding site overheads and profit	PC £	Labour hours	Labour £	Plant £	Material £	Unit	Total rate £
200 × 100 × 65 mm; 10 mm joints; 15.7 kg/m²	628.00	0.10	1.85	–	628.00	m²	629.85
215 × 112.5 × 65 mm; 10 mm joints; 14.3 kg/m²	572.00	0.08	1.54	–	572.00	m²	573.54
300 × 450 × 65 mm; 10 mm joints; 6.24 kg/m²	249.60	0.04	0.74	–	249.60	m²	250.34
300 × 450 × 65 mm; 20 mm joints; 11.98 kg/m²	479.20	0.10	1.85	–	479.20	m²	481.05
450 × 450 × 65 mm; 20 mm joints; 9.75 kg/m²	390.00	0.11	2.06	–	390.00	m²	392.06
450 × 600 × 65 mm; 20 mm joints; 8.6 kg/m²	344.00	0.09	1.68	–	344.00	m²	345.68
600 × 600 × 65 mm; 20 mm joints; 7.43 kg/m²	297.20	0.07	1.32	–	297.20	m²	298.52
Q26 SPECIAL SURFACINGS/PAVINGS FOR SPORT/GENERAL AMENITY							
Market prices of surfacing materials Surfacings; Melcourt Industries Ltd (items labelled FSC are Forest Stewardship Council certified)							
Playbark® 10/50; per 25 m³ load	–	–	–	–	59.60	m³	59.60
Playbark® 10/50; per 80 m³ load	–	–	–	–	46.60	m³	46.60
Playbark® 8/25; per 25 m³ load	–	–	–	–	58.10	m³	58.10
Playbark® 8/25; per 80 m³ load	–	–	–	–	45.10	m³	45.10
Playchips®; per 25 m³ load; FSC	–	–	–	–	37.75	m³	37.75
Playchips®; per 80 m³ load; FSC	–	–	–	–	26.75	m³	26.75
Kushyfall; per 25 m³ load; FSC	–	–	–	–	36.20	m³	36.20
Kushyfall; per 80 m³ load; FSC	–	–	–	–	23.20	m³	23.20
Softfall; per 25 m³ load	–	–	–	–	29.40	m³	29.40
Softfall; per 80 m³ load	–	–	–	–	16.40	m³	16.40
Playsand; per 10 t load	–	–	–	–	94.86	m³	94.86
Playsand; per 20 t load	–	–	–	–	84.42	m³	84.42
Walk Chips; per 25 m³ load; FSC	–	–	–	–	36.70	m³	36.70
Walk Chips; per 80 m³ load; FSC	–	–	–	–	23.70	m³	23.70
Woodfibre; per 25 m³ load; FSC	–	–	–	–	36.70	m³	36.70
Woodfibre; per 80 m³ load; FSC	–	–	–	–	23.70	m³	23.70
Natural sports pitches; Agripower Ltd; **sports pitches; gravel raft construction** Professional standard							
football pitch; 6500 m²	–	–	–	–	–	nr	150000.00
cricket wicket; county standard	–	–	–	–	–	nr	40000.00
Playing fields; Sport England Compliant							
football pitch; 6500 m²	–	–	–	–	–	nr	80000.00
cricket wicket; playing field standard	–	–	–	–	–	nr	25000.00
track and infield	–	–	–	–	–	nr	125000.00
Artificial surfaces and finishes – General Preamble: Advice should also be sought from the Technical Unit for Sport, the appropriate regional office of the Sports Council or the National Playing Fields Association. Some of the following prices include base work whereas others are for a specialist surface only on to a base prepared and costed separately.							

Q PAVING/PLANTING/FENCING/SITE FURNITURE

Item Excluding site overheads and profit	PC £	Labour hours	Labour £	Plant £	Material £	Unit	Total rate £
Q26 SPECIAL SURFACINGS/PAVINGS FOR SPORT/GENERAL AMENITY – cont							
Artificial sports pitches; Agripower Ltd; third generation rubber crumb							
Sports pitches to respective National governing body standards; inclusive of all excavation, drainage, lighting, fencing, etc.							
football pitch; 6500 m²	–	–	–	–	–	nr	490000.00
rugby union multi-use pitch; 120 × 75 m	–	–	–	–	–	nr	590000.00
hockey pitch; water-based; 101.4 × 63 m	–	–	–	–	–	nr	550000.00
athletic track; 8 lane; International amateur athletic federation	–	–	–	–	–	nr	750000.00
Multi-sport pitches; inclusive of all excavation, drainage, lighting, fencing, etc.							
sand dressed; 101.4 × 63 m	–	–	–	–	–	nr	390000.00
sand filled; 101.4 × 63 m	–	–	–	–	–	nr	340000.00
polymeric (synthetic bound rubber)	–	–	–	–	–	m²	130.00
macadam	–	–	–	–	–	m²	80.00
Sports areas							
Sports tracks; polyurethane rubber surfacing; on bitumen-macadam (not included); prices for 5500 m² minimum							
International	–	–	–	–	–	m²	40.00
Club Grade	–	–	–	–	–	m²	31.00
Sports areas; multi-component polyurethane rubber surfacing; on bitumen-macadam; finished with polyurethane or acrylic coat; green or red							
Permaprene	–	–	–	–	–	m²	41.00
Sports areas; Exclusive Leisure Ltd							
Cricketweave artificial wicket system; including proprietary subbase (all by specialist subcontractor)							
28 × 2.74 m; hard porous base	–	–	–	–	–	nr	6702.00
28 × 2.74 m; tarmac base	–	–	–	–	–	nr	8039.00
Cricketweave practice batting end							
11 × 2 .74 m; hard porous base	–	–	–	–	–	each	3515.00
11 × 2 .74 m; tarmac base	–	–	–	–	–	each	4378.00
11 × 3.65 m; hard porous base	–	–	–	–	–	each	4584.00
11 × 3.65 m; tarmac base	–	–	–	–	–	each	5732.00
Cricketweave practice bowling end							
8 × 2.74 m; hard porous base	–	–	–	–	–	each	2639.00
8 × 2.74 m; tarmac base	–	–	–	–	–	each	3265.00
Supply and erection of single bay cricket cage							
18.3 × 3.65 × 3.6 m high	–	–	–	–	–	each	2599.00

Q PAVING/PLANTING/FENCING/SITE FURNITURE

Item Excluding site overheads and profit	PC £	Labour hours	Labour £	Plant £	Material £	Unit	Total rate £
Tennis courts; Duracourt (Spadeoak) Ltd Hard playing surfaces to SAPCA Code of Practice minimum requirements; laid on 65 mm thick macadam base on 150 mm thick stone foundation; to include lines, nets, posts, brick edging, 2.75 m high fence and perimeter drainage (excluding excavation, levelling or additional foundation); based on court size 36.6 × 18.3 m (670 m²)							
Durapore; all weather porous macadam and acrylic colour coating	–	–	–	–	–	m²	**55.56**
Tiger Turf TT; sand filled artificial grass	–	–	–	–	–	m²	**72.39**
Tiger Turf Grand Prix; short pile sand filled artificial grass	–	–	–	–	–	m²	**75.76**
DecoColour; impervious acrylic hardcourt (200 mm foundation)	–	–	–	–	–	m²	**70.71**
DecoTurf; cushioned impervious acrylic tournament surface (200 mm foundation)	–	–	–	–	–	m²	**80.81**
Porous Kushion Kourt; porous cushioned acrylic surface	–	–	–	–	–	m²	**79.12**
EasiClay; synthetic clay system	–	–	–	–	–	m²	**82.49**
Canada Tenn; American green clay; fast dry surface (no macadam but including irrigation)	–	–	–	–	–	m²	**84.17**
Playgrounds; Rubaflex in situ playground **surfacing, porous; on prepared stone/** **granular base Type 1 (not included) and** **macadam base course (not included)** Black							
15 mm thick (0.50 m critical fall height)	–	–	–	–	–	m²	**27.15**
35 mm thick (1.00 m critical fall height)	–	–	–	–	–	m²	**41.85**
60 mm thick (1.50 m critical fall height)	–	–	–	–	–	m²	**55.45**
Playgrounds; Rubaflex in situ playground **surfacing, porous; on prepared stone/** **granular base Type 1 (not included) and** **macadam base course (not included)** Coloured							
15 mm thick (0.50 m critical fall height)	–	–	–	–	–	m²	**64.48**
35 mm thick (1.00 m critical fall height)	–	–	–	–	–	m²	**70.15**
60 mm thick (1.50 m critical fall height)	–	–	–	–	–	m²	**79.19**
Playgrounds; Wicksteed Leisure Ltd Safety tiles; on prepared base (not included)							
1000 × 1000 × 60 mm; red or green	75.00	0.13	2.31	–	75.00	m²	**77.31**
1000 × 1000 × 60 mm; black	72.00	0.13	2.31	–	72.00	m²	**74.31**
1000 × 1000 × 43 mm; red or green	72.00	0.13	2.31	–	72.00	m²	**74.31**
1000 × 1000 × 43 mm; black	66.00	0.13	2.31	–	66.00	m²	**68.31**
Playgrounds; SMP Playgrounds Ltd Tiles; on prepared base (not included)							
Premier 25; 1000 × 1000 × 25 mm; black; for general use; freefall height up to 0.8 m	48.00	0.20	3.70	–	57.00	m²	**60.70**
Premier 70; 1000 × 1000 × 70 mm; black; for higher equipment; freefall height up to 2.6 m	84.00	0.20	3.70	–	93.00	m²	**96.70**

Q PAVING/PLANTING/FENCING/SITE FURNITURE

Item Excluding site overheads and profit	PC £	Labour hours	Labour £	Plant £	Material £	Unit	Total rate £
Q26 SPECIAL SURFACINGS/PAVINGS FOR SPORT/GENERAL AMENITY – cont							
Playgrounds; Melcourt Industries Ltd; specifiers and users should contact the supplier for performance specifications of the materials below							
Play surfaces; on drainage layer (not included); minimum 300 mm settled depth; prices for 80 m³ loads unless specified							
Playbark® 8/25; 8–25 mm particles; red/ brown; 25 m³ loads	17.43	0.35	6.49	–	17.43	m²	23.92
Playbark® 8/25; 8–25 mm particles; red/ brown	14.88	0.35	6.49	–	14.88	m²	21.37
Playbark® 10/50; 10–50 mm particles; red/ brown	17.08	0.35	6.49	–	17.08	m²	23.57
Playchips®; FSC graded woodchips	8.92	0.35	6.49	–	8.92	m²	15.41
Kushyfall; 'fiberized' woodchips	7.73	0.35	6.49	–	7.73	m²	14.22
Softfall; conifer shavings	5.47	0.35	6.49	–	5.47	m²	11.96
Playgrounds; timber edgings							
Timber edging boards; 50 × 50 × 750 mm timber pegs at 1000 mm centres; excavations and hardcore under edgings (not included)							
50 × 150 mm; hardwood (iroko) edge boards	9.09	0.10	1.85	–	9.09	m	10.94
38 × 150 mm; hardwood (iroko) edge boards	6.17	0.10	1.85	–	6.17	m	8.02
50 × 150 mm; treated softwood edge boards	2.94	0.10	1.85	–	2.94	m	4.79
38 × 150 mm; treated softwood edge boards	2.19	0.10	1.85	–	2.19	m	4.04
Q30 SEEDING/TURFING							
Seeding/turfing – General							
Preamble: The following market prices generally reflect the manufacturer's recommended retail prices. Trade and bulk discounts are often available on the prices shown. The manufacturers of these products generally recommend application rates. Note: the following rates reflect the average rate for each product.							
Market prices of pre-seeding materials							
Rigby Taylor Ltd							
turf fertilizer; Mascot Outfield 16+6+6	–	–	–	–	2.49	100 m²	2.49
Boughton Loam							
screened topsoil; 100 mm	–	–	–	–	99.00	m³	99.00
screened Kettering loam; 3 mm	–	–	–	–	117.00	m³	117.00
screened Kettering loam; sterilized; 3 mm	–	–	–	–	120.60	m³	120.60
top dressing; sand soil mixtures; 90/10 to 50/50	–	–	–	–	93.60	m³	93.60

Q PAVING/PLANTING/FENCING/SITE FURNITURE

Item Excluding site overheads and profit	PC £	Labour hours	Labour £	Plant £	Material £	Unit	Total rate £
Market prices of turf fertilizers; **recommended average application rates** Scotts UK; turf fertilizers							
grass fertilizer; slow release; Sierraform GT Preseeder 18+22+05	–	–	–	–	4.66	100 m²	**4.66**
grass fertilizer; slow release; Sierraform GT All Season 18+06+18	–	–	–	–	3.88	100 m²	**3.88**
grass fertilizer; slow release; Sierraform GT Anti-Stress 15+00+26	–	–	–	–	4.66	100 m²	**4.66**
grass fertilizer; slow release; Sierraform GT Momentum 22+05+11	–	–	–	–	3.11	100 m²	**3.11**
grass fertilizer; slow release; Sierraform GT Spring Start 16+00+16	–	–	–	–	3.88	100 m²	**3.88**
grass fertilizer; water soluble; Sierrasol 28 +05+18+TE	–	–	–	–	11.68	100 m²	**11.68**
grass fertilizer; water soluble; Sierrasol 20 +05+30+TE	–	–	–	–	9.08	100 m²	**9.08**
grass fertilizer; controlled release; Sierrablen 27+05+05 (8–9 months)	–	–	–	–	5.41	100 m²	**5.41**
grass fertilizer; controlled release; Sierrablen 28+05+05 (5–6 months)	–	–	–	–	5.16	100 m²	**5.16**
grass fertilizer; controlled release; Sierrablen 15+00+22 (5–6 months)	–	–	–	–	6.45	100 m²	**6.45**
grass fertilizer; Sierrablen Fine 38+00+00 (2–3 months)	–	–	–	–	1.85	100 m²	**1.85**
grass fertilizer; Sierrablen Mini 00+00+37 (5–6 months)	–	–	–	–	3.52	100 m²	**3.52**
grass fertilizer; Greenmaster Pro-Lite Turf Tonic 08+00+00	–	–	–	–	2.30	100 m²	**2.30**
grass fertilizer; Greenmaster Pro-Lite Spring & Summer 14+05+10	–	–	–	–	3.29	100 m²	**3.29**
grass fertilizer; Greenmaster Pro-Lite Zero Phosphate 14+00+10	–	–	–	–	3.29	100 m²	**3.29**
grass fertilizer; Greenmaster Pro-Lite Mosskiller 14+00+00	–	–	–	–	3.66	100 m²	**3.66**
grass fertilizer; Greenmaster Pro-Lite Invigorator 04+00+08	–	–	–	–	2.49	100 m²	**2.49**
grass fertilizer; Greenmaster Pro-Lite Extra 14+02+04	–	–	–	–	4.16	100 m²	**4.16**
grass fertilizer; Greenmaster Pro-Lite Autumn 06+05+10	–	–	–	–	3.54	100 m²	**3.54**
grass fertilizer; Greenmaster Pro-Lite Double K 07+00+14	–	–	–	–	3.54	100 m²	**3.54**
grass fertilizer; Greenmaster Pro-Lite NK 12+00+12	–	–	–	–	3.54	100 m²	**3.54**
outfield turf fertilizer; Sportsmaster Standard Spring & Summer 09+07+07	–	–	–	–	2.69	100 m²	**2.69**
outfield turf fertilizer; Sportsmaster Standard Autumn 04+12+12	–	–	–	–	2.93	100 m²	**2.93**
outfield turf fertilizer; Sportsmaster Standard Zero Phosphate 12+00+09	–	–	–	–	3.06	100 m²	**3.06**
outfield turf fertilizer; Sportsmaster Specialities Municipal 15+05+15	–	–	–	–	5.59	100 m²	**5.59**
TPMC; tree planting and mulching compost	–	–	–	–	4.86	75 l	**4.86**

Q PAVING/PLANTING/FENCING/SITE FURNITURE

Item Excluding site overheads and profit	PC £	Labour hours	Labour £	Plant £	Material £	Unit	Total rate £
Q30 SEEDING/TURFING – cont							
Market prices of turf fertilizers – cont							
Rigby Taylor Ltd; 35 g/m²							
grass fertilizer; Mascot Microfine 14+4+14 + 2% Mg	–	–	–	–	6.26	100 m²	6.26
grass fertilizer; Mascot Microfine 12+0+10 + 2% Mg + 2% Fe	–	–	–	–	3.71	100 m²	3.71
grass fertilizer; Mascot Microfine 8+0+6 + 2% Mg + 4% Fe	–	–	–	–	5.00	100 m²	5.00
grass fertilizer; Mascot Microfine Organic OC1 8+0+0 + 2% Fe	–	–	–	–	5.72	100 m²	5.72
grass fertilizer; Mascot Microfine Organic OC2 5+2+10	–	–	–	–	6.46	100 m²	6.46
grass fertilizer; Mascot Guardian 6+1+12 + 2% Mg + 2% Fe + seaweed	–	–	–	–	7.65	100 m²	7.65
grass fertilizer; Mascot Delta Sport 12+4+8 + 0.5% Fe	–	–	–	–	3.83	100 m²	3.83
grass fertilizer; Mascot Fine Turf 11+5+5	–	–	–	–	4.50	100 m²	4.50
grass fertilizer; Mascot Fine Turf 12+0+9 + 1% Mg + 1% Fe + seaweed	–	–	–	–	4.96	100 m²	4.96
outfield fertilizer; Mascot Outfield 16+6+6	–	–	–	–	3.49	100 m²	3.49
outfield fertilizer; Mascot Outfield 4+10+10	–	–	–	–	2.86	100 m²	2.86
outfield fertilizer; Mascot Outfield 9+5+5	–	–	–	–	3.01	100 m²	3.01
outfield fertilizer; Mascot Outfield 12+4+4	–	–	–	–	3.00	100 m²	3.00
liquid fertilizer; Mascot Microflow-C 25+0+0 + trace elements; 200–1400 ml/100 m²	–	–	–	–	6.21	100 m²	6.21
liquid fertilizer; Mascot Microflow-CX 4+2 +18 + trace elements; 200–1400 ml/100 m²	–	–	–	–	6.18	100 m²	6.18
liquid fertilizer; Mascot Microflow-CX 12+0 +8 + trace elements; 200–1400 ml/100 m²	–	–	–	–	5.99	100 m²	5.99
liquid fertilizer; Mascot Microflow-CX 17+2 +5 + trace elements; 200–1400 ml/100 m²	–	–	–	–	5.91	100 m²	5.91
Market prices of grass seed							
Preamble: The prices shown are for supply only at one number 20 kg or 25 kg bag purchase price unless otherwise stated. Rates shown are based on the manufacturer's maximum recommendation for each seed type. Trade and bulk discounts are often available on the prices shown for quantities of more than one bag.							
Bowling greens; fine lawns; ornamental turf; croquet lawns							
British Seed Houses; A1 Greens; 35 g/m²	–	–	–	–	21.88	100 m²	21.88
DLF Trifolium; J Green; 34–50 g/m²	–	–	–	–	32.30	100 m²	32.30
DLF Trifolium; J Premier Green; 34–50 g/m²	–	–	–	–	55.85	100 m²	55.85
DLF Trifolium; Promaster 20; 35–50 g/m²	–	–	–	–	16.30	100 m²	16.30
DLF Trifolium; Promaster 10; 35–50 g/m²	–	–	–	–	31.05	100 m²	31.05
Tennis courts; cricket squares							
British Seed Houses; A2 Lawns & Tennis; 35 g/m²	–	–	–	–	14.44	100 m²	14.44
British Seed Houses; A5 Cricket Square; 35 g/m²	–	–	–	–	21.25	100 m²	21.25
DLF Trifolium; J Premier Green; 34–50 g/m²	–	–	–	–	55.85	100 m²	55.85
DLF Trifolium; J Court; 18–25 g/m²	–	–	–	–	8.65	100 m²	8.65
DLF Trifolium; Promaster 35; 35 g/m²	–	–	–	–	12.85	100 m²	12.85

Q PAVING/PLANTING/FENCING/SITE FURNITURE

Item Excluding site overheads and profit	PC £	Labour hours	Labour £	Plant £	Material £	Unit	**Total rate £**
Amenity grassed areas; general purpose lawns							
British Seed Houses; A3 Landscape;							
25–50 g/m^2	–	–	–	–	20.00	100 m^2	**20.00**
DLF Trifolium; J Rye Fairway; 18–25 g/m^2	–	–	–	–	11.50	100 m^2	**11.50**
DLF Trifolium; J Court; 18–25 g/m^2	–	–	–	–	8.65	100 m^2	**8.65**
DLF Trifolium; Promaster 50; 25–35 g/m^2	–	–	–	–	12.36	100 m^2	**12.36**
DLF Trifolium; Promaster 120; 25–35 g/m^2	–	–	–	–	12.74	100 m^2	**12.74**
Rigby Taylor; Mascot R15 General							
Landscape; 35 g/m^2	–	–	–	–	21.30	100 m^2	**21.30**
Conservation; country parks; slopes and banks							
British Seed Houses; A4							
Low-maintenance; 17–35 g/m^2	–	–	–	–	14.04	100 m^2	**14.04**
British Seed Houses; A16 Country Park;							
8–19 g/m^2	–	–	–	–	8.35	100 m^2	**8.35**
British Seed Houses; A17 Legume and							
Clover; 2 g/m^2	–	–	–	–	22.26	100 m^2	**22.26**
Shaded areas							
British Seed Houses; A6 Supra Shade;							
50 g/m^2	–	–	–	–	24.44	100 m^2	**24.44**
DLF Trifolium; J Green; 34–50 g/m^2	–	–	–	–	32.30	100 m^2	**32.30**
DLF Trifolium; Promaster 60; 35–50 g/m^2	–	–	–	–	22.35	100 m^2	**22.35**
Rigby Taylor; Mascot R18 Shade/Drought							
Tolerance; 35 g/m^2	–	–	–	–	34.52	100 m^2	**34.52**
Sports pitches; rugby; soccer pitches							
British Seed Houses; A7 Sportsground;							
20 g/m^2	–	–	–	–	9.75	100 m^2	**9.75**
DLF Trifolium; J Pitch; 18–30 g/m^2	–	–	–	–	9.09	100 m^2	**9.09**
DLF Trifolium; Promaster 70; 15–35 g/m^2	–	–	–	–	9.59	100 m^2	**9.59**
DLF Trifolium; Promaster 75; 15–35 g/m^2	–	–	–	–	15.40	100 m^2	**15.40**
DLF Trifolium; Promaster 80; 17–35 g/m^2	–	–	–	–	9.70	100 m^2	**9.70**
Rigby Taylor; Mascot R11 Football &							
Rugby; 35 g/m^2	–	–	–	–	17.11	100 m^2	**17.11**
Rigby Taylor; Mascot R12 General Playing							
Fields; 35 g/m^2	–	–	–	–	19.19	100 m^2	**19.19**
Outfields							
British Seed Houses; A7 Sportsground;							
20 g/m^2	–	–	–	–	9.75	100 m^2	**9.75**
British Seed Houses; A9 Outfield; 17–35 g/m^2	–	–	–	–	13.13	100 m^2	**13.13**
DLF Trifolium; J Fairway; 12–25 g/m^2	–	–	–	–	9.90	100 m^2	**9.90**
DLF Trifolium; Promaster 40; 35 g/m^2	–	–	–	–	12.63	100 m^2	**12.63**
DLF Trifolium; Promaster 70; 15–35 g/m^2	–	–	–	–	9.59	100 m^2	**9.59**
Rigby Taylor; Mascot R4 Cricket Outfields;							
35 g/m^2	–	–	–	–	25.37	100 m^2	**25.37**
Hockey pitches							
DLF Trifolium; J Fairway; 12–25 g/m^2	–	–	–	–	9.90	100 m^2	**9.90**
DLF Trifolium; Promaster 70; 15–35 g/m^2	–	–	–	–	9.59	100 m^2	**9.59**
Rigby Taylor; Mascot R10 Cricket &							
Hockey Outfield; 35 g/m^2	–	–	–	–	20.39	100 m^2	**20.39**
Parks							
British Seed Houses; A7 Sportsground							
20 g/m^2	–	–	–	–	9.75	100 m^2	**9.75**
British Seed Houses; A9 Outfield; 17–35 g/m^2	–	–	–	–	13.13	100 m^2	**13.13**
DLF Trifolium; Promaster 120; 25–35 g/m^2	–	–	–	–	12.74	100 m^2	**12.74**
Rigby Taylor; Mascot R16 Landscape &							
Ornamental; 35 g/m^2	–	–	–	–	21.26	100 m^2	**21.26**

Q PAVING/PLANTING/FENCING/SITE FURNITURE

Item Excluding site overheads and profit	PC £	Labour hours	Labour £	Plant £	Material £	Unit	Total rate £
Q30 SEEDING/TURFING – cont							
Market prices of grass seed – cont							
Informal playing fields							
DLF Trifolium; J Rye Fairway; 18–25 g/m²	–	–	–	–	11.50	100 m²	**11.50**
DLF Trifolium; Promaster 45; 35 g/m²	–	–	–	–	12.63	100 m²	**12.63**
Caravan sites							
British Seed Houses; A9 Outfield; 17–35 g/m²	–	–	–	–	13.13	100 m²	**13.13**
Sports pitch re-seeding and repair							
British Seed Houses; A8 Pitch Renovator; 20–35 g/m²	–	–	–	–	12.69	100 m²	**12.69**
British Seed Houses; A20 Ryesport; 20–35 g/m²	–	–	–	–	13.56	100 m²	**13.56**
DLF Trifolium; J Pitch; 18–30 g/m²	–	–	–	–	9.09	100 m²	**9.09**
DLF Trifolium; Promaster 80; 17–35 g/m²	–	–	–	–	9.70	100 m²	**9.70**
DLF Trifolium; Promaster 81; 17–35 g/m²	–	–	–	–	11.34	100 m²	**11.34**
Rigby Taylor; Mascot R14 Premier Winter Games Renovation + ESP coating; 35 g/m²	–	–	–	–	18.88	100 m²	**18.88**
Racecourses; gallops; polo grounds; horse rides							
British Seed Houses; Racecourse; 25–30 g/m²	–	–	–	–	12.75	100 m²	**12.75**
DLF Trifolium; J Court; 18–25 g/m²	–	–	–	–	8.65	100 m²	**8.65**
DLF Trifolium; Promaster 65; 17–35 g/m²	–	–	–	–	17.99	100 m²	**17.99**
Motorway and road verges							
British Seed Houses; A18 Road Verge; 6–15 g/m²	–	–	–	–	6.00	100 m²	**6.00**
DLF Trifolium; Promaster 85; 10 g/m²	–	–	–	–	4.03	100 m²	**4.03**
DLF Trifolium; Promaster 120; 25–35 g/m²	–	–	–	–	12.74	100 m²	**12.74**
Golf courses; tees							
British Seed Houses; A10 Golf Tee, 35–50 g/m²	–	–	–	–	22.50	100 m²	**22.50**
DLF Trifolium; J Premier Fairway; 18–30 g/m²	–	–	–	–	18.72	100 m²	**18.72**
DLF Trifolium; J Court; 18–25 g/m²	–	–	–	–	8.65	100 m²	**8.65**
DLF Trifolium; Promaster 40; 35 g/m²	–	–	–	–	12.63	100 m²	**12.63**
DLF Trifolium; Promaster 45; 35 g/m²	–	–	–	–	12.63	100 m²	**12.63**
Rigby Taylor; Mascot R4 Golf Tees & Fairways; 35 g/m²	–	–	–	–	51.45	100 m²	**51.45**
Golf courses; greens							
British Seed Houses; A11 Golf Green; 35 g/m²	–	–	–	–	21.88	100 m²	**21.88**
British Seed Houses; A13 Golf Roughs; 8 g/m²	–	–	–	–	3.04	100 m²	**3.04**
DLF Trifolium; J All Bent; 8 g/m²	–	–	–	–	11.57	100 m²	**11.57**
DLF Trifolium; J Green; 34–50 g/m²	–	–	–	–	32.30	100 m²	**32.30**
DLF Trifolium; Promaster 5; 35–50 g/m²	–	–	–	–	19.85	100 m²	**19.85**
Rigby Taylor; Mascot R1 Greenkeeper; 35 g/m²	–	–	–	–	65.29	100 m²	**65.29**
Golf courses; fairways							
British Seed Houses; A12 Golf Fairway; 15–25 g/m²	–	–	–	–	10.31	100 m²	**10.31**
DLF Trifolium; J Rye Fairway; 18–30 g/m²	–	–	–	–	13.80	100 m²	**13.80**
DLF Trifolium; J Premier Fairway; 18–30 g/m²	–	–	–	–	18.72	100 m²	**18.72**

Q PAVING/PLANTING/FENCING/SITE FURNITURE

Item Excluding site overheads and profit	PC £	Labour hours	Labour £	Plant £	Material £	Unit	Total rate £
DLF Trifolium; Promaster 40; 35 g/m²	–	–	–	–	12.63	100 m²	**12.63**
DLF Trifolium; Promaster 45; 35 g/m²	–	–	–	–	12.63	100 m²	**12.63**
Rigby Taylor; Mascot R6 Fescue Rye Fairway; 35 g/m²	–	–	–	–	50.81	100 m²	**50.81**
Golf courses; roughs							
DLF Trifolium; Promaster 25; 17–35 g/m²	–	–	–	–	13.34	100 m²	**13.34**
Rigby Taylor; Mascot R127 Golf Links & Rough; 35 g/m²	–	–	–	–	4.54	100 m²	**4.54**
Waste land; spoil heaps; quarries							
British Seed Houses; A15 Reclamation; 15–20 g/m²	–	–	–	–	8.10	100 m²	**8.10**
DLF Trifolium; Promaster 95; 12–35 g/m²	–	–	–	–	20.20	100 m²	**20.20**
DLF Trifolium; Promaster 105; 5 g/m²	–	–	–	–	4.05	100m2	**4.05**
Low maintenance; housing estates; amenity grassed areas							
British Seed Houses; A19 Housing Estate; 25–35 g/m²	–	–	–	–	12.69	100 m²	**12.69**
British Seed Houses; A22 Low Maintenance; 25–35 g/m²	–	–	–	–	14.70	100 m²	**14.70**
DLF Trifolium; Promaster 120; 25–35 g/m²	–	–	–	–	12.74	100 m²	**12.74**
Saline coastal; roadside areas							
British Seed Houses; A21 Coastal/Saline Restoration; 15–20 g/m²	–	–	–	–	8.75	100 m²	**8.75**
DLF Trifolium; Promaster 90; 15–35 g/m²	–	–	–	–	13.58	100 m²	**13.58**
Turf production							
British Seed Houses; A25 Ley Mixture; 160 kg/ha	–	–	–	–	744.00	ha	**744.00**
British Seed Houses; A24 Wear & Tear; 185 kg/ha	–	–	–	–	765.90	ha	**765.90**
Market prices of wild flora seed mixtures							
Acid soils							
British Seed Houses; WF1 (Annual Flowering); 1–2 g/m²	–	–	–	–	20.63	100 m²	**20.63**
Neutral soils							
British Seed Houses; WF3 (Neutral Soils); 0.5–1 g/m²	–	–	–	–	10.31	100 m²	**10.31**
Market prices of wild flora and grass seed mixtures							
General purpose							
DLF Trifolium; Pro Flora 8 Old English Country Meadow Mix; 5 g/m²	–	–	–	–	12.50	100 m²	**12.50**
DLF Trifolium; Pro Flora 9 General purpose; 5 g/m²	–	–	–	–	10.00	100 m²	**10.00**
Acid soils							
British Seed Houses; WFG2 (Annual Meadow); 5 g/m²	–	–	–	–	22.50	100 m²	**22.50**
DLF Trifolium; Pro Flora 2 Acidic soils; 5 g/m²	–	–	–	–	11.25	100 m²	**11.25**
Neutral soils							
British Seed Houses; WFG4 (Neutral Meadow); 5 g/m²	–	–	–	–	24.70	100 m²	**24.70**
British Seed Houses; WFG13 (Scotland); 5 g/m²	–	–	–	–	21.80	100 m²	**21.80**
DLF Trifolium; Pro Flora 3 Damp loamy soils; 5 g/m²	–	–	–	–	15.75	100 m²	**15.75**

Q PAVING/PLANTING/FENCING/SITE FURNITURE

Item Excluding site overheads and profit	PC £	Labour hours	Labour £	Plant £	Material £	Unit	Total rate £
Q30 SEEDING/TURFING – cont							
Market prices of wild flora and grass seed mixtures – cont							
Calcareous soils							
British Seed Houses; WFG5 (Calcareous Soils); 5 g/m²	–	–	–	–	23.10	100 m²	**23.10**
DLF Trifolium; Pro Flora 4 Calcareous soils; 5 g/m²	–	–	–	–	13.50	100 m²	**13.50**
Heavy clay soils							
British Seed Houses; WFG6 (Clay Soils); 5 g/m²	–	–	–	–	27.75	100 m²	**27.75**
British Seed Houses; WFG12 (Ireland); 5 g/m²	–	–	–	–	21.88	100 m²	**21.88**
DLF Trifolium; Pro Flora 5 Wet loamy soils; 5 g/m²	–	–	–	–	33.00	100 m²	**33.00**
Sandy soils							
British Seed Houses; WFG7 (Free Draining Soils); 5 g/m²	–	–	–	–	32.50	100 m²	**32.50**
British Seed Houses; WFG11 (Ireland); 5 g/m²	–	–	–	–	23.00	100 m²	**23.00**
British Seed Houses; WFG14 (Scotland); 5 g/m²	–	–	–	–	25.25	100 m²	**25.25**
DLF Trifolium; Pro Flora 6 Dry free draining loamy soils; 5 g/m²	–	–	–	–	29.25	100 m²	**29.25**
Shaded areas							
British Seed Houses; WFG8 (Woodland and Hedgerow); 5 g/m²	–	–	–	–	24.25	100 m²	**24.25**
DLF Trifolium; Pro Flora 7 Hedgerow and light shade; 5 g/m²	–	–	–	–	33.75	100 m²	**33.75**
Educational							
British Seed Houses; WFG15 (Schools and Colleges); 5 g/m²	–	–	–	–	34.00	100 m²	**34.00**
Wetlands							
British Seed Houses; WFG9 (Wetlands and Ponds); 5 g/m²	–	–	–	–	30.00	100 m²	**30.00**
DLF Trifolium; Pro Flora 5 Wet loamy soils; 5 g/m²	–	–	–	–	33.00	100 m²	**33.00**
Scrub and moorland							
British Seed Houses; WFG10 (Cornfield Annuals); 5 g/m²	–	–	–	–	33.05	100 m²	**33.05**
Hedgerow							
DLF Trifolium; Pro Flora 7 Hedgerow and light shade; 5 g/m²	–	–	–	–	33.75	100 m²	**33.75**
Vacant sites							
DLF Trifolium; Pro Flora 1 Cornfield annuals; 5 g/m²	–	–	–	–	30.00	100 m²	**30.00**
Regional Environmental mixes							
British Seed Houses; RE1 (Traditional Hay); 5 g/m²	–	–	–	–	24.95	100 m²	**24.95**
British Seed Houses; RE2 (Lowland Meadow); 5 g/m²	–	–	–	–	32.75	100 m²	**32.75**
British Seed Houses; RE3 (Riverflood Plain/Water Meadow); 5 g/m²	–	–	–	–	32.25	100 m²	**32.25**

Q PAVING/PLANTING/FENCING/SITE FURNITURE

Item Excluding site overheads and profit	PC £	Labour hours	Labour £	Plant £	Material £	Unit	Total rate £
British Seed Houses; RE4 (Lowland Limestone); 5 g/m²	–	–	–	–	31.75	100 m²	**31.75**
British Seed Houses; RE5 (Calcareous Sub-mountain Restoration); 5 g/m²	–	–	–	–	36.10	100 m²	**36.10**
British Seed Houses; RE6 (Upland Limestone); 5 g/m²	–	–	–	–	46.45	100 m²	**46.45**
British Seed Houses; RE7 (Acid Sub-mountain Restoration); 5 g/m²	–	–	–	–	31.25	100 m²	**31.25**
British Seed Houses; RE8 (Coastal Reclamation); 5 g/m²	–	–	–	–	40.45	100 m²	**40.45**
British Seed Houses; RE9 (Farmland Mixture); 5 g/m²	–	–	–	–	27.85	100 m²	**27.85**
British Seed Houses; RE10 (Marginal Land); 5 g/m²	–	–	–	–	41.75	100 m²	**41.75**
British Seed Houses; RE11 (Heath Scrubland); 5 g/m²	–	–	–	–	12.80	100 m²	**12.80**
British Seed Houses; RE12 (Drought Land); 5 g/m²	–	–	–	–	39.50	100 m²	**39.50**
Cultivation							
Ripping up subsoil; using approved subsoiling machine; minimum depth 250 mm below topsoil; at 1.20 m centres; in							
gravel or sandy clay	–	–	–	2.40	–	100 m²	**2.40**
soil compacted by machines	–	–	–	2.80	–	100 m²	**2.80**
clay	–	–	–	3.00	–	100 m²	**3.00**
chalk or other soft rock	–	–	–	6.01	–	100 m²	**6.01**
Extra for subsoiling at 1 m centres	–	–	–	0.60	–	100 m²	**0.60**
Breaking up existing ground; using pedestrian operated tine cultivator or rotavator							
100 mm deep	–	0.22	4.07	2.12	–	100 m²	**6.19**
150 mm deep	–	0.28	5.09	2.65	–	100 m²	**7.74**
200 mm deep	–	0.37	6.78	3.54	–	100 m²	**10.32**
As above but in heavy clay or wet soils							
100 mm deep	–	0.44	8.14	4.25	–	100 m²	**12.39**
150 mm deep	–	0.66	12.21	6.37	–	100 m²	**18.58**
200 mm deep	–	0.82	15.26	7.96	–	100 m²	**23.22**
Breaking up existing ground; using tractor drawn tine cultivator or rotavator							
Single pass							
100 mm deep	–	–	–	0.28	–	100 m²	**0.28**
150 mm deep	–	–	–	0.36	–	100 m²	**0.36**
200 mm deep	–	–	–	0.47	–	100 m²	**0.47**
600 mm deep	–	–	–	1.42	–	100 m²	**1.42**
Cultivating ploughed ground; using disc, drag or chain harrow							
4 passes	–	–	–	1.71	–	100 m²	**1.71**
Rolling cultivated ground lightly; using self-propelled agricultural roller	–	0.06	1.03	0.60	–	100 m²	**1.63**
Importing and storing selected and approved topsoil; from source not exceeding 13 km from site; inclusive of settlement							
small quantities (less than 15 m³)	60.00	–	–	–	60.00	m³	**60.00**
over 15 m³	26.40	–	–	–	26.40	m³	**26.40**

Q PAVING/PLANTING/FENCING/SITE FURNITURE

Item Excluding site overheads and profit	PC £	Labour hours	Labour £	Plant £	Material £	Unit	Total rate £
Q30 SEEDING/TURFING – cont							
Cultivation – cont							
Spreading and lightly consolidating approved topsoil (imported or from spoil heaps); in layers not exceeding 150 mm; travel distance from spoil heaps not exceeding 100 m; by machine (imported topsoil not included)							
minimum depth 100 mm	–	1.55	28.68	37.35	–	100 m²	66.03
minimum depth 150 mm	–	2.33	43.17	56.21	–	100 m²	99.38
minimum depth 300 mm	–	4.67	86.33	112.41	–	100 m²	198.74
minimum depth 450 mm	–	6.99	129.31	168.44	–	100 m²	297.75
Spreading and lightly consolidating approved topsoil (imported or from spoil heaps); in layers not exceeding 150 mm; travel distance from spoil heaps not exceeding 100 m; by hand (imported topsoil not included)							
minimum depth 100 mm	–	20.00	370.07	–	–	100 m²	370.07
minimum depth 150 mm	–	30.01	555.11	–	–	100 m²	555.11
minimum depth 300 mm	–	60.01	1110.22	–	–	100 m²	1110.22
minimum depth 450 mm	–	90.02	1665.33	–	–	100 m²	1665.33
Extra over for spreading topsoil to slopes 15–30°; by machine or hand	–	–	–	–	–	10%	–
Extra over for spreading topsoil to slopes over 30°; by machine or hand	–	–	–	–	–	25%	–
Extra over for spreading topsoil from spoil heaps; travel exceeding 100 m; by machine							
100–150 m	–	0.01	0.23	0.06	–	m³	0.29
150–200 m	–	0.02	0.34	0.09	–	m³	0.43
200–300 m	–	0.03	0.51	0.14	–	m³	0.65
Extra over spreading topsoil for travel exceeding 100 m; by hand							
100 m	–	0.83	15.42	–	–	m³	15.42
200 m	–	1.67	30.84	–	–	m³	30.84
300 m	–	2.50	46.25	–	–	m³	46.25
Evenly grading; to general surfaces to bring to finished levels							
by machine (tractor mounted rotavator)	–	–	–	0.01	–	m²	0.01
by pedestrian operated rotavator	–	–	0.07	0.04	–	m²	0.11
by hand	–	0.01	0.19	–	–	m²	0.19
Extra over grading for slopes 15–30°; by machine or hand	–	–	–	–	–	10%	–
Extra over grading for slopes over 30°; by machine or hand	–	–	–	–	–	25%	–
Apply screened topdressing to grass surfaces; spread using Tru-Lute							
sand soil mixes 90/10 to 50/50	–	–	0.04	0.03	0.14	m²	0.21
Spread only existing cultivated soil to final levels using Tru-Lute							
cultivated soil	–	–	0.04	0.03	–	m²	0.07
Clearing stones; disposing off site; to distance not exceeding 13 km							
by hand; stones not exceeding 50 mm in any direction; loading to skip 4.6 m³	–	0.01	0.19	0.04	–	m²	0.23
by mechanical stone rake; stones not exceeding 50 mm in any direction; loading to 15 m³ truck by mechanical loader	–	–	0.04	0.08	–	m²	0.12

Q PAVING/PLANTING/FENCING/SITE FURNITURE

Item Excluding site overheads and profit	PC £	Labour hours	Labour £	Plant £	Material £	Unit	Total rate £
Lightly cultivating; weeding; to fallow areas; disposing debris off site							
by hand	–	0.01	0.26	–	0.08	m²	**0.34**
Surface applications; soil additives;							
pre-seeding; material delivered to a							
maximum of 25 m from area of application;							
applied; by machine							
Soil conditioners; to cultivated ground; ground							
limestone; PC £29.00/tonne; including turning in							
0.25 kg/m² = 2.50 tonnes/ha	0.72	–	–	2.92	0.72	100 m²	**3.64**
0.50 kg/m² = 5.00 tonnes/ha	1.45	–	–	2.92	1.45	100 m²	**4.37**
0.75 kg/m² = 7.50 tonnes/ha	2.17	–	–	2.92	2.17	100 m²	**5.09**
1.00 kg/m² = 10.00 tonnes/ha	2.90	–	–	2.92	2.90	100 m²	**5.82**
Soil conditioners; to cultivated ground;							
medium bark; based on deliveries of 25 m³							
loads; PC £37.15/m³; including turning in							
1 m³ per 40 m² = 25 mm thick	0.93	–	–	0.09	0.93	m²	**1.02**
1 m³ per 20 m² = 50 mm thick	1.86	–	–	0.14	1.86	m²	**2.00**
1 m³ per 13.33 m² = 75 mm thick	2.79	–	–	0.20	2.79	m²	**2.99**
1 m³ per 10 m² = 100 mm thick	3.71	–	–	0.26	3.71	m²	**3.97**
Soil conditioners; to cultivated ground;							
mushroom compost; delivered in 25 m³ loads;							
PC £21.20/m³; including turning in							
1 m³ per 40 m² = 25 mm thick	0.53	0.02	0.31	0.09	0.53	m²	**0.93**
1 m³ per 20 m² = 50 mm thick	1.06	0.03	0.57	0.14	1.06	m²	**1.77**
1 m³ per 13.33 m² = 75 mm thick	1.59	0.04	0.74	0.20	1.59	m²	**2.53**
1 m³ per 10 m² = 100 mm thick	2.12	0.05	0.93	0.26	2.12	m²	**3.31**
Soil conditioners; to cultivated ground;							
mushroom compost; delivered in 35 m³ loads;							
PC £11.75/m³; including turning in							
1 m³ per 40 m² = 25 mm thick	0.29	0.02	0.31	0.09	0.29	m²	**0.69**
1 m³ per 20 m² = 50 mm thick	0.59	0.03	0.57	0.14	0.59	m²	**1.30**
1 m³ per 13.33 m² = 75 mm thick	0.88	0.04	0.74	0.20	0.88	m²	**1.82**
1 m³ per 10 m² = 100 mm thick	1.18	0.05	0.93	0.26	1.18	m²	**2.37**
Surface applications and soil additives;							
pre-seeding; material delivered to a							
maximum of 25 m from area of application;							
applied; by hand							
Soil conditioners; to cultivated ground; ground							
limestone; PC £29.00/tonne; including turning in							
0.25 kg/m² = 2.50 tonnes/ha	0.72	1.20	22.20	–	0.72	100 m²	**22.92**
0.50 kg/m² = 5.00 tonnes/ha	1.45	1.33	24.66	–	1.45	100 m²	**26.11**
0.75 kg/m² = 7.50 tonnes/ha	2.17	1.50	27.75	–	2.17	100 m²	**29.92**
1.00 kg/m² = 10.00 tonnes/ha	2.90	1.71	31.71	–	2.90	100 m²	**34.61**
Soil conditioners; to cultivated ground;							
medium bark; based on deliveries of 25 m³							
loads; PC £37.15/m³; including turning in							
1 m³ per 40 m² = 25 mm thick	0.93	0.02	0.41	–	0.93	m²	**1.34**
1 m³ per 20 m² = 50 mm thick	1.86	0.04	0.82	–	1.86	m²	**2.68**
1 m³ per 13.33 m² = 75 mm thick	2.79	0.07	1.23	–	2.79	m²	**4.02**
1 m³ per 10 m² = 100 mm thick	3.71	0.08	1.48	–	3.71	m²	**5.19**

Q PAVING/PLANTING/FENCING/SITE FURNITURE

Item Excluding site overheads and profit	PC £	Labour hours	Labour £	Plant £	Material £	Unit	Total rate £
Q30 SEEDING/TURFING – cont							
Surface applications and soil additives –							
cont							
Soil conditioners; to cultivated ground;							
mushroom compost; delivered in 25 m³ loads;							
PC £21.20/m³; including turning in							
1 m³ per 40 m² = 25 mm thick	0.53	0.02	0.41	–	0.53	m²	**0.94**
1 m³ per 20 m² = 50 mm thick	1.06	0.04	0.82	–	1.06	m²	**1.88**
1 m³ per 13.33 m² = 75 mm thick	1.59	0.07	1.23	–	1.59	m²	**2.82**
1 m³ per 10 m² = 100 mm thick	2.12	0.08	1.48	–	2.12	m²	**3.60**
Soil conditioners; to cultivated ground;							
mushroom compost; delivered in 60 m³ loads;							
PC £9.50/m³; including turning in							
1 m³ per 40 m² = 25 mm thick	0.24	0.02	0.41	–	0.24	m²	**0.65**
1 m³ per 20 m² = 50 mm thick	0.42	0.04	0.82	–	0.42	m²	**1.24**
1 m³ per 13.33 m² = 75 mm thick	0.71	0.07	1.23	–	0.71	m²	**1.94**
1 m³ per 10 m² = 100 mm thick	0.95	0.08	1.48	–	0.95	m²	**2.43**
Preparation of seedbeds – General							
Preamble: For preliminary operations see							
'Cultivation' section.							
Preparation of seedbeds; soil preparation							
Lifting selected and approved topsoil from							
spoil heaps; passing through 6 mm screen;							
removing debris	–	0.08	1.54	4.90	0.01	m³	**6.45**
Topsoil; supply only; PC £22.00/m³; allowing							
for 20% settlement							
25 mm	–	–	–	–	0.66	m²	**0.66**
50 mm	–	–	–	–	1.32	m²	**1.32**
100 mm	–	–	–	–	2.64	m²	**2.64**
150 mm	–	–	–	–	3.96	m²	**3.96**
200 mm	–	–	–	–	5.28	m²	**5.28**
250 mm	–	–	–	–	6.60	m²	**6.60**
300 mm	–	–	–	–	7.92	m²	**7.92**
400 mm	–	–	–	–	10.56	m²	**10.56**
450 mm	–	–	–	–	11.88	m²	**11.88**
Spreading topsoil to form seedbeds (topsoil							
not included); by machine							
25 mm deep	–	–	0.05	0.10	–	m²	**0.15**
50 mm deep	–	–	0.06	0.13	–	m²	**0.19**
75 mm deep	–	–	0.07	0.15	–	m²	**0.22**
100 mm deep	–	0.01	0.09	0.19	–	m²	**0.28**
150 mm deep	–	0.01	0.14	0.29	–	m²	**0.43**
Spreading only topsoil to form seedbeds							
(topsoil not included); by hand							
25 mm deep	–	0.03	0.46	–	–	m²	**0.46**
50 mm deep	–	0.03	0.62	–	–	m²	**0.62**
75 mm deep	–	0.04	0.79	–	–	m²	**0.79**
100 mm deep	–	0.05	0.93	–	–	m²	**0.93**
150 mm deep	–	0.08	1.39	–	–	m²	**1.39**
Bringing existing topsoil to a fine tilth for							
seeding; by raking or harrowing; stones not to							
exceed 6 mm; by machine	–	–	0.07	0.04	–	m²	**0.11**
Bringing existing topsoil to a fine tilth for							
seeding; by raking or harrowing; stones not to							
exceed 6 mm; by hand	–	0.01	0.16	–	–	m²	**0.16**

Q PAVING/PLANTING/FENCING/SITE FURNITURE

Item Excluding site overheads and profit	PC £	Labour hours	Labour £	Plant £	Material £	Unit	Total rate £
Preparation of seedbeds; soil treatments							
For the following operations add or subtract							
the following amounts for every £0.10							
difference in the material cost price							
35 g/m²	–	–	–	–	0.35	100 m²	0.35
50 g/m²	–	–	–	–	0.50	100 m²	0.50
70 g/m²	–	–	–	–	0.70	100 m²	0.70
100 g/m²	–	–	–	–	1.00	100 m²	1.00
125 kg/ha	–	–	–	–	12.50	ha	12.50
150 kg/ha	–	–	–	–	15.00	ha	15.00
175 kg/ha	–	–	–	–	17.50	ha	17.50
200 kg/ha	–	–	–	–	20.00	ha	20.00
225 kg/ha	–	–	–	–	22.50	ha	22.50
250 kg/ha	–	–	–	–	25.00	ha	25.00
300 kg/ha	–	–	–	–	30.00	ha	30.00
350 kg/ha	–	–	–	–	35.00	ha	35.00
400 kg/ha	–	–	–	–	40.00	ha	40.00
500 kg/ha	–	–	–	–	50.00	ha	50.00
700 kg/ha	–	–	–	–	70.00	ha	70.00
1000 kg/ha	–	–	–	–	100.00	ha	100.00
1250 kg/ha	–	–	–	–	125.00	ha	125.00
Pre-seeding fertilizers (12+00+09); PC £0.58/							
kg; to seedbeds; by machine							
35 g/m²	2.04	–	–	0.19	2.04	100 m²	2.23
50 g/m²	2.92	–	–	0.19	2.92	100 m²	3.11
70 g/m²	4.09	–	–	0.19	4.09	100 m²	4.28
100 g/m²	5.84	–	–	0.19	5.84	100 m²	6.03
Pre-seeding fertilizers (12+00+09); PC £0.58/							
kg; to seedbeds; by hand							
35 g/m²	2.04	0.17	3.08	–	2.04	100 m²	5.12
50 g/m²	2.92	0.17	3.08	–	2.92	100 m²	6.00
70 g/m²	4.09	0.17	3.08	–	4.09	100 m²	7.17
100 g/m²	5.84	0.20	3.70	–	5.84	100 m²	9.54
Seeding							
Seeding labours only in two operations; by							
machine (for seed prices see above)							
35 g/m²	–	–	–	0.51	–	100 m²	0.51
Grass seed; spreading in two operations; PC							
£4.50/kg (for changes in material prices							
please refer to table above); by machine							
35 g/m²	–	–	–	0.51	15.75	100 m²	16.26
50 g/m²	–	–	–	0.51	22.50	100 m²	23.01
70 g/m²	–	–	–	0.51	31.50	100 m²	32.01
100 g/m²	–	–	–	0.51	45.00	100 m²	45.51
125 kg/ha	–	–	–	50.95	562.50	ha	613.45
150 kg/ha	–	–	–	50.95	675.00	ha	725.95
200 kg/ha	–	–	–	50.95	900.00	ha	950.95
250 kg/ha	–	–	–	50.95	1125.00	ha	1175.95
300 kg/ha	–	–	–	50.95	1350.00	ha	1400.95
350 kg/ha	–	–	–	50.95	1575.00	ha	1625.95
400 kg/ha	–	–	–	50.95	1800.00	ha	1850.95
500 kg/ha	–	–	–	50.95	2250.00	ha	2300.95
700 kg/ha	–	–	–	50.95	3150.00	ha	3200.95

Q PAVING/PLANTING/FENCING/SITE FURNITURE

Item Excluding site overheads and profit	PC £	Labour hours	Labour £	Plant £	Material £	Unit	Total rate £
Q30 SEEDING/TURFING – cont							
Seeding – cont							
Extra over seeding by machine for slopes over 30° (allowing for the actual area but measured in plan)							
35 g/m²	–	–	–	0.08	2.36	100 m²	2.44
50 g/m²	–	–	–	0.08	3.38	100 m²	3.46
70 g/m²	–	–	–	0.08	4.72	100 m²	4.80
100 g/m²	–	–	–	0.08	6.75	100 m²	6.83
125 kg/ha	–	–	–	7.64	84.38	ha	92.02
150 kg/ha	–	–	–	7.64	101.25	ha	108.89
200 kg/ha	–	–	–	7.64	135.00	ha	142.64
250 kg/ha	–	–	–	7.64	168.75	ha	176.39
300 kg/ha	–	–	–	7.64	202.50	ha	210.14
350 kg/ha	–	–	–	7.64	236.25	ha	243.89
400 kg/ha	–	–	–	7.64	270.00	ha	277.64
500 kg/ha	–	–	–	7.64	337.50	ha	345.14
700 kg/ha	–	–	–	7.64	472.50	ha	480.14
Seeding labours only in two operations; by machine (for seed prices see above)							
35 g/m²	–	0.17	3.08	–	–	100 m²	3.08
Grass seed; spreading in two operations; PC £4.50/kg (for changes in material prices please refer to table above); by hand							
35 g/m²	–	0.17	3.08	–	15.75	100 m²	18.83
50 g/m²	–	0.17	3.08	–	22.50	100 m²	25.58
70 g/m²	–	0.17	3.08	–	31.50	100 m²	34.58
100 g/m²	–	0.20	3.70	–	45.00	100 m²	48.70
125 g/m²	–	0.20	3.70	–	56.25	100 m²	59.95
Extra over seeding by hand for slopes over 30° (allowing for the actual area but measured in plan)							
35 g/m²	2.34	–	0.07	–	2.34	100 m²	2.41
50 g/m²	3.38	–	0.07	–	3.38	100 m²	3.45
70 g/m²	4.72	–	0.07	–	4.72	100 m²	4.79
100 g/m²	6.75	–	0.08	–	6.75	100 m²	6.83
125 g/m²	8.41	–	0.08	–	8.41	100 m²	8.49
Harrowing seeded areas; light chain harrow	–	–	–	0.09	–	100 m²	0.09
Raking over seeded areas							
by mechanical stone rake	–	–	–	2.18	–	100 m²	2.18
by hand	–	0.80	14.80	–	–	100 m²	14.80
Rolling seeded areas; light roller							
by tractor drawn roller	–	–	–	0.54	–	100 m²	0.54
by pedestrian operated mechanical roller	–	0.08	1.54	0.56	–	100 m²	2.10
by hand drawn roller	–	0.17	3.08	–	–	100 m²	3.08
Extra over harrowing, raking or rolling seeded areas for slopes over 30°; by machine or hand	–	–	–	–	–	25%	–
Turf edging; to seeded areas; 300 mm wide	–	0.05	0.88	–	1.85	m²	2.73
Liquid sod; Turf Management Systems							
Spray on grass system of grass plantlets fertilizer, bio-degradable mulch carrier, root enhancer and water							
to prepared ground	–	–	–	–	–	m²	1.80

Q PAVING/PLANTING/FENCING/SITE FURNITURE

Item Excluding site overheads and profit	PC £	Labour hours	Labour £	Plant £	Material £	Unit	Total rate £
Preparation of turf beds							
Rolling turf to be lifted; lifting by hand or							
mechanical turf stripper; stacks to be not more							
than 1 m high							
cutting only preparing to lift; pedestrian turf							
cutter	–	0.75	13.88	9.96	–	100 m²	23.84
lifting and stacking; by hand	–	8.33	154.17	–	–	100 m²	154.17
Rolling up; moving to stacks							
distance not exceeding 100 m	–	2.50	46.25	–	–	100 m²	46.25
extra over rolling and moving turf to stacks							
to transport per additional 100 m	–	0.83	15.42	–	–	100 m²	15.42
Lifting selected and approved topsoil from							
spoil heaps							
passing through 6 mm screen; removing							
debris	–	0.17	3.08	9.80	–	m³	12.88
Extra over lifting topsoil and passing through							
screen for imported topsoil; plus 20%							
allowance for settlement	–	–	–	–	26.40	m³	26.40
Topsoil; PC £22.00/m³; plus 20% allowance							
for settlement							
25 mm deep	–	–	–	–	0.66	m²	0.66
50 mm deep	–	–	–	–	1.32	m²	1.32
100 mm deep	–	–	–	–	2.64	m²	2.64
150 mm deep	–	–	–	–	3.96	m²	3.96
200 mm deep	–	–	–	–	5.28	m²	5.28
250 mm deep	–	–	–	–	6.60	m²	6.60
300 mm deep	–	–	–	–	7.92	m²	7.92
400 mm deep	–	–	–	–	10.56	m²	10.56
450 mm deep	–	–	–	–	11.88	m²	11.88
Spreading topsoil to form turfbeds (topsoil not							
included); by machine							
25 mm deep	–	–	0.05	0.10	–	m²	0.15
50 mm deep	–	–	0.06	0.13	–	m²	0.19
75 mm deep	–	–	0.07	0.15	–	m²	0.22
100 mm deep	–	0.01	0.09	0.19	–	m²	0.28
150 mm deep	–	0.01	0.14	0.29	–	m²	0.43
Spreading topsoil to form turfbeds (topsoil not							
included); by hand							
25 mm deep	–	0.03	0.46	–	–	m²	0.46
50 mm deep	–	0.03	0.62	–	–	m²	0.62
75 mm deep	–	0.04	0.79	–	–	m²	0.79
100 mm deep	–	0.05	0.93	–	–	m²	0.93
150 mm deep	–	0.08	1.39	–	–	m²	1.39
Bringing existing topsoil to a fine tilth for turfing							
by raking or harrowing; stones not to exceed							
6 mm; by machine	–	–	0.07	0.04	–	m²	0.11
Bringing existing topsoil to a fine tilth for turfing							
by raking or harrowing; stones not to exceed							
6 mm; by hand	–	0.01	0.17	–	–	m²	0.17

Q PAVING/PLANTING/FENCING/SITE FURNITURE

Item Excluding site overheads and profit	PC £	Labour hours	Labour £	Plant £	Material £	Unit	Total rate £
Q30 SEEDING/TURFING – cont							
Turfing							
Turfing; laying only; to stretcher bond; butt							
joints; including providing and working from							
barrow plank runs where necessary to							
surfaces not exceeding 30° from horizontal							
specially selected lawn turves from							
previously lifted stockpile	–	0.08	1.39	–	–	m²	1.39
cultivated lawn turves; to large open areas	–	0.06	1.08	–	–	m²	1.08
cultivated lawn turves; to domestic or							
garden areas	–	0.08	1.44	–	–	m²	1.44
road verge quality turf	–	0.04	0.74	–	–	m²	0.74
Industrially grown turf; PC prices listed							
represent the general range of industrial							
turf prices for sportsfields and amenity							
purposes; prices will vary with quantity							
and site location							
Rolawn							
RB Medallion; sports fields, domestic							
lawns, general landscape; full loads							
1720 m²	1.69	0.07	1.29	–	1.69	m²	2.98
RB Medallion; sports fields, domestic							
lawns, general landscape; part loads	1.85	0.07	1.29	–	1.85	m²	3.14
Tensar Ltd							
Tensar Mat 400 reinforced turf for							
embankments; laid to embankments 3.3 m²							
turves	3.50	0.11	2.04	–	3.50	m²	5.54
Inturf							
Inturf Masters; formal lawns, golf greens,							
bowling greens and low maintenance areas	2.42	0.10	1.76	–	2.42	m²	4.18
Inturf Lawn; garden areas and open							
spaces, racecourses and hockey fields	1.67	0.05	0.97	–	1.67	m²	2.64
Inturf Classic; golf tees, surrounds, lawns,							
parks, general purpose landscaping areas							
and winter sports	1.92	0.05	0.97	–	1.92	m²	2.89
Inturf Ornamental; medium fine turf for							
ornamental lawns, fairways and green							
surrounds	1.92	0.05	1.00	–	1.92	m²	2.92
Inturf Custom Grown Turf; tailor made for							
any turfgrass specification	6.42	0.08	1.48	–	6.42	m²	7.90
RTF Rhizomatous Tall Fescue; deep							
rooted self repairing for our changing							
climate	2.42	0.08	1.48	–	2.42	m²	3.90
Reinforced turf; Boddingtons Netlon							
Advanced Turf; Boddingtons Ltd; blended							
mesh elements incorporated into root							
zone; root zone spread and levelled over							
cultivated, prepared and reduced and							
levelled ground (not included); compacted							
with vibratory roller							
Boddingtons ATS 400/B with selected turf and							
fertilizer 100 mm thick							
100–500 m²	25.30	0.03	0.62	0.73	25.30	m²	26.65
over 500 m²	23.65	0.03	0.62	0.73	23.65	m²	25.00

Q PAVING/PLANTING/FENCING/SITE FURNITURE

Item Excluding site overheads and profit	PC £	Labour hours	Labour £	Plant £	Material £	Unit	Total rate £
Boddingtons ATS 400/B with selected turf and fertilizer 150 mm thick							
100–500 m²	31.90	0.04	0.74	0.82	31.90	m²	**33.46**
over 500 m²	30.80	0.04	0.74	0.82	30.80	m²	**32.36**
Boddingtons ATS 400/B with selected turf and fertilizer 200 mm thick							
100–500 m²	35.20	0.05	0.93	0.94	35.20	m²	**37.07**
over 500 m²	33.00	0.05	0.93	0.94	33.00	m²	**34.87**
Firming turves with wooden beater	–	0.01	0.19	–	–	m²	**0.19**
Rolling turfed areas; light roller							
by tractor with turf tyres and roller	–	–	–	0.54	–	100 m²	**0.54**
by pedestrian operated mechanical roller	–	0.08	1.54	0.56	–	100 m²	**2.10**
by hand drawn roller	–	0.17	3.08	–	–	100 m²	**3.08**
Dressing with finely sifted topsoil; brushing into joints	0.03	0.05	0.93	–	0.03	m²	**0.96**
Turfing; laying only							
to slopes over 30°; to diagonal bond (measured as plan area – add 15% to these rates for the incline area of 30° slopes)	–	0.12	2.22	–	–	m²	**2.22**
Extra over laying turfing for pegging down turves							
wooden or galvanized wire pegs; 200 mm long; 2 pegs per 0.50 m²	1.46	0.01	0.25	–	1.46	m²	**1.71**
Artificial grass; Artificial Lawn Company; laid to sharp sand bed priced separately 15 kg kiln sand brushed in per m²							
Leisure Lawn; 24 mm thick artificial sports turf; sand filled	–	–	–	–	–	m²	**27.80**
Budget Grass; for general use; budget surface; sand filled	–	–	–	–	–	m²	**21.95**
Multi Grass; patios, conservatories and pool surrounds; sand filled	–	–	–	–	–	m²	**23.90**
Premier; lawns and patios	–	–	–	–	–	m²	**30.00**
Play Lawn; grass/sand and rubber filled	–	–	–	–	–	m²	**33.85**
Grassflex; safety surfacing for play areas	–	–	–	–	–	m²	**49.50**
Maintenance operations (Note: the following rates apply to aftercare maintenance executed as part of a landscaping contract only)							
Initial cutting; to turfed areas							
20 mm high; using pedestrian guided power driven cylinder mower; including boxing off cuttings (stone picking and rolling not included)	–	0.18	3.33	0.29	–	100 m²	**3.62**
Repairing damaged grass areas							
scraping out; removing slurry; from ruts and holes; average 100 mm deep	–	0.13	2.47	–	–	m²	**2.47**
100 mm topsoil	–	0.13	2.47	–	2.64	m²	**5.11**
Repairing damaged grass areas; sowing grass seed to match existing or as specified; to individually prepared worn patches							
35 g/m²	0.18	0.01	0.19	–	0.18	m²	**0.37**
50 g/m²	0.26	0.01	0.19	–	0.26	m²	**0.45**

Q PAVING/PLANTING/FENCING/SITE FURNITURE

Item Excluding site overheads and profit	PC £	Labour hours	Labour £	Plant £	Material £	Unit	Total rate £
Q30 SEEDING/TURFING – cont							
Maintenance operations (Note: the following rates apply to aftercare maintenance executed as part of a landscaping contract only) – cont							
Sweeping leaves; disposing off site; motorized vacuum sweeper or rotary brush sweeper							
areas of maximum 2500 m² with occasional large tree and established boundary planting; 4.6 m³ (1 skip of material to be removed)	–	0.40	7.40	3.29	–	100 m²	**10.69**
Leaf clearance; clearing grassed area of leaves and other extraneous debris							
Using equipment towed by tractor							
large grassed areas with perimeters of mature trees such as sports fields and amenity areas	–	0.01	0.23	0.04	–	100 m²	**0.27**
large grassed areas containing ornamental trees and shrub beds	–	0.03	0.46	0.06	–	100 m²	**0.52**
Using pedestrian operated mechanical equipment and blowers							
grassed areas with perimeters of mature trees such as sports fields and amenity areas	–	0.04	0.74	0.05	–	100 m²	**0.79**
grassed areas containing ornamental trees and shrub beds	–	0.10	1.85	0.13	–	100 m²	**1.98**
verges	–	0.07	1.23	0.09	–	100 m²	**1.32**
By hand							
grassed areas with perimeters of mature trees such as sports fields and amenity areas	–	0.05	0.93	0.10	–	100 m²	**1.03**
grassed areas containing ornamental trees and shrub beds	–	0.08	1.54	0.17	–	100 m²	**1.71**
verges	–	1.00	18.50	1.99	–	100 m²	**20.49**
Removal of arisings							
areas with perimeters of mature trees	–	0.01	0.10	0.09	1.20	100 m²	**1.39**
areas containing ornamental trees and shrub beds	–	0.02	0.31	0.34	3.00	100 m²	**3.65**
Cutting grass to specified height; per cut							
multi-unit gang mower	–	0.59	10.88	18.31	–	ha	**29.19**
ride-on triple cylinder mower	–	0.01	0.26	0.14	–	100 m²	**0.40**
ride-on triple rotary mower	–	0.01	0.26	–	–	100 m²	**0.26**
pedestrian mower	–	0.18	3.33	0.69	–	100 m²	**4.02**
Cutting grass to banks; per cut							
side arm cutter bar mower	–	0.02	0.43	0.34	–	100 m²	**0.77**
Cutting rough grass; per cut							
power flail or scythe cutter	–	0.04	0.65	–	–	100 m²	**0.65**
Extra over cutting grass for slopes not exceeding 30°	–	–	–	–	–	10%	**–**
Extra over cutting grass for slopes exceeding 30°	–	–	–	–	–	40%	**–**
Cutting fine sward							
pedestrian operated seven-blade cylinder lawn mower	–	0.14	2.59	0.22	–	100 m²	**2.81**
Extra over cutting fine sward for boxing off cuttings							
pedestrian mower	–	0.03	0.52	0.04	–	100 m²	**0.56**

Q PAVING/PLANTING/FENCING/SITE FURNITURE

Item Excluding site overheads and profit	PC £	Labour hours	Labour £	Plant £	Material £	Unit	Total rate £
Cutting areas of rough grass							
scythe	–	1.00	18.50	–	–	100 m²	18.50
sickle	–	2.00	37.00	–	–	100 m²	37.00
petrol operated strimmer	–	0.30	5.56	0.42	–	100 m²	5.98
Cutting areas of rough grass which contain trees or whips							
petrol operated strimmer	–	0.40	7.40	0.56	–	100 m²	7.96
Extra over cutting rough grass for on site raking up and dumping	–	0.33	6.17	–	–	100 m²	6.17
Trimming edge of grass areas; edging tool							
with petrol powered strimmer	–	0.13	2.47	0.19	–	100 m	2.66
by hand	–	0.67	12.33	–	–	100 m	12.33
Marking out pitches using approved line marking compound; including initial setting out and marking							
discus, hammer, javelin or shot put area	4.68	2.00	37.00	–	4.68	nr	41.68
cricket square	3.12	2.00	37.00	–	3.12	nr	40.12
cricket boundary	10.92	8.00	148.00	–	10.92	nr	158.92
grass tennis court	4.68	4.00	74.00	–	4.68	nr	78.68
hockey pitch	15.60	8.00	148.00	–	15.60	nr	163.60
football pitch	15.60	8.00	148.00	–	15.60	nr	163.60
rugby pitch	15.60	8.00	148.00	–	15.60	nr	163.60
eight lane running track; 400 m	31.20	16.00	296.00	–	31.20	nr	327.20
Re-marking out pitches using approved line marking compound							
discus, hammer, javelin or shot put area	3.12	0.50	9.25	–	3.12	nr	12.37
cricket square	3.12	0.50	9.25	–	3.12	nr	12.37
grass tennis court	10.92	1.00	18.50	–	10.92	nr	29.42
hockey pitch	10.92	1.00	18.50	–	10.92	nr	29.42
football pitch	10.92	1.00	18.50	–	10.92	nr	29.42
rugby pitch	10.92	1.00	18.50	–	10.92	nr	29.42
eight lane running track; 400 m	31.20	2.50	46.25	–	31.20	nr	77.45
Rolling grass areas; light roller							
by tractor drawn roller	–	–	–	0.54	–	100 m²	0.54
by pedestrian operated mechanical roller	–	0.08	1.54	0.56	–	100 m²	2.10
by hand drawn roller	–	0.17	3.08	–	–	100 m²	3.08
Aerating grass areas; to a depth of 100 mm							
using tractor-drawn aerator	–	0.06	1.08	1.01	–	100 m²	2.09
using pedestrian-guided motor powered solid or slitting tine turf aerator	–	0.18	3.24	2.96	–	100 m²	6.20
using hollow tine aerator; including sweeping up and dumping corings	–	0.50	9.25	5.92	–	100 m²	15.17
using hand aerator or fork	–	1.67	30.83	–	–	100 m²	30.83
Extra over aerating grass areas for on site sweeping up and dumping corings	–	0.17	3.08	–	–	100 m²	3.08
Switching off dew; from fine turf areas	–	0.20	3.70	–	–	100 m²	3.70
Scarifying grass areas to break up thatch; removing dead grass							
using tractor-drawn scarifier	–	0.07	1.29	0.31	–	100 m²	1.60
using self-propelled scarifier; including removing and disposing of grass on site	–	0.33	6.17	0.15	–	100 m²	6.32
Harrowing grass areas							
using drag mat	–	0.03	0.52	0.30	–	100 m²	0.82
using chain harrow	–	0.04	0.65	0.38	–	100 m²	1.03
using drag mat	–	2.80	51.81	30.25	–	ha	82.06
using chain harrow	–	3.50	64.75	37.81	–	ha	102.56

Q PAVING/PLANTING/FENCING/SITE FURNITURE

Item Excluding site overheads and profit	PC £	Labour hours	Labour £	Plant £	Material £	Unit	Total rate £
Q30 SEEDING/TURFING – cont							
Maintenance operations (Note: the							
following rates apply to aftercare							
maintenance executed as part of a							
landscaping contract only) – cont							
Extra for scarifying and harrowing grass areas							
for disposing excavated material off site; to tip							
not exceeding 13 km; loading by machine							
slightly contaminated	–	–	–	1.68	24.00	m³	**25.68**
rubbish	–	–	–	1.68	24.00	m³	**25.68**
inert material	–	–	–	1.12	8.00	m³	**9.12**
For the following topsoil improvement and							
seeding operations add or subtract the							
following amounts for every £0.10 difference in							
the material cost price							
35 g/m²	–	–	–	–	0.35	100 m²	**0.35**
50 g/m²	–	–	–	–	0.50	100 m²	**0.50**
70 g/m²	–	–	–	–	0.70	100 m²	**0.70**
100 g/m²	–	–	–	–	1.00	100 m²	**1.00**
125 kg/ha	–	–	–	–	12.50	ha	**12.50**
150 kg/ha	–	–	–	–	15.00	ha	**15.00**
175 kg/ha	–	–	–	–	17.50	ha	**17.50**
200 kg/ha	–	–	–	–	20.00	ha	**20.00**
225 kg/ha	–	–	–	–	22.50	ha	**22.50**
250 kg/ha	–	–	–	–	25.00	ha	**25.00**
300 kg/ha	–	–	–	–	30.00	ha	**30.00**
350 kg/ ha	–	–	–	–	35.00	ha	**35.00**
400 kg/ha	–	–	–	–	40.00	ha	**40.00**
500 kg/ha	–	–	–	–	50.00	ha	**50.00**
700 kg/ha	–	–	–	–	70.00	ha	**70.00**
1000 kg/ha	–	–	–	–	100.00	ha	**100.00**
1250 kg/ha	–	–	–	–	125.00	ha	**125.00**
Top dressing fertilizers (7+7+7); PC £1.06/kg;							
to seedbeds; by machine							
35 g/m²	3.71	–	–	0.19	3.71	100 m²	**3.90**
50 g/m²	5.30	–	–	0.19	5.30	100 m²	**5.49**
300 kg/ha	318.00	–	–	18.76	318.00	ha	**336.76**
350 kg/ha	371.00	–	–	18.76	371.00	ha	**389.76**
400 kg/ha	424.00	–	–	18.76	424.00	ha	**442.76**
500 kg/ha	530.00	–	–	30.01	530.00	ha	**560.01**
Top dressing fertilizers (7+7+7); PC £1.06/kg;							
to seedbeds; by hand							
35 g/m²	3.71	0.17	3.08	–	3.71	100 m²	**6.79**
50 g/m²	5.30	0.17	3.08	–	5.30	100 m²	**8.38**
70 g/m²	7.42	0.17	3.08	–	7.42	100 m²	**10.50**
Watering turf; evenly; at a rate of 5 l/m²							
using movable spray lines powering 3 nr							
sprinkler heads with a radius of 15 m and							
allowing for 60% overlap (irrigation							
machinery costs not included)	–	0.02	0.29	–	–	100 m²	**0.29**
using sprinkler equipment and with							
sufficient water pressure to run 1 nr 15 m							
radius sprinkler	–	0.02	0.37	–	–	100 m²	**0.37**
using hand-held watering equipment	–	0.25	4.63	–	–	100 m²	**4.63**

Q PAVING/PLANTING/FENCING/SITE FURNITURE

Item Excluding site overheads and profit	PC £	Labour hours	Labour £	Plant £	Material £	Unit	Total rate £
Q31 PLANTING							
Planting – General							
Preamble: Prices for all planting work are							
deemed to include carrying out planting in							
accordance with all good horticultural practice.							
Q31 PLANTING – MARKET PRICES OF							
PLANTING MATERIALS							
Note: For market prices of landscape							
chemicals please see sections Q30 or Q35.							
Market prices of planting materials (Note:							
the rates shown generally reflect the							
manufacturer's recommended retail prices;							
trade and bulk discounts are often							
available on the prices shown)							
Topsoil							
general purpose screened	–	–	–	–	22.00	m³	**22.00**
50/50; as dug/10 mm screened	–	–	–	–	33.00	m³	**33.00**
Market prices of mulching materials							
Melcourt Industries Ltd; 25 m³ loads (items							
labelled FSC are Forest Stewardship Council							
certified, items marked FT are certified fire							
tested)							
Ornamental Bark Mulch FT	–	–	–	–	54.40	m³	**54.40**
Bark Nuggets® FT	–	–	–	–	51.70	m³	**51.70**
Graded Bark Flakes FT	–	–	–	–	54.30	m³	**54.30**
Amenity Bark Mulch FSC FT	–	–	–	–	37.15	m³	**37.15**
Contract Bark Mulch FSC	–	–	–	–	34.40	m³	**34.40**
Spruce Ornamental FSC FT	–	–	–	–	37.90	m³	**37.90**
Decorative Biomulch®	–	–	–	–	35.55	m³	**35.55**
Rustic Biomulch®	–	–	–	–	39.30	m³	**39.30**
Mulch 2000	–	–	–	–	28.70	m³	**28.70**
Forest BioMulch®	–	–	–	–	33.05	m³	**33.05**
Melcourt Industries Ltd; 50 m³ loads							
Mulch 2000	–	–	–	–	18.20	m³	**18.20**
Melcourt Industries Ltd; 70 m³ loads							
Contract Bark Mulch FSC	–	–	–	–	22.00	m³	**22.00**
Melcourt Industries Ltd; 80 m³ loads							
Ornamental Bark Mulch FT	–	–	–	–	41.40	m³	**41.40**
Bark Nuggets® FT	–	–	–	–	38.70	m³	**38.70**
Graded Bark Flakes FT	–	–	–	–	41.30	m³	**41.30**
Amenity Bark Mulch FSC	–	–	–	–	24.15	m³	**24.15**
Spruce Ornamental FSC FT	–	–	–	–	24.90	m³	**24.90**
Decorative Biomulch®	–	–	–	–	22.55	m³	**22.55**
Rustic Biomulch®	–	–	–	–	26.30	m³	**26.30**
Forest BioMulch®	–	–	–	–	20.05	m³	**20.05**
Mulch; Melcourt Industries Ltd; 25 m³ loads							
Composted Fine Bark FSC	–	–	–	–	33.15	m³	**33.15**
Humus 2000	–	–	–	–	27.85	m³	**27.85**
Spent Mushroom Compost	–	–	–	–	21.20	m³	**21.20**
Topgrow	–	–	–	–	32.05	m³	**32.05**

Q PAVING/PLANTING/FENCING/SITE FURNITURE

Item Excluding site overheads and profit	PC £	Labour hours	Labour £	Plant £	Material £	Unit	Total rate £
Q31 PLANTING – MARKET PRICES OF PLANTING MATERIALS – cont							
Market prices of mulching materials – cont							
Mulch; Melcourt Industries Ltd; 50 m³ loads							
Humus 2000	–	–	–	–	17.35	m³	**17.35**
Mulch; Melcourt Industries Ltd; 60 m³ loads							
Spent Mushroom Compost	–	–	–	–	9.50	m³	**9.50**
Mulch; Melcourt Industries Ltd; 65 m³ loads							
Composted Fine Bark	–	–	–	–	21.15	m³	**21.15**
Super Humus	–	–	–	–	18.60	m³	**18.60**
Topgrow	–	–	–	–	21.55	m³	**21.55**
Market prices of fertilizers							
Fertilizers; British Seed Houses							
Floranid Permanent; 16+7+15+te; slow release fertilizer	39.50	–	–	–	39.50	kg	**39.50**
Floranid Eagle NK; 20+0+18+te; slow release fertilizer	46.75	–	–	–	46.75	kg	**46.75**
Floranid Master Extra; 19+5+10+te; slow release fertilizer	43.75	–	–	–	43.75	kg	**43.75**
Basatop Fair; 23+6+10+te; slow release fertilizer	40.50	–	–	–	40.50	kg	**40.50**
Fertilizers; Scotts UK							
Enmag; controlled release fertilizer (8–9 months); 11+22+09; 70 g/m²	–	–	–	–	14.56	100 m²	**14.56**
Fertilizers; Scotts UK; granular Osmocote Flora; 15+09+11; controlled release fertilizer; costs for recommended application rates							
transplant	–	–	–	–	0.09	nr	**0.09**
whip	–	–	–	–	0.13	nr	**0.13**
feathered	–	–	–	–	0.18	nr	**0.18**
light standard	–	–	–	–	0.18	nr	**0.18**
standard	–	–	–	–	0.22	nr	**0.22**
selected standard	–	–	–	–	0.31	nr	**0.31**
heavy standard	–	–	–	–	0.35	nr	**0.35**
extra heavy standard	–	–	–	–	0.44	nr	**0.44**
16–18 cm girth	–	–	–	–	0.48	nr	**0.48**
18–20 cm girth	–	–	–	–	0.53	nr	**0.53**
20–22 cm girth	–	–	–	–	0.61	nr	**0.61**
22–24 cm girth	–	–	–	–	0.66	nr	**0.66**
24–26 cm girth	–	–	–	–	0.70	nr	**0.70**
Fertilizers; Farmura Environmental Ltd; Seanure Soilbuilder							
soil amelioration; 70 g/m²	–	–	–	–	7.75	100 m²	**7.75**
to plant pits; 300 × 300 × 300 mm	–	–	–	–	0.03	nr	**0.03**
to plant pits; 600 × 600 × 600 mm	–	–	–	–	0.36	nr	**0.36**
to tree pits; 1.00 × 1.00 × 1.00	–	–	–	–	1.66	nr	**1.66**
Fertilizers; Scotts UK							
TPMC tree planting and mulching compost	–	–	–	–	4.86	bag	**4.86**
Fertilizers; Rigby Taylor Ltd; fertilizer application rates 35 g/m² unless otherwise shown							
straight fertilizer; Bone Meal; at 70 g/m²	–	–	–	–	7.76	100 m²	**7.76**
straight fertilizer; Sulphate of Ammonia	–	–	–	–	2.77	100 m²	**2.77**
straight fertilizer; Sulphate of Iron	–	–	–	–	2.68	100 m²	**2.68**

Q PAVING/PLANTING/FENCING/SITE FURNITURE

Item Excluding site overheads and profit	PC £	Labour hours	Labour £	Plant £	Material £	Unit	Total rate £
straight fertilizer; Sulphate of Potash	–	–	–	–	6.20	100 m²	6.20
straight fertilizer; Super Phosphate Powder	–	–	–	–	1.22	100 m²	1.22
liquid fertilizer; Mascot Microflow-C; 25+0 +0 + trace elements; 200–1400 ml/100 m²	–	–	–	–	6.21	100 m²	6.21
liquid fertilizer; Mascot Microflow-CX; 4+2 +18 + trace elements; 200–1400 ml/100 m²	–	–	–	–	6.18	100 m²	6.18
liquid fertilizer; Mascot Microflow-CX; 12+0 +8 + trace elements; 200–1400 ml/100 m²	–	–	–	–	5.99	100 m²	5.99
liquid fertilizer; Mascot Microflow-CX; 17+2 +5 + trace elements; 200–1400 ml/100 m²	–	–	–	–	5.91	100 m²	5.91
Wetting agents; Rigby Taylor Ltd							
wetting agent; Breaker Advance Liquid; 10 l	–	–	–	–	1.77	100 m²	1.77
wetting agent; Breaker Advance Granules; 20 kg	–	–	–	–	14.17	100 m²	14.17

Q31 PLANTING PREPARATION

Site protection; temporary protective fencing
Cleft chestnut rolled fencing; to 100 mm diameter chestnut posts; driving into firm ground at 3 m centres; pales at 50 mm centres

900 mm high	3.12	0.11	1.97	–	3.64	m	5.61
1100 mm high	4.07	0.11	1.97	–	4.59	m	6.56
1500 mm high	5.71	0.11	1.97	–	6.24	m	8.21
Extra over temporary protective fencing for removing and making good (no allowance for re-use of material)	–	0.07	1.23	0.20	–	m	1.43

Cultivation
Ripping up subsoil; using approved subsoiling machine; minimum depth 250 mm below topsoil; at 1.20 m centres; in

gravel or sandy clay	–	–	–	2.40	–	100 m²	2.40
soil compacted by machines	–	–	–	2.80	–	100 m²	2.80
clay	–	–	–	3.00	–	100 m²	3.00
chalk or other soft rock	–	–	–	6.01	–	100 m²	6.01
Extra for subsoiling at 1 m centres	–	–	–	0.60	–	100 m²	0.60
Breaking up existing ground; using pedestrian operated tine cultivator or rotavator							
100 mm deep	–	0.22	4.07	2.12	–	100 m²	6.19
150 mm deep	–	0.28	5.09	2.65	–	100 m²	7.74
200 mm deep	–	0.37	6.78	3.54	–	100 m²	10.32
As above but in heavy clay or wet soils							
100 mm deep	–	0.44	8.14	4.25	–	100 m²	12.39
150 mm deep	–	0.66	12.21	6.37	–	100 m²	18.58
200 mm deep	–	0.82	15.26	7.96	–	100 m²	23.22
Breaking up existing ground; using tractor drawn tine cultivator or rotavator							
100 mm deep	–	–	–	0.28	–	100 m²	0.28
150 mm deep	–	–	–	0.36	–	100 m²	0.36
200 mm deep	–	–	–	0.47	–	100 m²	0.47
600 mm deep	–	–	–	1.42	–	100 m²	1.42
Cultivating ploughed ground; using disc, drag or chain harrow							
4 passes	–	–	–	1.71	–	100 m²	1.71
Rolling cultivated ground lightly; using self-propelled agricultural roller	–	0.06	1.03	0.60	–	100 m²	1.63

Q PAVING/PLANTING/FENCING/SITE FURNITURE

Item Excluding site overheads and profit	PC £	Labour hours	Labour £	Plant £	Material £	Unit	Total rate £
Q31 PLANTING PREPARATION – cont							
Cultivation – cont							
Importing only selected and approved topsoil							
1–14 m³	26.40	–	–	–	26.40	m³	**26.40**
over 15 m³	22.00	–	–	–	22.00	m³	**22.00**
Spreading and lightly consolidating approved topsoil (imported or from spoil heaps); in layers not exceeding 150 mm; travel distance from spoil heaps not exceeding 100 m; by machine (imported topsoil not included)							
minimum depth 100 mm	–	1.55	28.68	37.35	–	100 m²	**66.03**
minimum depth 150 mm	–	2.33	43.17	56.21	–	100 m²	**99.38**
minimum depth 300 mm	–	4.67	86.33	112.41	–	100 m²	**198.74**
minimum depth 450 mm	–	6.99	129.31	168.44	–	100 m²	**297.75**
Spreading and lightly consolidating approved topsoil (imported or from spoil heaps); in layers not exceeding 150 mm; travel distance from spoil heaps not exceeding 100 m; by hand (imported topsoil not included)							
minimum depth 100 mm	–	20.00	370.07	–	–	100 m²	**370.07**
minimum depth 150 mm	–	30.01	555.11	–	–	100 m²	**555.11**
minimum depth 300 mm	–	60.01	1110.22	–	–	100 m²	**1110.22**
minimum depth 450 mm	–	90.02	1665.33	–	–	100 m²	**1665.33**
Extra over for spreading topsoil to slopes 15–30°; by machine or hand	–	–	–	–	–	10%	**–**
Extra over for spreading topsoil to slopes over 30°; by machine or hand	–	–	–	–	–	25%	**–**
Extra over spreading topsoil for travel exceeding 100 m; by machine							
100–150 m	–	0.02	0.37	0.10	–	m³	**0.47**
150–200 m	–	0.03	0.49	0.13	–	m³	**0.62**
200–300 m	–	0.04	0.66	0.18	–	m³	**0.84**
Extra over spreading topsoil for travel exceeding 100 m; by hand							
100 m	–	2.50	46.25	–	–	m³	**46.25**
200 m	–	3.50	64.75	–	–	m³	**64.75**
300 m	–	4.50	83.25	–	–	m³	**83.25**
Evenly grading; to general surfaces to bring to finished levels							
by machine (tractor mounted rotavator)	–	–	–	0.01	–	m²	**0.01**
by pedestrian operated rotavator	–	–	0.07	0.04	–	m²	**0.11**
by hand	–	0.01	0.19	–	–	m²	**0.19**
Extra over grading for slopes 15–30°; by machine or hand	–	–	–	–	–	10%	**–**
Extra over grading for slopes over 30°; by machine or hand	–	–	–	–	–	25%	**–**
Clearing stones; disposing off site; to distance not exceeding 13 km							
by hand; stones not exceeding 50 mm in any direction; loading to skip 4.6 m³	–	0.01	0.19	0.04	–	m²	**0.23**
by mechanical stone rake; stones not exceeding 50 mm in any direction; loading to 15 m³ truck by mechanical loader	–	–	0.04	0.08	–	m²	**0.12**
Lightly cultivating; weeding; to fallow areas; disposing debris off site; to distance not exceeding 13 km							
by hand	–	0.01	0.26	–	0.12	m²	**0.38**

Q PAVING/PLANTING/FENCING/SITE FURNITURE

Item Excluding site overheads and profit	PC £	Labour hours	Labour £	Plant £	Material £	Unit	Total rate £
Preparation of planting operations							
For the following topsoil improvement and							
planting operations add or subtract the							
following amounts for every £0.10 difference in							
the material cost price							
35 g/m²	–	–	–	–	0.35	100 m²	0.35
50 g/m²	–	–	–	–	0.50	100 m²	0.50
70 g/m²	–	–	–	–	0.70	100 m²	0.70
100 g/m²	–	–	–	–	1.00	100 m²	1.00
150 kg/ha	–	–	–	–	15.00	ha	15.00
200 kg/ha	–	–	–	–	20.00	ha	20.00
250 kg/ha	–	–	–	–	25.00	ha	25.00
300 kg/ha	–	–	–	–	30.00	ha	30.00
400 kg/ha	–	–	–	–	40.00	ha	40.00
500 kg/ha	–	–	–	–	50.00	ha	50.00
700 kg/ha	–	–	–	–	70.00	ha	70.00
1000 kg/ha	–	–	–	–	100.00	ha	100.00
1250 kg/ha	–	–	–	–	125.00	ha	125.00
Fertilizers; in top 150 mm of topsoil; at 35 g/m²							
Mascot Microfine; controlled release turf							
fertilizer; 8+0+6 + 2% Mg + 4% Fe	5.00	0.12	2.27	–	5.00	100 m²	7.27
Enmag; controlled release fertilizer							
(8–9 months); 11+22+09	7.64	0.12	2.27	–	7.64	100 m²	9.91
Mascot Outfield; turf fertilizer; 4+10+10	2.86	0.12	2.27	–	2.86	100 m²	5.13
Super Phosphate Powder	4.48	0.12	2.27	–	4.48	100 m²	6.75
Mascot Outfield; turf fertilizer; 9+5+5	3.01	0.12	2.27	–	3.01	100 m²	5.28
Bone meal	4.07	0.12	2.27	–	4.07	100 m²	6.34
Fertilizers; in top 150 mm of topsoil at 70 g/m²							
Mascot Microfine; controlled release turf							
fertilizer; 8+0+6 + 2% Mg + 4% Fe	10.01	0.12	2.27	–	10.01	100 m²	12.28
Enmag; controlled release fertilizer							
(8–9 months); 11+22+09	15.29	0.12	2.27	–	15.29	100 m²	17.56
Mascot Outfield; turf fertilizer; 4+10+10	5.71	0.12	2.27	–	5.71	100 m²	7.98
Super Phosphate Powder	8.97	0.12	2.27	–	8.97	100 m²	11.24
Mascot Outfield; turf fertilizer; 9+5+5	6.02	0.12	2.27	–	6.02	100 m²	8.29
Bone meal	8.14	0.12	2.27	–	8.14	100 m²	10.41
Spreading topsoil; from dump not exceeding							
100 m distance; by machine (topsoil not							
included)							
at 1 m³ per 13 m²; 75 mm thick	–	0.56	10.28	28.01	–	100 m²	38.29
at 1 m³ per 10 m²; 100 mm thick	–	0.89	16.44	28.62	–	100 m²	45.06
at 1 m³ per 6.50 m²; 150 mm thick	–	1.11	20.54	50.35	–	100 m²	70.89
at 1 m³ per 5 m²; 200 mm thick	–	1.48	27.38	66.94	–	100 m²	94.32
Spreading topsoil; from dump not exceeding							
100 m distance; by hand (topsoil not included)							
at 1 m³ per 13 m²; 75 mm thick	–	15.00	277.56	–	–	100 m²	277.56
at 1 m³ per 10 m²; 100 mm thick	–	20.00	370.07	–	–	100 m²	370.07
at 1 m³ per 6.50 m²; 150 mm thick	–	30.01	555.11	–	–	100 m²	555.11
at 1 m³ per 5 m²; 200 mm thick	–	36.67	678.47	–	–	100 m²	678.47
Imported topsoil; tipped 100 m from area of							
application; by machine							
at 1 m³ per 13 m²; 75 mm thick	198.00	0.56	10.28	28.01	198.00	100 m²	236.29
at 1 m³ per 10 m²; 100 mm thick	264.00	0.89	16.44	28.62	264.00	100 m²	309.06
at 1 m³ per 6.50 m²; 150 mm thick	396.00	1.11	20.54	50.35	396.00	100 m²	466.89
at 1 m³ per 5 m²; 200 mm thick	528.00	1.48	27.38	66.94	528.00	100 m²	622.32

Q PAVING/PLANTING/FENCING/SITE FURNITURE

Item Excluding site overheads and profit	PC £	Labour hours	Labour £	Plant £	Material £	Unit	Total rate £
Q31 PLANTING PREPARATION – cont							
Preparation of planting operations – cont							
Imported topsoil; tipped 100 m from area of							
application; by hand							
at 1 m³ per 13 m²; 75 mm thick	198.00	15.00	277.56	–	198.00	100 m²	475.56
at 1 m³ per 10 m²; 100 mm thick	264.00	20.00	370.07	–	264.00	100 m²	634.07
at 1 m³ per 6.50 m²; 150 mm thick	396.00	30.01	555.11	–	396.00	100 m²	951.11
at 1 m³ per 5 m²; 200 mm thick	528.00	36.67	678.47	–	528.00	100 m²	1206.47
Composted bark and manure soil conditioner							
(20 m³ loads); from not further than 25 m from							
location; cultivating into topsoil by machine							
50 mm thick	159.07	2.86	52.86	1.77	159.07	100 m²	213.70
100 mm thick	318.15	6.05	111.91	1.77	318.15	100 m²	431.83
150 mm thick	477.23	8.90	164.74	1.77	477.23	100 m²	643.74
200 mm thick	636.30	12.90	238.74	1.77	636.30	100 m²	876.81
Composted bark soil conditioner (20 m³ loads);							
placing on beds by mechanical loader;							
spreading and rotavating into topsoil by							
machine							
50 mm thick	151.50	–	–	6.29	151.50	100 m²	157.79
100 mm thick	318.15	–	–	10.62	318.15	100 m²	328.77
150 mm thick	477.23	–	–	15.13	477.23	100 m²	492.36
200 mm thick	636.30	–	–	19.55	636.30	100 m²	655.85
Mushroom compost (20 m³ loads); from not							
further than 25 m from location; cultivating into							
topsoil by machine							
50 mm thick	106.00	2.86	52.86	1.77	106.00	100 m²	160.63
100 mm thick	212.00	6.05	111.91	1.77	212.00	100 m²	325.68
150 mm thick	318.00	8.90	164.74	1.77	318.00	100 m²	484.51
200 mm thick	424.00	12.90	238.74	1.77	424.00	100 m²	664.51
Mushroom compost (20 m³ loads); placing on							
beds by mechanical loader; spreading and							
rotavating into topsoil by machine							
50 mm thick	106.00	–	–	6.29	106.00	100 m²	112.29
100 mm thick	212.00	–	–	10.62	212.00	100 m²	222.62
150 mm thick	318.00	–	–	15.13	318.00	100 m²	333.13
200 mm thick	424.00	–	–	19.55	424.00	100 m²	443.55
Manure (60 m³ loads); from not further than							
25 m from location; cultivating into topsoil by							
machine							
50 mm thick	114.90	2.86	52.86	1.77	114.90	100 m²	169.53
100 mm thick	241.29	6.05	111.91	1.77	241.29	100 m²	354.97
150 mm thick	361.94	8.90	164.74	1.77	361.94	100 m²	528.45
200 mm thick	482.58	12.90	238.74	1.77	482.58	100 m²	723.09
Surface applications and soil additives;							
pre-planting; from not further than 25 m							
from location; by machine							
Medium bark soil conditioner; AHS Ltd;							
including turning in to cultivated ground;							
delivered in 15 m³ loads							
1 m³ per 40 m² = 25 mm thick	0.93	–	–	0.09	0.93	m²	1.02
1 m³ per 20 m² = 50 mm thick	1.86	–	–	0.14	1.86	m²	2.00
1 m³ per 13.33 m² = 75 mm thick	2.79	–	–	0.20	2.79	m²	2.99
1 m³ per 10 m² = 100 mm thick	3.71	–	–	0.26	3.71	m²	3.97

Q PAVING/PLANTING/FENCING/SITE FURNITURE

Item Excluding site overheads and profit	PC £	Labour hours	Labour £	Plant £	Material £	Unit	Total rate £
Mushroom compost soil conditioner; AHS Ltd; including turning in to cultivated ground; delivered in 20 m³ loads							
1 m³ per 40 m² = 25 mm thick	0.53	0.02	0.31	–	0.53	m²	**0.84**
1 m³ per 20 m² = 50 mm thick	1.06	0.03	0.57	–	1.06	m²	**1.63**
1 m³ per 13.33 m² = 75 mm thick	1.59	0.04	0.74	–	1.59	m²	**2.33**
1 m³ per 10 m² = 100 mm thick	2.12	0.05	0.93	–	2.12	m²	**3.05**
Mushroom compost soil conditioner; AHS Ltd; including turning in to cultivated ground; delivered in 35 m³ loads							
1 m³ per 40 m² = 25 mm thick	0.29	0.02	0.31	–	0.29	m²	**0.60**
1 m³ per 20 m² = 50 mm thick	0.59	0.03	0.57	–	0.59	m²	**1.16**
1 m³ per 13.33 m² = 75 mm thick	0.88	0.04	0.74	–	0.88	m²	**1.62**
1 m³ per 10 m² = 100 mm thick	1.18	0.05	0.93	–	1.18	m²	**2.11**
Surface applications and soil additives; **pre-planting; from not further than 25 m** **from location; by hand**							
Ground limestone soil conditioner; including turning in to cultivated ground							
0.25 kg/m² = 2.50 tonnes/ha	0.72	1.20	22.20	–	0.72	100 m²	**22.92**
0.50 kg/m² = 5.00 tonnes/ha	1.45	1.33	24.66	–	1.45	100 m²	**26.11**
0.75 kg/m² = 7.50 tonnes/ha	2.17	1.50	27.75	–	2.17	100 m²	**29.92**
1.00 kg/m² = 10.00 tonnes/ha	2.90	1.71	31.71	–	2.90	100 m²	**34.61**
Medium bark soil conditioner; AHS Ltd; including turning in to cultivated ground; delivered in 15 m³ loads							
1 m³ per 40 m² = 25 mm thick	0.93	0.02	0.41	–	0.93	m²	**1.34**
1 m³ per 20 m² = 50 mm thick	1.86	0.04	0.82	–	1.86	m²	**2.68**
1 m³ per 13.33 m² = 75 mm thick	2.79	0.07	1.23	–	2.79	m²	**4.02**
1 m³ per 10 m² = 100 mm thick	3.71	0.08	1.48	–	3.71	m²	**5.19**
Mushroom compost soil conditioner; Melcourt Industries Ltd; including turning in to cultivated ground; delivered in 25 m³ loads							
1 m³ per 40 m² = 25 mm thick	0.53	0.02	0.41	–	0.53	m²	**0.94**
1 m³ per 20 m² = 50 mm thick	1.06	0.04	0.82	–	1.06	m²	**1.88**
1 m³ per 13.33 m² = 75 mm thick	1.59	0.07	1.23	–	1.59	m²	**2.82**
1 m³ per 10 m² = 100 mm thick	2.12	0.08	1.48	–	2.12	m²	**3.60**
Mushroom compost soil conditioner; AHS Ltd; including turning in to cultivated ground; delivered in 35 m³ loads							
1 m³ per 40 m² = 25 mm thick	0.29	0.02	0.41	–	0.29	m²	**0.70**
1 m³ per 20 m² = 50 mm thick	0.59	0.04	0.82	–	0.59	m²	**1.41**
1 m³ per 13.33 m² = 75 mm thick	0.88	0.07	1.23	–	0.88	m²	**2.11**
1 m³ per 10 m² = 100 mm thick	1.18	0.08	1.48	–	1.18	m²	**2.66**
Super Humus mixed bark and manure conditioner; Melcourt Industries Ltd; delivered in 25 m³ loads							
1 m³ per 40 m² = 25 mm thick	0.80	0.02	0.41	–	0.80	m²	**1.21**
1 m³ per 20 m² = 50 mm thick	1.59	0.04	0.82	–	1.59	m²	**2.41**
1 m³ per 13.33 m² = 75 mm thick	2.39	0.07	1.23	–	2.39	m²	**3.62**
1 m³ per 10 m² = 100 mm thick	3.18	0.08	1.48	–	3.18	m²	**4.66**

Q PAVING/PLANTING/FENCING/SITE FURNITURE

Item Excluding site overheads and profit	PC £	Labour hours	Labour £	Plant £	Material £	Unit	Total rate £
Q31 TREE PLANTING							
Tree planting; pre-planting operations							
Excavating tree pits; depositing soil alongside pits; by machine							
600 × 600 × 600 mm deep	–	0.15	2.72	0.65	–	nr	3.37
900 × 900 × 600 mm deep	–	0.33	6.11	1.45	–	nr	7.56
1.00 × 1.00 m × 600 mm deep	–	0.61	11.35	1.80	–	nr	13.15
1.25 × 1.25 m × 600 mm deep	–	0.96	17.77	2.81	–	nr	20.58
1.00 × 1.00 × 1.00 m deep	–	1.02	18.92	3.00	–	nr	21.92
1.50 × 1.50 m × 750 mm deep	–	1.73	31.93	5.06	–	nr	36.99
1.50 × 1.50 × 1.00 m deep	–	2.30	42.46	6.73	–	nr	49.19
1.75 × 1.75 × 1.00 m deep	–	3.13	57.95	9.19	–	nr	67.14
2.00 × 2.00 × 1.00 m deep	–	4.09	75.68	12.00	–	nr	87.68
Excavating tree pits; depositing soil alongside pits; by hand							
600 × 600 × 600 mm deep	–	0.44	8.14	–	–	nr	8.14
900 × 900 × 600 mm deep	–	1.00	18.50	–	–	nr	18.50
1.00 × 1.00 m × 600 mm deep	–	1.13	20.81	–	–	nr	20.81
1.25 × 1.25 m × 600 mm deep	–	1.93	35.70	–	–	nr	35.70
1.00 × 1.00 × 1.00 m deep	–	2.06	38.11	–	–	nr	38.11
1.50 × 1.50 m × 750 mm deep	–	3.47	64.19	–	–	nr	64.19
1.75 × 1.50 m × 750 mm deep	–	4.05	74.92	–	–	nr	74.92
1.50 × 1.50 × 1.00 m deep	–	4.63	85.66	–	–	nr	85.66
2.00 × 2.00 m × 750 mm deep	–	6.17	114.14	–	–	nr	114.14
2.00 × 2.00 × 1.00 m deep	–	8.23	152.26	–	–	nr	152.26
Breaking up subsoil in tree pits; to a depth of 200 mm	–	0.03	0.62	–	–	m²	0.62
Spreading and lightly consolidating approved topsoil (imported or from spoil heaps); in layers not exceeding 150 mm; distance from spoil heaps not exceeding 100 m (imported topsoil not included); by machine							
minimum depth 100 mm	–	1.55	28.68	37.35	–	100 m²	66.03
minimum depth 150 mm	–	2.33	43.17	56.21	–	100 m²	99.38
minimum depth 300 mm	–	4.67	86.33	112.41	–	100 m²	198.74
minimum depth 450 mm	–	6.99	129.31	168.44	–	100 m²	297.75
Spreading and lightly consolidating approved topsoil (imported or from spoil heaps); in layers not exceeding 150 mm; distance from spoil heaps not exceeding 100 m (imported topsoil not included); by hand							
minimum depth 100 mm	–	20.00	370.07	–	–	100 m²	370.07
minimum depth 150 mm	–	30.01	555.11	–	–	100 m²	555.11
minimum depth 300 mm	–	60.01	1110.22	–	–	100 m²	1110.22
minimum depth 450 mm	–	90.02	1665.33	–	–	100 m²	1665.33
Extra for filling tree pits with imported topsoil; PC £22.00/m³; plus allowance for 20% settlement							
depth 100 mm	–	–	–	–	2.64	m²	2.64
depth 150 mm	–	–	–	–	3.96	m²	3.96
depth 200 mm	–	–	–	–	5.28	m²	5.28
depth 300 mm	–	–	–	–	7.92	m²	7.92
depth 400 mm	–	–	–	–	10.56	m²	10.56
depth 450 mm	–	–	–	–	11.88	m²	11.88
depth 500 mm	–	–	–	–	13.20	m²	13.20
depth 600 mm	–	–	–	–	15.84	m²	15.84

Q PAVING/PLANTING/FENCING/SITE FURNITURE

Item Excluding site overheads and profit	PC £	Labour hours	Labour £	Plant £	Material £	Unit	Total rate £
Add or deduct the following amounts for every £0.50 change in the material price of topsoil							
depth 100 mm	–	–	–	–	0.06	m²	0.06
depth 150 mm	–	–	–	–	0.09	m²	0.09
depth 200 mm	–	–	–	–	0.12	m²	0.12
depth 300 mm	–	–	–	–	0.18	m²	0.18
depth 400 mm	–	–	–	–	0.24	m²	0.24
depth 450 mm	–	–	–	–	0.27	m²	0.27
depth 500 mm	–	–	–	–	0.30	m²	0.30
depth 600 mm	–	–	–	–	0.36	m²	0.36
Structural soils (Amsterdam tree sand); Heicom; non compressive soil mixture for tree planting in areas to receive compressive surface treatments							
Backfilling and lightly compacting in layers; excavation, disposal, moving of material from delivery position and surface treatments not included; by machine							
individual treepits	61.05	0.50	9.25	2.20	61.05	m³	72.50
in trenches	61.05	0.42	7.71	1.83	61.05	m³	70.59
Backfilling and lightly compacting in layers; excavation, disposal, moving of material from delivery position and surface treatments not included; by hand							
individual treepits	61.05	1.60	29.60	–	61.05	m³	90.65
in trenches	61.05	1.33	24.61	–	61.05	m³	85.66
Tree staking							
J Toms Ltd; extra over trees for tree stake(s); driving 500 mm into firm ground; trimming to approved height; including two tree ties to approved pattern							
one stake; 1.52 m long × 32 × 32 mm	1.47	0.20	3.70	–	1.47	nr	5.17
two stakes; 1.21 m long × 25 × 25 mm	1.62	0.30	5.55	–	1.62	nr	7.17
two stakes; 1.52 m long × 32 × 32 mm	2.16	0.30	5.55	–	2.16	nr	7.71
three stakes; 1.52 m long × 32 × 32 mm	3.24	0.36	6.66	–	3.24	nr	9.90
Tree anchors							
Platipus Anchors Ltd; extra over trees for tree anchors							
RF1P rootball kit; for 75–220 mm girth × 2–4.5 m high; inclusive of Plati-Mat PM1	28.25	1.00	18.50	–	28.25	nr	46.75
RF2P; rootball kit; for 220–450 mm girth × 4.5-7.5 m high; inclusive of Plati-Mat PM2	48.09	1.33	24.61	–	48.09	nr	72.70
RF3P; rootball kit; for 450–750 mm girth × 7.5 to 12 m high; inclusive of Plati-Mat PM3	101.82	1.50	27.75	–	101.82	nr	129.57
CG1; guy fixing kit; 75–220 mm girth × 2–4.5 m high	18.17	1.67	30.83	–	18.17	nr	49.00
CG2; guy fixing kit; 220–450 mm girth × 4.5–7.5 m high	36.75	2.00	37.00	–	36.75	nr	73.75
installation tools; drive rod for RF1/CG1 kits	–	–	–	–	59.80	nr	59.80
installation tools; drive rod for RF2/CG2 kits	–	–	–	–	88.76	nr	88.76
Extra over trees for land drain to tree pits; 100 mm diameter perforated flexible agricultural drain; including excavating drain trench; laying pipe; backfilling	1.02	1.00	18.50	–	1.02	m	19.52

Q PAVING/PLANTING/FENCING/SITE FURNITURE

Item Excluding site overheads and profit	PC £	Labour hours	Labour £	Plant £	Material £	Unit	Total rate £
Q31 TREE PLANTING – cont							
Platipus anchors; deadman anchoring							
Deadman kits; anchoring to concrete kerbs							
placed in base of tree pit							
Trees 12–25 cm girth; 2.5–4.0 m high	26.62	0.75	13.88	–	41.24	nr	**55.12**
Trees 25–45 cm girth; 4.0–7.5 m high	45.82	1.50	27.75	–	60.44	nr	**88.19**
Trees 45–75 cm girth; 7.5–12.0 m high	150.86	2.00	37.00	–	165.48	nr	**202.48**
Tree planting; tree pit additives							
Melcourt Industries Ltd; Topgrow;							
incorporating into topsoil at 1 part Topgrow to							
3 parts excavated topsoil; supplied in 75 l							
bags; pit size							
600 × 600 × 600 mm	1.51	0.02	0.37	–	1.51	nr	**1.88**
900 × 900 × 900 mm	5.10	0.06	1.11	–	5.10	nr	**6.21**
1.00 × 1.00 × 1.00 m	6.99	0.24	4.44	–	6.99	nr	**11.43**
1.25 × 1.25 × 1.25 m	13.67	0.40	7.40	–	13.67	nr	**21.07**
1.50 × 1.50 × 1.50 m	23.63	0.90	16.65	–	23.63	nr	**40.28**
Melcourt Industries Ltd; Topgrow;							
incorporating into topsoil at 1 part Topgrow to							
3 parts excavated topsoil; supplied in 60 m³							
loose loads; pit size							
600 × 600 × 600 mm	1.16	0.02	0.31	–	1.16	nr	**1.47**
900 × 900 × 900 mm	3.93	0.05	0.93	–	3.93	nr	**4.86**
1.00 × 1.00 × 1.00 m	5.39	0.20	3.70	–	5.39	nr	**9.09**
1.25 × 1.25 × 1.25 m	10.52	0.33	6.17	–	10.52	nr	**16.69**
1.50 × 1.50 × 1.50 m	18.18	0.75	13.88	–	18.18	nr	**32.06**
Tree planting; root barriers							
English Woodlands; Root Director; one-piece							
root control planters for installation at time of							
planting to divert root growth down away from							
pavements and out for anchorage; excavation							
measured separately							
RD 1050; 1050 × 1050 mm to 1300 ×							
1300 mm at base	94.90	0.25	4.63	–	94.90	nr	**99.53**
RD 640; 640 × 640 mm to 870 × 870 mm at							
base	61.64	0.25	4.63	–	61.64	nr	**66.27**
Greenleaf Horticulture; linear root deflection							
barriers; installed to trench measured							
separately							
Re-Root 2000; 1.5 mm thick	8.20	0.05	0.93	–	8.20	m	**9.13**
Re-Root 2000; 1.0 mm thick	4.62	0.05	0.93	–	4.62	m	**5.55**
Re-Root 600; 1.0 mm thick	7.11	0.05	0.93	–	7.11	m	**8.04**
Greenleaf Horticulture; irrigation systems;							
Root Rain tree pit irrigation systems							
Metro; small; 35 mm pipe diameter × 1.25 m							
long; for specimen shrubs and standard trees							
plastic	6.48	0.25	4.63	–	6.48	nr	**11.11**
plastic; with chain	8.49	0.25	4.63	–	8.49	nr	**13.12**
metal; with chain	10.52	0.25	4.63	–	10.52	nr	**15.15**
Metro; medium; 35 mm pipe diameter × 1.75 m							
long; for standard and selected standard trees							
plastic	6.73	0.29	5.29	–	6.73	nr	**12.02**
plastic; with chain	8.91	0.29	5.29	–	8.91	nr	**14.20**
metal; with chain	10.90	0.29	5.29	–	10.90	nr	**16.19**

Q PAVING/PLANTING/FENCING/SITE FURNITURE

Item Excluding site overheads and profit	PC £	Labour hours	Labour £	Plant £	Material £	Unit	Total rate £
Metro; large; 35 mm pipe diameter × 2.50 m long; for selected standards and extra heavy standards							
plastic	7.96	0.33	6.17	–	7.96	nr	**14.13**
plastic; with chain	9.97	0.33	6.17	–	9.97	nr	**16.14**
metal; with chain	13.33	0.33	6.17	–	13.33	nr	**19.50**
Urban; for large capacity general purpose irrigation to parkland and street verge planting							
RRUrb1; 3.0 m pipe	15.98	0.29	5.29	–	15.98	nr	**21.27**
RRUrb2; 5.0 m pipe	18.35	0.33	6.17	–	18.35	nr	**24.52**
RRUrb3; 8.0 m pipe	21.95	0.40	7.40	–	21.95	nr	**29.35**
Civic; heavy cast aluminium inlet; for heavily trafficked locations							
5.0 m pipe	31.55	0.33	6.17	–	31.55	nr	**37.72**
8.0 m pipe	35.05	0.40	7.40	–	35.05	nr	**42.45**
Tree irrigation kits; direct water delivery to the rootball area of trees							
Piddler tree irrigation system; permeable membrane delivering targeted irrigation to rootball surround							
root balls up to 550 mm diameter; Irrigation Kit PID0	7.00	0.25	4.63	–	7.00	nr	**11.63**
root balls up to 900 mm diameter; Irrigation Kit PID1	12.50	0.30	5.55	–	12.50	nr	**18.05**
root balls up to 1.55 m diameter; Irrigation Kit PID2	16.50	0.40	7.40	–	16.50	nr	**23.90**
root balls up to 2.40 m diameter; Irrigation Kit PID3	20.50	0.45	8.32	–	20.50	nr	**28.82**
root balls up to 3.10 m diameter; Irrigation Kit PID4	25.50	0.50	9.25	–	25.50	nr	**34.75**
Mulching of tree pits; Melcourt Industries Ltd; FSC (Forest Stewardship Council) and FT (fire tested) certified							
Spreading mulch; to individual trees; maximum distance 25 m (mulch not included)							
50 mm thick	–	0.05	0.90	–	–	m²	**0.90**
75 mm thick	–	0.07	1.35	–	–	m²	**1.35**
100 mm thick	–	0.10	1.80	–	–	m²	**1.80**
Mulch; Bark Nuggets®; to individual trees; delivered in 80 m³ loads; maximum distance 25 m							
50 mm thick	2.03	0.05	0.90	–	2.03	m²	**2.93**
75 mm thick	3.05	0.07	1.35	–	3.05	m²	**4.40**
100 mm thick	4.06	0.10	1.80	–	4.06	m²	**5.86**
Mulch; Bark Nuggets®; to individual trees; delivered in 25 m³ loads; maximum distance 25 m							
50 mm thick	2.71	0.05	0.90	–	2.71	m²	**3.61**
75 mm thick	4.07	0.05	0.93	–	4.07	m²	**5.00**
100 mm thick	5.43	0.07	1.23	–	5.43	m²	**6.66**
Mulch; Amenity Bark Mulch; to individual trees; delivered in 80 m³ loads; maximum distance 25 m							
50 mm thick	1.27	0.05	0.90	–	1.27	m²	**2.17**
75 mm thick	1.90	0.07	1.35	–	1.90	m²	**3.25**
100 mm thick	2.54	0.10	1.80	–	2.54	m²	**4.34**

Q PAVING/PLANTING/FENCING/SITE FURNITURE

Item Excluding site overheads and profit	PC £	Labour hours	Labour £	Plant £	Material £	Unit	Total rate £
Q31 TREE PLANTING – cont							
Mulching of tree pits – cont							
Mulch; Amenity Bark Mulch; to individual trees; delivered in 25 m³ loads; maximum distance 25 m							
50 mm thick	1.95	0.05	0.90	–	1.95	m²	2.85
75 mm thick	2.93	0.07	1.35	–	2.93	m²	4.28
100 mm thick	3.90	0.07	1.23	–	3.90	m²	5.13
Trees; planting labours only							
Bare root trees; including backfilling with previously excavated material (all other operations and materials not included)							
light standard; 6–8 cm girth	–	0.35	6.47	–	–	nr	6.47
standard; 8–10 cm girth	–	0.40	7.40	–	–	nr	7.40
selected standard; 10–12 cm girth	–	0.58	10.73	–	–	nr	10.73
heavy standard; 12–14 cm girth	–	0.83	15.42	–	–	nr	15.42
extra heavy standard; 14–16 cm girth	–	1.00	18.50	–	–	nr	18.50
Root balled trees; including backfilling with previously excavated material (all other operations and materials not included)							
standard; 8–10 cm girth	–	0.50	9.25	–	–	nr	9.25
selected standard; 10–12 cm girth	–	0.60	11.10	–	–	nr	11.10
heavy standard; 12–14 cm girth	–	0.80	14.80	–	–	nr	14.80
extra heavy standard; 14–16 cm girth	–	1.50	27.75	–	–	nr	27.75
16–18 cm girth	–	1.30	23.98	21.81	–	nr	45.79
18–20 cm girth	–	1.60	29.60	26.93	–	nr	56.53
20–25 cm girth	–	4.50	83.25	86.86	–	nr	170.11
25–30 cm girth	–	6.00	111.00	114.33	–	nr	225.33
30–35 cm girth	–	11.00	203.50	201.82	–	nr	405.32
Tree planting; containerized trees; nursery stock; James Coles & Sons (Nurseries) Ltd							
Acer platanoides 'Emerald Queen'; including backfillling with excavated material (other operations not included)							
standard; 8–10 cm girth	44.75	0.48	8.88	–	44.75	nr	53.63
selected standard; 10–12 cm girth	70.00	0.56	10.37	–	70.00	nr	80.37
heavy standard; 12–14 cm girth	91.00	0.76	14.14	–	91.00	nr	105.14
extra heavy standard; 14–16 cm girth	105.00	1.20	22.20	–	105.00	nr	127.20
Carpinus betulus; including backfillling with excavated material (other operations not included)							
standard; 8–10 cm girth	44.75	0.48	8.88	–	44.75	nr	53.63
selected standard; 10–12 cm girth	70.00	0.56	10.37	–	70.00	nr	80.37
heavy standard; 12–14 cm girth	91.00	0.76	14.14	–	91.00	nr	105.14
extra heavy standard; 14–16 cm girth	112.00	1.20	22.20	–	112.00	nr	134.20
Fraxinus excelsior 'Altena'; including backfillling with excavated material (other operations not included)							
standard; 8–10 cm girth	42.00	0.48	8.88	–	42.00	nr	50.88
selected standard; 10–12 cm girth	67.25	0.56	10.37	–	67.25	nr	77.62
heavy standard; 12–14 cm girth	84.00	0.76	14.14	–	84.00	nr	98.14
extra heavy standard; 14–16 cm girth	98.00	1.20	22.20	–	98.00	nr	120.20

Q PAVING/PLANTING/FENCING/SITE FURNITURE

Item Excluding site overheads and profit	PC £	Labour hours	Labour £	Plant £	Material £	Unit	Total rate £
Prunus avium 'Plena'; including backfillling with excavated material (other operations not included)							
standard; 8–10 cm girth	42.00	0.40	7.40	–	42.00	nr	49.40
selected standard; 10–12 cm girth	70.00	0.56	10.37	–	70.00	nr	80.37
heavy standard; 12–14 cm girth	84.00	0.76	14.14	–	84.00	nr	98.14
extra heavy standard; 14–16 cm girth	98.00	1.20	22.20	–	98.00	nr	120.20
Quercus robur; including backfillling with excavated material (other operations not included)							
standard; 8–10 cm girth	49.00	0.48	8.88	–	49.00	nr	57.88
selected standard; 10–12 cm girth	77.00	0.56	10.37	–	77.00	nr	87.37
heavy standard; 12–14 cm girth	105.00	0.76	14.14	–	105.00	nr	119.14
extra heavy standard; 14–16 cm girth	119.00	1.20	22.20	–	119.00	nr	141.20
Betula utilis jaquemontii; multistemmed; including backfillling with excavated material (other operations not included)							
175/200 mm high	70.00	0.48	8.88	–	70.00	nr	78.88
200/250 mm high	98.00	0.56	10.37	–	98.00	nr	108.37
250/300 mm high	119.00	0.76	14.14	–	119.00	nr	133.14
300/350 mm high	196.00	1.20	22.20	–	196.00	nr	218.20
Tree planting; root balled trees; advanced nursery stock and semi-mature – General							
Preamble: The cost of planting semi-mature trees will depend on the size and species, and on the access to the site for tree handling machines. Prices should be obtained for individual trees and planting.							
Tree planting; bare root trees; nursery stock; James Coles & Sons (Nurseries) Ltd							
Acer platanoides; including backfillling with excavated material (other operations not included)							
light standard; 6–8 cm girth	9.00	0.35	6.47	–	9.00	nr	15.47
standard; 8–10 cm girth	12.00	0.40	7.40	–	12.00	nr	19.40
selected standard; 10–12 cm girth	19.50	0.58	10.73	–	19.50	nr	30.23
heavy standard; 12–14 cm girth	42.00	0.83	15.42	–	42.00	nr	57.42
extra heavy standard; 14–16 cm girth	56.00	1.00	18.50	–	56.00	nr	74.50
Carpinus betulus; including backfillling with excavated material (other operations not included)							
light standard; 6–8 cm girth	13.25	0.35	6.47	–	13.25	nr	19.72
standard; 8–10 cm girth	26.50	0.40	7.40	–	26.50	nr	33.90
selected standard; 10–12 cm girth	37.75	0.58	10.73	–	37.75	nr	48.48
heavy standard; 12–14 cm girth	39.25	0.83	15.42	–	39.25	nr	54.67
extra heavy standard; 14–16 cm girth	46.25	1.00	18.50	–	46.25	nr	64.75
Fraxinus excelsior; including backfillling with excavated material (other operations not included)							
light standard; 6-8 cm girth	9.75	0.35	6.47	–	9.75	nr	16.22
standard; 8-10 cm girth	16.00	0.40	7.40	–	16.00	nr	23.40
selected standard; 10–12 cm girth	22.50	0.58	10.73	–	22.50	nr	33.23
heavy standard; 12–14 cm girth	39.25	0.83	15.36	–	39.25	nr	54.61
extra heavy standard; 14–16 cm girth	47.50	1.00	18.50	–	47.50	nr	66.00

Q PAVING/PLANTING/FENCING/SITE FURNITURE

Item Excluding site overheads and profit	PC £	Labour hours	Labour £	Plant £	Material £	Unit	Total rate £
Q31 TREE PLANTING – cont							
Tree planting – cont							
Prunus avium 'Plena'; including backfillling with excavated material (other operations not included)							
light standard; 6–8 cm girth	9.75	0.36	6.73	–	9.75	nr	16.48
standard; 8–10 cm girth	16.00	0.40	7.40	–	16.00	nr	23.40
selected standard; 10–12 cm girth	29.50	0.58	10.73	–	29.50	nr	40.23
heavy standard; 12–14 cm girth	47.50	0.83	15.42	–	47.50	nr	62.92
extra heavy standard; 14–16 cm girth	63.00	1.00	18.50	–	63.00	nr	81.50
Quercus robur; including backfillling with excavated material (other operations not included)							
light standard; 6–8 cm girth	21.00	0.35	6.47	–	21.00	nr	27.47
standard; 8–10 cm girth	30.75	0.40	7.40	–	30.75	nr	38.15
selected standard; 10–12 cm girth	42.00	0.58	10.73	–	42.00	nr	52.73
heavy standard; 12–14 cm girth	58.75	0.83	15.42	–	58.75	nr	74.17
Robinia pseudoacacia 'Frisia'; including backfillling with excavated material (other operations not included)							
light standard; 6–8 cm girth	21.00	0.35	6.47	–	21.00	nr	27.47
standard; 8–10 cm girth	28.00	0.40	7.40	–	28.00	nr	35.40
selected standard; 10–12 cm girth	44.75	0.58	10.73	–	44.75	nr	55.48
Tree planting; root balled trees; nursery stock; James Coles & Sons (Nurseries) Ltd							
Acer platanoides; including backfillling with excavated material (other operations not included)							
standard; 8–10 cm girth	19.50	0.48	8.88	–	19.50	nr	28.38
selected standard; 10–12 cm girth	29.50	0.56	10.37	–	29.50	nr	39.87
heavy standard; 12–14 cm girth	57.00	0.76	14.14	–	57.00	nr	71.14
extra heavy standard; 14–16 cm girth	71.00	1.20	22.20	–	71.00	nr	93.20
Carpinus betulus; including backfillling with excavated material (other operations not included)							
standard; 8–10 cm girth	34.00	0.48	8.88	–	34.00	nr	42.88
selected standard; 10–12 cm girth	47.75	0.56	10.37	–	47.75	nr	58.12
heavy standard; 12–14 cm girth	79.50	0.76	14.14	–	79.50	nr	93.64
extra heavy standard; 14–16 cm girth	106.00	1.20	22.20	–	106.00	nr	128.20
Fraxinus excelsior; including backfillling with excavated material (other operations not included)							
standard; 8–10 cm girth	23.50	0.48	8.88	–	23.50	nr	32.38
selected standard; 10–12 cm girth	32.50	0.56	10.37	–	32.50	nr	42.87
heavy standard; 12–14 cm girth	54.25	0.76	14.14	–	54.25	nr	68.39
extra heavy standard; 14–16 cm girth	62.50	1.20	22.20	–	62.50	nr	84.70
Prunus avium 'Plena'; including backfillling with excavated material (other operations not included)							
standard; 8–10 cm girth	30.50	0.40	7.40	–	30.50	nr	37.90
selected standard; 10–12 cm girth	30.50	0.56	10.37	–	30.50	nr	40.87
heavy standard; 12–14 cm girth	62.50	0.76	14.14	–	62.50	nr	76.64
extra heavy standard; 14–16 cm girth	78.00	1.20	22.20	–	78.00	nr	100.20

Q PAVING/PLANTING/FENCING/SITE FURNITURE

Item Excluding site overheads and profit	PC £	Labour hours	Labour £	Plant £	Material £	Unit	Total rate £
Quercus robur; including backfilling with excavated material (other operations not included)							
standard; 8–10 cm girth	38.25	0.48	8.88	–	38.25	nr	**47.13**
selected standard; 10–12 cm girth	52.00	0.56	10.37	–	52.00	nr	**62.37**
heavy standard; 12–14 cm girth	73.75	0.76	14.14	–	73.75	nr	**87.89**
extra heavy standard; 14–16 cm girth	112.00	1.20	22.20	–	112.00	nr	**134.20**
Robinia pseudoacacia 'Frisia'; including backfilling with excavated material (other operations not included)							
standard; 8–10 cm girth	35.50	0.48	8.88	–	35.50	nr	**44.38**
selected standard; 10–12 cm girth	54.75	0.56	10.37	–	54.75	nr	**65.12**
heavy standard; 12–14 cm girth	98.00	0.76	14.14	–	98.00	nr	**112.14**
Tree planting; Airpot container grown trees; advanced nursery stock and semi-mature; Deepdale Trees Ltd							
Acer platanoides 'Emerald Queen'; including backfilling with excavated material (other operations not included)							
16–18 cm girth	95.00	1.98	36.63	39.67	95.00	nr	**171.30**
18–20 cm girth	130.00	2.18	40.29	42.52	130.00	nr	**212.81**
20–25 cm girth	190.00	2.38	43.96	47.60	190.00	nr	**281.56**
25–30 cm girth	250.00	2.97	54.95	72.25	250.00	nr	**377.20**
30–35 cm girth	450.00	3.96	73.26	79.34	450.00	nr	**602.60**
Aesculus briotti; including backfilling with excavated material (other operations not included)							
16–18 cm girth	110.00	1.98	36.63	39.67	110.00	nr	**186.30**
18–20 cm girth	130.00	1.60	29.60	42.52	130.00	nr	**202.12**
20–25 cm girth	200.00	2.38	43.96	47.60	200.00	nr	**291.56**
25–30 cm girth	300.00	2.97	54.95	72.25	300.00	nr	**427.20**
30–35 cm girth	450.00	3.96	73.26	79.34	450.00	nr	**602.60**
Prunus avium 'Flora Plena'; including backfilling with excavated material (other operations not included)							
16–18 cm girth	95.00	1.98	36.63	39.67	95.00	nr	**171.30**
18–20 cm girth	130.00	1.60	29.60	42.52	130.00	nr	**202.12**
20–25 cm girth	190.00	2.38	43.96	47.60	190.00	nr	**281.56**
25–30 cm girth	250.00	2.97	54.95	72.25	250.00	nr	**377.20**
30–35 cm girth	350.00	3.96	73.26	79.34	350.00	nr	**502.60**
Quercus palustris 'Pin Oak'; including backfilling with excavated material (other operations not included)							
16–18 cm girth	100.00	1.98	36.63	39.67	100.00	nr	**176.30**
18–20 cm girth	140.00	1.60	29.60	42.52	140.00	nr	**212.12**
20–25 cm girth	210.00	2.38	43.96	47.60	210.00	nr	**301.56**
25–30 cm girth	275.00	2.97	54.95	72.25	275.00	nr	**402.20**
30–35 cm girth	400.00	3.96	73.26	79.34	400.00	nr	**552.60**

Q PAVING/PLANTING/FENCING/SITE FURNITURE

Item Excluding site overheads and profit	PC £	Labour hours	Labour £	Plant £	Material £	Unit	Total rate £
Q31 TREE PLANTING – cont							
Tree planting; Airpot container grown trees;							
semi-mature and mature trees; Deepdale							
Trees Ltd; planting and back filling; planted							
by telehandler or by crane; delivery included;							
all other operations priced separately							
Semi-mature trees; indicative prices							
40–45 cm girth	550.00	4.00	74.00	49.03	550.00	nr	673.03
45–50 cm girth	750.00	4.00	74.00	49.03	750.00	nr	873.03
55–60 cm girth	1350.00	6.00	111.00	49.03	1350.00	nr	1510.03
60–70 cm girth	2500.00	7.00	129.50	66.34	2500.00	nr	2695.84
70–80 cm girth	3500.00	7.50	138.75	83.65	3500.00	nr	3722.40
80–90 cm girth	4500.00	8.00	148.00	98.06	4500.00	nr	4746.06
Tree planting; rootballed trees; advanced							
nursery stock and semi-mature; Lorenz							
von Ehren							
Acer platanoides 'Emerald Queen'; including							
backfilling with excavated material (other							
operations not included)							
16–18 cm girth	85.00	1.30	23.98	3.67	85.00	nr	112.65
18–20 cm girth	105.00	1.60	29.60	3.67	105.00	nr	138.27
20–25 cm girth	130.00	4.50	83.25	16.62	130.00	nr	229.87
25–30 cm girth	170.00	6.00	111.00	20.68	170.00	nr	301.68
30–35 cm girth	305.00	11.00	203.50	27.68	305.00	nr	536.18
Aesculus carnea 'Briotti'; including backfilling							
with excavated material (other operations not							
included)							
16–18 cm girth	150.00	1.30	23.98	3.67	150.00	nr	177.65
18–20 cm girth	180.00	1.60	29.60	3.67	180.00	nr	213.27
20–25 cm girth	215.00	4.50	83.25	16.62	215.00	nr	314.87
25–30 cm girth	290.00	6.00	111.00	20.68	290.00	nr	421.68
30–35 cm girth	390.00	11.00	203.50	27.68	390.00	nr	621.18
Prunus avium 'Plena'; including backfilling with							
excavated material (other operations not							
included)							
16–18 cm girth	110.00	1.30	23.98	3.67	110.00	nr	137.65
18–20 cm girth	130.00	1.60	29.60	3.67	130.00	nr	163.27
20–25 cm girth	150.00	4.50	83.25	16.62	150.00	nr	249.87
25–30 cm girth	155.00	6.00	111.00	20.68	155.00	nr	286.68
30–35 cm girth	250.00	11.00	203.50	24.35	250.00	nr	477.85
Quercus palustris 'Pin Oak'; including							
backfilling with excavated material (other							
operations not included)							
16–18 cm girth	125.00	1.30	23.98	3.67	125.00	nr	152.65
18–20 cm girth	150.00	1.60	29.60	3.67	150.00	nr	183.27
20–25 cm girth	190.00	4.50	83.25	16.62	190.00	nr	289.87
25–30 cm girth	225.00	6.00	111.00	20.68	225.00	nr	356.68
30–35 cm girth	340.00	11.00	203.50	27.68	340.00	nr	571.18
Tilia cordata 'Greenspire'; including backfilling							
with excavated material (other operations not							
included)							
16–18 cm girth	95.00	1.30	23.98	3.67	95.00	nr	122.65
18–20 cm girth	115.00	1.60	29.60	3.67	115.00	nr	148.27
20–25 cm girth	140.00	4.50	83.25	16.62	140.00	nr	239.87
25–30 cm girth; 5 × transplanted; 4.0–5.0 m tall	150.00	6.00	111.00	20.68	150.00	nr	281.68
30–35 cm girth; 5 × transplanted; 5.0–7.0 m tall	220.00	11.00	203.50	27.68	220.00	nr	451.18

Q PAVING/PLANTING/FENCING/SITE FURNITURE

Item Excluding site overheads and profit	PC £	Labour hours	Labour £	Plant £	Material £	Unit	**Total rate £**
Betula pendula (3 stems); including backfilling with excavated material (other operations not included)							
3.0–3.5 m high	60.00	1.98	36.63	39.67	60.00	nr	**136.30**
3.5–4.0 m high	90.00	1.60	29.60	42.52	90.00	nr	**162.12**
4.0–4.5 m high	130.00	2.38	43.96	47.60	130.00	nr	**221.56**
4.5–5.0 m high	150.00	2.97	54.95	72.25	150.00	nr	**277.20**
5.0–6.0 m high	210.00	3.96	73.26	79.34	210.00	nr	**362.60**
6.0–7.0 m high	350.00	4.50	83.25	93.92	350.00	nr	**527.17**
Pinus sylvestris; including backfilling with excavated material (other operations not included)							
3.0–3.5 m high	400.00	1.98	36.63	39.67	400.00	nr	**476.30**
3.5–4.0 m high	500.00	1.60	29.60	42.52	500.00	nr	**572.12**
4.0–4.5 m high	690.00	2.38	43.96	47.60	690.00	nr	**781.56**
4.5–5.0 m high	950.00	2.97	54.95	72.25	950.00	nr	**1077.20**
5.0–6.0 m high	1500.00	3.96	73.26	79.34	1500.00	nr	**1652.60**
6.0–7.0 m high	2500.00	4.50	83.25	96.80	2500.00	nr	**2680.05**
Pleached trees; Carpinus betulus; specimen trees; Lorenz Von Ehren; supply and planting only; excavation, support and treepit additives not included							
Box shaped trees to provide 'floating hedge' or screen effect; planting distance 1 tree per m run; trunk and box alignment; wire root balls							
4 × transplanted; 20–25 cm; 1.00 m centres	300.00	5.00	92.50	16.62	300.00	m	**409.12**
5 × transplanted; 25–30 cm; 1.00 m centres	430.00	6.50	120.25	20.68	430.00	m	**570.93**
5 × transplanted; 30–35 cm; 1.20 m centres	720.00	11.00	203.50	27.68	720.00	m	**951.18**
Pleached trees; Tilia europaea 'Pallida'; specimen trees; Lorenz Von Ehren; supply and planting only; excavation, support and treepit additives not included							
Box pleached trees to provide 'floating hedge' or screen effect; wire root balls							
4 × transplanted; 20–25 cm; 1.00 m centres	300.00	5.00	92.50	16.62	300.00	m	**409.12**
4 × transplanted; 25–30 cm; 1.00 m centres	370.00	5.00	92.50	16.62	370.00	m	**479.12**
5 × transplanted; 30–35 cm; 1.20 m centres	358.32	5.00	92.50	16.62	358.32	m	**467.44**
5 × transplanted; 35–40 cm; 1.50 m centres	380.02	5.00	92.50	16.62	380.02	m	**489.14**
6 × transplanted; 40–45 cm; 1.80 m centres	444.48	5.00	92.50	16.62	444.48	m	**553.60**
6 × transplanted; 45–50 cm; 1.80 m centres	550.04	5.00	92.50	16.62	550.04	m	**659.16**
6 × transplanted; 50–60 cm; 2.00 m centres	700.00	5.00	92.50	16.62	700.00	m	**809.12**
Espalier pleached trees; specimen trees; Lorenz Von Ehren; supply and planting only; excavation, support and treepit additives not included							
Frame pleached trees; to provide floating screen effects; wire root balls							
Carpinus betulus; 20–25 cm; 1.00 m centres	420.00	5.00	92.50	16.62	420.00	m	**529.12**
Carpinus betulus; 25–30 cm; 1.00 m centres	550.00	5.00	92.50	16.62	550.00	m	**659.12**
Tilia europaea 'Pallida'; 20–25 cm; 1.00 m centres	420.00	5.00	92.50	16.62	420.00	m	**529.12**
Tilia europaea 'Pallida'; 25–30 cm; 1.00 m centres	550.00	5.00	92.50	16.62	550.00	m	**659.12**
Tilia europaea 'Pallida'; 30–35 cm; 1.20 m centres	658.31	5.00	92.50	16.62	658.31	m	**767.43**

Q PAVING/PLANTING/FENCING/SITE FURNITURE

Item Excluding site overheads and profit	PC £	Labour hours	Labour £	Plant £	Material £	Unit	Total rate £
Q31 TREE PLANTING – cont							
Umbrella shaped pleaches							
Umbrella or roof pleached trees to provide							
umbrella effect; wire root balls							
Tilia euchlora; specimen; 4 × transplanted;							
20–25 cm	390.00	5.00	92.50	16.62	390.00	ea	499.12
Tilia euchlora; specimen; 5 × transplanted;							
25–30 cm	480.00	5.00	92.50	16.62	480.00	ea	589.12
Platanus acerifolia; clear stem;							
4 × transplanted; 20–25 cm	390.00	5.00	92.50	16.62	390.00	ea	499.12
Platanus acerifolia; specimen;							
5 × transplanted; 30–35 cm	700.00	5.00	92.50	16.62	700.00	ea	809.12
Tilia europaea; specimen; 5 × transplanted;							
30–35 cm	700.00	5.00	92.50	16.62	700.00	ea	809.12
Platanus acerifolia; specimen;							
5 × transplanted; 25–30 cm	480.00	5.00	92.50	16.62	480.00	ea	589.12
Tree planting; containerized trees; Lorenz							
von Ehren; to the tree prices above add for							
Airpot containerization only							
Tree size							
20–25 cm	–	–	–	–	45.00	nr	45.00
25–30 cm	–	–	–	–	75.00	nr	75.00
30–35 cm	–	–	–	–	95.00	nr	95.00
35–40 cm	–	–	–	–	130.00	nr	130.00
40–45 cm	–	–	–	–	150.00	nr	150.00
45–50 cm	–	–	–	–	190.00	nr	190.00
50–60 cm	–	–	–	–	250.00	nr	250.00
60–70 cm	–	–	–	–	320.00	nr	320.00
70–80 cm	–	–	–	–	390.00	nr	390.00
80–90 cm	–	–	–	–	450.00	nr	450.00
Tree planting; rootballed trees;							
semi-mature and mature trees; Lorenz von							
Ehren; planting and back filling; planted by							
telehandler or by crane; delivery included;							
all other operations priced separately							
Semi-mature trees							
40–45 cm girth	600.00	8.00	148.00	63.41	600.00	nr	811.41
45–50 cm girth	850.00	8.00	148.00	85.02	850.00	nr	1083.02
50–60 cm girth	1200.00	10.00	185.00	102.55	1200.00	nr	1487.55
60–70 cm girth	2200.00	15.00	277.50	196.94	2200.00	nr	2674.44
70–80 cm girth	3200.00	18.00	333.00	188.52	3200.00	nr	3721.52
80–90 cm girth	4800.00	18.00	333.00	188.52	4800.00	nr	5321.52
90–100 cm girth	5900.00	18.00	333.00	188.52	5900.00	nr	6421.52
Q31 HEDGE/SHRUB/TOPIARY PLANTING							
Hedges							
Excavating trench for hedges; depositing soil							
alongside trench; by machine							
300 mm deep × 300 mm wide	–	0.03	0.56	0.26	–	m	0.82
300 mm deep × 450 mm wide	–	0.05	0.83	0.40	–	m	1.23
Excavating trench for hedges; depositing soil							
alongside trench; by hand							
300 mm deep × 300 mm wide	–	0.12	2.22	–	–	m	2.22
300 mm deep × 450 mm wide	–	0.23	4.17	–	–	m	4.17

Q PAVING/PLANTING/FENCING/SITE FURNITURE

Item Excluding site overheads and profit	PC £	Labour hours	Labour £	Plant £	Material £	Unit	Total rate £
Setting out; notching out; excavating trench; breaking up subsoil to minimum depth 300 mm							
minimum 400 mm deep	–	0.25	4.63	–	–	m	**4.63**
Hedge planting; including backfill with excavated topsoil; PC £0.33/nr							
single row; 200 mm centres	1.65	0.06	1.16	–	1.65	m	**2.81**
single row; 300 mm centres	1.10	0.06	1.03	–	1.10	m	**2.13**
single row; 400 mm centres	0.82	0.04	0.77	–	0.82	m	**1.59**
single row; 500 mm centres	0.66	0.03	0.62	–	0.66	m	**1.28**
double row; 200 mm centres	3.30	0.17	3.08	–	3.30	m	**6.38**
double row; 300 mm centres	2.20	0.13	2.47	–	2.20	m	**4.67**
double row; 400 mm centres	1.65	0.08	1.54	–	1.65	m	**3.19**
double row; 500 mm centres	1.32	0.07	1.23	–	1.32	m	**2.55**
Extra over hedges for incorporating manure; at 1 m³ per 30 m	0.77	0.03	0.46	–	0.77	m	**1.23**
Green Screen; Mobilane Ltd Fully installed vertical green screen of Hedera hibernica; 65 plants per linear m; all on 5 mm galvanized steel weldmesh; attached to existing wall fence or hoarding							
screen 1.80 m high	–	–	–	–	170.00	m	**170.00**
screen 2.20 m high	–	–	–	–	230.00	m	**230.00**
screen 4.00 m high	–	–	–	–	500.00	m	**500.00**
Living wall; Mobilane Ltd LivePanel; fully installed living wall adjusted to fit modules with developed substrate; panels are stacked onto building or wall face of the building and secured by an aluminium frame	–	–	–	–	600.00	m²	**600.00**
Topiary							
Clipped topiary; Lorenz von Ehren; German field grown clipped and transplanted as detailed; planted to plantpit; including backfilling with excavated material and TPMC Buxus sempervirens (box); balls							
300 mm diameter; 3 × transplanted; container grown or rootballed	15.00	0.25	4.63	–	24.22	nr	**28.85**
500 mm diameter; 4 × transplanted; wire rootballed	45.00	1.50	27.75	5.58	113.86	nr	**147.19**
900 mm diameter; 5 × transplanted; wire rootballed	205.00	2.75	50.88	5.58	319.02	nr	**375.48**
1300 mm diameter; 6 × transplanted; wire rootballed	650.00	3.10	57.35	6.69	761.58	nr	**825.62**
Buxus sempervirens (box); pyramids							
500 mm high; 3 × transplanted; container grown or rootballed	30.00	1.50	27.75	5.58	98.86	nr	**132.19**
900 mm high; 4 × transplanted; wire rootballed	95.00	2.75	50.88	5.58	209.02	nr	**265.48**
1300 mm high; 5 × transplanted; wire rootballed	390.00	3.10	57.35	6.69	501.58	nr	**565.62**

Q PAVING/PLANTING/FENCING/SITE FURNITURE

Item Excluding site overheads and profit	PC £	Labour hours	Labour £	Plant £	Material £	Unit	Total rate £
Q31 HEDGE/SHRUB/TOPIARY PLANTING – **cont**							
Clipped topiary – cont							
Buxus sempervirens (box); truncated pyramids							
500 mm high; 3 × transplanted; container grown or rootballed	80.00	1.50	27.75	5.58	148.86	nr	**182.19**
900 mm high; 4 × transplanted; wire rootballed	300.00	2.75	50.88	5.58	414.02	nr	**470.48**
Buxus sempervirens (box); truncated cone							
1300 mm high; 5 × transplanted; wire rootballed	950.00	3.10	57.35	6.69	1061.58	nr	**1125.62**
Buxus sempervirens (box); cubes							
500 mm square; 4 × transplanted; rootballed	80.00	1.50	27.75	5.58	148.86	nr	**182.19**
900 mm square; 5 × transplanted; wire rootballed	360.00	2.75	50.88	5.58	474.02	nr	**530.48**
Taxus baccata (yew); balls							
500 mm diameter; 4 × transplanted; rootballed	55.00	1.50	27.75	5.58	123.86	nr	**157.19**
900 mm diameter; 5 × transplanted; wire rootballed	180.00	2.75	50.88	5.58	294.02	nr	**350.48**
1300 mm diameter; 6 × transplanted; wire rootballed	560.00	3.10	57.35	6.69	671.58	nr	**735.62**
Taxus baccata (yew); cones							
800 mm high; 4 × transplanted; wire rootballed	55.00	1.50	27.75	5.58	123.86	nr	**157.19**
1500 mm high; 5 × transplanted; wire rootballed	140.00	2.00	37.00	5.58	224.86	nr	**267.44**
2500 mm high; 6 × transplanted; wire rootballed	470.00	3.10	57.35	6.69	581.58	nr	**645.62**
Taxus baccata (yew); cubes							
500 mm square; 4 × transplanted; rootballed	65.00	1.50	27.75	5.58	133.86	nr	**167.19**
900 mm square; 5 × transplanted; wire rootballed	240.00	2.75	50.88	5.58	354.02	nr	**410.48**
Taxus baccata (yew); pyramids							
900 mm high; 4 × transplanted; wire rootballed	120.00	1.50	27.75	5.58	188.86	nr	**222.19**
1500 mm high; 5 × transplanted; wire rootballed	240.00	3.10	57.35	6.69	351.58	nr	**415.62**
2500 mm high; 6 × transplanted; wire rootballed	690.00	4.00	74.00	11.16	833.01	nr	**918.17**
Carpinus betulus (common hornbeam); columns; round base							
800 mm wide × 2000 mm high; 4 × transplanted; wire rootballed	150.00	3.10	57.35	6.69	261.58	nr	**325.62**
800 mm wide × 2750 mm high; 4 × transplanted; wire rootballed	300.00	4.00	74.00	11.16	443.01	nr	**528.17**
Carpinus betulus (common hornbeam); cones							
3 m high; 5 × transplanted; wire rootballed	270.00	4.00	74.00	11.16	413.01	nr	**498.17**
4 m high; 5 × transplanted; wire rootballed	560.00	5.00	92.50	11.16	729.73	nr	**833.39**
Carpinus betulus 'Fastigiata'; pyramids							
4 m high; 6 × transplanted; wire rootballed	960.00	5.00	92.50	11.16	1129.73	nr	**1233.39**
7 m high; 7 × transplanted; wire rootballed	2600.00	7.50	138.75	16.73	2809.73	nr	**2965.21**

Q PAVING/PLANTING/FENCING/SITE FURNITURE

Item Excluding site overheads and profit	PC £	Labour hours	Labour £	Plant £	Material £	Unit	Total rate £
Shrub planting – General Preamble: For preparation of planting areas see 'Cultivation' at the beginning of the section on planting.							
Shrub planting Setting out; selecting planting from holding area; loading to wheelbarrows; planting as plan or as directed; distance from holding area maximum 50 m; plants 2–3 litre containers							
single plants not grouped	–	0.04	0.74	–	–	nr	**0.74**
plants in groups of 3–5 nr	–	0.03	0.46	–	–	nr	**0.46**
plants in groups of 10–100 nr	–	0.02	0.31	–	–	nr	**0.31**
plants in groups of 100 nr minimum	–	0.01	0.21	–	–	nr	**0.21**
Forming planting holes; in cultivated ground (cultivating not included); by mechanical auger; trimming holes by hand; depositing excavated material alongside holes							
250 mm diameter	–	0.03	0.62	0.04	–	nr	**0.66**
250 × 250 mm	–	0.04	0.74	0.08	–	nr	**0.82**
300 × 300 mm	–	0.08	1.39	0.10	–	nr	**1.49**
Hand excavation; forming planting holes; in cultivated ground (cultivating not included); depositing excavated material alongside holes							
100 × 100 × 100 mm deep; with mattock or hoe	–	0.01	0.12	–	–	nr	**0.12**
250 × 250 × 300 mm deep	–	0.04	0.74	–	–	nr	**0.74**
300 × 300 × 300 mm deep	–	0.06	1.03	–	–	nr	**1.03**
400 × 400 × 400 mm deep	–	0.13	2.31	–	–	nr	**2.31**
500 × 500 × 500 mm deep	–	0.25	4.63	–	–	nr	**4.63**
600 × 600 × 600 mm deep	–	0.43	8.01	–	–	nr	**8.01**
900 × 900 × 600 mm deep	–	1.00	18.50	–	–	nr	**18.50**
1.00 × 1.00 m × 600 mm deep	–	1.23	22.75	–	–	nr	**22.75**
1.25 × 1.25 m × 600 mm deep	–	1.93	35.70	–	–	nr	**35.70**
Hand excavation; forming planting holes; in uncultivated ground; depositing excavated material alongside holes							
100 × 100 × 100 mm deep; with mattock or hoe	–	0.03	0.46	–	–	nr	**0.46**
250 × 250 × 300 mm deep	–	0.06	1.03	–	–	nr	**1.03**
300 × 300 × 300 mm deep	–	0.06	1.16	–	–	nr	**1.16**
400 × 400 × 400 mm deep	–	0.25	4.63	–	–	nr	**4.63**
500 × 500 × 500 mm deep	–	0.33	6.02	–	–	nr	**6.02**
600 × 600 × 600 mm deep	–	0.55	10.18	–	–	nr	**10.18**
900 × 900 × 600 mm deep	–	1.25	23.13	–	–	nr	**23.13**
1.00 × 1.00 m × 600 mm deep	–	1.54	28.44	–	–	nr	**28.44**
1.25 × 1.25 m × 600 mm deep	–	2.41	44.63	–	–	nr	**44.63**
Bare root planting; to planting holes (forming holes not included); including backfilling with excavated material (bare root plants not included)							
bare root 1+1; 30–90 mm high	–	0.02	0.31	–	–	nr	**0.31**
bare root 1+2; 90–120 mm high	–	0.02	0.31	–	–	nr	**0.31**

Q PAVING/PLANTING/FENCING/SITE FURNITURE

Item Excluding site overheads and profit	PC £	Labour hours	Labour £	Plant £	Material £	Unit	Total rate £
Q31 HEDGE/SHRUB/TOPIARY PLANTING – **cont**							
Shrub planting – cont							
Containerized planting; to planting holes							
(forming holes not included); including							
backfilling with excavated material (shrub or							
ground cover not included)							
9 cm pot	–	0.01	0.19	–	–	nr	0.19
2 litre container	–	0.02	0.37	–	–	nr	0.37
3 litre container	–	0.02	0.41	–	–	nr	0.41
5 litre container	–	0.03	0.62	–	–	nr	0.62
10 litre container	–	0.05	0.93	–	–	nr	0.93
15 litre container	–	0.07	1.23	–	–	nr	1.23
20 litre container	–	0.08	1.54	–	–	nr	1.54
Shrub planting; 2 litre containerized plants; in							
cultivated ground (cultivating not included); PC							
£2.65/nr							
average 2 plants per m²	–	0.06	1.04	–	5.30	m²	6.34
average 3 plants per m²	–	0.08	1.55	–	7.95	m²	9.50
average 4 plants per m²	–	0.11	2.07	–	10.60	m²	12.67
average 6 plants per m²	–	0.17	3.11	–	15.90	m²	19.01
Extra over shrubs for stakes	0.70	0.02	0.31	–	0.70	nr	1.01
Composted bark soil conditioners; 20 m³ loads;							
on beds by mechanical loader; spreading and							
rotavating into topsoil; by machine							
50 mm thick	151.50	–	–	6.29	151.50	100 m²	157.79
100 mm thick	318.15	–	–	10.62	318.15	100 m²	328.77
150 mm thick	477.23	–	–	15.13	477.23	100 m²	492.36
200 mm thick	636.30	–	–	19.55	636.30	100 m²	655.85
Mushroom compost; 25 m³ loads; delivered							
not further than 25 m from location; cultivating							
into topsoil by pedestrian operated machine							
50 mm thick	106.00	2.86	52.86	1.77	106.00	100 m²	160.63
100 mm thick	212.00	6.05	111.91	1.77	212.00	100 m²	325.68
150 mm thick	318.00	8.90	164.74	1.77	318.00	100 m²	484.51
200 mm thick	424.00	12.90	238.74	1.77	424.00	100 m²	664.51
Mushroom compost; 25 m³ loads; on beds by							
mechanical loader; spreading and rotavating							
into topsoil by tractor drawn rotavator							
50 mm thick	106.00	–	–	6.29	106.00	100 m²	112.29
100 mm thick	212.00	–	–	10.62	212.00	100 m²	222.62
150 mm thick	318.00	–	–	15.13	318.00	100 m²	333.13
200 mm thick	424.00	–	–	19.55	424.00	100 m²	443.55
Manure; 20 m³ loads; delivered not further							
than 25 m from location; cultivating into topsoil							
by pedestrian operated machine							
50 mm thick	225.00	2.86	52.86	1.77	225.00	100 m²	279.63
100 mm thick	241.29	6.05	111.91	1.77	241.29	100 m²	354.97
150 mm thick	361.94	8.90	164.74	1.77	361.94	100 m²	528.45
200 mm thick	482.58	12.90	238.74	1.77	482.58	100 m²	723.09
Fertilizer (7+7+7); PC £1.06/kg; to beds; by hand							
35 g/m²	3.71	0.17	3.08	–	3.71	100 m²	6.79
50 g/m²	5.30	0.17	3.08	–	5.30	100 m²	8.38
70 g/m²	7.42	0.17	3.08	–	7.42	100 m²	10.50

Q PAVING/PLANTING/FENCING/SITE FURNITURE

Item Excluding site overheads and profit	PC £	Labour hours	Labour £	Plant £	Material £	Unit	Total rate £
Fertilizers; Enmag; PC £2.08/kg; controlled release fertilizer; to beds; by hand							
35 g/m²	7.28	0.17	3.08	–	7.28	100 m²	10.36
50 g/m²	10.40	0.17	3.08	–	10.40	100 m²	13.48
70 g/m²	15.60	0.17	3.08	–	15.60	100 m²	18.68
Note: For machine incorporation of fertilizers and soil conditioners see 'Cultivation'.							
Q31 HERBACEOUS/GROUNDCOVER/ BULB/BEDDING PLANTING							
Herbaceous and groundcover planting Herbaceous plants; PC £1.20/nr; including forming planting holes in cultivated ground (cultivating not included); backfilling with excavated material; 1 litre containers							
average 4 plants per m²; 500 mm centres	–	0.09	1.73	–	4.80	m²	6.53
average 6 plants per m²; 408 mm centres	–	0.14	2.59	–	7.20	m²	9.79
average 8 plants per m²; 354 mm centres	–	0.19	3.46	–	9.60	m²	13.06
Note: For machine incorporation of fertilizers and soil conditioners see 'Cultivation'.							
Plant support netting; Bridport Gundry; on 50 mm diameter stakes; 750 mm long; driving into ground at 1.50 m centres							
green extruded plastic mesh; 125 mm square	0.53	0.04	0.74	–	0.53	m²	1.27
Bulb planting Bulbs; including forming planting holes in cultivated area (cultivating not included); backfilling with excavated material							
small	13.00	0.83	15.42	–	13.00	100 nr	28.42
medium	22.00	0.83	15.42	–	22.00	100 nr	37.42
large	25.00	0.91	16.82	–	25.00	100 nr	41.82
Bulbs; in grassed area; using bulb planter; including backfilling with screened topsoil or peat and cut turf plug							
small	13.00	1.67	30.83	–	13.00	100 nr	43.83
medium	22.00	1.67	30.83	–	22.00	100 nr	52.83
large	25.00	2.00	37.00	–	25.00	100 nr	62.00
Aquatic planting Aquatic plants; in prepared growing medium in pool; plant size 2–3 litre containerized (plants not included)	–	0.04	0.74	–	–	nr	0.74

Q PAVING/PLANTING/FENCING/SITE FURNITURE

Item Excluding site overheads and profit	PC £	Labour hours	Labour £	Plant £	Material £	Unit	Total rate £
Q31 HERBACEOUS/GROUNDCOVER/ **BULB/BEDDING PLANTING – cont**							
MARKET PRICES OF NATIVE SPECIES							
Oakover Nurseries Ltd; The following **prices are typically some of the more** **popular species used in native plantations;** **prices vary between species; readers** **should check the catalogue of the supplier** 1+1; 1 year seedling transplanted and grown for a year							
30–40 cm	–	–	–	–	0.45	nr	**0.45**
40–60 cm	–	–	–	–	0.55	nr	**0.55**
1+2; 1 year seedling transplanted and grown for 2 years							
60–80 cm	–	–	–	–	0.65	nr	**0.65**
Whips							
80–100 cm	–	–	–	–	0.90	nr	**0.90**
100–125 cm	–	–	–	–	1.30	nr	**1.30**
Feathered trees							
125–150 cm	–	–	–	–	2.10	nr	**2.10**
150–180 cm	–	–	–	–	2.50	nr	**2.50**
180–200 cm	–	–	–	–	4.00	nr	**4.00**
200–250 cm	–	–	–	–	6.50	nr	**6.50**
250–300 cm	–	–	–	–	8.00	nr	**8.00**
Native species planting Forming planting holes; in cultivated ground (cultivating not included); by mechanical auger; trimming holes by hand; depositing excavated material alongside holes							
250 mm diameter	–	0.03	0.62	0.04	–	nr	**0.66**
250 × 250 mm	–	0.04	0.74	0.08	–	nr	**0.82**
300 × 300 mm	–	0.08	1.39	0.10	–	nr	**1.49**
Hand excavation; forming planting holes; in cultivated ground (cultivating not included); depositing excavated material alongside holes 100 × 100 × 100 mm deep; with mattock or							
hoe	–	0.01	0.12	–	–	nr	**0.12**
250 × 250 × 300 mm deep	–	0.04	0.74	–	–	nr	**0.74**
300 × 300 × 300 mm deep	–	0.06	1.03	–	–	nr	**1.03**
Q31 TREE/SHRUB PROTECTION							
Tree planting; tree protection – General Preamble: Care must be taken to ensure that tree grids and guards are removed when trees grow beyond the specified diameter of guard.							
Tree planting; tree protection Crowders Nurseries; Crowders Tree Tube; olive green							
1200 mm high × 80 × 80 mm	0.95	0.07	1.23	–	0.95	nr	**2.18**
stakes; 1500 mm high for Crowders Tree Tube; driving into ground	0.70	0.05	0.93	–	0.70	nr	**1.63**

Q PAVING/PLANTING/FENCING/SITE FURNITURE

Item Excluding site overheads and profit	PC £	Labour hours	Labour £	Plant £	Material £	Unit	Total rate £
Crowders Nurseries; expandable plastic tree guards; including 25 mm softwood stakes							
500 mm high	0.68	0.17	3.08	–	0.68	nr	**3.76**
1.00 m high	0.72	0.17	3.08	–	0.72	nr	**3.80**
English Woodlands; Weldmesh Tree Guards; nailing to tree stakes (tree stakes not included)							
1800 mm high × 200 mm diameter	12.50	0.33	6.16	–	12.50	nr	**18.66**
1800 mm high × 250 mm diameter	14.50	0.25	4.63	–	14.50	nr	**19.13**
1800 mm high × 300 mm diameter	15.50	0.33	6.16	–	15.50	nr	**21.66**
J. Toms Ltd; spiral rabbit guards; clear or brown							
450 × 38 mm	0.19	0.03	0.62	–	0.19	nr	**0.81**
610 × 38 mm	0.21	0.03	0.62	–	0.21	nr	**0.83**
English Woodlands; Plastic Mesh Tree Guards; black; supplied in 50 m rolls							
13 × 13 mm small mesh; roll width 60 cm	0.97	0.04	0.66	–	1.47	nr	**2.13**
13 × 13 mm small mesh; roll width 120 cm	1.90	0.06	1.09	–	2.40	nr	**3.49**
English Woodlands; Plastic Mesh Tree Guard in pre-cut pieces; 600 × 150 mm; supplied flat packed	0.81	0.06	1.09	–	1.31	nr	**2.40**
Tree guards of 3 nr 2.40 m × 100 mm stakes; driving 600 mm into firm ground; bracing with timber braces at top and bottom; including 3 strands barbed wire	0.38	1.00	18.50	–	0.38	nr	**18.88**
English Woodlands; strimmer guard in heavy duty black plastic; 225 mm high	2.58	0.07	1.23	–	2.58	nr	**3.81**
Tubex Ltd; Standard Treeshelter inclusive of 25 mm stake; prices shown for quantities of 500 nr							
0.6 m high	0.66	0.05	0.93	–	0.87	each	**1.80**
0.75 m high	0.77	0.05	0.93	–	1.09	each	**2.02**
1.2 m high	0.99	0.05	0.93	–	1.33	each	**2.26**
1.5 m high	1.35	0.05	0.93	–	1.76	each	**2.69**
Tubex Ltd; Shrubshelter inclusive of 25 mm stake; prices shown for quantities of 500 nr							
Ecostart shelter for forestry transplants and seedlings	0.72	0.07	1.23	–	0.72	nr	**1.95**
0.6 m high	1.38	0.07	1.23	–	1.59	nr	**2.82**
0.75 m high	1.50	0.07	1.23	–	1.75	nr	**2.98**
Extra over trees for spraying with antidesiccant spray; Wiltpruf							
selected standards; standards; light standards	2.62	0.20	3.70	–	2.62	nr	**6.32**
standards; heavy standards	4.36	0.25	4.63	–	4.36	nr	**8.99**

Q31 FORESTRY PLANTING

Forestry planting

Deep ploughing rough ground to form planting ridges at							
2.00 m centres	–	0.63	11.56	7.51	–	100 m²	**19.07**
3.00 m centres	–	0.59	10.88	7.07	–	100 m²	**17.95**
4.00 m centres	–	0.40	7.40	4.81	–	100 m²	**12.21**
Notching plant forestry seedlings; T or L notch	33.00	0.75	13.88	–	33.00	100 nr	**46.88**
Turf planting forestry seedlings	33.00	2.00	37.00	–	33.00	100 nr	**70.00**
Tree tubes; to young trees	165.00	0.30	5.55	–	165.00	100 nr	**170.55**
Cleaning and weeding around seedlings; once	–	0.50	9.25	–	–	100 nr	**9.25**

Q PAVING/PLANTING/FENCING/SITE FURNITURE

Item Excluding site overheads and profit	PC £	Labour hours	Labour £	Plant £	Material £	Unit	Total rate £
Q31 FORESTRY PLANTING – cont							
Forestry planting – cont							
Treading in and firming ground around							
seedlings planted; at 2500 per ha after frost or							
other ground disturbance; once	–	0.33	6.17	–	–	100 nr	**6.17**
Beating up initial planting; once (including							
supply of replacement seedlings at 10% of							
original planting)	3.30	0.25	4.63	–	3.30	100 nr	**7.93**
Q31 OPERATIONS AFTER PLANTING							
Operations after planting							
Initial cutting back to shrubs and hedge plants;							
including disposal of all cuttings	–	1.00	18.50	–	–	100 m²	**18.50**
Mulch; Melcourt Industries Ltd; Bark							
Nuggets®; to plant beds; delivered in 80 m³							
loads; maximum distance 25 m							
50 mm thick	2.03	0.04	0.82	–	2.03	m²	**2.85**
75 mm thick	3.05	0.07	1.23	–	3.05	m²	**4.28**
100 mm thick	4.06	0.09	1.64	–	4.06	m²	**5.70**
Mulch; Melcourt Industries Ltd; Bark							
Nuggets®; to plant beds; delivered in 25 m³							
loads; maximum distance 25 m							
50 mm thick	2.41	0.03	0.54	–	2.41	m²	**2.95**
75 mm thick	4.07	0.07	1.23	–	4.07	m²	**5.30**
100 mm thick	5.43	0.09	1.64	–	5.43	m²	**7.07**
Mulch; Melcourt Industries Ltd; Amenity Bark							
Mulch FSC; to plant beds; delivered in 80 m³							
loads; maximum distance 25 m							
50 mm thick	1.27	0.04	0.82	–	1.27	m²	**2.09**
75 mm thick	1.90	0.07	1.23	–	1.90	m²	**3.13**
100 mm thick	2.54	0.09	1.64	–	2.54	m²	**4.18**
Mulch; Melcourt Industries Ltd; Amenity Bark							
Mulch FSC; to plant beds; delivered in 25 m³							
loads; maximum distance 25 m							
50 mm thick	1.95	0.04	0.82	–	1.95	m²	**2.77**
75 mm thick	2.93	0.07	1.23	–	2.93	m²	**4.16**
100 mm thick	3.90	0.09	1.64	–	3.90	m²	**5.54**
Mulch mats; English Woodlands; lay mulch							
mat to planted area or plant station; mat							
secured with metal J pins 240 mm × 3 mm							
Mats to individual plants and plant stations							
Hemcore Biodegradable; 50 × 50 cm;							
square mat	0.80	0.05	0.93	–	1.08	ea	**2.01**
Woven Polypropylene; 50 × 50 cm; square							
mat	0.25	0.05	0.93	–	0.53	ea	**1.46**
Woven Polypropylene; 1 × 1 m; square mat	0.69	0.07	1.23	–	0.97	ea	**2.20**
Hedging mats							
Woven Polypropylene; 1 × 100 m roll;							
hedge planting	0.39	0.01	0.23	–	0.53	m²	**0.76**
General planting areas							
Plantex membrane; 1 × 14 m roll; weed							
control	1.00	0.01	0.19	–	1.14	m²	**1.33**

Q PAVING/PLANTING/FENCING/SITE FURNITURE

Item Excluding site overheads and profit	PC £	Labour hours	Labour £	Plant £	Material £	Unit	Total rate £
Fertilizers; application during aftercare period							
Fertilizers; in top 150 mm of topsoil at 35 g/m²							
Mascot Microfine; turf fertilizer; 8+0+6 + 2% Mg + 4% Fe	5.00	0.12	2.27	–	5.00	100 m²	7.27
Enmag; controlled release fertilizer (8-9 months); 11+22+09	7.64	0.12	2.27	–	7.64	100 m²	9.91
Mascot Outfield; turf fertilizer; 8+12+8	3.71	0.12	2.27	–	3.71	100 m²	5.98
Mascot Outfield; turf fertilizer; 9+5+5	3.01	0.12	2.27	–	3.01	100 m²	5.28
Super Phosphate Powder	4.48	0.12	2.27	–	4.48	100 m²	6.75
Bone Meal	4.07	0.12	2.27	–	4.07	100 m²	6.34
Fertilizers; in top 150 mm of topsoil at 70 g/m²							
Mascot Microfine; turf fertilizer; 8+0+6 + 2% Mg + 4% Fe	10.01	0.12	2.27	–	10.01	100 m²	12.28
Enmag; controlled release fertilizer (8-9 months); 11+22+09	15.29	0.12	2.27	–	15.29	100 m²	17.56
Mascot Outfield; turf fertilizer; 8+12+8	7.42	0.12	2.27	–	7.42	100 m²	9.69
Mascot Outfield; turf fertilizer; 9+5+5	6.02	0.12	2.27	–	6.02	100 m²	8.29
Super Phosphate Powder	8.97	0.12	2.27	–	8.97	100 m²	11.24
Bone meal	8.14	0.12	2.27	–	8.14	100 m²	10.41
Q31 AFTERCARE AS PART OF A LANDSCAPE CONTRACT							
Maintenance operations (Note: the following rates apply to aftercare maintenance executed as part of a landscaping contract only – Please see section Q35 of this publication for further maintenance rates)							
Weeding and hand forking planted areas; including disposing weeds and debris on site; areas maintained weekly	–	–	0.07	–	–	m²	0.07
Weeding and hand forking planted areas; including disposing weeds and debris on site; areas maintained monthly	–	0.01	0.19	–	–	m²	0.19
Extra over weeding and hand forking planted areas for disposing excavated material off site; to tip not exceeding 13 km; mechanically loaded							
slightly contaminated	–	–	–	1.68	24.00	m³	25.68
rubbish	–	–	–	1.68	24.00	m³	25.68
inert material	–	–	–	1.12	8.00	m³	9.12
Mulch; Melcourt Industries Ltd; Bark Nuggets®; to plant beds; delivered in 80 m³ loads; maximum distance 25 m							
50 mm thick	2.03	0.04	0.82	–	2.03	m²	2.85
75 mm thick	3.05	0.07	1.23	–	3.05	m²	4.28
100 mm thick	4.06	0.09	1.64	–	4.06	m²	5.70
Mulch; Melcourt Industries Ltd; Bark Nuggets®; to plant beds; delivered in 25 m³ loads; maximum distance 25 m							
50 mm thick	2.71	0.04	0.82	–	2.71	m²	3.53
75 mm thick	4.07	0.07	1.23	–	4.07	m²	5.30
100 mm thick	5.43	0.09	1.64	–	5.43	m²	7.07

Q PAVING/PLANTING/FENCING/SITE FURNITURE

Item Excluding site overheads and profit	PC £	Labour hours	Labour £	Plant £	Material £	Unit	Total rate £
Q31 AFTERCARE AS PART OF A **LANDSCAPE CONTRACT – cont**							
Maintenance operations (Note: the **following rates apply to aftercare** **maintenance executed as part of a** **landscaping contract only – Please see** **section Q35 of this publication for further** **maintenance rates) – cont**							
Mulch; Melcourt Industries Ltd; Amenity Bark Mulch FSC; to plant beds; delivered in 80 m³ loads; maximum distance 25 m							
50 mm thick	1.27	0.04	0.82	–	1.27	m²	2.09
75 mm thick	1.90	0.07	1.23	–	1.90	m²	3.13
100 mm thick	2.54	0.09	1.65	–	2.54	m²	4.19
Mulch; Melcourt Industries Ltd; Amenity Bark Mulch FSC; to plant beds; delivered in 25 m³ loads; maximum distance 25 m							
50 mm thick	1.95	0.04	0.82	–	1.95	m²	2.77
75 mm thick	2.93	0.07	1.23	–	2.93	m²	4.16
100 mm thick	3.90	0.09	1.64	–	3.90	m²	5.54
Fertilizers; at 35 g/m²							
Mascot Microfine; turf fertilizer; 8+0+6 + 2% Mg + 4% Fe	5.00	0.12	2.27	–	5.00	100 m²	7.27
Enmag; controlled release fertilizer (8–9 months); 11+22+09	7.64	0.12	2.27	–	7.64	100 m²	9.91
Mascot Outfield; turf fertilizer; 8+12+8	3.71	0.12	2.27	–	3.71	100 m²	5.98
Mascot Outfield; turf fertilizer; 9+5+5	3.01	0.12	2.27	–	3.01	100 m²	5.28
Super Phosphate Powder	4.48	0.12	2.27	–	4.48	100 m²	6.75
Bone meal	4.07	0.12	2.27	–	4.07	100 m²	6.34
Fertilizers; at 70 g/m²							
Mascot Microfine; turf fertilizer; 8+0+6 + 2% Mg + 4% Fe	10.01	0.12	2.27	–	10.01	100 m²	12.28
Enmag; controlled release fertilizer (8–9 months); 11+22+09	15.29	0.12	2.27	–	15.29	100 m²	17.56
Mascot Outfield; turf fertilizer; 8+12+8	7.42	0.12	2.27	–	7.42	100 m²	9.69
Mascot Outfield; turf fertilizer; 9+5+5	6.02	0.12	2.27	–	6.02	100 m²	8.29
Super Phosphate Powder	8.97	0.12	2.27	–	8.97	100 m²	11.24
Bone meal	8.14	0.12	2.27	–	8.14	100 m²	10.41
Watering planting; evenly; at a rate of 5 l/m²							
using hand-held watering equipment	–	0.25	4.63	–	–	100 m²	4.63
using sprinkler equipment and with sufficient water pressure to run one 15 m radius sprinkler	–	0.14	2.57	–	–	100 m²	2.57
using movable spray lines powering three sprinkler heads with a radius of 15 m and allowing for 60% overlap (irrigation machinery costs not included)	–	0.02	0.29	–	–	100 m²	0.29

Q PAVING/PLANTING/FENCING/SITE FURNITURE

Item Excluding site overheads and profit	PC £	Labour hours	Labour £	Plant £	Material £	Unit	Total rate £
Q35 LANDSCAPE MAINTENANCE							
Preamble: Long-term landscape maintenance							
Maintenance on long-term contracts differs in cost from that of maintenance as part of a landscape contract. In this section the contract period is generally 3-5 years. Staff are generally allocated to a single project only and therefore productivity is higher whilst overhead costs are lower. Labour costs in this section are lower than the costs used in other parts of the book. Machinery is assumed to be leased over a five year period and written off over the same period. The costs of maintenance and consumables for the various machinery types have been included in the information that follows. Finance costs for the machinery have not been allowed for. The rates shown below are for machines working in unconfined contiguous areas. Users should adjust the times and rates if working in smaller spaces or spaces with obstructions.							
LANDSCAPE CHEMICALS							
Chemical applications labours only							
Controlled droplet application (CDA)							
plants at 1.50 m centres; 4444 nr/ha	–	18.52	342.62	–	–	ha	**342.62**
plants at 1.50 m centres; 44.44 nr/100 m²	–	1.85	34.26	–	–	100 m²	**34.26**
plants at 1.75 m centres; 3265 nr/ha	–	13.60	251.60	–	–	ha	**251.60**
plants at 1.75 m centres; 3265 nr/ha	–	1.36	25.16	–	–	100 m²	**25.16**
plants at 2.00 m centres; 2500 nr/ha	–	10.42	192.77	–	–	ha	**192.77**
plants at 2.00 m centres; 2500 nr/ha	–	1.04	19.28	–	–	100 m²	**19.28**
spot spraying to shrub beds	–	0.40	7.40	–	–	100 m²	**7.40**
spot spraying to hard surfaces	–	0.25	4.63	–	–	100 m²	**4.63**
mass spraying	–	5.00	92.50	–	–	ha	**92.50**
mass spraying	–	0.50	9.25	–	–	100 m²	**9.25**
Power spraying							
mass spraying to areas containing vegetation	–	4.00	74.00	60.02	–	ha	**134.02**
mass spraying to hard surfaces	–	0.01	0.15	0.12	–	100 m²	**0.27**
spraying to kerbs and channels along roadways 300 mm wide	–	2.50	46.25	37.51	–	1 km	**83.76**
General herbicides; in accordance with manufacturer's instructions; Knapsack spray application; see 'Market rates of chemicals' for costs of specific applications							
knapsack sprayer; selective spraying around bases of plants	–	0.35	6.49	–	–	100 m²	**6.49**
knapsack sprayer; spot spraying	–	0.20	3.70	–	–	100 m²	**3.70**
knapsack sprayer; mass spraying	–	0.20	3.70	–	–	100 m²	**3.70**
granular distribution by hand or hand applicator	–	0.40	7.40	–	–	100 m²	**7.40**
Backpack spraying; keeping weed free at bases of plants							
plants at 1.50 m centres; 4444 nr/ha	–	27.78	513.93	–	–	ha	**513.93**
plants at 1.75 m centres; 3265 nr/ha	–	20.40	377.40	–	–	ha	**377.40**
plants at 2.00 m centres; 2500 nr/ha	–	15.63	289.15	–	–	ha	**289.15**
mass spraying	–	15.00	277.50	–	–	ha	**277.50**

Q PAVING/PLANTING/FENCING/SITE FURNITURE

Item Excluding site overheads and profit	PC £	Labour hours	Labour £	Plant £	Material £	Unit	Total rate £
Q35 LANDSCAPE MAINTENANCE – cont							
The following table provides the areas of spraying in each planting scenario for plants which require 1.00 m diameter weed free circles							
Area of weed free circles required in each planting density							
plants at 500 mm centres; 400 nr/100 m²	–	–	–	–	400.00	m²	**400.00**
plants at 600 m centres; 278 nr/100 m²	–	–	–	–	278.00	m²	**278.00**
plants at 750 mm centres; 178 nr/100 m²	–	–	–	–	178.00	m²	**178.00**
plants at 1.00 m centres; 100 nr/100 m²	–	–	–	–	100.00	m²	**100.00**
plants at 1.50 m centres; 4444 nr/ha	–	–	–	–	3490.00	m²	**3490.00**
plants at 1.75 m centres; 3265 nr/ha	–	–	–	–	2564.00	m²	**2564.00**
plants at 2.00 m centres; 2500 nr/ha	–	–	–	–	1963.00	m²	**1963.00**
MARKET PRICES OF LANDSCAPE CHEMICALS AT SUGGESTED APPLICATION RATES							
Note: All chemicals are standard knapsack or backpack applied unless specifically stated as CDA or TDA.							
Total herbicides							
Herbicide applications; CDA; chemical application via low pressure specialized wands to landscape planting; application to maintain 1.00 m diameter clear circles (0.79 m²) around new planting							
Vanquish Biactive; ALS Ltd; Bayer; enhanced movement glyphosate; application rate 15 litre/ha							
plants at 1.50 m centres; 4444 nr/ha	–	–	–	–	52.74	ha	**52.74**
plants at 1.75 m centres; 3265 nr/ha	–	–	–	–	38.93	ha	**38.93**
plants at 2.00 m centres; 2500 nr/ha	–	–	–	–	29.85	ha	**29.85**
mass spraying	–	–	–	–	151.27	ha	**151.27**
spot spraying	–	–	–	–	1.51	ha	**1.51**
Total herbicides; CDA; low pressure sustainable applications							
Hilite; Nomix Enviro Ltd; glyphosate oil-emulsion; Total Droplet Control (TDC formerly CDA); low volume sustainable chemical applications; formulation controls annual and perennial grasses and non-woody broad-leaved weeds in amenity and industrial areas, parks, highways, paved areas and factory sites; also effective for clearing vegetation prior to planting of ornamental species, around the bases of trees, shrubs and roses, as well as in forestry areas							
10 l/ha	–	–	–	–	1.63	100 m²	**1.63**
7.5 l/ ha	–	–	–	–	1.22	100 m²	**1.22**

Q PAVING/PLANTING/FENCING/SITE FURNITURE

Item Excluding site overheads and profit	PC £	Labour hours	Labour £	Plant £	Material £	Unit	Total rate £
Tribute Plus; Nomix Enviro Ltd; TDA; MCPA/ dicamba/mecaprop-p; TDC; for selective control of broad-leaved weeds in turf							
6 l/ha; white clover, plantains, creeping buttercup, thistle	–	–	–	–	0.66	100 m²	**0.66**
8 l/ha; yarrow, ragwort, parsley, piert, cinquefoil, mouse eared chickweed, selfheal, pearlwort, sorrel, black meddick, dandelion and related weeds, bulbous buttercup	–	–	–	–	0.87	100 m²	**0.87**
Stirrup; Rigby Taylor Ltd; glyphosate; CDA; 192 g							
10 l/ha	–	–	–	–	1.63	100 m²	**1.63**
7.5 l/ha	–	–	–	–	1.63	100 m²	**1.63**
Gallup Biograde Amenity; Rigby Taylor Ltd; glyphosate 360 g/l formulation							
general use; woody weeds; ash, beech, bracken, bramble; 3 l/ha	–	–	–	–	0.38	100 m²	**0.38**
annual and perennial grasses; heather (peat soils); 4 l/ha	–	–	–	–	0.51	100 m²	**0.51**
pre-planting; general clearance; 5 l/ha	–	–	–	–	0.63	100 m²	**0.63**
heather; mineral soils; 6 l/ha	–	–	–	–	0.76	100 m²	**0.76**
rhododendron; 10 l/ha	–	–	–	–	1.26	100 m²	**1.26**
Gallup Hi-aktiv Amenity; Rigby Taylor Ltd; 490 g/l formulation							
general use; woody weeds; ash, beech, bracken, bramble; 2.2 l/ha	–	–	–	–	0.35	100 m²	**0.35**
annual and perennial grasses; heather (peat soils); 2.9 l/ha	–	–	–	–	0.46	100 m²	**0.46**
pre-planting; general clearance; 3.7 l/ha	–	–	–	–	0.59	100 m²	**0.59**
heather; mineral soils; 4.4 l/ha	–	–	–	–	0.70	100 m²	**0.70**
rhododendron; 7.3 l/ha	–	–	–	–	1.16	100 m²	**1.16**
Discman CDA Biograde; Rigby Taylor Ltd							
application rate; 7 l/ha; light vegetation and annual weeds	–	–	–	–	1.42	100 m²	**1.42**
application rate; 8.5 l/ha; forestry/ woodlands; before planting, to control broad-leaved and grass weeds	–	–	–	–	1.72	100 m²	**1.72**
application rate; 10 l/ha; established deep-rooted annuals, perennial grasses and broad-leaved weeds	–	–	–	–	2.02	100 m²	**2.02**
Total herbicides; standard backpack or **power sprayer applications**							
Roundup Pro Biactive 360; Scotts UK							
application rate; 5 l/ha; land not intended for cropping	–	–	–	–	0.61	100 m²	**0.61**
application rate; 10 l/ha; forestry; weed control	–	–	–	–	1.21	100 m²	**1.21**
Roundup Pro Biactive 450; Scotts UK; 450 g/ litre glyphosate							
application rate 4 l/ha; natural surfaces not intended to bear vegetation; amenity vegetation control; permeable surfaces overlaying soil; hard surfaces	–	–	–	–	0.58	100 m²	**0.58**
application rate 8 l/ha; forest; forest nursery weed control	–	–	–	–	1.15	100 m²	**1.15**
application rate 16 mm/l; stump application	–	–	–	–	0.23	litre	**0.23**

Q PAVING/PLANTING/FENCING/SITE FURNITURE

Item Excluding site overheads and profit	PC £	Labour hours	Labour £	Plant £	Material £	Unit	Total rate £
Q35 LANDSCAPE MAINTENANCE – cont							
Total herbicides – cont							
Roundup Pro Biactive 450; Scotts UK;							
enhanced movement glyphosate; application							
rate 4 litre/ha							
plants at 1.50 m centres; 4444 nr/ha	83.32	–	–	–	83.32	ha	**83.32**
plants at 1.75 m centres; 3265 nr/ha	61.28	–	–	–	61.28	ha	**61.28**
plants at 2.00 m centres; 2500 nr/ha	46.89	–	–	–	46.89	ha	**46.89**
mass spraying	63.36	–	–	–	63.36	ha	**63.36**
spot spraying	2.38	–	–	–	2.38	100 m²	**2.38**
Kerb							
Kerb Flowable; propyzamide	–	–	–	–	2.43	100 m²	**2.43**
Kerb Granules; 2 × 2000 tree pack	–	–	–	–	0.08	tree	**0.08**
Dual; Nomix Enviro Ltd; glyphosate/							
sulfosulfuron; TDC; low volume sustainable							
chemical applications; residual effect lasting							
up to 6 months; shrub beds, gravelled areas,							
tree bases, parks, industrial areas and around							
obstacles							
application rate 9 l/ha	–	–	–	–	2.03	100 m²	**2.03**
Glyfos Dakar Pro; Headland Amenity Ltd; 68%							
glyphosate as water dispersible granules;							
broad-leaved weeds and grasses							
application rate 2.5 kg/ha	–	–	–	–	0.32	100 m²	**0.32**
Total herbicides with residual pre-emergent							
action							
Pistol; ALS; Bayer; barrier prevention for							
emergence control and glyphosate for control							
of emerged weeds at a single application per							
season; application rate 4.5 litre/ha; 40 g/l							
diflufenican and 250 g/l glyphosate							
plants at 1.50 m centres; 4444 nr/ha	–	–	–	–	36.53	ha	**36.53**
plants at 1.75 m centres; 3265 nr/ha	–	–	–	–	26.75	ha	**26.75**
plants at 2.00 m centres; 2500 nr/ha	–	–	–	–	26.17	ha	**26.17**

Q PAVING/PLANTING/FENCING/SITE FURNITURE

Item Excluding site overheads and profit	PC £	Labour hours	Labour £	Plant £	Material £	Unit	Total rate £
Residual herbicides Chikara; Rigby Taylor Ltd; Belchim Crop Protection; flazasulfuron; systemic pre-emergent and early post-emergent herbicide for the control of annual and perennial weeds on natural surfaces not intended to bear vegetation and permeable surfaces over-lying soil							
150 g/ha	–	–	–	–	2.41	100 m²	**2.41**
Contact herbicides Jewel; Scotts UK; carfentrazone and mecoprop-p; selective weed and mosskiller							
application rate; 1.5 kg/ha	–	–	–	–	1.28	100 m²	**1.28**
Natural Weed and Moss Spray; Headland Amenity Ltd; acetic acid							
application rate; 100 l/ha	–	–	–	–	4.50	100 m²	**4.50**
Selective herbicides CDA Hilite; Nomix Enviro; glyphosate oil-emulsion; TD; low volume sustainable chemical applications; formulation controls annual and perennial grasses and non-woody broad-leaved weeds in amenity and industrial areas, parks, highways, paved areas and factory sites; also effective for clearing vegetation prior to planting of ornamental species, around the bases of trees, shrubs and roses, as well as in forestry areas							
10 l/ha	–	–	–	–	1.63	100 m²	**1.63**
7.5 l/ha	–	–	–	–	1.22	100 m²	**1.22**
Tribute Plus; Nomix Enviro; MCPA/dicamba/mecaprop-p; TDC; for selective control of broad-leaved weeds in turf							
6 l/ha; white clover, plantains, creeping buttercup, thistle	–	–	–	–	0.66	100 m²	**0.66**
8 l/ha; yarrow, ragwort, parsley, piert, cinquefoil, mouse eared chickweed, selfheal, pearlwort, sorrel, black meddick, dandelion and related weeds, bulbous buttercup	–	–	–	–	0.87	100 m²	**0.87**
Contact and systemic combined herbicides Jewel; Scotts UK; carfentrazone and mecoprop-p; selective weed and mosskiller							
application rate; 1.5 kg/ha	–	–	–	–	1.28	100 m²	**1.28**

Q PAVING/PLANTING/FENCING/SITE FURNITURE

Item Excluding site overheads and profit	PC £	Labour hours	Labour £	Plant £	Material £	Unit	Total rate £
Q35 LANDSCAPE MAINTENANCE – cont							
Contact and systemic combined herbicides – cont							
Finale; ALS; Bayer; glufosinate-ammonium; controls grasses and broad-leaved weeds; it can also be used for weed control around non-edible ornamental crops including trees and shrubs and in tree nurseries; may be used to 'burn off' weeds in shrub beds prior to treatment with a residual herbicide; is also ideal for preparing sports turf for linemarking							
application rate; 5 l/ha; seedlings of all species; established annual weeds and grasses as specified	–	–	–	–	0.82	100 m²	0.82
application rate; 8 l/ha; established annual and perennial weeds and grasses as specified in the product guide	–	–	–	–	1.32	100 m²	1.32
Selective herbicides; power or knapsack sprayer							
Junction; Rigby Taylor Ltd; selective post-emergence turf herbicide containing 6.25 g/litre florasulam plus 452 g/litre 2,4-D; specially designed for use by contractors for maintenance treatments of a wide range of broad-leaved weeds whilst giving excellent safety to turfgrass							
application rate; 1.2 l/ha	–	–	–	–	0.55	100 m²	0.55
Greenor; Rigby Taylor Ltd; fluroxypyr, clopyralid and MCPA; systemic selective treatment of weeds in turfgrass							
application rate; 4 l/ha	–	–	–	–	1.00	100 m²	1.00
Super Selective Plus; Rigby Taylor Ltd; combination of three active ingredients (mecoprop-p, MCPA and dicamba) enables this formulation to give good control of most commonly occurring broad-leaved weeds found in sports and amenity turf							
application rate; 3.5 l/ha	–	–	–	–	0.20	100 m²	0.20
Crossbar Herbicide; Rigby Taylor Ltd; fluroxypyr, 2,4-D and dicamba for control of weeds in amenity turf including fine turf							
application rate; 2 l/ha	–	–	–	–	0.93	100 m²	0.93
Barclay Greengard; Rigby Taylor Ltd; 2,4-D, triclopyr, and dicamba; selective herbicide for the control of certain broad-leaved weeds in amenity grassland and scrub control in forestry and uncropped areas including Japanese Knotweed and weeds listed under the Weeds Act of 1959							
application rate; 5 l/ha	–	–	–	–	1.66	100 m²	1.66
application rate; 2 l/ha	–	–	–	–	0.66	100 m²	0.66

Q PAVING/PLANTING/FENCING/SITE FURNITURE

Item Excluding site overheads and profit	PC £	Labour hours	Labour £	Plant £	Material £	Unit	**Total rate £**
Intrepid 2; Scotts UK; for broad-leaved weed control in grass							
application rate; 7700 m²/5 l	–	–	–	–	0.51	100 m²	**0.51**
Re-Act; Scotts UK; MCPA, mecoprop-p and dicamba; selective herbicide; established managed amenity turf and newly seeded grass; controls many annual and perennial weeds; will not vaporize in hot conditions							
application rate; 9090–14,285 m²/5 l	–	–	–	–	0.46	100 m²	**0.46**
Praxys; Scotts UK; fluroxypyr-meptyl, clopyralid and florasulam; selective, systemic post-emergence herbicide for managed amenity turf, lawns and amenity grassland with high selectivity to established and young turf							
application rate; 9090–14,285 m²/5 l	–	–	–	–	1.10	100 m²	**1.10**
Asulox; ALS; Bayer; post-emergence translocated herbicide for the control of bracken and docks							
application rate; docks; 17857 m²/5 l	–	–	–	–	0.37	100 m²	**0.37**
application rate; bracken; 4545 m²/5 l	–	–	–	–	1.45	100 m²	**1.45**
Spearhead; ALS; Bayer; selective herbicide for the control of broad-leaved weeds in established turf							
application rate; 4.5 l/ha	–	–	–	–	0.85	100 m²	**0.85**
Cabadex; Headland Amenity Ltd; fluroxypyr and florasulam; newly sown and established turf (particularly speedwell, yarrow and yellow suckling clover)							
application rate 2 l/ha	–	–	–	–	0.67	100 m²	**0.67**
Relay Turf; Headland Amenity Ltd; mecoprop-p, MCPA and dicamba; many broad-leaved weeds in managed amenity turf							
application rate; 5 l/ha	–	–	–	–	0.45	100 m²	**0.45**
Blaster; Headland Amenity Ltd; triclopyr clopyralid; nettles, docks, thistles, brambles, woody weeds in amenity grassland							
application rate; 4 l/ha	–	–	–	–	1.08	100 m²	**1.08**
Woody weed herbicides; backpack or power sprayers							
Timbrel; Rigby Taylor Ltd; Bayer; summer applied; water-based; selective scrub and brushwood herbicide							
2.0 l/ha; bramble, briar, broom, gorse, nettle	–	–	–	–	1.33	100 m²	**1.33**
4.0 l/ha; alder, birch, blackthorn, dogwood, elder, poplar, rosebay willowherb, sycamore	–	–	–	–	2.66	100 m²	**2.66**
6.0 l/ha; beech, box, buckthorn, elm, hazel, hornbeam, horse chestnut, lime, maple, privet, rowan, Spanish chestnut, willow, wild pear	–	–	–	–	3.98	100 m²	**3.98**
8.0 l/ha; ash, oak, rhododendron	–	–	–	–	5.31	100 m²	**5.31**

Q PAVING/PLANTING/FENCING/SITE FURNITURE

Item Excluding site overheads and profit	PC £	Labour hours	Labour £	Plant £	Material £	Unit	Total rate £
Q35 LANDSCAPE MAINTENANCE – cont							
Woody weed herbicides – cont							
Timbrel; Rigby Taylor Ltd; Bayer; winter							
applied; paraffin or diesel based; selective							
scrub and brushwood herbicide							
3.0 l/ha; bramble, briar, broom, gorse,							
nettle	–	–	–	–	1.99	100 m²	**1.99**
6.0 l/ha; alder, ash, beech, birch,							
blackthorn, box, buckthorn, dogwood,							
elder, elm, hazel, hornbeam, horse							
chestnut, lime, maple, oak, poplar, privet,							
rowan, Spanish chestnut, sycamore,							
willow, wild pear	–	–	–	–	3.98	100 m²	**3.98**
10.0 l/ha; hawthorn, laurel, rhododendron	–	–	–	–	6.64	100 m²	**6.64**
Garlon 4; Nomix Enviro Ltd; 667 g/litre							
triclopyr; control of brushwood, scrub, woody							
and perennial herbaceous weeds in amenity,							
forestry, rail and industrial areas; also suitable							
for forestry production							
2 l/ha; bramble, briar, broom, gorse, nettle	–	–	–	–	1.02	100 m²	**1.02**
4 l/ha; alder, birch, blackthorn,dogwood,							
elder, poplar, rosebay, willow-herb,							
sycamore	–	–	–	–	2.05	100 m²	**2.05**
6 l/ha; beech, box, elm, buckthorn, hazel,							
hornbeam, horse chestnut, lime, maple,							
privet, rowan, Spanish chestnut, willow,							
wild pear	–	–	–	–	3.08	100 m²	**3.08**
8 l/ha; rhododendron, ash, oak	–	–	–	–	4.10	100 m²	**4.10**
Tordon 22K; Nomix Enviro; picloram; selective							
action; control of tough woody weeds, invasive							
species, scrub and deep-rooted perennials;							
residual action grass cover is maintained							
2.8 l/ha; white clover, creeping thistle,							
hogweed, plantains	–	–	–	–	1.22	100 m²	**1.22**
4.3–5.6 l/ha; common bird's-foot trefoil,							
field bindweed, creeping buttercup,							
burdock, cat's ear, common chickweed,							
colt's foot, creeping cinquefoil, dandelion,							
docks, hawkweed, Japanese knotweed,							
mugwort, common nettle, silverweed,							
tansy, cotton thistle, rosebay willowherb,							
yarrow	–	–	–	–	2.17	100 m²	**2.17**
8.4 l/ha; wild mignonettte, common ragwort	–	–	–	–	3.65	100 m²	**3.65**
11.2 l/ha; bracken, brambles	–	–	–	–	7.91	100 m²	**7.91**
Japanese knotweed and ragwort control							
Loram 24; ALS Ltd; Bayer; picloram							
4.2 l/ha; Japanese Knotweed and other							
weeds as listed in the suppliers literature	–	–	–	–	1.64	100 m²	**1.64**
5.6 l/ha; common ragwort, wild mignonette.	–	–	–	–	2.18	100 m²	**2.18**
Aquatic herbicides; backpack or power							
sprayers							
Roundup Pro Biactive 360; Scotts UK							
application rate; 6 l/ha; aquatic weed							
control	–	–	–	–	0.73	100 m²	**0.73**

Q PAVING/PLANTING/FENCING/SITE FURNITURE

Item Excluding site overheads and profit	PC £	Labour hours	Labour £	Plant £	Material £	Unit	Total rate £
Roundup Pro Biactive 450; Scotts UK; 450 g/litre glyphosate							
application rate 4.8 l/ha; enclosed waters, open waters, land immediately adjacent to aquatic areas	–	–	–	–	0.69	100 m²	**0.69**
Aquatic herbicides; TDC							
Conqueror; Nomix Enviro; low volume sustainable chemical applications glyphosate with aquatic approval							
10 l/ha; established annuals, deep-rooted perennials and broad-leaved weeds	–	–	–	–	0.18	100 m²	**0.18**
7.5 l/ha; light vegetation and small annual weeds aquatic situations	–	–	–	–	1.32	100 m²	**1.32**
12.5 l/ha; aquatic situations emergent weeds e.g. reed, grasses and watercress	–	–	–	–	0.22	100 m²	**0.22**
15 l/ha; aquatic situations floating weeds such as water lilies	–	–	–	–	0.26	100 m²	**0.26**
Fungicides							
Masalon; Rigby Taylor Ltd; systemic							
application rate; 8 l/ha	–	–	–	–	7.86	100 m²	**7.86**
Defender; Rigby Taylor Ltd; broad spectrum fungicide displaying protective and curative activity against fusarium patch and red thread							
application rate; 700 g/ha	–	–	–	–	6.43	100 m²	**6.43**
Fusion; Rigby Taylor Ltd; contact and systemic fungicide with curative and preventative activity against major turf diseases such as fusarium, red thread, dollar spot, anthracnose and rust							
application rate; 700 g/ha	–	–	–	–	6.24	100 m²	**6.24**
Rayzor; Rigby Taylor Ltd; broad spectrum contact fungicide with both protectant and eradicant activity for the control of fusarium patch and red thread in managed amenity turf							
application rate; 5 l/2000 m²	–	–	–	–	8.69	100 m²	**8.69**
Mascot Contact; multi-site protectant fungicide for the control of a range of diseases on turf							
application rate; 5 l/1667 m²	–	–	–	–	6.60	100 m²	**6.60**
Moss control							
Jewel; Scotts UK; carfentrazone and mecoprop-p; selective weed and moss killer							
application rate; 1.5 kg/ha	–	–	–	–	1.28	100 m²	**1.28**
Insect control							
Crossfire 480; Rigby Taylor Ltd; control of leatherjackets and fruit fly in sports and amenity turf							
application rate; 1.5 l/ha	–	–	–	–	0.58	100 m²	**0.58**
Merit Turf; Bayer; insecticide for the control of chafer grubs and leatherjackets							
application rate; 10 kg/3333 m²	–	–	–	–	5.18	100 m²	**5.18**

Q PAVING/PLANTING/FENCING/SITE FURNITURE

Item Excluding site overheads and profit	PC £	Labour hours	Labour £	Plant £	Material £	Unit	Total rate £
Q35 LANDSCAPE MAINTENANCE – cont							
Harrowing							
Harrowing grassed area with							
drag harrow	–	0.01	0.19	0.14	–	100 m²	0.33
chain or light flexible spiked harrow	–	0.02	0.26	0.14	–	100 m²	0.40
Scarifying							
Mechanical							
Sisis ARP4; including grass collection box;							
towed by tractor; area scarified annually	–	0.02	0.36	0.10	–	100 m²	0.46
Sisis ARP4; including grass collection box;							
towed by tractor; area scarified two years							
previously	–	0.03	0.43	0.12	–	100 m²	0.55
pedestrian operated self-powered							
equipment	–	0.07	1.08	0.22	–	100 m²	1.30
add for disposal of arisings	–	0.03	0.39	1.02	14.98	100 m²	16.39
By hand							
hand implement	–	0.50	7.75	–	–	100 m²	7.75
add for disposal of arisings	–	0.03	0.39	1.02	14.98	100 m²	16.39
Rolling							
Rolling grassed area; equipment towed by							
tractor; once over; using							
smooth roller	–	0.01	0.22	0.15	–	100 m²	0.37
Turf aeration							
By machine							
Vertidrain turf aeration equipment towed by							
tractor to effect a minimum penetration of							
100–250 mm at 100 mm centres	–	0.04	0.60	1.29	–	100 m²	1.89
Ryan GA 30; self-propelled turf aerating							
equipment; to effect a minimum penetration							
of 100 mm at varying centres	–	0.04	0.60	1.51	–	100 m²	2.11
Groundsman; pedestrian operated;							
self-powered solid or slitting tine turf							
aerating equipment to effect a minimum							
penetration of 100 mm	–	0.14	2.17	1.43	–	100 m²	3.60
Cushman core harvester; self-propelled; for							
collection of arisings	–	0.02	0.30	0.66	–	100 m²	0.96
By hand							
hand fork; to effect a minimum penetration							
of 100 mm and spaced 150 mm apart	–	1.33	20.67	–	–	100 m²	20.67
hollow tine hand implement; to effect a							
minimum penetration of 100 mm and							
spaced 150 mm apart	–	2.00	31.00	–	–	100 m²	31.00
collection of arisings by hand	–	3.00	46.50	–	–	100 m²	46.50
Turf areas; surface treatments and top							
dressing; Boughton Loam Ltd							
Apply screened topdressing to grass surfaces;							
spread using Tru-Lute							
Sand soil mixes 90/10 to 50/50	0.14	–	0.03	0.03	0.14	m²	0.20
Apply screened soil 3 mm Kettering loam to							
goal mouths and worn areas							
20 mm thick	1.56	0.01	0.16	–	1.56	m²	1.72
10 mm thick	0.78	0.01	0.16	–	0.78	m²	0.94

Q PAVING/PLANTING/FENCING/SITE FURNITURE

Item Excluding site overheads and profit	PC £	Labour hours	Labour £	Plant £	Material £	Unit	Total rate £
Leaf clearance; clearing grassed area of leaves and other extraneous debris							
Using equipment towed by tractor							
large grassed areas with perimeters of mature trees such as sports fields and amenity areas	–	0.01	0.19	0.07	–	100 m²	**0.26**
large grassed areas containing ornamental trees and shrub beds	–	–	0.07	0.98	–	100 m²	**1.05**
Using pedestrian operated mechanical equipment and blowers							
grassed areas with perimeters of mature trees such as sports fields and amenity areas	–	0.02	0.26	1.57	–	100 m²	**1.83**
grassed areas containing ornamental trees and shrub beds	–	0.10	1.55	0.67	–	100 m²	**2.22**
verges	–	0.07	1.03	0.09	–	100 m²	**1.12**
By hand							
grassed areas with perimeters of mature trees such as sports fields and amenity areas	–	0.10	1.55	0.20	–	100 m²	**1.75**
grassed areas containing ornamental trees and shrub beds	–	0.17	2.58	0.33	–	100 m²	**2.91**
verges	–	0.33	5.17	0.66	–	100 m²	**5.83**
Removal of arisings							
areas with perimeters of mature trees	–	0.01	0.10	0.09	1.20	100 m²	**1.39**
areas containing ornamental trees and shrub beds	–	0.02	0.31	0.34	3.00	100 m²	**3.65**
Litter clearance							
Collection and disposal of litter from grassed area	–	0.01	0.16	–	0.07	100 m²	**0.23**
Collection and disposal of litter from isolated grassed area not exceeding 1000 m²	–	0.04	0.62	–	0.07	100 m²	**0.69**
Edge maintenance							
Maintain edges where lawn abuts pathway or hard surface using							
strimmer	–	0.01	0.08	0.01	–	m	**0.09**
shears	–	0.02	0.26	–	–	m	**0.26**
Maintain edges where lawn abuts plant bed using							
mechanical edging tool	–	0.01	0.10	0.03	–	m	**0.13**
shears	–	0.01	0.17	–	–	m	**0.17**
half moon edging tool	–	0.02	0.31	–	–	m	**0.31**
Tree guards, stakes and ties							
Adjusting existing tree tie	–	0.03	0.52	–	–	nr	**0.52**
Taking up single or double tree stake and ties; removing and disposing	–	0.05	0.93	–	–	nr	**0.93**
Pruning shrubs							
Trimming ground cover planting							
soft groundcover; vinca ivy and the like	–	1.00	15.50	–	–	100 m²	**15.50**
woody groundcover; cotoneaster and the like	–	1.50	23.25	–	–	100 m²	**23.25**

Q PAVING/PLANTING/FENCING/SITE FURNITURE

Item Excluding site overheads and profit	PC £	Labour hours	Labour £	Plant £	Material £	Unit	Total rate £
Q35 LANDSCAPE MAINTENANCE – cont							
Pruning shrubs – cont							
Pruning massed shrub border (measure ground area)							
shrub beds pruned annually	–	0.01	0.16	–	–	m²	0.16
shrub beds pruned hard every 3 years	–	0.03	0.43	–	–	m²	0.43
Cutting off dead heads							
bush or standard rose	–	0.05	0.78	–	–	nr	0.78
climbing rose	–	0.08	1.29	–	–	nr	1.29
Pruning roses							
bush or standard rose	–	0.05	0.78	–	–	nr	0.78
climbing rose or rambling rose; tying in as required	–	0.07	1.03	–	–	nr	1.03
Pruning ornamental shrub; height before pruning (increase these rates by 50% if pruning work has not been executed during the previous two years)							
not exceeding 1m	–	0.04	0.62	–	–	nr	0.62
1 to 2m	–	0.06	0.86	–	–	nr	0.86
exceeding 2m	–	0.13	1.94	–	–	nr	1.94
Removing excess growth etc from face of building etc.; height before pruning							
not exceeding 2m	–	0.03	0.44	–	–	nr	0.44
2 to 4m	–	0.05	0.78	–	–	nr	0.78
4 to 6m	–	0.08	1.29	–	–	nr	1.29
6 to 8m	–	0.13	1.94	–	–	nr	1.94
8 to 10m	–	0.14	2.21	–	–	nr	2.21
Removing epicormic growth from base of shrub or trunk and base of tree (any height, any diameter); number of growths							
not exceeding 10	–	0.05	0.78	–	–	nr	0.78
10 to 20	–	0.07	1.03	–	–	nr	1.03
Beds, borders and planters							
Lifting							
bulbs	–	0.50	7.75	–	–	100 nr	7.75
tubers or corms	–	0.40	6.20	–	–	100 nr	6.20
established herbaceous plants; hoeing and depositing for replanting	–	2.00	31.00	–	–	100 nr	31.00
Temporary staking and tying in herbaceous plant	–	0.03	0.52	–	0.12	nr	0.64
Cutting down spent growth of herbaceous plant; clearing arisings							
unstaked	–	0.02	0.31	–	–	nr	0.31
staked; not exceeding 4 stakes per plant; removing stakes and putting into store	–	0.03	0.39	–	–	nr	0.39
Hand weeding							
newly planted areas	–	2.00	31.00	–	–	100 m²	31.00
established areas	–	0.50	7.75	–	–	100 m²	7.75
Removing grasses from groundcover areas	–	3.00	46.55	–	–	100 m²	46.55
Hand digging with fork; not exceeding 150 mm deep; breaking down lumps; leaving surface with a medium tilth	–	1.33	20.67	–	–	100 m²	20.67

Q PAVING/PLANTING/FENCING/SITE FURNITURE

Item Excluding site overheads and profit	PC £	Labour hours	Labour £	Plant £	Material £	Unit	Total rate £
Hand digging with fork or spade to an average depth of 230 mm; breaking down lumps; leaving surface with a medium tilth	–	2.00	31.00	–	–	100 m²	31.00
Hand hoeing; not exceeding 50 mm deep; leaving surface with a medium tilth	–	0.40	6.20	–	–	100 m²	6.20
Hand raking to remove stones etc.; breaking down lumps: leaving surface with a fine tilth prior to planting	–	0.67	10.33	–	–	100 m²	10.33
Hand weeding; planter, window box; not exceeding 1.00 m²							
ground level box	–	0.05	0.78	–	–	nr	0.78
box accessed by stepladder	–	0.08	1.29	–	–	nr	1.29
Spreading only compost, mulch or processed bark to a depth of 75 mm							
on shrub bed with existing mature planting	–	0.09	1.41	–	–	m²	1.41
recently planted areas	–	0.07	1.03	–	–	m²	1.03
groundcover and herbaceous areas	–	0.08	1.16	–	–	m²	1.16
Clearing cultivated area of leaves, litter and other extraneous debris; using hand implement							
weekly maintenance	–	0.13	1.94	–	–	100 m²	1.94
daily maintenance	–	0.02	0.26	–	–	100 m²	0.26
Bedding							
Lifting							
bedding plants; hoeing and depositing for disposal	–	3.00	46.50	–	–	100 m²	46.50
Hand digging with fork; not exceeding 150 mm deep: breaking down lumps; leaving surface with a medium tilth	–	0.75	11.63	–	–	100 m²	11.63
Hand weeding							
newly planted areas	–	2.00	31.00	–	–	100 m²	31.00
established areas	–	0.50	7.75	–	–	100 m²	7.75
Hand digging with fork or spade to an average depth of 230 mm; breaking down lumps; leaving surface with a medium tilth	–	0.50	7.75	–	–	100 m²	7.75
Hand hoeing: not exceeding 50 mm deep; leaving surface with a medium tilth	–	0.40	6.20	–	–	100 m²	6.20
Hand raking to remove stones etc.; breaking down lumps; leaving surface with a fine tilth prior to planting	–	0.67	10.33	–	–	100 m²	10.33
Hand weeding; planter, window box; not exceeding 1.00 m²							
ground level box	–	0.05	0.78	–	–	nr	0.78
box accessed by stepladder	–	0.08	1.29	–	–	nr	1.29
Spreading only; compost, mulch or processed bark to a depth of 75 mm							
on shrub bed with existing mature planting	–	0.09	1.41	–	–	m²	1.41
recently planted areas	–	0.07	1.03	–	–	m²	1.03
groundcover and herbaceous areas	–	0.08	1.16	–	–	m²	1.16
Collecting bedding from nursery	–	3.00	46.50	13.64	–	100 m²	60.14
Setting out							
mass planting single variety	–	0.13	1.94	–	–	m²	1.94
pattern	–	0.33	5.17	–	–	m²	5.17
Planting only							
massed bedding plants	–	0.20	3.10	–	–	m²	3.10

Q PAVING/PLANTING/FENCING/SITE FURNITURE

Item Excluding site overheads and profit	PC £	Labour hours	Labour £	Plant £	Material £	Unit	Total rate £
Q35 LANDSCAPE MAINTENANCE – cont							
Bedding – cont							
Clearing cultivated area of leaves, litter and other extraneous debris; using hand implement							
weekly maintenance	–	0.13	1.94	–	–	100 m²	1.94
daily maintenance	–	0.02	0.26	–	–	100 m²	0.26
Irrigation and watering							
Hand held hosepipe; flow rate 25 litres per minute; irrigation requirement							
10 litres/m²	–	0.74	11.42	–	–	100 m²	11.42
15 litres/m²	–	1.10	17.05	–	–	100 m²	17.05
20 litres/m²	–	1.46	22.68	–	–	100 m²	22.68
25 litres/m²	–	1.84	28.47	–	–	100 m²	28.47
Hand held hosepipe; flow rate 40 litres per minute; irrigation requirement							
10 litres/m²	–	0.46	7.16	–	–	100 m²	7.16
15 litres/m²	–	0.69	10.74	–	–	100 m²	10.74
20 litres/m²	–	0.91	14.15	–	–	100 m²	14.15
25 litres/m²	–	1.15	17.77	–	–	100 m²	17.77
Hedge cutting; field hedges cut once or twice annually							
Trimming sides and top using hand tool or hand held mechanical tools							
not exceeding 2 m high	–	0.10	1.55	0.14	–	10 m²	1.69
2 to 4 m high	–	0.33	5.17	0.47	–	10 m²	5.64
Hedge cutting; ornamental							
Trimming sides and top using hand tool or hand held mechanical tools							
not exceeding 2 m high	–	0.13	1.94	0.18	–	10 m²	2.12
2 to 4 m high	–	0.50	7.75	0.70	–	10 m²	8.45
Hedge cutting; reducing width; hand tool or hand held mechanical tools							
Not exceeding 2 m high							
average depth of cut not exceeding 300 mm	–	0.03	0.39	0.04	–	m²	0.43
average depth of cut 300–600 mm	–	0.06	0.86	0.08	–	m²	0.94
average depth of cut 600–900 mm	–	0.06	0.97	0.09	–	m²	1.06
2–4 m high							
average depth of cut not exceeding 300 mm	–	0.04	0.62	0.06	–	m²	0.68
average depth of cut 300–600 mm	–	0.06	0.97	0.09	–	m²	1.06
average depth of cut 600–900 mm	–	0.10	1.55	0.14	–	m²	1.69
4–6 m high							
average depth of cut not exceeding 300 mm	–	0.10	1.55	0.14	–	m²	1.69
average depth of cut 300–600 mm	–	0.17	2.58	0.23	–	m²	2.81
average depth of cut 600–900 mm	–	0.50	7.75	0.70	–	m²	8.45

Q PAVING/PLANTING/FENCING/SITE FURNITURE

Item Excluding site overheads and profit	PC £	Labour hours	Labour £	Plant £	Material £	Unit	Total rate £
Hedge cutting; reducing width; tractor **mounted hedge cutting equipment**							
Not exceeding 2 m high							
average depth of cut not exceeding							
300 mm	–	0.04	0.62	0.67	–	10 m²	**1.29**
average depth of cut 300–600 mm	–	0.05	0.78	0.83	–	10 m²	**1.61**
average depth of cut 600–900 mm	–	0.20	3.10	3.33	–	10 m²	**6.43**
2–4 m high							
average depth of cut not exceeding							
300 mm	–	0.01	0.19	0.21	–	10 m²	**0.40**
average depth of cut 300–600 mm	–	0.03	0.39	0.42	–	10 m²	**0.81**
average depth of cut 600–900 mm	–	0.02	0.31	0.33	–	10 m²	**0.64**
Hedge cutting; reducing height; hand tool **or hand held mechanical tools**							
Not exceeding 2 m high							
average depth of cut not exceeding							
300 mm	–	0.07	1.03	0.09	–	10 m²	**1.12**
average depth of cut 300–600 mm	–	0.13	2.07	0.19	–	10 m²	**2.26**
average depth of cut 600–900 mm	–	0.40	6.20	0.56	–	10 m²	**6.76**
2–4 m high							
average depth of cut not exceeding							
300 mm	–	0.03	0.52	0.05	–	m²	**0.57**
average depth of cut 300–600 mm	–	0.07	1.03	0.09	–	m²	**1.12**
average depth of cut 600–900 mm	–	0.20	3.10	0.28	–	m²	**3.38**
4–6 m high							
average depth of cut not exceeding							
300 mm	–	0.07	1.03	0.09	–	m²	**1.12**
average depth of cut 300–600 mm	–	0.13	1.94	0.18	–	m²	**2.12**
average depth of cut 600–900 mm	–	0.25	3.88	0.35	–	m²	**4.23**
Hedge cutting; removal and disposal of **arisings**							
Sweeping up and depositing arisings							
300 mm cut	–	0.05	0.78	–	–	10 m²	**0.78**
600 mm cut	–	0.20	3.10	–	–	10 m²	**3.10**
900 mm cut	–	0.40	6.20	–	–	10 m²	**6.20**
Chipping arisings							
300 mm cut	–	0.02	0.31	0.11	–	10 m²	**0.42**
600 mm cut	–	0.08	1.29	0.48	–	10 m²	**1.77**
900 mm cut	–	0.20	3.10	1.14	–	10 m²	**4.24**
Disposal of unchipped arisings							
300 mm cut	–	0.02	0.26	0.51	2.00	10 m²	**2.77**
600 mm cut	–	0.03	0.52	1.02	3.00	10 m²	**4.54**
900 mm cut	–	0.08	1.29	2.56	7.49	10 m²	**11.34**
Disposal of chipped arisings							
300 mm cut	–	–	0.05	0.17	3.00	10 m²	**3.22**
600 mm cut	–	0.02	0.26	0.17	5.99	10 m²	**6.42**
900 mm cut	–	0.03	0.52	0.34	14.98	10 m²	**15.84**

Q PAVING/PLANTING/FENCING/SITE FURNITURE

Item Excluding site overheads and profit	PC £	Labour hours	Labour £	Plant £	Material £	Unit	Total rate £
Q35 LANDSCAPE MAINTENANCE – cont							
Herbicide applications; standard backpack **spray applicators; application to maintain** **1.00 m diameter clear circles (0.79m²)** **around new planting**							
Scotts UK; Roundup Pro Biactive 360; glyphosate; enhanced movement glyphosate; application rate 5 litre/ha							
plants at 1.50 m centres; 4444 nr/ha	23.33	24.69	456.76	–	23.33	ha	480.0█
plants at 1.75 m centres; 3265 nr/ha	17.07	18.14	335.59	–	17.07	ha	352.6█
plants at 2.00 m centres; 2500 nr/ha	13.17	13.89	256.96	–	13.17	ha	270.1█
mass spraying	66.66	13.00	240.50	–	66.66	ha	307.1█
Japanese Knotweed Management Plan; **PBA Solutions Ltd**							
Formulate Japanese Knotweed Management Plan (KMP) as recommended by the Environment Agency Code of Practice 2006 (EA CoP)	–	–	–	–	–	nr	1200.0█
Treatment of Japanese Knotweed; PBA **Solutions Ltd; removal and/or treatment in** **accordance with 'The Knotweed Code of** **Practice' published by the Environment** **Agency 2006 (EACoP); Japanese knotweed** **must be treated by qualified practitioners**							
Commercial chemical treatments based on areas less than 50 m² knotweed area; allows for the provision of contractual arrangements including method statements and spraying records; also allows for the provision of risk, environment and COSHH assessments							
rate per visit	–	–	–	–	–	nr	225.0█
rate per year	–	–	–	–	–	nr	900.0█
rate for complete treatment programme lasting 3 years	–	–	–	–	–	nr	2700.0█
Commercial chemical treatments based on areas between 50–300 m² knotweed area; allows for the provision of contractual arrangements including method statements and spraying recordsl also allows for the provision of risk, environment and COSHH assessments							
rate per visit	–	–	–	–	–	nr	300.0█
rate per year	–	–	–	–	–	nr	1200.0█
rate for complete treatment programme lasting 3 years	–	–	–	–	–	nr	3600.0█
Extra over minimum rate above; for areas 300–900 m²							
rate per visit	–	–	–	–	–	m²	0.4█
rate per year	–	–	–	–	–	m²	1.9█
rate for complete treatment programme lasting 3 years	–	–	–	–	–	m²	5.7█

Q PAVING/PLANTING/FENCING/SITE FURNITURE

Item excluding site overheads and profit	PC £	Labour hours	Labour £	Plant £	Material £	Unit	Total rate £
Extra over minimum rate above; for areas over 100 m²							
rate per visit	–	–	–	–	–	m²	0.40
rate per year	–	–	–	–	–	m²	1.60
rate for complete treatment programme							
lasting 3 years	–	–	–	–	–	m²	4.80

Note: Taking waste off site will require pre-treatment. Landfill operators are required by the Environment Agency to obtain evidence of pretreatment to show that the waste has been treated to either reduce its volume or hazardousness nature, or facilitate its handling, or enhance its recovery by the waste produce (site of origin) prior being taken to landfill. Generally pretreatment of Japanese knotweed can include chemical treatment, cutting or burning. The cost to implement this operation would be similar to item 4/5 with a minimum charge of £300.00. To enable waste to go to landfill soil testing would need to be completed and consultation with a landfill operator or invasive weed specialist is recommended in the first instance.

Japanese Knotweed contaminated waste removal; PBA Solutions Ltd (see note above) Removal off site to a licensed landfill site using 30 tonne gross vehicle weight lorries; price is for haulage and tipping only; the mileage stated below is the distance from point of collection to licensed landfill site; price excludes landfill tax and loading of material into lorries; rates will fluctuate between landfill operators and the figures given are for initial guidance only							
less than 20 miles	–	–	–	–	–	load	575.00
20–40 miles	–	–	–	–	–	load	637.00
40–60 miles	–	–	–	–	–	load	835.00
Rootbarriers; PBA Solutions Ltd; used in conjunction with on site control methods such as bund treatment, cell burial, vertical boundary protection and capping. The following barriers have been reliably used in connection with Japanese Knotweed control and as such are effective when correctly installed as part of a control programme which would usually include or chemical control Linear root deflection barriers; installed to trench measured separately							
Flexiroot non permeable; reinforced LDPE coated polypropylene geomembrane	5.89	0.05	0.93	–	5.89	m	6.82
Bioroot-barrier × permeable root barrier; mechanically bonded geocomposite consisting of a cooper-foil bonded between 2 layers of geotextile	10.76	0.05	0.93	–	10.76	m²	11.69

Q PAVING/PLANTING/FENCING/SITE FURNITURE

Item Excluding site overheads and profit	PC £	Labour hours	Labour £	Plant £	Material £	Unit	Total rate £
Q35 LANDSCAPE MAINTENANCE – cont							
Landfill tax rates; based on waste being **within landfill operator thresholds for** **controlled waste and with Japanese** **Knotweed being the only form of** **contamination; the rates below are based** **on visual inspections of the load at the** **gate of the tip**							
Rates to April 2010							
higher rate (where knotweed is visible at more than 5% on inspection at landfill)	–	–	–	–	40.00	tonne	**40.00**
lower rate (where knotweed is visible at less than 5% on inspection at landfill)	–	–	–	–	2.50	tonne	**2.50**
Rates from April 2010 – April 2011							
higher rate (where knotweed is visible at more than 5% on inspection at landfill)	–	–	–	–	48.00	tonne	**48.00**
lower rate (where knotweed is visible at less than 5% on inspection at landfill)	–	–	–	–	2.50	tonne	**2.50**
Q35 LANDSCAPE MAINTENANCE/GRASS **CUTTING**							
Grass cutting only; ride-on or tractor drawn **equipment; Norris & Gardiner Ltd; works** **carried out on one site only such as large** **fields or amenity areas where labour and** **machinery is present for a full day; grass** **cutting only; strimming not included**							
Using multiple-gang mower with cylindrical cutters; contiguous areas such as playing fields and the like larger than 3000 m²							
3 gang; 2.13 m cutting width	–	0.01	0.16	0.11	–	100 m²	**0.27**
5 gang; 3.40 m cutting width	–	0.01	0.15	0.13	–	100 m²	**0.28**
7 gang; 4.65 m cutting width	–	0.01	0.09	0.10	–	100 m²	**0.19**
Using multiple-gang mower with cylindrical cutters; non-contiguous areas such as verges and general turf areas							
3 gang	–	0.02	0.32	0.11	–	100 m²	**0.43**
5 gang	–	0.01	0.21	0.13	–	100 m²	**0.34**
Using multiple-rotary mower with vertical drive shaft and horizontally rotating bar or disc cutters; contiguous areas larger than 3000 m²							
cutting grass, overgrowth or the like using flail mower or reaper	–	0.02	0.31	0.25	–	100 m²	**0.56**

Q PAVING/PLANTING/FENCING/SITE FURNITURE

Item Excluding site overheads and profit	PC £	Labour hours	Labour £	Plant £	Material £	Unit	Total rate £
Grass cutting with strimming; ride-on or tractor drawn equipment; Norris & Gardiner Ltd; works carried out on one site only such as large fields or amenity areas where labour and machinery is present for a full day; includes accompanying strimmer operative							
Using multiple-gang mower with cylindrical cutters; contiguous areas such as playing fields and the like larger than 3000 m²							
3 gang; 2.13 m cutting width	–	0.02	0.32	0.12	–	100 m²	**0.44**
5 gang; 3.40 m cutting width	–	0.02	0.29	0.14	–	100 m²	**0.43**
7 gang; 4.65 m cutting width	–	0.01	0.18	0.13	–	100 m²	**0.31**
Using multiple-gang mower with cylindrical cutters; non-contiguous areas such as verges and general turf areas							
3 gang	–	0.04	0.65	0.17	–	100 m²	**0.82**
5 gang	–	0.03	0.41	0.15	–	100 m²	**0.56**
Using multiple-rotary mower with vertical drive shaft and horizontally rotating bar or disc cutters; contiguous areas larger than 3000 m²							
cutting grass, overgrowth or the like using flail mower or reaper	–	0.04	0.62	0.31	–	100 m²	**0.93**
Grass cutting only; ride-on or tractor drawn equipment; Norris & Gardiner Ltd; works carried out on multiple sites; machinery moved by trailer between sites; Grass cutting only, strimming not included							
Using multiple-gang mower with cylindrical cutters; contiguous areas such as playing fields and the like larger than 3000 m²							
3 gang; 2.13 m cutting width	–	0.02	0.26	0.11	–	100 m²	**0.37**
5 gang; 3.40 m cutting width	–	0.02	0.24	0.13	–	100 m²	**0.37**
7 gang; 4.65 m cutting width	–	0.01	0.15	0.10	–	100 m²	**0.25**
Using multiple-gang mower with cylindrical cutters; non-contiguous areas such as verges and general turf areas							
3 gang	–	0.03	0.52	0.11	–	100 m²	**0.63**
5 gang	–	0.02	0.33	0.13	–	100 m²	**0.46**
Using multiple-rotary mower with vertical drive shaft and horizontally rotating bar or disc cutters; contiguous areas larger than 3000 m²							
cutting grass, overgrowth or the like using flail mower or reaper	–	0.03	0.50	0.25	–	100 m²	**0.75**
Cutting grass, overgrowth or the like; using tractor-mounted side-arm flail mower; in areas inaccessible to alternative machine; on surface							
not exceeding 30° from horizontal	–	0.02	0.36	0.32	–	100 m²	**0.68**
30° to 50° from horizontal	–	0.05	0.72	0.32	–	100 m²	**1.04**

Q PAVING/PLANTING/FENCING/SITE FURNITURE

Item Excluding site overheads and profit	PC £	Labour hours	Labour £	Plant £	Material £	Unit	Total rate £
Q35 LANDSCAPE MAINTENANCE/GRASS CUTTING – cont							
Grass cutting and strimming; ride-on or tractor drawn equipment; Norris & Gardiner Ltd; works carried out on multiple sites; machinery moved by trailer between sites; includes for accompanying strimmer operative							
Using multiple-gang mower with cylindrical cutters; contiguous areas such as playing fields and the like larger than 3000 m^2							
3 gang; 2.13 m cutting width	–	0.03	0.53	0.13	–	100 m^2	0.6●
5 gang; 3.40 m cutting width	–	0.03	0.48	0.15	–	100 m^2	0.6:
7 gang; 4.65 m cutting width	–	0.02	0.29	0.11	–	100 m^2	0.4●
Using multiple-gang mower with cylindrical cutters; non-contiguous areas such as verges and general turf areas							
3 gang	–	0.07	1.05	0.16	–	100 m^2	1.2*
5 gang	–	0.04	0.67	0.19	–	100 m^2	0.8●
Using multiple-rotary mower with vertical drive shaft and horizontally rotating bar or disc cutters; contiguous areas larger than 3000 m^2							
cutting grass, overgrowth or the like using flail mower or reaper	–	0.07	1.01	0.30	–	100 m^2	1.3*
Cutting grass, overgrowth or the like; using tractor-mounted side-arm flail mower; in areas inaccessible to alternative machine; on surface							
not exceeding 30° from horizontal	–	0.05	0.74	0.35	–	100 m^2	1.0●
30° to 50° from horizontal	–	0.09	1.45	0.38	–	100 m^2	1.8:
Grass cutting; pedestrian operated equipment; Norris & Gardiner Ltd; the following rates are for grass cutting only; there is no allowance for accompanying strimming operations							
Using cylinder lawn mower fitted with not less than five cutting blades, front and rear rollers; on surface not exceeding 30° from horizontal; arisings let fly; width of cut							
51 cm	–	0.06	0.94	0.19	–	100 m	1.1:
61 cm	–	0.05	0.79	0.17	–	100 m	0.9●
71 cm	–	0.04	0.68	0.17	–	100 m	0.8:
91 cm	–	0.03	0.53	0.20	–	100 m	0.7:
Using rotary self-propelled mower; width of cut							
45 cm	–	0.04	0.61	0.08	–	100 m^2	0.6●
81 cm	–	0.02	0.34	0.08	–	100 m^2	0.4:
91 cm	–	0.02	0.30	0.11	–	100 m^2	0.4*
120 cm	–	0.01	0.23	0.36	–	100 m^2	0.5●
Add for using grass box for collecting and depositing arisings							
removing and depositing arisings	–	0.05	0.78	–	–	100 m^2	0.7●

Q PAVING/PLANTING/FENCING/SITE FURNITURE

Item Excluding site overheads and profit	PC £	Labour hours	Labour £	Plant £	Material £	Unit	Total rate £
Grass cutting; pedestrian operated equipment; Norris & Gardiner Ltd; the following rates are for grass cutting with accompanying strimming operations							
Using cylinder lawn mower fitted with not less than five cutting blades, front and rear rollers; on surface not exceeding 30° from horizontal; arisings let fly; width of cut							
51 cm	–	0.12	1.89	0.28	–	100 m	**2.17**
61 cm	–	0.10	1.58	0.24	–	100 m	**1.82**
71 cm	–	0.09	1.35	0.23	–	100 m	**1.58**
91 cm	–	0.07	1.06	0.25	–	100 m	**1.31**
Using rotary self-propelled mower; width of cut							
45 cm	–	0.08	1.23	0.14	–	100 m^2	**1.37**
81 cm	–	0.04	0.68	0.11	–	100 m^2	**0.79**
91 cm	–	0.04	0.60	0.14	–	100 m^2	**0.74**
120 cm	–	0.03	0.46	0.11	–	100 m^2	**0.57**
Add for using grass box for collecting and depositing arisings							
removing and depositing arisings	–	0.05	0.78	–	–	100 m^2	**0.78**
Add for 30° to 50° from horizontal	–	–	–	–	–	33%	**–**
Add for slopes exceeding 50°	–	–	–	–	–	100%	**–**
Cutting grass or light woody undergrowth; using trimmer with nylon cord or metal disc cutter; on surface							
not exceeding 30° from horizontal	–	0.20	3.10	0.28	–	100 m^2	**3.38**
30–50° from horizontal	–	0.40	6.20	0.56	–	100 m^2	**6.76**
exceeding 50° from horizontal	–	0.50	7.75	0.70	–	100 m^2	**8.45**
Grass cutting; collecting arisings; Norris & Gardener Ltd							
Extra over for tractor drawn and self-propelled machinery using attached grass boxes; depositing arisings							
22 cuts per year	–	0.05	0.78	–	–	100 m^2	**0.78**
18 cuts per year	–	0.08	1.16	–	–	100 m^2	**1.16**
12 cuts per year	–	0.10	1.55	–	–	100 m^2	**1.55**
4 cuts per year	–	0.25	3.88	–	–	100 m^2	**3.88**
bedding plant	–	–	–	–	0.29	each	**0.29**
Arisings collected by trailed sweepers							
22 cuts per year	–	0.02	0.26	0.24	–	100 m^2	**0.50**
18 cuts per year	–	0.02	0.34	0.24	–	100 m^2	**0.58**
12 cuts per year	–	0.03	0.52	0.24	–	100 m^2	**0.76**
4 cuts per year	–	0.07	1.03	0.24	–	100 m^2	**1.27**
Disposing arisings on site; 100 m distance maximum							
22 cuts per year	–	0.01	0.12	0.02	–	100 m^2	**0.14**
18 cuts per year	–	0.01	0.15	0.03	–	100 m^2	**0.18**
12 cuts per year	–	0.01	0.22	0.04	–	100 m^2	**0.26**
4 cuts per year	–	0.04	0.68	0.13	–	100 m^2	**0.81**
Disposal of arisings off site							
22 cuts per year	–	0.01	0.13	0.04	0.75	100 m^2	**0.92**
18 cuts per year	–	0.01	0.16	0.05	1.00	100 m^2	**1.21**
12 cuts per year	–	0.01	0.10	0.14	1.80	100 m^2	**2.04**
4 cuts per year	–	0.02	0.31	0.82	3.00	100 m^2	**4.13**

Q PAVING/PLANTING/FENCING/SITE FURNITURE

Item Excluding site overheads and profit	PC £	Labour hours	Labour £	Plant £	Material £	Unit	Total rate £
Q40 FENCING							
Temporary security fence; HSS Hire; mesh							
framed unclimbable fencing; including							
precast concrete supports and couplings							
Weekly hire; 2.85 × 2.00 m high							
weekly hire rate	–	–	–	2.11	–	m	**2.11**
erection of fencing; labour only	–	0.10	1.85	–	–	m	**1.85**
removal of fencing loading to collection							
vehicle	–	0.07	1.23	–	–	m	**1.23**
delivery charge	–	–	–	0.80	–	m	**0.80**
return haulage charge	–	–	–	0.60	–	m	**0.60**
Protective fencing; AVS Fencing Supplies							
Ltd							
Cleft chestnut rolled fencing; fixing to 100 mm							
diameter chestnut posts; driving into firm							
ground at 3 m centres							
900 mm high	3.12	0.11	1.97	–	3.64	m	**5.61**
1200 mm high	4.27	0.16	2.96	–	4.80	m	**7.76**
1500 mm high; 3 strand	5.71	0.21	3.95	–	6.24	m	**10.19**
Enclosures; Earth Anchors							
Rootfast anchored galvanized steel							
enclosures post ADP 20–1000; 1000 mm							
high × 20 mm diameter with AA 25–750							
socket and padlocking ring	35.00	0.10	1.85	–	37.10	nr	**38.95**
steel cable, orange plastic coated	1.25	–	0.04	–	1.25	m	**1.29**
Timber fencing; AVS Fencing Supplies Ltd;							
tanalized softwood fencing; all timber							
posts are kiln dried redwood with 15 year							
guarantee							
Timber lap panels; pressure treated; fixed to							
timber posts 75 × 75 mm in 1:3:6 concrete; at							
1.90 m centres							
900 mm high	7.86	0.67	12.33	–	14.49	m	**26.82**
1200 mm high	8.17	0.75	13.88	–	14.39	m	**28.27**
1500 mm high	8.17	0.80	14.80	–	14.70	m	**29.50**
1800 mm high	8.91	0.90	16.65	–	16.38	m	**33.03**
Timber lap panels; fixed to slotted concrete							
posts 100 × 100 mm 1:3:6 concrete at 1.88 m							
centres							
900 mm high	7.86	0.75	13.88	–	17.50	m	**31.38**
1200 mm high	8.12	0.80	14.80	–	17.76	m	**32.56**
1500 mm high	9.20	0.85	15.72	–	20.39	m	**36.11**
1800 mm high	9.95	1.00	18.50	–	21.32	m	**39.82**
extra for corner posts	19.55	0.90	16.65	–	25.76	nr	**42.41**
extra over panel fencing for 300 mm high							
trellis tops; slats at 100 mm centres;							
including additional length of posts	4.29	1.00	18.50	–	4.29	m	**22.79**

Q PAVING/PLANTING/FENCING/SITE FURNITURE

Item Excluding site overheads and profit	PC £	Labour hours	Labour £	Plant £	Material £	Unit	Total rate £
Close boarded fencing; to concrete posts 100 × 100 mm; 2 nr softwood arris rails; 100 × 22 mm softwood pales lapped 13 mm; including excavating and backfilling into firm ground at 3.00 m centres; setting in concrete 1:3:6; concrete gravel board 150 × 150 mm							
900 mm high	9.88	1.00	18.50	–	12.21	m	**30.71**
1050 mm high	10.12	1.10	20.35	–	12.45	m	**32.80**
1500 mm high	12.16	1.15	21.27	–	14.49	m	**35.76**
Close boarded fencing; to concrete posts 100 × 100 mm; 3 nr softwood arris rails; 100 × 22 mm softwood pales lapped 13 mm; including excavating and backfilling into firm ground at 3.00 m centres; setting in concrete 1:3:6							
1650 mm high	13.46	1.25	23.13	–	15.79	m	**38.92**
1800 mm high	13.58	1.30	24.05	–	15.91	m	**39.96**
Close boarded fencing; to timber redwood posts 100 × 100 mm; 2 nr softwood arris rails; 100 × 22 mm softwood pales lapped 13 mm; including excavating and backfilling into firm ground at 3.00 m centres; setting in concrete 1:3:6							
1350 mm high	11.76	1.00	18.50	–	14.08	m	**32.58**
1650 mm high	12.72	1.25	23.13	–	15.04	m	**38.17**
1800 mm high	13.21	1.30	24.05	–	15.54	m	**39.59**
Close boarded fencing; to timber posts 100 × 100 mm; 3 nr softwood arris rails; 100 × 22 mm softwood pales lapped 13 mm; including excavating and backfilling into firm ground at 3.00 m centres; setting in concrete 1:3:6							
1350 mm high	12.70	0.95	17.62	–	15.02	m	**32.64**
1650 mm high	13.66	1.30	24.05	–	15.98	m	**40.03**
1800 mm high	14.15	1.35	24.98	–	16.48	m	**41.46**
extra over for post 125 × 100 mm; for 1800 mm high fencing	8.44	–	–	–	8.44	m	**8.44**
extra over to the above for counter rail	0.59	–	–	–	0.59	m	**0.59**
extra over to the above for capping rail	0.59	0.10	1.85	–	0.59	m	**2.44**
extra over to the above for concrete gravel board 150 × 50 mm	4.97	–	–	–	4.97	m	**4.97**
Feather edge board fencing; to timber posts 100 × 100 mm; 2 nr softwood cant rails; 125 × 22 mm feather edge boards lapped 13 mm; including excavating and backfilling into firm ground at 3.00 m centres; setting in concrete 1:3:6							
1350 mm high	10.62	0.56	10.36	–	11.67	m	**22.03**
1650 mm high	12.04	0.56	10.36	–	13.09	m	**23.45**
1800 mm high	12.40	0.56	10.36	–	13.45	m	**23.81**

Q PAVING/PLANTING/FENCING/SITE FURNITURE

Item Excluding site overheads and profit	PC £	Labour hours	Labour £	Plant £	Material £	Unit	Total rate £
Q40 FENCING – cont							
Timber fencing – cont							
Feather edge board fencing; to timber posts 100 × 100 mm; 3 nr softwood cant rails; 125 × 22 mm feather edge boards lapped 13 mm; including excavating and backfilling into firm ground at 3.00 m centres; setting in concrete 1:3:6							
1350 mm high	11.64	0.57	10.57	–	12.69	m	**23.26**
1650 mm high	13.06	0.57	10.57	–	14.11	m	**24.68**
1800 mm high	12.35	0.57	10.57	–	14.47	m	**25.04**
extra over for post 125 × 100 mm; for							
1800 mm high fencing	8.44	–	–	–	8.44	nr	**8.44**
extra over to the above for counter rail and							
capping	1.81	0.10	1.85	–	1.81	m	**3.66**
Palisade fencing; 22 × 75 mm softwood vertical palings with pointed tops; nailing to two 50 × 100 mm horizontal softwood arris rails; morticed to 100 × 100 mm softwood posts with weathered tops at 3.00 m centres; setting in concrete							
900 mm high	12.20	0.90	16.65	–	14.52	m	**31.17**
1050 mm high	9.73	0.90	16.65	–	12.06	m	**28.71**
1200 mm high	10.13	0.95	17.57	–	12.46	m	**30.03**
extra over for rounded tops	0.66	–	–	–	0.66	m	**0.66**
Post-and-rail fencing; 90 × 38 mm softwood horizontal rails; fixing with galvanized nails to 150 × 75 mm softwood posts; including excavating and backfilling into firm ground at 1.80 m centres; all treated timber							
1200 mm high; 3 horizontal rails	8.25	0.35	6.48	–	8.33	m	**14.81**
1200 mm high; 4 horizontal rails	9.54	0.35	6.48	–	9.54	m	**16.02**
Cleft rail fencing; oak or chestnut adze tapered rails 2.80 m long; morticed into joints; to 125 × 100 mm softwood posts 1.95 m long; including excavating and backfilling into firm ground at 2.80 m centres							
two rails	9.48	0.25	4.63	–	9.48	m	**14.11**
three rails	11.83	0.28	5.18	–	11.83	m	**17.01**
four rails	14.18	0.35	6.48	–	14.18	m	**20.66**
Hit and Miss horizontal rail fencing; 87 × 38 mm top and bottom rails; 100 × 22 mm vertical boards arranged alternately on opposite side of rails; to 100 × 100 mm posts; including excavating and backfilling into firm ground; setting in concrete at 1.8 m centres							
treated softwood; 1600 mm high	25.66	1.20	22.20	–	27.76	m	**49.96**
treated softwood; 1800 mm high	28.18	1.33	24.66	–	30.49	m	**55.15**
treated softwood; 2000 mm high	30.71	1.40	25.90	–	33.02	m	**58.92**
Trellis tops to fencing (see more detailed items in the Minor Works section of this book)							
Extra over screen fencing for 300 mm high trellis tops; slats at 100 mm centres; including additional length of posts	4.29	0.10	1.85	–	4.29	m	**6.14**

Q PAVING/PLANTING/FENCING/SITE FURNITURE

Item Excluding site overheads and profit	PC £	Labour hours	Labour £	Plant £	Material £	Unit	Total rate £
Boundary fencing; strained wire and wire mesh; AVS Fencing Supplies Ltd							
Strained wire fencing; concrete inter posts only at 2750 mm centres; 610 mm below ground; excavating holes; filling with concrete; replacing topsoil; disposing surplus soil off site							
900 mm high	2.89	0.56	10.28	0.61	4.10	m	**14.99**
1200 mm high	3.62	0.56	10.28	0.61	4.83	m	**15.72**
1800 mm high	4.95	0.78	14.39	1.53	6.16	m	**22.08**
Extra over strained wire fencing for concrete straining posts with one strut; posts and struts 610 mm below ground; struts, cleats, stretchers, winders, bolts and eye bolts; excavating holes; filling to within 150 mm of ground level with concrete (1:12) – 40 mm aggregate; replacing topsoil; disposing surplus soil off site							
900 mm high	23.35	0.67	12.39	1.38	49.04	nr	**62.81**
1200 mm high	25.97	0.67	12.34	1.38	52.60	nr	**66.32**
1800 mm high	36.96	0.67	12.34	1.38	65.93	nr	**79.65**
Extra over strained wire fencing for concrete straining posts with two struts; posts and struts 610 mm below ground; excavating holes; filling to within 150 mm of ground level with concrete (1:12) – 40 mm aggregate; replacing topsoil; disposing surplus soil off site							
900 mm high	33.68	0.91	16.84	1.77	72.05	nr	**90.66**
1200 mm high	37.54	0.85	15.72	1.38	75.90	nr	**93.00**
1800 mm high	53.66	0.85	15.72	1.38	101.41	nr	**118.51**
Strained wire fencing; galvanized steel angle posts only at 2750 mm centres; 610 mm below ground; driving in							
900 mm high; 40 × 40 × 5 mm	2.65	0.03	0.56	–	2.65	m	**3.21**
1200 mm high; 40 × 40 × 5 mm	3.01	0.03	0.62	–	3.01	m	**3.63**
1500 mm high; 40 × 40 × 5 mm	4.01	0.06	1.12	–	4.01	m	**5.13**
1800 mm high; 40 × 40 × 5 mm	4.32	0.07	1.34	–	4.32	m	**5.66**
1800 mm high; 45 × 45 × 5 mm with extention for three rows barbed wire	5.54	0.12	2.24	–	5.54	m	**7.78**
Galvanized steel straining posts with two struts for strained wire fencing; setting in concrete							
900 mm high; 50 × 50 × 6 mm	47.20	1.00	18.50	–	49.55	nr	**68.05**
1200 mm high; 50 × 50 × 6 mm	57.01	1.00	18.50	–	59.36	nr	**77.86**
1500 mm high; 50 × 50 × 6 mm	71.23	1.00	18.50	–	73.58	nr	**92.08**
1800 mm high; 50 × 50 × 6 mm	73.83	1.00	18.50	–	76.18	nr	**94.68**
2400 mm high; 60 × 60 × 6 mm	83.59	1.00	18.50	–	85.94	nr	**104.44**
Strained wire; to posts (posts not included); 3 mm galvanized wire; fixing with galvanized stirrups							
900 mm high; 2 wire	0.16	0.03	0.61	–	0.31	m	**0.92**
1200 mm high; 3 wire	0.24	0.05	0.86	–	0.39	m	**1.25**
1400 mm high; 3 wire	0.24	0.05	0.86	–	0.39	m	**1.25**
1800 mm high; 3 wire	0.24	0.05	0.86	–	0.39	m	**1.25**

Q PAVING/PLANTING/FENCING/SITE FURNITURE

Item Excluding site overheads and profit	PC £	Labour hours	Labour £	Plant £	Material £	Unit	Total rate £
Q40 FENCING – cont							
Boundary fencing – cont							
Barbed wire; to posts (posts not included); 3 mm galvanized wire; fixing with galvanized stirrups							
900 mm high; 2 wire	0.25	0.07	1.23	–	0.40	m	**1.63**
1200 mm high; 3 wire	0.37	0.09	1.72	–	0.52	m	**2.24**
1400 mm high; 3 wire	0.37	0.09	1.72	–	0.52	m	**2.24**
1800 mm high; 3 wire	0.37	0.09	1.72	–	0.52	m	**2.24**
Chain link fencing; AVS Fencing Supplies Ltd; to strained wire and posts priced separately; 3 mm galvanized wire; 51 mm mesh; galvanized steel components; fixing to line wires threaded through posts and strained with eye-bolts; posts (not included)							
900 mm high	3.89	0.07	1.23	–	4.04	m	**5.27**
1200 mm high	5.44	0.07	1.23	–	5.59	m	**6.82**
1800 mm high	7.54	0.10	1.85	–	7.69	m	**9.54**
Chain link fencing; to strained wire and posts priced separately; 3.15 mm plastic coated galvanized wire (wire only 2.5 mm); 51 mm mesh; galvanized steel components; fencing with line wires threaded through posts and strained with eye-bolts; posts (not included) (Note: plastic coated fencing can be cheaper than galvanized finish as wire of a smaller cross-sectional area can be used)							
900 mm high	2.75	0.07	1.23	–	3.05	m	**4.28**
1200 mm high	3.79	0.07	1.23	–	4.09	m	**5.32**
1800 mm high	5.40	0.13	2.31	–	6.60	m	**8.91**
Extra over strained wire fencing for cranked arms and galvanized barbed wire							
1 row	3.47	0.02	0.31	–	3.47	m	**3.78**
2 row	3.60	0.05	0.93	–	3.60	m	**4.53**
3 row	3.72	0.05	0.93	–	3.72	m	**4.65**
Field fencing; AVS Fencing Supplies Ltd; welded wire mesh; fixed to posts and straining wires measured separately							
Cattle fence; 1200 m high; 114 × 300 mm at bottom to 230 × 300 mm at top	0.76	0.10	1.85	–	0.96	m	**2.81**
Sheep fence; 800 mm high; 140 × 300 mm at bottom to 230 × 300 mm at top	0.65	0.10	1.85	–	0.85	m	**2.70**
Deer fence; 2.0 m high; 89 × 150 mm at bottom to 267 × 300 mm at top	1.53	0.13	2.31	–	1.73	m	**4.04**
Extra for concreting in posts	–	0.50	9.25	–	3.72	nr	**12.97**
Extra for straining post	8.25	0.75	13.88	–	8.25	nr	**22.13**
Rabbit netting; AVS Fencing Supplies Ltd							
Timber stakes; peeled kiln dried pressure treated; pointed; 1.8 m posts driven 900 mm into ground at 3 m centres (line wires and netting priced separately)							
75–100 mm stakes	1.92	0.25	4.63	–	1.92	m	**6.55**
Corner posts or straining posts 150 mm diameter × 2.3 m high set in concrete; centres to suit local conditions or changes of direction							
1 strut	10.19	1.00	18.50	3.94	15.91	each	**38.35**
2 strut	12.51	1.00	18.50	3.94	18.23	each	**40.67**

Q PAVING/PLANTING/FENCING/SITE FURNITURE

Item Excluding site overheads and profit	PC £	Labour hours	Labour £	Plant £	Material £	Unit	Total rate £
Strained wire; to posts (posts not included); 3 mm galvanized wire; fixing with galvanized stirrups							
900 mm high; 2 wire	0.16	0.03	0.61	–	0.31	m	0.92
1200 mm high; 3 wire	0.24	0.05	0.86	–	0.39	m	1.25
Rabbit netting; 31 mm 19 gauge 1050 mm high netting fixed to posts line wires and straining posts or corner posts all priced separately							
900 mm high; turned in	0.94	0.04	0.74	–	0.94	m	1.68
900 mm high; buried 150 mm in trench	0.94	0.08	1.54	–	0.94	m	2.48
Boundary fencing; strained wire and wire **mesh; Jacksons Fencing**							
Tubular chain link fencing; galvanized; plastic coated; 60.3 mm diameter posts at 3.0 m centres; setting 700 mm into ground; choice of 10 mesh colours; including excavating holes, backfilling and removing surplus soil; with top rail only							
900 mm high	–	–	–	–	–	m	33.68
1200 mm high	–	–	–	–	–	m	35.78
1800 mm high	–	–	–	–	–	m	39.89
2000 mm high	–	–	–	–	–	m	42.10
Tubular chain link fencing; galvanized; plastic coated; 60.3 mm diameter posts at 3.0 m centres; cranked arms and three lines barbed wire; setting 700 mm into ground; including excavating holes; backfilling and removing surplus soil; with top rail only							
2000 mm high	–	–	–	–	–	m	42.68
1800 mm high	–	–	–	–	–	m	41.47
Boundary fencing; Steelway-Fensecure Ltd							
Classic two rail tubular fencing; top and bottom with stretcher bars and straining wires in between; comprising 60.3 mm tubular posts at 3.00 m centres; setting in concrete; 35 mm top rail tied with aluminium and steel fittings; 50 × 50 × 355/2.5 mm PVC coated chain link; all components galvanized and coated in green nylon							
964 mm high	–	–	–	–	–	m	36.90
1269 mm high	–	–	–	–	–	m	40.63
1574 mm high	–	–	–	–	–	m	46.19
1878 mm high	–	–	–	–	–	m	48.51
2188 mm high	–	–	–	–	–	m	51.71
2458 mm high	–	–	–	–	–	m	57.82
2948 mm high	–	–	–	–	–	m	73.28
3562 mm high	–	–	–	–	–	m	73.09
End posts; Classic range; 60.3 mm diameter; setting in concrete							
964 mm high	–	–	–	–	–	nr	51.19
1269 mm high	–	–	–	–	–	nr	55.19
1574 mm high	–	–	–	–	–	nr	57.83
1878 mm high	–	–	–	–	–	nr	67.19
2188 mm high	–	–	–	–	–	nr	73.08
2458 mm high	–	–	–	–	–	nr	83.92
2948 mm high	–	–	–	–	–	nr	91.24
3562 mm high	–	–	–	–	–	nr	109.80

Q PAVING/PLANTING/FENCING/SITE FURNITURE

Item Excluding site overheads and profit	PC £	Labour hours	Labour £	Plant £	Material £	Unit	Total rate £
Q40 FENCING – cont							
Boundary fencing – cont							
Corner posts; 60.3 mm diameter; setting in concrete							
964 mm high	–	–	–	–	–	nr	41.42
1269 mm high	–	–	–	–	–	nr	24.83
1574 mm high	–	–	–	–	–	nr	48.45
1878 mm high	–	–	–	–	–	nr	50.77
2188 mm high	–	–	–	–	–	nr	59.05
2458 mm high	–	–	–	–	–	nr	62.90
2948 mm high	–	–	–	–	–	nr	75.54
3562 mm high	–	–	–	–	–	nr	80.43
Boundary fencing; McArthur Group							
Paladin welded mesh colour coated green fencing; fixing to metal posts at 2.975 m centres with manufacturer's fixings; setting 600 mm deep in firm ground; including excavating holes; backfilling and removing surplus excavated material; includes 10 year product guarantee							
1800 mm high	53.29	0.66	12.21	–	53.29	m	65.50
2000 mm high	54.40	0.66	12.21	–	62.45	m	74.66
2400 mm high	62.69	0.66	12.21	–	72.06	m	84.27
Extra over welded galvanized plastic coated mesh fencing for concreting in posts	7.45	0.11	2.06	–	7.45	m	9.51
Boundary fencing							
Orsogril rectangular steel bar mesh fence panels; pleione pattern; bolting to 60 × 8 mm uprights at 2 m centres; mesh 62 × 66 mm; setting in concrete							
930 mm high panels	–	–	–	–	–	m	118.00
1326 mm high panels	–	–	–	–	–	m	151.00
1722 mm high panels	–	–	–	–	–	m	198.00
Orsogril rectangular steel bar mesh fence panels; pleione pattern; bolting to 8 × 80 mm uprights at 2 m centres; mesh 62 × 66 mm; setting in concrete							
930 mm high panel	–	–	–	–	–	m	122.00
1326 mm high panel	–	–	–	–	–	m	157.00
1722 mm high panel	–	–	–	–	–	m	202.00
Orsogril rectangular steel bar mesh fence panels; sterope pattern; bolting to 60 × 80 mm uprights at 2.00 m centres; mesh 62 × 132 mm; setting in concrete paving (paving not included)							
930 mm high panels	–	–	–	–	–	m	149.03
1326 mm high panels	–	–	–	–	–	m	195.34
1722 mm high panels	–	–	–	–	–	m	245.68
Security fencing; Jacksons Fencing							
Barbican galvanized steel paling fencing; on 60 × 60 mm posts at 3 m centres; setting in concrete							
1250 mm high	–	–	–	–	–	m	67.27
1500 mm high	–	–	–	–	–	m	73.04
2000 mm high	–	–	–	–	–	m	90.30
2500 mm high	–	–	–	–	–	m	109.34

Q PAVING/PLANTING/FENCING/SITE FURNITURE

Item Excluding site overheads and profit	PC £	Labour hours	Labour £	Plant £	Material £	Unit	Total rate £
Gates; to match Barbican galvanized steel paling fencing							
width 1 m	–	–	–	–	–	nr	1250.00
width 2 m	–	–	–	–	–	nr	1280.00
width 3 m	–	–	–	–	–	nr	1352.00
width 4 m	–	–	–	–	–	nr	1370.00
width 8 m	–	–	–	–	–	pair	2185.00
width 9 m	–	–	–	–	–	pair	2780.00
width 10 m	–	–	–	–	–	pair	3160.00
Security fencing; Jacksons Fencing; intruder guards							
Viper Spike Intruder Guards; to existing structures; including fixing bolts							
Viper 1; 40 × 5 mm × 1.1 m long; with base plate	31.05	1.00	18.50	–	31.05	nr	49.55
Viper 3; 160 × 190 mm wide; U shape; to prevent intruders climbing pipes	35.05	0.50	9.25	–	35.05	nr	44.30
Security fencing; Jacksons Fencing							
Razor Barb Concertina; spiral wire security barriers; fixing to 600 mm steel ground stakes							
Ref 3275; 450 mm diameter roll; medium barb; galvanized	2.14	0.02	0.37	–	2.29	m	2.66
Ref 3276; 730 mm diameter roll; medium barb; galvanized	2.91	0.02	0.37	–	3.06	m	3.43
Ref 3277; 950 mm diameter roll; medium barb; galvanized	2.90	0.02	0.37	–	3.05	m	3.42
Three lines Barbed Tape; medium barb; on 50 × 50 mm mild steel angle posts; setting in concrete							
Ref 3283; barbed tape; medium barb; galvanized	0.85	0.21	3.90	–	10.29	m	14.19
Five lines Barbed Tape; medium barb; on 50 × 50 mm mild steel angle posts; setting in concrete							
Ref 3283; barbed tape; medium barb; galvanized	1.42	0.22	4.11	–	10.86	m	14.97
Trip rails; metal; Broxap Ltd							
Double row 48.3 mm internal diameter mild steel tubular rails with sleeved joints; to cast iron posts 525 mm above ground; setting in concrete at 1.20 m centres; all standard painted							
BX 201; cast iron with single rail; 42 mm diameter rail	65.66	1.00	18.50	–	70.84	m	89.34
Heavy duty; Anti Ram 540 mm high with 100 × 100 mm rail; Birdsmouth style	46.06	1.00	18.50	–	46.35	m	64.85

Q PAVING/PLANTING/FENCING/SITE FURNITURE

Item Excluding site overheads and profit	PC £	Labour hours	Labour £	Plant £	Material £	Unit	Total rate £
Q40 FENCING – cont							
Trip rails; Birdsmouth; Jacksons Fencing							
Diamond rail fencing; Birdsmouth; posts 100 ×							
100 mm softwood planed at 1.35 m centres set							
in 1:3:6 concrete; rail 75 × 75 mm nominal							
secured with galvanized straps nailed to posts							
posts 900 mm (600 mm above ground); at							
1.35 m centres	7.98	0.50	9.25	0.26	9.54	m	19.05
posts 1.20 m (900 mm above ground); at							
1.35 m centres	9.12	0.50	9.25	0.26	10.67	m	20.18
posts 900 mm (600 mm above ground); at							
1.80 m centres	6.73	0.38	6.95	0.20	7.90	m	15.05
posts 1.20 m (900 mm above ground); at							
1.80 m centres	7.58	0.38	6.95	0.20	8.74	m	15.89
Trip rails; Townscape Products Ltd							
Hollow steel section knee rails; galvanized;							
500 mm high; setting in concrete							
1000 mm bays	99.02	2.00	37.00	–	105.20	m	142.20
1200 mm bays	23.38	2.00	37.00	–	91.76	m	128.76
Concrete fencing							
Panel fencing; to precast concrete posts; in							
2 m bays; setting posts 600 mm into ground;							
sandfaced finish							
900 mm high	11.74	0.25	4.63	–	13.21	m	17.84
1200 mm high	15.68	0.25	4.63	–	17.15	m	21.78
Panel fencing; to precast concrete posts; in							
2 m bays; setting posts 750 mm into ground;							
sandfaced finish							
1500 mm high	20.46	0.33	6.17	–	21.93	m	28.10
1800 mm high	24.52	0.36	6.73	–	25.98	m	32.71
2100 mm high	28.39	0.40	7.40	–	29.86	m	37.26
2400 mm high	32.63	0.40	7.40	–	34.10	m	41.50
extra over capping panel to all of the above	6.72	–	–	–	6.72	m	6.72
Windbreak fencing							
Fencing; English Woodlands; Shade and							
Shelter Netting windbreak fencing; green; to							
100 mm diameter treated softwood posts;							
setting 450 mm into ground; fixing with 50 ×							
25 mm treated softwood battens nailed to							
posts; including excavating and backfilling into							
firm ground; setting in concrete at 3 m centres							
1200 mm high	1.24	0.16	2.88	–	5.10	m	7.98
1800 mm high	1.76	0.16	2.88	–	5.62	m	8.50
Ball stop fencing; Steelway-Fensecure Ltd							
Ball stop net; 30 × 30 mm netting fixed to							
60.3 mm diameter 12 mm solid bar lattice;							
galvanized; dual posts; top, middle and bottom							
rails							
4.5 m high	–	–	–	–	–	m	102.56
5.0 m high	–	–	–	–	–	m	129.49

Q PAVING/PLANTING/FENCING/SITE FURNITURE

Item Excluding site overheads and profit	PC £	Labour hours	Labour £	Plant £	Material £	Unit	Total rate £
6.0 m high	–	–	–	–	–	m	144.27
7.0 m high	–	–	–	–	–	m	159.07
8.0 m high	–	–	–	–	–	m	173.83
9.0 m high	–	–	–	–	–	m	188.63
10.0 m high	–	–	–	–	–	m	172.36
Ball stop net; corner posts							
4.5 m high	–	–	–	–	–	m	70.37
5.0 m high	–	–	–	–	–	m	77.02
6.0 m high	–	–	–	–	–	m	119.65
7.0 m high	–	–	–	–	–	m	106.59
8.0 m high	–	–	–	–	–	m	121.39
9.0 m high	–	–	–	–	–	m	126.17
10 m high	–	–	–	–	–	m	150.95
Ball stop net; end posts							
4.5 m high	–	–	–	–	–	m	58.54
5.0 m high	–	–	–	–	–	m	65.20
6.0 m high	–	–	–	–	–	m	79.98
7.0 m high	–	–	–	–	–	m	94.78
8.0 m high	–	–	–	–	–	m	109.54
9.0 m high	–	–	–	–	–	m	124.35
10 m high	–	–	–	–	–	m	126.17
Railings; Steelway-Fensecure Ltd							
Mild steel bar railings of balusters at 115 mm centres welded to flat rail top and bottom; bays 2.00 m long; bolting to 51 × 51 mm hollow square section posts; setting in concrete							
galvanized; 900 mm high	–	–	–	–	–	m	74.80
galvanized; 1200 mm high	–	–	–	–	–	m	90.60
galvanized; 1500 mm high	–	–	–	–	–	m	100.69
galvanized; 1800 mm high	–	–	–	–	–	m	120.79
primed; 900 mm high	–	–	–	–	–	m	70.50
primed; 1200 mm high	–	–	–	–	–	m	85.60
primed; 1500 mm high	–	–	–	–	–	m	94.63
primed; 1800 mm high	–	–	–	–	–	m	112.78
Mild steel blunt top railings of 19 mm balusters at 130 mm centres welded to bottom rail; passing through holes in top rail and welded; top and bottom rails 40 × 10 mm; bolting to 51 × 51 mm hollow square section posts; setting in concrete							
galvanized; 800 mm high	–	–	–	–	–	m	80.54
galvanized; 1000 mm high	–	–	–	–	–	m	84.58
galvanized; 1300 mm high	–	–	–	–	–	m	100.69
galvanized; 1500 mm high	–	–	–	–	–	m	108.74
primed; 800 mm high	–	–	–	–	–	m	77.53
primed; 1000 mm high	–	–	–	–	–	m	80.54
primed; 1300 mm high	–	–	–	–	–	m	95.65
primed; 1500 mm high	–	–	–	–	–	m	102.72
Railings; traditional pattern; 16 mm diameter verticals at 127 mm intervals with horizontal bars near top and bottom; balusters with spiked tops; 51 × 20 mm standards; including setting 520 mm into concrete at 2.75 m centres							
primed; 1200 mm high	–	–	–	–	–	m	90.60
primed; 1500 mm high	–	–	–	–	–	m	102.72
primed; 1800 mm high	–	–	–	–	–	m	110.75

Q PAVING/PLANTING/FENCING/SITE FURNITURE

Item Excluding site overheads and profit	PC £	Labour hours	Labour £	Plant £	Material £	Unit	Total rate £
Q40 FENCING – cont							
Railings – cont							
Interlaced bow-top mild steel railings;							
traditional park type; 16 mm diameter verticals							
at 80 mm intervals; welded at bottom to 50 ×							
10 mm flat and slotted through 38 × 8 mm top							
rail to form hooped top profile; 50 × 10 mm							
standards; setting 560 mm into concrete at							
2.75 m centres							
galvanized; 900 mm high	–	–	–	–	–	m	136.94
galvanized; 1200 mm high	–	–	–	–	–	m	167.15
galvanized; 1500 mm high	–	–	–	–	–	m	199.36
galvanized; 1800 mm high	–	–	–	–	–	m	231.58
primed; 900 mm high	–	–	–	–	–	m	132.90
primed; 1200 mm high	–	–	–	–	–	m	161.10
primed; 1500 mm high	–	–	–	–	–	m	191.32
primed; 1800 mm high	–	–	–	–	–	m	221.50
Metal estate fencing; Jacksons Fencing;							
mild steel flat bar angular fencing;							
galvanized; main posts at 5.00 m centres;							
fixed with 1:3:6 concrete 430 mm deep;							
intermediate posts fixed with fixing claw at							
1.00 m centres							
Five plain, flat or round bars							
1200 high; under 500 m	60.00	0.28	5.14	–	62.10	m	67.24
1200 high; over 500 m	55.00	0.28	5.14	–	57.10	m	62.24
Five bar horizontal steel fencing; 1.20 m high;							
38 × 10 mm joiner standards at 4.50 m							
centres; 38 × 8 mm intermediate standards at							
900 mm centres; including excavating and							
backfilling into firm ground; setting in concrete							
galvanized	–	–	–	–	–	m	106.94
primed	–	–	–	–	–	m	87.91
Extra over five bar horizontal steel fencing for							
76 mm diameter end and corner posts							
galvanized	–	–	–	–	–	each	116.92
primed	–	–	–	–	–	each	68.41
Five bar horizontal steel fencing with mild							
steel copings; 38 mm square hollow							
section mild steel standards at 1.80 m							
centres; including excavating and							
backfilling into firm ground; setting in							
concrete							
Galvanized							
900 mm high	–	–	–	–	–	m	139.96
1200 mm high	–	–	–	–	–	m	147.23
1500 mm high	–	–	–	–	–	m	157.62
1800 mm high	–	–	–	–	–	m	168.01
Primed							
900 mm high	–	–	–	–	–	m	124.19
1200 mm high	–	–	–	–	–	m	127.48
1500 mm high	–	–	–	–	–	m	134.40
1800 mm high	–	–	–	–	–	m	144.64

Q PAVING/PLANTING/FENCING/SITE FURNITURE

Item Excluding site overheads and profit	PC £	Labour hours	Labour £	Plant £	Material £	Unit	Total rate £
Traditional trellis panels; The Garden Trellis Company; bespoke trellis panels for decorative, screening or security applications; timber planed all round; height of trellis 1800 mm							
Freestanding panels; posts 70 × 70 mm set in concrete; timber frames mitred and grooved 45 × 34 mm; heights and widths to suit; slats 32 × 10 mm joinery quality tanalized timber at 100 mm ccs; elements fixed by galvanized staples; capping rail 70 × 34 mm							
HV68 horizontal and vertical slats;							
softwood	79.20	1.00	18.50	–	84.77	m	103.27
D68 diagonal slats; softwood	93.60	1.00	18.50	–	99.17	m	117.67
HV68 horizontal and vertical slats;							
hardwood iroko	190.80	1.00	18.50	–	196.37	m	214.87
D68 diagonal slats; hardwood iroko or							
Western Red Cedar	216.00	1.00	18.50	–	221.57	m	240.07
Trellis panels fixed to face of existing wall or railings; timber frames mitred and grooved 45 × 34 mm; heights and widths to suit; slats 32 × 10 mm joinery quality tanalized timber at 100 mm ccs; elements fixed by galvanized staples; capping rail 70 × 34 mm							
HV68 horizontal and vertical slats;							
softwood	75.60	0.50	9.25	–	77.37	m	86.62
D68 diagonal slats; softwood	88.20	0.50	9.25	–	89.97	m	99.22
HV68 horizontal and vertical slats; iroko or							
Western Red Cedar	189.00	0.50	9.25	–	190.77	m	200.02
D68 diagonal slats; iroko or Western Red							
Cedar	198.00	0.50	9.25	–	199.77	m	209.02
Contemporary style trellis panels; The Garden Trellis Company; bespoke trellis panels for decorative, screening or security applications; timber planed all round							
Freestanding panels 30/15; posts 70 × 70 mm set in concrete with 90 × 30 mm top capping; slats 30 × 14 mm with 15 mm gaps; vertical support at 450 mm ccs							
joinery treated softwood	126.00	1.00	18.50	–	131.57	m	150.07
hardwood iroko or Western Red Cedar	207.00	1.00	18.50	–	212.57	m	231.07
Panels 30/15 face fixed to existing wall or fence; posts 70 × 70 mm set in concrete with 90 × 30 mm top capping; slats 30 × 14 mm with 15 mm gaps; vertical support at 450 mm ccs							
joinery treated softwood	117.00	0.50	9.25	–	118.77	m	128.02
hardwood iroko or Western Red Cedar	189.00	0.50	9.25	–	190.77	m	200.02
Integral arches to ornamental trellis panels; The Garden Trellis Company							
Arches to trellis panels in 45 × 34 mm grooved timbers to match framing; fixed to posts							
R450 1/4 circle; joinery treated softwood	38.00	1.50	27.75	–	43.57	nr	71.32
R450 1/4 circle; hardwood iroko or Western Red Cedar	49.00	1.50	27.75	–	54.57	nr	82.32

Q PAVING/PLANTING/FENCING/SITE FURNITURE

Item Excluding site overheads and profit	PC £	Labour hours	Labour £	Plant £	Material £	Unit	Total rate £
Q40 FENCING – cont							
Integral arches to ornamental trellis panels – cont							
Arches to span 1800 mm wide							
joinery treated softwood	49.00	1.50	27.75	–	54.57	nr	**82.32**
hardwood iroko or Western Red Cedar	65.00	1.50	27.75	–	70.57	nr	**98.32**
Painting or staining of trellis panels; high quality coatings							
microporous opaque paint or spirit based stain	–	–	–	–	24.00	m²	**24.00**
Gates – General							
Preamble: Gates in fences; see specification for fencing as gates in traditional or proprietary fencing systems are usually constructed of the same materials and finished as the fencing itself.							
Gates; hardwood; AVS Fencing Supplies Ltd							
Hardwood entrance gate; five bar diamond braced; curved hanging stile; planed iroko; fixed to 150 × 150 mm softwood posts; inclusive of hinges and furniture							
Ref 1100 040; 0.9 m wide	189.97	5.00	92.50	–	239.77	nr	**332.27**
Ref 1100 041; 1.2 m wide	202.55	5.00	92.50	–	252.35	nr	**344.85**
Ref 1100 042; 1.5 m wide	264.02	5.00	92.50	–	313.82	nr	**406.32**
Ref 1100 043; 1.8 m wide	279.57	5.00	92.50	–	329.37	nr	**421.87**
Ref 1100 044; 2.1 m wide	306.11	5.00	92.50	–	355.91	nr	**448.41**
Ref 1100 045; 2.4 m wide	288.09	5.00	92.50	–	337.89	nr	**430.39**
Ref 1100 047; 3.0 m wide	351.86	5.00	92.50	–	401.66	nr	**494.16**
Ref 1100 048; 3.3 m wide	367.51	5.00	92.50	–	417.31	nr	**509.81**
Ref 1100 049; 3.6 m wide	380.05	5.00	92.50	–	429.85	nr	**522.35**
Hardwood field gate; five bar diamond braced; planed iroko; fixed to 150 × 150 mm softwood posts; inclusive of hinges and furniture							
Ref 1100 100; 0.9 m wide	129.10	4.00	74.00	–	178.90	nr	**252.90**
Ref 1100 101; 1.2 m wide	140.26	4.00	74.00	–	190.06	nr	**264.06**
Ref 1100 102; 1.5 m wide	167.22	4.00	74.00	–	217.02	nr	**291.02**
Ref 1100 103; 1.8 m wide	179.74	4.00	74.00	–	229.54	nr	**303.54**
Ref 1100 104; 2.1 m wide	221.32	4.00	74.00	–	271.12	nr	**345.12**
Ref 1100 105; 2.4 m wide	234.49	4.00	74.00	–	284.29	nr	**358.29**
Ref 1100 106; 2.7 m wide	247.69	4.00	74.00	–	297.49	nr	**371.49**
Ref 1100 108; 3.3 m wide	274.06	4.00	74.00	–	323.86	nr	**397.86**
Ref 1100 109; 3.6 m wide	287.92	4.00	74.00	–	337.72	nr	**411.72**
Gates; treated softwood; Jacksons Fencing							
Timber field gates; including wrought iron ironmongery; five bar type; diamond braced; 1.80 m high; to 200 × 200 mm posts; setting 750 mm into firm ground							
width 2400 mm	81.63	10.00	185.00	–	200.63	nr	**385.63**
width 2700 mm	90.63	10.00	185.00	–	209.63	nr	**394.63**
width 3000 mm	97.56	10.00	185.00	–	216.56	nr	**401.56**
width 3300 mm	103.32	10.00	185.00	–	222.32	nr	**407.32**

Q PAVING/PLANTING/FENCING/SITE FURNITURE

Item Excluding site overheads and profit	PC £	Labour hours	Labour £	Plant £	Material £		
Featherboard garden gates; including ironmongery; to 100 × 120 mm posts; one diagonal brace							
1.0 × 1.2 m high	25.11	3.00	55.50	–	105.70	nr	**161.20**
1.0 × 1.5 m high	32.94	3.00	55.50	–	116.59	nr	**172.09**
1.0 × 1.8 m high	83.03	3.00	55.50	–	127.17	nr	**182.67**
Picket garden gates; including ironmongery; to match picket fence; width 1000 mm; to 100 × 120 mm posts; one diagonal brace							
950 mm high	68.22	3.00	55.50	–	100.57	nr	**156.07**
1200 mm high	71.91	3.00	55.50	–	104.26	nr	**159.76**
1800 mm high	84.38	3.00	55.50	–	128.52	nr	**184.02**
Gates; tubular mild steel; Jacksons Fencing							
Field gates; galvanized; including ironmongery; diamond braced; 1.80 m high; to tubular steel posts; setting in concrete							
width 3000 mm	110.25	5.00	92.50	–	206.74	nr	**299.24**
width 3300 mm	116.50	5.00	92.50	–	213.00	nr	**305.50**
width 3600 mm	122.89	5.00	92.50	–	219.39	nr	**311.89**
width 4200 mm	135.68	5.00	92.50	–	232.17	nr	**324.67**
Gates; sliding; Jacksons Fencing							
Sliding Gate; including all galvanized rails and vertical rail infill panels; special guide and shutting frame posts (Note: foundations installed by suppliers)							
access width 4.00 m; 1.5 m high gates	–	–	–	–	–	nr	**4381.28**
access width 4.00 m; 2.0 m high gates	–	–	–	–	–	nr	**4502.00**
access width 4.00 m; 2.5 m high gates	–	–	–	–	–	nr	**4621.28**
access width 6.00 m; 1.5 m high gates	–	–	–	–	–	nr	**5154.28**
access width 6.00 m; 2.0 m high gates	–	–	–	–	–	nr	**5309.00**
access width 6.00 m; 2.5 m high gates	–	–	–	–	–	nr	**5465.28**
access width 8.00 m; 1.5 m high gates	–	–	–	–	–	nr	**5875.28**
access width 8.00 m; 2.0 m high gates	–	–	–	–	–	nr	**6068.28**
access width 8.00 m; 2.5 m high gates	–	–	–	–	–	nr	**6260.28**
access width 10.00 m; 2.0 m high gates	–	–	–	–	–	nr	**7450.00**
access width 10.00 m; 2.5 m high gates	–	–	–	–	–	nr	**7649.00**
Kissing gates							
Kissing gates; Jacksons Fencing; in galvanized metal bar; fixing to fencing posts (posts not included); 1.65 × 1.30 × 1.00 m high	221.40	5.00	92.50	–	221.40	nr	**313.90**
Stiles							
Stiles; two posts; setting into firm ground; three rails; two treads	72.00	3.00	55.50	–	81.93	nr	**137.43**
Pedestrian guard rails and barriers							
Mild steel pedestrian guard rails; Broxap Street Furniture; 1.00 m high with 150 mm toe space; to posts at 2.00 m centres; galvanized finish							
vertical in-line bar infill panel	44.00	2.00	37.00	–	49.72	m	**86.72**
vertical staggered bar infill with 230 mm visibility gap at top	46.00	2.00	37.00	–	51.72	m	**88.72**

Q PAVING/PLANTING/FENCING/SITE FURNITURE

Item Excluding site overheads and profit	PC £	Labour hours	Labour £	Plant £	Material £	Unit	Total rate £
Q50 SITE/STREET FURNITURE/ EQUIPMENT							
Note: The costs of delivery from the supplier are generally not included in these items. Readers should check with the supplier to verify the delivery costs.							
Furniture/equipment – General Preamble: The following items include fixing to manufacturer's instructions; holding down bolts or other fittings and making good (excavating, backfilling and tarmac, concrete or paving bases not included).							
Q50 STONES AND BOULDERS							
Standing stones; CED Ltd; erect standing stones; vertical height above ground; in concrete base; including excavation setting in concrete to ⅓ depth and crane offload into position Purple schist							
1.00 m high	90.16	1.50	27.75	14.42	104.28	nr	**146.45**
1.25 m high	114.21	1.50	27.75	14.42	123.18	nr	**165.35**
1.50 m high	182.72	2.00	37.00	17.30	194.02	nr	**248.32**
2.00 m high	300.54	3.00	55.50	28.83	333.20	nr	**417.53**
2.50 m high	456.81	1.50	27.75	57.66	520.09	nr	**605.50**
Rockery stone – General Preamble: Rockery stone prices vary considerably with source, carriage, distance and load. Typical PC prices are in the range of £ 60–80 per tonne collected.							
Rockery stone; CED Ltd Boulders; maximum distance 25 m; by machine							
750 mm diameter	78.14	0.90	16.65	6.66	78.14	nr	**101.45**
1 m diameter	241.55	2.00	37.00	13.32	241.55	nr	**291.87**
1.5 m diameter	803.30	2.00	37.00	47.98	803.30	nr	**888.28**
2 m diameter	1949.27	2.00	37.00	47.98	1949.27	nr	**2034.25**
Boulders; maximum distance 25 m; by hand							
750 mm diameter	78.14	0.75	13.88	–	78.14	nr	**92.02**
1 m diameter	241.55	1.89	34.88	–	241.55	nr	**276.43**

Q PAVING/PLANTING/FENCING/SITE FURNITURE

Item Excluding site overheads and profit	PC £	Labour hours	Labour £	Plant £	Material £	Unit	Total rate £
Q50 BARRIERS							
Barriers – General							
Preamble: The provision of car and lorry							
control barriers may form part of the							
landscape contract. Barriers range from							
simple manual counterweighted poles to fully							
automated remote-control security gates, and							
the exact degree of control required must be							
specified. Complex barriers may need special							
maintenance and repair.							
Barriers; Autopa Ltd							
Manually operated pole barriers;							
counterbalance; to tubular steel supports;							
bolting to concrete foundation (foundation not							
included); aluminium boom; various finishes							
clear opening up to 3.00 m	817.00	6.00	111.00	–	836.32	nr	947.32
clear opening up to 4.00 m	879.00	6.00	111.00	–	898.32	nr	1009.32
clear opening 5.00 m	941.00	6.00	111.00	–	960.32	nr	1071.32
clear opening 6.00 m	1003.00	6.00	111.00	–	1022.32	nr	1133.32
clear opening 7.00 m	1063.00	6.00	111.00	–	1082.32	nr	1193.32
catch pole; arm rest for all manual barriers	116.00	–	–	–	118.41	nr	118.41
Electrically operated pole barriers; enclosed							
fan-cooled motor; double worm reduction gear;							
overload clutch coupling with remote controls;							
aluminium boom; various finishes (exclusive of							
electrical connections by electrician)							
Autopa AU 3.0; 3 m boom	2365.00	6.00	111.00	–	2384.32	nr	2495.32
Autopa AU 4.5; 4.5 m boom with catchpost	2465.00	6.00	111.00	–	2484.32	nr	2595.32
Autopa AU 6.0; 6.0 m boom with catchpost	2565.00	6.00	111.00	–	2584.32	nr	2695.32
Vehicle crash barriers – General							
Preamble: See Department of Environment							
Technical Memorandum BE5.							
Vehicle crash barriers							
Steel corrugated beams; untensioned; Broxap							
Street Furniture; effective length 3.20 m ×							
310 mm deep × 85 mm corrugations							
steel posts; Z-section; roadside posts	75.94	0.33	6.17	0.69	83.55	m	90.41
steel posts; Z-section; off highway	68.44	0.33	6.17	0.69	76.05	m	82.91
steel posts RSJ 760 high; for anchor fixing	88.44	0.33	6.17	–	102.50	m	108.67
steel posts RSJ 560 high; for anchor fixing							
to car parks	85.31	0.33	6.17	–	99.38	m	105.55
extra over for curved rail 6.00 m radius	67.19	–	–	–	67.19	m	67.19
Q50 BOLLARDS							
Excavating; for bollards and barriers; by							
hand							
Holes for bollards							
400 × 400 × 400 mm; disposing off site	–	0.42	7.71	–	0.61	nr	8.32
600 × 600 × 600 mm; disposing off site	–	0.97	17.95	–	1.73	nr	19.68

Q PAVING/PLANTING/FENCING/SITE FURNITURE

Item Excluding site overheads and profit	PC £	Labour hours	Labour £	Plant £	Material £	Unit	Total rate £
Q50 BOLLARDS – cont							
Concrete bollards – General							
Preamble: Precast concrete bollards are							
available in a very wide range of shapes and							
sizes. The bollards listed here are the most							
commonly used sizes and shapes;							
manufacturer's catalogues should be							
consulted for the full range. Most							
manufacturers produce bollards to match their							
suites of street furniture, which may include							
planters, benches, litter bins and cycle stands.							
Most parallel sided bollards can be supplied in							
removable form, with a reduced shank,							
precast concrete socket and lifting hole to							
permit removal with a bar.							
Concrete bollards							
Marshalls Plc; cylinder; straight or tapered;							
200–400 mm diameter; plain grey concrete;							
setting into firm ground (excavating and							
backfilling not included)							
Bridgford; 915 mm high above ground	70.00	2.00	37.00	–	79.70	nr	**116.70**
Marshalls Plc; cylinder; straight or tapered;							
200–400 mm diameter; Beadalite reflective							
finish; setting into firm ground (excavating and							
backfilling not included)							
Wexham concrete bollard; exposed silver							
grey	153.00	2.00	37.00.	–	158.21	nr	**195.21**
Wexham Major concrete bollard; exposed							
silver grey	250.00	2.00	37.00	–	259.70	nr	**296.70**
Precast concrete verge markers; various							
shapes; 450 mm high							
plain grey concrete	49.47	1.00	18.50	–	53.19	nr	**71.69**
white concrete	51.86	1.00	18.50	–	55.58	nr	**74.08**
exposed aggregate	54.92	1.00	18.50	–	58.64	nr	**77.14**
Bollards							
Recycled plastic bollards; Furnitubes							
International Ltd							
Aberdeen Circular ABR150; 1000 ×							
150 mm diameter	56.00	2.00	37.00	–	61.21	ea	**98.21**
Aberdeen Square ABR140; 1000 × 140 ×							
140 mm	35.00	2.00	37.00	–	40.21	ea	**77.21**
Service bollards; Furnitubes International Ltd;							
stainless steel; for housing services							
connection points to electricity, water etc.							
(services connections not included)							
Kenton KEN717; 900 × 250 mm diameter	595.00	2.00	37.00	–	600.21	ea	**637.21**
Zenith ZEN707; 900 × 250 mm diameter	550.00	2.00	37.00	–	555.21	ea	**592.21**
Other bollards							
Removable parking posts; Marshalls Plc							
RT\RD4; domestic telescopic bollard	160.00	2.00	37.00	–	165.21	nr	**202.21**
RT\R8; heavy duty telescopic bollard	214.00	2.00	37.00	–	219.21	nr	**256.21**
Plastic bollards; Marshalls Plc							
Lismore; three ring recycled plastic bollard	75.00	2.00	37.00	–	80.21	nr	**117.21**

Q PAVING/PLANTING/FENCING/SITE FURNITURE

Item Excluding site overheads and profit	PC £	Labour hours	Labour £	Plant £	Material £	Unit	Total rate £
Cast iron bollards – General							
Preamble: The following bollards are particularly suitable for conservation areas. Logos for civic crests can be incorporated to order.							
Cast iron bollards							
Bollards; Furnitubes International Ltd (excavating and backfilling not included)							
Doric Round; 920 mm high × 170 mm diameter	87.00	2.00	37.00	–	91.97	nr	**128.97**
Gunner Round; 750 mm high × 165 mm diameter	62.00	2.00	37.00	–	66.97	nr	**103.97**
Manchester Round; 975 mm high; 225 mm square base	107.00	2.00	37.00	–	111.97	nr	**148.97**
Cannon; 1140 mm × 210 mm diameter	132.00	2.00	37.00	–	136.97	nr	**173.97**
Kenton; heavy duty galvanized steel; 900 mm high; 350 mm diameter	171.00	2.00	37.00	–	175.97	nr	**212.97**
Bollards; Marshalls Plc (excavating and backfilling not included)							
MSF103; cast iron bollard; Small Manchester	115.00	2.00	37.00	–	120.21	nr	**157.21**
MSF102; cast iron bollard; Manchester	120.00	2.00	37.00	–	125.21	nr	**162.21**
Cast iron bollards with rails – General							
Preamble: The following cast iron bollards are suitable for conservation areas.							
Cast iron bollards with rails							
Cast iron posts with steel tubular rails; Broxap Street Furniture; setting into firm ground (excavating not included)							
Sheffield Short; 420 mm high; one rail, type A	70.00	1.50	27.75	–	72.41	nr	**100.16**
Mersey; 1085 mm high above ground; two rails, type D	60.00	1.50	27.75	–	62.41	nr	**90.16**
Promenade; 1150 mm high above ground; square; three rails, type C	124.00	1.50	27.75	–	126.41	nr	**154.16**
Type A mild steel tubular rail; including connector	6.05	0.03	0.62	–	7.44	m	**8.06**
Type C mild steel tubular rail; including connector	8.50	0.03	0.62	–	10.05	m	**10.67**
Steel bollards							
Steel bollards; Marshalls Plc (excavating and backfilling not included)							
SSB01; stainless steel bollard; 101 × 1250 mm	105.00	2.00	37.00	–	110.21	nr	**147.21**
RS001; stainless steel bollard; 114 × 1500 mm	120.00	2.00	37.00	–	125.21	nr	**162.21**
RB119 Brunel; steel bollard; 168 × 1500 mm	129.00	2.00	37.00	–	138.70	nr	**175.70**

Q PAVING/PLANTING/FENCING/SITE FURNITURE

Item Excluding site overheads and profit	PC £	Labour hours	Labour £	Plant £	Material £	Unit	Total rate £
Q50 BOLLARDS – cont							
Timber bollards							
Woodscape Ltd; durable hardwood							
RP 250/1500; 250 mm diameter × 1500 mm long	223.75	1.00	18.50	–	223.97	nr	**242.47**
SP 250/1500; 250 mm square × 1500 mm long	223.75	1.00	18.50	–	223.97	nr	**242.47**
SP 150/1200; 150 mm square × 1200 mm long	80.65	1.00	18.50	–	80.87	nr	**99.37**
SP 125/750; 125 mm square × 750 mm long	44.50	1.00	18.50	–	44.72	nr	**63.22**
RP 125/750; 125 mm diameter × 750 mm long	44.50	1.00	18.50	–	44.72	nr	**63.22**
Deterrent bollards							
Semi-mountable vehicle deterrent and kerb protection bollards; Furnitubes International Ltd (excavating and backfilling not included)							
Bell decorative	417.00	2.00	37.00	–	424.07	nr	**461.07**
Half bell	395.00	2.00	37.00	–	402.07	nr	**439.07**
Full bell	542.00	2.00	37.00	–	549.07	nr	**586.07**
Three quarter bell	419.00	2.00	37.00	–	426.07	nr	**463.07**
Security bollards							
Security bollards; Furnitubes International Ltd (excavating and backfilling not included)							
Gunner; reinforced with steel insert and tie bars; 750 mm high above ground; 600 mm below ground	93.00	2.00	37.00	–	100.45	nr	**137.45**
Burr Bloc Type 6; removable steel security bollard; 750 mm high above ground; 410 × 285 mm	487.00	2.00	37.00	–	503.76	nr	**540.76**
Q50 BASES FOR STREET FURNITURE							
Bases for street furniture							
Excavating; filling with concrete 1:3:6; bases for street furniture							
300 × 450 × 500 mm deep	–	0.75	13.88	–	5.89	nr	**19.77**
300 × 600 × 500 mm deep	–	1.11	20.54	–	7.84	nr	**28.38**
300 × 900 × 500 mm deep	–	1.67	30.89	–	11.78	nr	**42.67**
1750 × 900 × 300 mm deep	–	2.33	43.10	–	41.21	nr	**84.31**
2000 × 900 × 300 mm deep	–	2.67	49.40	–	47.09	nr	**96.49**
2400 × 900 × 300 mm deep	–	3.20	59.20	–	56.51	nr	**115.71**
2400 × 1000 × 300 mm deep	–	3.56	65.86	–	62.78	nr	**128.64**
Precast concrete flags; to concrete bases (not included); bedding and jointing in cement: mortar (1:4)							
450 × 600 × 50 mm	10.44	1.17	21.58	–	15.45	m²	**37.03**
Precast concrete paving blocks; to concrete bases (not included); bedding in sharp sand; butt joints							
200 × 100 × 65 mm	7.82	0.50	9.25	–	9.57	m²	**18.82**
200 × 100 × 80 mm	8.66	0.50	9.25	–	10.48	m²	**19.73**

Q PAVING/PLANTING/FENCING/SITE FURNITURE

Item Excluding site overheads and profit	PC £	Labour hours	Labour £	Plant £	Material £	Unit	Total rate £
Engineering paving bricks; to concrete bases (not included); bedding and jointing in sulphate-resisting cement: lime: sand mortar (1:1:6)							
over 300 mm wide	9.90	0.56	10.39	–	20.66	m²	31.05
Edge restraints to pavings; haunching in concrete (1:3:6)							
200 × 300 mm	4.66	0.10	1.85	–	4.66	m	6.51
Bases; Earth Anchors Ltd							
Rootfast ancillary anchors; A1; 500 mm long × 25 mm diameter and top strap F2; including bolting to site furniture (site furniture not included)	9.50	0.50	9.25	–	9.50	set	18.75
installation tool for above	23.50	–	–	–	23.50	nr	23.50
Rootfast ancillary anchors; A4; heavy duty 40 mm square fixed head anchors; including bolting to site furniture (site furniture not included)	29.00	0.33	6.17	–	29.00	set	35.17
Rootfast ancillary anchors; F4; vertical socket; including bolting to site furniture (site furniture not included)	12.75	0.05	0.93	–	12.75	set	13.68
Rootfast ancillary anchors; F3; horizontal socket; including bolting to site furniture (site furniture not included)	14.00	0.05	0.93	–	14.00	set	14.93
installation tools for the above	66.00	–	–	–	66.00	nr	66.00
Rootfast ancillary anchors; A3; anchored bases; including bolting to site furniture (site furniture not included)	46.00	0.33	6.17	–	46.00	set	52.17
installation tools for the above	66.00	–	–	–	66.00	nr	66.00
Q50 BINS – LITTER/WASTE/DOG WASTE/ SALT AND GRIT BINS							
Dog waste bins							
Earth Anchors Ltd							
HG45A; steel; 45 l; earth anchored; post mounted	176.00	0.33	6.17	–	176.00	nr	182.17
HG45A; steel; 45 l; as above with pedal operation	206.00	0.33	6.17	–	206.00	nr	212.17
Furnitubes International Ltd							
PED 701; Pedigree; post mounted cast iron dog waste bins; 1250 mm total height above ground; 400 mm square bin	417.00	0.75	13.88	–	422.72	nr	436.60
LUK745 P; Lucky; steel dog bin; post mounted; 47 l	287.00	1.50	27.75	–	292.72	ea	320.47
LUK745 W; Lucky; steel dog bin; wall mounted; 47 l	262.00	0.75	13.88	–	267.72	ea	281.60
TER801 P; Terrier; polythene dog bin; post mounted; 40 l	139.00	1.50	27.75	–	144.72	ea	172.47
TER801 W; Terrier; polythene dog bin; wall mounted; 40 l	93.00	0.75	13.88	–	98.72	ea	112.60

Q PAVING/PLANTING/FENCING/SITE FURNITURE

Item Excluding site overheads and profit	PC £	Labour hours	Labour £	Plant £	Material £	Unit	Total rate £
Q50 BINS – LITTER/WASTE/DOG WASTE/ SALT AND GRIT BINS – cont							
Litter bins; precast concrete in textured white or exposed aggregate finish; with wire baskets and drainage holes							
Bins; Marshalls Plc							
Boulevard 700; concrete circular litter bin	415.00	0.50	9.25	–	415.00	nr	**424.25**
Bins; Neptune Outdoor Furniture Ltd							
SF16; 42 l	222.00	0.50	9.25	–	222.00	nr	**231.25**
SF14; 100 l	310.00	0.50	9.25	–	310.00	nr	**319.25**
Bins; Townscape Products Ltd							
Sutton; 750 mm high × 500 mm diameter; 70 l capacity; including GRP canopy	242.33	0.33	6.17	–	415.49	nr	**421.66**
Braunton; 750 mm high × 500 mm diameter; 70 l capacity; including GRP canopy	256.80	1.50	27.75	–	429.96	nr	**457.71**
Litter bins; metal; stove-enamelled perforated metal for holder and container							
Bins; Townscape Products Ltd							
Metro; 440 × 420 × 800 mm high; 62 l capacity	733.98	0.33	6.17	–	733.98	nr	**740.15**
Voltan; large round; 460 mm diameter × 780 mm high; 56 l capacity	657.15	0.33	6.17	–	657.15	nr	**663.32**
Voltan; small round with pedestal; 410 mm diameter × 760 mm high; 31 l capacity	586.58	0.33	6.17	–	586.58	nr	**592.75**
Litter bins; all-steel							
Bins; Marshall Plc							
MSF Central; steel litter bin	250.00	1.00	18.50	–	250.00	nr	**268.50**
MSF Central; stainless steel litter bin	500.00	0.67	12.33	–	500.00	nr	**512.33**
Bins; Furnitubes International Ltd							
Wave Bin; WVB 440; free standing 55 l liners; 440 mm diameter × 850 mm high	352.00	0.50	9.25	–	356.06	nr	**365.31**
Wave Bin WVB 520; free standing 85 l liners; 520 mm diameter × 850 mm high; cast iron plinth	436.00	0.50	9.25	–	440.06	nr	**449.31**
Wave Bin; WVB 440 S304; free standing; steel with satin finish; cast bronze plinth; open top; 55 l	688.00	0.50	9.25	–	688.00	ea	**697.25**
Wave Bin; WVB 520 S304; free standing; steel with satin finish; cast bronze plinth; open top; 80 l	760.00	0.50	9.25	–	760.00	ea	**769.25**
Liverpool Bin; LVR520 S304; free standing; stainless steel; side opening; 965 mm high; 125 l	1200.00	0.50	9.25	–	1200.00	ea	**1209.25**
Bins; Earth Anchors Ltd							
Ranger; 100 l; pedestal mounted	427.00	0.50	9.25	–	427.00	nr	**436.25**
Big Ben; 82 l; steel frame and liner; colour coated finish; earth anchored	281.00	1.00	18.50	–	281.00	nr	**299.50**
Beau; 42 l; steel frame and liner; colour coated finish; earth anchored	229.00	1.00	18.50	–	229.00	nr	**247.50**
Bins; Townscape Products Ltd							
Baltimore Major with GRP canopy; 560 mm diameter × 960 mm high; 140 l capacity	1060.57	1.00	18.50	–	1060.57	nr	**1079.07**

Q PAVING/PLANTING/FENCING/SITE FURNITURE

Item Excluding site overheads and profit	PC £	Labour hours	Labour £	Plant £	Material £	Unit	Total rate £
Litter bins; cast iron							
Bins; Marshalls Plc							
MSF5501 Heritage; cast iron litter bin	490.00	1.00	18.50	3.72	490.00	nr	512.22
Bins; Furnitubes International Ltd							
Covent Garden COV 702; side opening, 500 mm square × 1050 mm high; 105 l capacity	236.50	0.50	9.25	–	242.97	nr	252.22
Covent Garden COV 803; side opening, 500 mm diameter × 1025 mm high; 85 l capacity	244.00	0.50	9.25	–	250.47	nr	259.72
Covent Garden COV 912; open top; 500 mm A/F octagonal × 820 mm high; 85 l capacity	284.00	0.50	9.25	–	290.47	nr	299.72
Albert ALB 800; open top; 400 mm diameter × 845 mm high; 55 l capacity	238.00	0.50	9.25	–	244.47	nr	253.72
Bins; Broxap Street Furniture							
Derby Hercules; post mounted; steel; 40 l	160.00	1.00	18.50	–	166.47	nr	184.97
Bins; Townscape Products Ltd							
York Major; 650 mm diameter × 1060 mm high; 140 l capacity	1030.55	1.00	18.50	–	1032.45	nr	1050.95
Litter bins; timber faced; hardwood slatted casings with removable metal litter containers; ground or wall fixing							
Bins; Lister Lutyens Co Ltd							
Monmouth; 675 mm high × 450 mm wide; freestanding	123.00	0.33	6.17	–	123.00	nr	129.17
Monmouth; 675 mm high × 450 mm wide; bolting to ground (without legs)	114.00	1.00	18.50	–	122.50	nr	141.00
Bins; Woodscape Ltd							
square; 580 × 580 × 950 mm high; with lockable lid	660.00	0.50	9.25	–	660.00	nr	669.25
round; 580 mm diameter × 950 mm high; with lockable lid	660.00	0.50	9.25	–	660.00	nr	669.25
Plastic litter and grit bins; glassfibre reinforced polyester grit bins; yellow body; hinged lids							
Bins; Wybone Ltd; Victoriana glass fibre; cast iron effect litter bins; including lockable liner							
LBV/2; 521 × 521 × 673 mm high; open top; square shape; 0.078 m^3 capacity	204.90	1.00	18.50	–	211.37	nr	229.87
LVC/3; 457 mm diameter × 648 mm high; open top; drum shape; 0.084 m^3 capacity; with lockable liner	223.52	1.00	18.50	–	229.99	nr	248.49
Grit bins; Furnitubes International Ltd							
Grit and salt bin; yellow glass fibre; hinged lid; 170 l	180.00	–	–	–	180.00	ea	180.00
Cigarette bins							
Furnitubes International Ltd							
ZEN 275 Zenith cigarette bin; stainless steel; wall or post mounted; 1.7 L	29.99	0.75	13.88	–	35.71	ea	49.59
LVR250 PC Liverpool cigarette bin; steel; wall mounted; 1.9 L	16.75	0.50	9.25	–	16.75	ea	26.00
SMK500F Smoke King cigarette bin; cast aluminium; bolt down; 3.5 L	199.99	0.75	13.88	–	205.71	ea	219.59

Q PAVING/PLANTING/FENCING/SITE FURNITURE

Item Excluding site overheads and profit	PC £	Labour hours	Labour £	Plant £	Material £	Unit	Total rate £
Q50 SMOKING SHELTERS							
Smoking shelters Furnitubes International Ltd; Ashby freestanding bolt down; aluminium frame; PET glazing; 2050 mm length × 2330 mm high							
3290 mm depth; maximum 3 people	2067.00	4.00	74.00	–	2078.45	ea	**2152.45**
6050 mm depth; maximum 6 people	2597.00	5.00	92.50	–	2608.45	ea	**2700.95**
Q50 SEATING							
Outdoor seats							
CED Ltd; stone bench; Sinuous bench; stone type bench to organic S pattern; laid to concrete base; 500 mm high × 500 mm wide (not included)							
2.00 m long	1235.85	4.00	74.00	–	1235.85	nr	**1309.85**
5.00 m long	3033.45	8.00	148.00	–	3033.45	nr	**3181.45**
10.00 m long	6066.90	11.00	203.50	–	6066.90	nr	**6270.40**
Outdoor seats; concrete framed – General Preamble: Prices for the following concrete framed seats and benches with hardwood slats include for fixing (where necessary) by bolting into existing paving or concrete bases (bases not included) or building into walls or concrete foundations (walls and foundations not included). Delivery generally not included.							
Outdoor seats; concrete framed Outdoor seats; Townscape Products Ltd							
Oxford benches; 1800 × 430 × 440 mm	353.56	2.00	37.00	–	365.01	nr	**402.01**
Maidstone seats; 1800 × 610 × 785 mm	528.59	2.00	37.00	–	540.04	nr	**577.04**
Outdoor seats; concrete Outdoor seats; Marshalls Ltd							
Boulevard 2000 concrete seat	725.00	2.00	37.00	–	744.57	nr	**781.57**
Outdoor seats; Furnitubes International Ltd							
Amesbury concrete seat, 2.2 m long	525.00	2.00	37.00	–	544.57	ea	**581.57**
Marlborough concrete seat, 1.8 m long	488.00	2.00	37.00	–	507.57	ea	**544.57**
Outdoor seats; metal framed – General Preamble: Metal framed seats with hardwood backs to various designs can be bolted to ground anchors.							
Outdoor seats; metal framed Outdoor seats; Furnitubes International Ltd							
NS6 Newstead; steel standards with iroko slats; 1.80 m long	301.00	2.00	37.00	–	320.57	nr	**357.57**
NEB6 New Forest Single Bench; cast iron standards with iroko slats; 1.83 m long	248.00	2.00	37.00	–	267.57	nr	**304.57**
EA6 Eastgate; cast iron standards with iroko slats; 1.86 m long	274.00	2.00	37.00	–	293.57	nr	**330.57**

Q PAVING/PLANTING/FENCING/SITE FURNITURE

Item Excluding site overheads and profit	PC £	Labour hours	Labour £	Plant £	Material £	Unit	Total rate £
NE6 New Forest Seat; cast iron standards with iroko slats; 1.83 m long	328.00	2.00	37.00	–	347.57	nr	**384.57**
Zenith long bench, satin stainless steel, iroko slats, 2.43m long	1034.00	2.00	37.00	–	1053.57	ea	**1090.57**
Ashburton; steel with recycled plastic slats; 1.8m long	318.00	2.00	37.00	–	337.57	ea	**374.57**
Outdoor seats; Orchard Street Furniture Ltd; Bramley; broad iroko slats to steel frame							
1.20 m long	172.90	2.00	37.00	–	185.85	nr	**222.85**
1.80 m long	213.05	2.00	37.00	–	226.00	nr	**263.00**
2.40 m long	246.09	2.00	37.00	–	259.04	nr	**296.04**
Outdoor seats; Orchard Street Furniture Ltd; Laxton; narrow iroko slats to steel frame							
1.20 m long	195.40	2.00	37.00	–	208.35	nr	**245.35**
1.80 m long	240.11	2.00	37.00	–	253.06	nr	**290.06**
2.40 m long	272.87	2.00	37.00	–	285.82	nr	**322.82**
Outdoor seats; Orchard Street Furniture Ltd; Lambourne; iroko slats to cast iron frame							
1.80 m long	507.58	2.00	37.00	–	520.53	nr	**557.53**
2.40 m long	597.30	2.00	37.00	–	610.25	nr	**647.25**
Outdoor seats; Broxap Street Furniture; metal and timber							
Eastgate	384.00	2.00	37.00	–	396.95	nr	**433.95**
Rotherham	639.00	2.00	37.00	–	651.95	nr	**688.95**
Outdoor seats; Earth Anchors Ltd; Forest-Saver; steel frame and recycled slats							
bench; 1.8 m	199.00	1.00	18.50	–	199.00	nr	**217.50**
seat; 1.8 m	318.00	1.00	18.50	–	318.00	nr	**336.50**
Outdoor seats; Earth Anchors Ltd; Evergreen; cast iron frame and recycled slats							
bench; 1.8 m	426.00	1.00	18.50	–	426.00	nr	**444.50**
seat; 1.8 m	532.00	1.00	18.50	–	532.00	nr	**550.50**
Outdoor seats; all steel							
Outdoor seats; Marshalls Plc							
MSF Central; stainless steel seat	1000.00	0.50	9.25	–	1000.00	nr	**1009.25**
MSF Central; steel seat	500.00	2.00	37.00	–	500.00	nr	**537.00**
MSF 502; cast iron Heritage seat	495.00	2.00	37.00	–	495.00	nr	**532.00**
Outdoor seats; Earth Anchors Ltd; Ranger							
bench; 1.8 m	275.00	1.00	18.50	–	275.00	nr	**293.50**
seat; 1.8 m	431.00	1.00	18.50	–	431.00	nr	**449.50**
Outdoor seats; all timber							
Outdoor seats; Broxap Street Furniture; teak seats with back and armrests							
Cambridge seat; three seater; 1.8m	390.00	2.00	37.00	–	409.57	nr	**446.57**
Milano seat; three seater; 1.8 m	420.00	2.00	37.00	–	439.57	nr	**476.57**
Outdoor seats; Lister Lutyens Co Ltd; Mendip; teak							
1.524 m long	465.00	1.00	18.50	–	488.14	nr	**506.64**
1.829 m long	561.00	1.00	18.50	–	584.14	nr	**602.64**
2.438 m long (including centre leg)	703.00	1.00	18.50	–	726.14	nr	**744.64**
Outdoor seats; Lister Lutyens Co Ltd; Sussex; hardwood							
1.5 m	185.00	2.00	37.00	–	208.14	nr	**245.14**

Q PAVING/PLANTING/FENCING/SITE FURNITURE

Item Excluding site overheads and profit	PC £	Labour hours	Labour £	Plant £	Material £	Unit	Total rate £
Q50 SEATING – cont							
Outdoor seats – cont							
Outdoor seats; Woodscape Ltd; solid hardwood							
seat type 3 with back; 2.00 m long; freestanding	790.00	2.00	37.00	–	813.14	nr	850.14
seat type 3 with back; 2.00 m long; building in	830.00	4.00	74.00	–	853.14	nr	927.14
seat type 4; 2.00 m long; fixing to wall	365.00	2.00	37.00	–	388.14	nr	425.14
seat type 4 with back; 2.00 m long; fixing to wall	485.00	3.00	55.50	–	497.18	nr	552.68
seat type 4 with back; 2.50 m long; fixing to wall	565.00	3.00	55.50	–	577.18	nr	632.68
seat type 5 with back; 2.00 m long; freestanding	805.00	2.00	37.00	–	828.14	nr	865.14
seat type 5 with back; 2.50 m long; freestanding	965.00	2.00	37.00	–	988.14	nr	1025.14
seat type 5 with back; 2.00 m long; building in	845.00	2.00	37.00	–	868.14	nr	905.14
seat type 5 with back; 2.50 m long; building in	1005.00	3.00	55.50	–	1028.14	nr	1083.64
bench type 1; 2.00 m long; freestanding	625.00	2.00	37.00	–	648.14	nr	685.14
bench type 1; 2.00 m long; building in	665.00	4.00	74.00	–	688.14	nr	762.14
bench type 2; 2.00 m long; freestanding	600.00	2.00	37.00	–	623.14	nr	660.14
bench type 2; 2.50 m long; freestanding	680.00	2.00	37.00	–	703.14	nr	740.14
bench type 2; 2.00 m long; building in	640.00	4.00	74.00	–	663.14	nr	737.14
bench type 2; 2.50 m long; building in	720.00	4.00	74.00	–	743.14	nr	817.14
bench type 2; 2.00 m long overall; curved to 5 m radius; building in	820.00	4.00	74.00	–	843.14	nr	917.14
Outdoor seats; Orchard Street Furniture Ltd; Allington; all iroko							
1.20 m long	272.30	2.00	37.00	–	295.44	nr	332.44
1.80 m long	313.03	2.00	37.00	–	336.17	nr	373.17
2.40 m long	380.82	2.00	37.00	–	403.96	nr	440.96
Outdoor seats; tree benches/seats							
Tree bench; Neptune Outdoor Furniture Ltd; Beaufort; hexagonal; timber							
SF 34–15A; 1500 mm diameter	1071.00	0.50	9.25	–	1071.00	nr	1080.25
Tree seat; Neptune Outdoor Furniture Ltd; Beaufort; hexagonal; timber; with back							
SF 32–10A; 720 mm diameter	1285.00	0.50	9.25	–	1285.00	nr	1294.25
SF 32–20A; 1720 mm diameter	1504.00	0.50	9.25	–	1504.00	nr	1513.25
Outdoor seats; recycled plastic							
Furnitubes International Ltd							
Aberdeen seat, brown recycled plastic; 1 slat; 2 m long	296.00	2.00	37.00	–	315.57	ea	352.57
Aberdeen seat, brown recycled plastic; 2 slats; 2 m long	401.00	2.00	37.00	–	420.57	ea	457.57
Anti-skateboard devices							
Stainless steel; Furnitubes International Ltd							
for fitting to seats	18.00	0.20	3.70	–	18.00	ea	21.70
for fitting to benches	36.00	0.20	3.70	–	36.00	ea	39.70

Q PAVING/PLANTING/FENCING/SITE FURNITURE

Item Excluding site overheads and profit	PC £	Labour hours	Labour £	Plant £	Material £	Unit	Total rate £
Q50 PLANT CONTAINERS							
Market prices of containers							
Plant containers; terracotta							
Capital Garden Products Ltd							
Large Pot LP63; weathered terracotta; 1170 × 1600 mm diameter	–	–	–	–	795.30	nr	795.30
Large Pot LP38; weathered terracotta; 610 × 970 mm diameter	–	–	–	–	341.00	nr	341.00
Large Pot LP23; weathered terracotta; 480 × 580 mm diameter	–	–	–	–	182.60	nr	182.60
Indian style Shimmer Pot 2322; 585 × 560 mm diameter	–	–	–	–	167.75	nr	167.75
Indian style Shimmer Pot 1717; 430 × 430 mm diameter	–	–	–	–	107.75	nr	107.75
Indian style Shimmer Pot 1314; 330 × 355 mm diameter	–	–	–	–	97.75	nr	97.75
Plant containers; faux lead							
Capital Garden Products Ltd							
Trough 2508 Tudor Rose; 620 × 220 × 230 mm high	–	–	–	–	61.75	nr	61.75
Tub 2004 Elizabethan; 510 mm square	–	–	–	–	97.75	nr	97.75
Tub 1513 Elizabethan; 380 mm square	–	–	–	–	65.75	nr	65.75
Tub 1601 Tudor Rose; 420 × 400 mm diameter	–	–	–	–	68.75	nr	68.75
Plant containers; window boxes							
Capital Garden Products Ltd							
Adam 5401; faux lead; 1370 × 270 × 210 mm high	–	–	–	–	107.75	nr	107.75
Oakleaf OAK24; terracotta; 610 × 230 × 240 mm high	–	–	–	–	86.75	nr	86.75
Swag 2402; faux lead; 610 × 200 × 210 mm high	–	–	–	–	53.75	nr	53.75
Plant containers; timber							
Plant containers; hardwood; Neptune Outdoor Furniture Ltd							
Beaufort T38-4D; 1500 × 1500 × 900 mm high	–	–	–	–	1224.00	nr	1224.00
Beaufort T38-3C; 1000 × 1500 × 700 mm high	–	–	–	–	909.00	nr	909.00
Beaufort T38-2A; 1000 × 500 × 500 mm high	–	–	–	–	475.00	nr	475.00
Kara T42-4D; 1500 × 1500 × 900 mm high	–	–	–	–	1320.00	nr	1320.00
Kara T42-3C; 1000 × 1500 × 700 mm high	–	–	–	–	985.00	nr	985.00
Kara T42-2A; 1000 × 500 × 500 mm high	–	–	–	–	513.00	nr	513.00
Plant containers; hardwood; Woodscape Ltd							
square; 900 × 900 × 420 mm high	–	–	–	–	342.60	nr	342.60
Measured works							
Plant containers; precast concrete							
Plant containers; Marshalls Plc							
Boulevard 700; circular base and ring	560.00	2.00	37.00	16.83	560.00	nr	613.83
Boulevard 1200; circular base and ring	705.00	1.00	18.50	16.83	705.00	nr	740.33

Q PAVING/PLANTING/FENCING/SITE FURNITURE

Item Excluding site overheads and profit	PC £	Labour hours	Labour £	Plant £	Material £	Unit	Total rate £
Q50 CYCLE HOLDERS/CYCLE SHELTERS							
Cycle holders							
Cycle stands; Marshalls Plc							
Sheffield; steel cycle stand; RCS1	44.00	0.50	9.25	–	44.00	nr	**53.25**
Sheffield; stainless steel cycle stand;							
RSCS1	121.00	0.50	9.25	–	121.00	nr	**130.25**
Cycle holders; Autopa; VELOPA; galvanized							
steel							
R; fixing to wall or post; making good	34.00	1.00	18.50	–	46.95	nr	**65.45**
SR(V); fixing in ground; making good	43.00	1.00	18.50	–	55.95	nr	**74.45**
Sheffield cycle stands; ragged steel	54.00	1.00	18.50	–	66.95	nr	**85.45**
Cycle holders; Townscape Products Ltd							
Guardian cycle holders; tubular steel							
frame; setting in concrete; 1250 × 550 ×							
775 mm high; making good	348.21	1.00	18.50	–	361.16	nr	**379.66**
Penny cycle stands; 600 mm diameter;							
exposed aggregate bollards with 8 nr cycle							
holders; in galvanized steel; setting in							
concrete; making good	788.45	1.00	18.50	–	801.40	nr	**819.90**
Cycle holders; Broxap Street Furniture							
Neath cycle rack; BX/MW/AG; for 6 nr							
cycles; semi-vertical; galvanized and							
polyester powder coated; 1.32 m wide ×							
2.542 m long × 1.80 m high	435.00	10.00	185.00	–	460.90	nr	**645.90**
Premier Senior economy combined shelter							
and rack; BX/MW/AW; for 10 nr cycles;							
horizontal; galvanized only; 2.13 m wide ×							
3.05 m long × 2.15 m high	1200.00	10.00	185.00	–	1225.90	nr	**1410.90**
Toast Rack double sided free standing							
cycle rack; BX/MW/GH; for 10 nr cycles;							
galvanized and polyester coated; 3.25 m							
long	375.00	4.00	74.00	–	384.66	nr	**458.66**
Cycle shelters							
Furnitubes International Ltd; Academy							
freestanding bolt down shelter, galvanized							
steel frame and roof; 3450 mm overall width;							
2150 mm maximum height; 2670 mm depth							
corrugated roof	1873.00	6.00	111.00	–	1884.45	ea	**1995.45**
clear polycarbonate roof	2024.00	6.00	111.00	–	2035.45	ea	**2146.45**
Q50 SIGNAGE							
Directional signage; cast aluminium							
Signage; Furnitubes International Ltd							
FFL1 Lancer; cast aluminium finials	68.25	0.07	1.23	–	68.25	nr	**69.48**
FAAIS; arrow end type cast aluminium							
directional arms; single line; 90 mm wide	122.85	2.00	37.00	–	122.85	nr	**159.85**
FAAID; arrow end type cast aluminium							
directional arms; double line; 145 mm wide	149.10	0.13	2.47	–	149.10	nr	**151.57**
FAAIS; arrow end type cast aluminium							
directional arms; treble line; 200 mm wide	184.80	0.20	3.70	–	184.80	nr	**188.50**
FCK1211G Kingston; composite standard							
root columns	403.20	2.00	37.00	–	418.83	nr	**455.83**

Q PAVING/PLANTING/FENCING/SITE FURNITURE

Item Excluding site overheads and profit	PC £	Labour hours	Labour £	Plant £	Material £	Unit	Total rate £
Park signage							
Entrance signs and map boards; vitreous enamel							
entrance map board; 1250 × 1000 mm high with two support posts; all associated works to post bases	–	–	–	–	–	nr	2159.00
information board with two locking cabinets; 1250 × 1000 mm high with two support posts; all associated works to post bases	–	–	–	–	–	nr	2061.20
Miscellaneous park signage; vitreous enamel							
No dogs' sign; 200 × 150 mm; fixing to fencing or gates	–	–	–	–	–	nr	319.58
Dog exercise area' sign; 300 × 400 mm; fixing to fencing or gates	–	–	–	–	–	nr	397.92
Nature conservation area' sign; 900 × 400 mm high with two support posts; all associated works to post bases	–	–	–	–	–	nr	1140.00
Monolith signage boards							
Furnitubes International Ltd; Fulham monolith signage board; stainless steel frame; vinyl graphics; 2600 mm above ground; 600 mm below ground; 350 mm width × 120 mm depth							
without baseplate	3550.00	2.00	37.00	–	3554.83	ea	3591.83
with baseplate	3700.00	2.00	37.00	–	3704.83	ea	3741.83
Q50 TREE GRILLES/PROTECTION							
Tree grilles; cast iron							
Cast iron tree grilles; Furnitubes International Ltd							
GS 1070 Greenwich; two part; 1000 mm square × 700 mm diameter tree hole	74.00	2.00	37.00	–	74.00	nr	111.00
GC 1270 Greenwich; two part, 1200 mm diameter × 700 mm diameter tree hole	93.00	2.00	37.00	–	93.00	nr	130.00
Cast iron tree grilles; Marshalls Plc							
Heritage; cast iron grille plus frame; 1 × 1 m	370.00	3.00	55.50	–	380.40	nr	435.90
Cast iron tree grilles; Townscape Products Ltd							
Baltimore; 1200 mm square × 460 mm diameter tree hole	501.44	2.00	37.00	–	501.44	nr	538.44
Baltimore; hexagonal; maximum width 1440 mm nominal × 600 mm diameter tree hole	806.84	2.00	37.00	–	806.84	nr	843.84
Tree grilles; steel							
Steel tree grilles; Furnitubes International Ltd							
GSF 102G Greenwich; steel tree grille frame for GS 1045; one part	123.00	2.00	37.00	–	123.00	nr	160.00
GSF 122G Greenwich; steel tree grille frame for GS 1270 and GC 1245; one part	131.00	2.00	37.00	–	131.00	nr	168.00
Note: Care must be taken to ensure that tree grids and guards are removed when trees grow beyond the specified diameter of guard.							

Q PAVING/PLANTING/FENCING/SITE FURNITURE

Item Excluding site overheads and profit	PC £	Labour hours	Labour £	Plant £	Material £	Unit	Total rate £
Q50 PLAYGROUND EQUIPMENT							
Playground equipment – General							
Preamble: The range of equipment							
manufactured or available in the UK is so							
great that comprehensive coverage would be							
impossible, especially as designs,							
specifications and prices change fairly							
frequently. The following information should be							
sufficient to give guidance to anyone							
designing or equipping a playground. In							
comparing prices note that only outline							
specification details are given here and that							
other refinements which are not mentioned							
may be the reason for some difference in price							
between two apparently identical elements.							
The fact that a particular manufacturer does							
not appear under one item heading does not							
necessarily imply that they do not make it.							
Landscape designers are advised to check							
that equipment complies with ever more							
stringent safety standards before specifying.							
Playground equipment – Installation							
The rates below include for installation of the							
specified equipment by the manufacturers.							
Most manufacturers will offer an option to							
install the equipment they have supplied.							
Play systems; Kompan Ltd; Galaxy;							
multiple play activity systems for							
non-prescribed play; for children 6–14							
years; galvanized steel and high density							
polyethylene							
GXY906 Adara; 14 different play activities	–	–	–	–	–	nr	**15390.90**
GXY8011 Sirius; 10 different play activities	–	–	–	–	–	nr	**9630.60**
Sports and social areas; Freegame							
multi-use games areas; Kompan Ltd;							
enclosed sports areas; complete with							
surfacing boundary and goals and targets;							
galvanized steel framework with high							
density polyethylene panels, galvanized							
steel goals and equipment; surfacing							
priced separately							
FRE1211 Classic Multigoal; 7 m	–	–	–	–	–	nr	**5028.45**
Pitch; complete; suitable for multiple ball							
sports; suitable for use with natural artificial or							
hard landscape surfaces; fully enclosed							
including two end sports walls							
FRE2110 Cosmos; 12 × 20 m	–	–	–	–	–	nr	**29362.20**
FRE2116 Cosmos; 19 × 36 m	–	–	–	–	–	nr	**43483.65**
FRE3000 Meeting Point; shelter or social							
area	–	–	–	–	–	nr	**4228.35**

Q PAVING/PLANTING/FENCING/SITE FURNITURE

Item Excluding site overheads and profit	PC £	Labour hours	Labour £	Plant £	Material £	Unit	Total rate £
Swings – General							
Preamble: Prices for the following vary							
considerably. Those given represent the							
middle of the range and include multiple							
swings with tubular steel frames and timber or							
tyre seats; ground fixing and priming only.							
Swings							
Swings; Wicksteed Ltd							
traditional swings; 1850 mm high; 1 bay;							
2 seat	–	–	–	–	–	nr	2338.00
traditional swings; 1850 mm high; 2 bay;							
4 seat	–	–	–	–	–	nr	3757.00
traditional swings; 2450 mm high; 1 bay;							
2 seat	–	–	–	–	–	nr	2259.00
traditional swings; 2450 mm high; 2 bay;							
4 seat	–	–	–	–	–	nr	3566.00
traditional swings; 3050 mm high; 1 bay;							
2 seat	–	–	–	–	–	nr	2398.00
traditional swings; 3050 mm high; 2 bay;							
4 seat	–	–	–	–	–	nr	3738.00
double arch swing; cradle safety seats;							
1850 mm high	–	–	–	–	–	nr	2372.00
twin double arch swing; cradle safety seat;							
1850 mm high	–	–	–	–	–	nr	3973.00
single arch swing; flat rubber safety seat;							
2450 mm high	–	–	–	–	–	nr	1826.00
double arch swing; flat rubber safety seats;							
2450 mm high	–	–	–	–	–	nr	2240.00
Swings; Lappset UK Ltd							
020414M; swing frame with two flat seats	–	–	–	–	–	nr	1520.00
Swings; Kompan Ltd							
M951P Sunflower swing; 1–3 years	–	–	–	–	–	nr	1556.10
M947P double swings; 1–6 years	–	–	–	–	–	nr	2071.65
M961P double swings; 6–12 years	–	–	–	–	–	nr	1792.35
Slides							
Slides; Wicksteed Ltd							
Pedestal slides; 3.40 m	–	–	–	–	–	nr	4223.00
Pedestal slides; 4.40 m	–	–	–	–	–	nr	3244.00
Pedestal slides; 5.80 m	–	–	–	–	–	nr	4914.00
Embankment slides; 3.40 m	–	–	–	–	–	nr	2485.00
Embankment slides; 4.40 m	–	–	–	–	–	nr	3231.00
Embankment slides; 5.80 m	–	–	–	–	–	nr	4112.00
Embankment slides; 7.30 m	–	–	–	–	–	nr	5225.00
Embankment slides; 9.10 m	–	–	–	–	–	nr	6545.00
Embankment slides; 11.00 m	–	–	–	–	–	nr	7827.00
Slides; Lappset UK Ltd							
142015M slide	–	–	–	–	–	nr	3682.00
141115M Jumbo slide	–	–	–	–	–	nr	5518.00
Slides; Kompan Ltd							
M351P slide	–	–	–	–	–	nr	2886.45
M326P Aladdin's Cave slide	–	–	–	–	–	nr	2655.45

Q PAVING/PLANTING/FENCING/SITE FURNITURE

Item Excluding site overheads and profit	PC £	Labour hours	Labour £	Plant £	Material £	Unit	Total rate £
Q50 PLAYGROUND EQUIPMENT – cont							
Moving equipment – General							
Preamble: The following standard items of							
playground equipment vary considerably in							
quality and price; the following prices are							
middle of the range.							
Moving equipment							
Roundabouts; Wicksteed Ltd							
Turnstile	–	–	–	–	–	nr	904.00
Speedway (without restrictor)	–	–	–	–	–	nr	3562.00
Spiro Whirl (without restrictor)	–	–	–	–	–	nr	4059.00
Roundabouts; Kompan Ltd							
Supernova GXY 916; multifunctional spinning							
and balancing disc; capacity approximately 15							
children	–	–	–	–	–	nr	4006.80
Seesaws							
Seesaws; Lappset UK Ltd							
010300; seesaws	–	–	–	–	–	nr	962.00
010237; seesaws	–	–	–	–	–	nr	2439.00
Seesaws; Wicksteed Ltd							
Seesaw; non-bump	–	–	–	–	–	nr	2631.00
Jolly Gerald; non-bump	–	–	–	–	–	nr	2837.00
Rocking Rockette; with motion restrictor	–	–	–	–	–	nr	3806.00
Rocking Horse; with motion restrictor	–	–	–	–	–	nr	4341.00
Play sculptures – General							
Preamble: Many variants on the shapes of							
playground equipment are available,							
simulating spacecraft, trains, cars, houses							
etc., and these designs are frequently							
changed. The basic principles remain constant							
but manufacturer's catalogues should be							
checked for the latest styles.							
Climbing equipment and play structures –							
General							
Preamble: Climbing equipment generally							
consists of individually designed modules.							
Play structures generally consist of interlinked							
and modular pieces of equipment and							
sculptures. These may consist of climbing,							
play and skill based modules, nets and							
various other activities. Both are set into either							
safety surfacing or defined sand pit areas. The							
equipment below outlines a range from							
various manufacturers. Individual catalogues							
should be consulted in each instance. Safety							
areas should be allowed round all equipment.							
Climbing equipment and play structures							
Climbing equipment and play structures;							
Kompan Ltd							
M623P Fairy Castle	–	–	–	–	–	nr	4764.90

Q PAVING/PLANTING/FENCING/SITE FURNITURE

Item Excluding site overheads and profit	PC £	Labour hours	Labour £	Plant £	Material £	Unit	Total rate £
Climbing equipment; Wicksteed Ltd							
Funrun Fitness Trail; Under Starter's							
Orders; set of 12 units	–	–	–	–	–	nr	16679.00
Climbing equipment; Lappset UK Ltd							
138401M Storks Nest	–	–	–	–	–	nr	4368.00
122457M Playhouse	–	–	–	–	–	nr	3765.00
120100M Activity Tower	–	–	–	–	–	nr	13853.00
120124M Tower and Climbing Frame	–	–	–	–	–	nr	13479.00
SMP Playgrounds Ltd							
Nexus – The Core; multi-play structure	–	–	–	–	–	nr	11961.00
Spring equipment							
Spring based equipment for 1–8 year olds;							
Kompan Ltd							
M101P Crazy Hen	–	–	–	–	–	nr	574.35
M128P Crazy Daisy	–	–	–	–	–	nr	835.80
M141P Spring Seesaw	–	–	–	–	–	nr	2001.30
M155P Quartet Seesaw	–	–	–	–	–	nr	1390.20
Spring based equipment for under 12s;							
Lappset UK Ltd							
010501 Horse Springer	–	–	–	–	–	nr	1113.00
Sandpits							
Market prices							
play pit sand; Boughton Loam Ltd	–	–	–	–	122.40	m³	122.40
Kompan Ltd							
Basic550; 276 × 154 × 31 cm deep	–	–	–	–	–	nr	1076.25
Q50 PERGOLAS							
Pergolas; AVS Fencing Supplies Ltd;							
construct timber pergola; posts 150 ×							
150 mm × 2.40 m finished height in 600 mm							
deep minimum concrete pits; beams of							
200 × 50 mm × 2.00 m wide; fixed to posts							
with dowels; rafters 200 × 38 mm notched							
to beams; inclusive of all mechanical							
excavation and disposal off site							
Pergola 2.00 m wide in green oak; posts at							
2.00 m centres; dowel fixed							
rafters at 600 mm centres	–	8.40	173.60	3.67	241.48	m	418.75
rafters at 400 mm centres	–	10.80	223.20	3.67	291.91	m	518.78
Pergola 3.00 m wide in green oak; posts at							
1.50 m centres							
rafters at 600 mm centres	–	7.20	156.60	4.40	320.36	m	481.36
rafters at 400 mm centres	–	8.00	174.00	4.40	395.86	m	574.26
Pergola 2.00 wide in prepared softwood;							
beams fixed with bolts; posts at 2.00 m centres							
rafters at 600 mm centres	–	5.00	108.75	3.67	142.82	m	255.24
rafters at 400 mm centres	–	6.00	130.50	3.67	164.28	m	298.45
Pergola 3.00 m wide in prepared softwood;							
posts at 1.50 m centres							
rafters at 600 mm centres	–	7.20	156.60	4.40	142.50	m	303.50
rafters at 400 mm centres	–	8.00	174.00	4.40	162.93	m	341.33

Q PAVING/PLANTING/FENCING/SITE FURNITURE

Item Excluding site overheads and profit	PC £	Labour hours	Labour £	Plant £	Material £	Unit	Total rate £
Q50 SPORTS EQUIPMENT							
Sports equipment							
Tennis posts; steel; suitable for hard or grass							
tennis courts; including winder, sockets and							
dust cap							
round	257.99	1.00	18.50	–	257.99	set	**276.49**
square	273.99	1.00	18.50	–	273.99	set	**292.49**
Tennis nets; not including posts or fixings							
Tournament	114.99	3.00	55.50	–	121.43	set	**176.93**
Match	89.99	3.00	55.50	–	96.43	set	**151.93**
Club	69.99	3.00	55.50	–	76.43	set	**131.93**
Football goals; full size; socketed; including							
international net supports and nets							
aluminium	1380.00	4.00	74.00	–	1386.44	set	**1460.44**
steel; heavyweight	1040.00	0.14	2.66	–	1218.84	set	**1221.50**
Football goals; full size; freestanding; including							
nets							
aluminium	1420.00	4.00	74.00	–	1420.00	set	**1494.00**
steel	1110.00	4.00	74.00	–	1110.00	set	**1184.00**
Mini-soccer goals; freestanding; including nets							
aluminium	645.00	4.00	74.00	–	645.00	set	**719.00**
steel	525.00	4.00	74.00	–	525.00	set	**599.00**
Rugby posts; socketed							
aluminium; 10 m high	1350.50	6.00	111.00	–	1367.67	set	**1478.67**
aluminium; 12 m high	1450.50	6.00	111.00	–	1467.67	set	**1578.67**
steel; 12 m high	1750.00	6.00	111.00	–	1767.17	set	**1878.17**
Hockey goals; steel; including backboards and							
nets							
socketed	950.00	1.00	18.50	–	950.00	set	**968.50**
freestanding	1130.00	1.00	18.50	–	1130.00	set	**1148.50**
Cricket cages; steel; including netting							
freestanding	895.50	12.00	222.00	–	895.50	set	**1117.50**
wheelaway	1195.50	12.00	222.00	–	1195.50	set	**1417.50**
Q50 FLAGPOLES							
Flagpoles							
Flagpoles; ground mounted; Harrison External							
Display Systems; in glass fibre; smooth white							
finish; including hinged baseplate, nylon							
halyard system and revolving gold onion finial;							
setting in concrete; to manufacturer's							
recommendations (excavating not included)							
6 m high; external halyard system	177.00	4.00	74.00	–	196.31	nr	**270.31**
6 m high; internal halyard system	279.99	4.00	74.00	–	299.31	nr	**373.31**
10 m high; external halyard system	322.00	5.00	92.50	–	341.31	nr	**433.81**
10 m high; internal halyard system	440.99	5.00	92.50	–	460.31	nr	**552.81**
12 m high; external halyard system	390.00	6.00	111.00	–	409.31	nr	**520.31**
12 m high; internal halyard system	502.99	6.00	111.00	–	522.31	nr	**633.31**
Flagpoles; wall mounted; Harrison External							
Display Systems; vertical or angled poles; in							
glass fibre; smooth white finish; including							
base, top bracket, external nylon halyard rope,							
cleat and revolving gold onion finial							
3 m pole	187.00	2.00	37.00	–	187.00	nr	**224.00**

Q PAVING/PLANTING/FENCING/SITE FURNITURE

Item Excluding site overheads and profit	PC £	Labour hours	Labour £	Plant £	Material £	Unit	Total rate £
Banner flagpoles; Harrison External Display Systems; 90 mm diameter aluminium pole; suitable for 1 × 2 m banner							
6 m high	458.62	4.00	74.00	–	477.94	nr	**551.94**
Q50 PICNIC TABLES/SUNDRY FURNITURE							
Lifebuoy stations							
Lifebuoy stations; Earth Anchors Ltd							
Rootfast lifebuoy station complete with post AP44; SOLAS approved lifebouys 590 mm and lifeline	165.00	2.00	37.00	–	173.94	nr	**210.94**
installation tool for above	–	–	–	–	81.00	nr	**81.00**
Street furniture ranges							
Townscape Products Ltd; Belgrave; natural grey concrete							
bollards; 250 mm diameter × 500 mm high	99.89	1.00	18.50	–	102.89	nr	**121.39**
seats; 1800 × 600 × 736 mm high	436.75	2.00	37.00	–	436.75	nr	**473.75**
Picnic benches – General							
Preamble: The following items include for fixing to ground to manufacturer's instructions or concreting in							
Picnic tables and benches							
Picnic tables and benches; Broxap Street Furniture							
Eastgate picnic unit	629.00	2.00	37.00	–	641.95	nr	**678.95**
Picnic tables and benches; Woodscape Ltd							
table and benches built in; 2 m long	1945.00	2.00	37.00	–	1957.95	nr	**1994.95**
Picnic table; recycled plastic							
Furnitubes International Ltd							
Dundee picnic table; brown recycled plastic; 1.8 m long	643.00	2.00	37.00	–	662.57	ea	**699.57**

R DISPOSAL SYSTEMS

Item Excluding site overheads and profit	PC £	Labour hours	Labour £	Plant £	Material £	Unit	Total rate £
R12 DRAINAGE BELOW GROUND							
Silt pits and inspection chambers							
Excavating pits; starting from ground level; by machine							
maximum depth not exceeding 1.00 m	–	0.50	9.25	16.83	–	m³	**26.08**
maximum depth not exceeding 2.00 m	–	0.50	9.25	25.25	–	m³	**34.50**
maximum depth not exceeding 4.00 m	–	0.50	9.25	50.49	–	m³	**59.74**
Disposal of excavated material; depositing on site in permanent spoil heaps; average 50 m	–	0.04	0.77	1.59	–	m³	**2.36**
Filling to excavations; obtained from on site spoil heaps; average thickness not exceeding 0.25 m	–	0.13	2.47	4.49	–	m³	**6.96**
Surface treatments; compacting; bottoms of excavations	–	0.05	0.93	–	–	m²	**0.93**
Earthwork support; distance between opposing faces not exceeding 2.00 m							
maximum depth not exceeding 1.00 m	–	0.20	3.70	–	5.97	m³	**9.67**
maximum depth not exceeding 2.00 m	–	0.30	5.55	–	5.97	m³	**11.52**
maximum depth not exceeding 4.00 m	–	0.67	12.33	–	2.75	m³	**15.08**
Silt pits and inspection chambers; in situ concrete							
Beds; plain in situ concrete; 11.50 N/mm² – 40 mm aggregate							
thickness not exceeding 150 mm	–	1.00	18.50	–	89.42	m³	**107.92**
thickness 150–450 mm	–	0.67	12.33	–	89.42	m³	**101.75**
Benchings in bottoms; plain in situ concrete; 25.50 N/mm² – 20 mm aggregate							
thickness 150–450 mm	–	2.00	37.00	–	89.42	m³	**126.42**
Isolated cover slabs; reinforced In situ concrete; 21.00 N/mm² – 20 mm aggregate							
thickness not exceeding 150 mm	89.42	4.00	74.00	–	89.42	m³	**163.42**
Fabric reinforcement; A193 (3.02 kg/m²) in cover slabs	3.50	0.06	1.16	–	3.50	m²	**4.66**
Formwork to reinforced in situ concrete; isolated cover slabs							
soffits; horizontal	–	3.28	60.68	–	5.25	m²	**65.93**
height not exceeding 250 mm	–	0.97	17.95	–	2.99	m	**20.94**
Silt pits and inspection chambers; precast concrete units							
Precast concrete inspection chamber units; FP McCann Ltd; bedding, jointing and pointing in cement mortar (1:3); 600 × 450 mm internally							
600 mm deep	51.82	6.00	111.00	–	55.58	nr	**166.58**
900 mm deep	64.78	7.00	129.50	–	68.54	nr	**198.04**
Drainage chambers; FP McCann Ltd; 1200 × 750 mm reducing to 600 × 600 mm; no base unit; depth of invert							
1050 mm deep	392.57	9.00	166.50	–	398.83	nr	**565.33**
1650 mm deep	581.17	11.00	203.50	–	591.19	nr	**794.69**
2250 mm deep	769.77	12.50	231.25	–	782.29	nr	**1013.54**

R DISPOSAL SYSTEMS

Item Excluding site overheads and profit	PC £	Labour hours	Labour £	Plant £	Material £	Unit	Total rate £
Cover slabs for chambers or shaft sections; FP McCann Ltd; heavy duty							
900 mm diameter internally	62.00	0.67	12.32	–	62.00	nr	74.32
1050 mm diameter internally	68.00	2.00	37.00	11.22	68.00	nr	116.22
1200 mm diameter internally	82.00	1.00	18.50	11.22	82.00	nr	111.72
1500 mm diameter internally	139.04	1.00	18.50	11.22	139.04	nr	168.76
1800 mm diameter internally	208.47	2.00	37.00	25.25	208.47	nr	270.72
Brickwork							
Walls to manholes; bricks; PC £300.00/1000; in cement mortar (1:3)							
one brick thick	37.80	3.00	55.50	–	45.31	m²	100.81
one and a half brick thick	56.70	4.00	74.00	–	67.97	m²	141.97
two brick thick projection of footing or the like	75.60	4.80	88.80	–	90.62	m²	179.42
Walls to manholes; engineering bricks; PC £239.00/1000; in cement mortar (1:3)							
one brick thick	30.11	3.00	55.50	–	37.63	m²	93.13
one and a half brick thick	45.17	4.00	74.00	–	56.44	m²	130.44
two brick thick projection of footing or the like	60.23	4.80	88.80	–	75.25	m²	164.05
Extra over common or engineering bricks in any mortar for fair face; flush pointing as work proceeds; English bond walls or the like	–	0.13	2.47	–	–	m²	2.47
In situ finishings; cement: sand mortar (1:3); steel trowelled; 13 mm one coat work to manhole walls; to brickwork or blockwork base; over 300 mm wide	–	0.80	14.80	–	2.50	m²	17.30
Building into brickwork; ends of pipes; making good facings or renderings							
small	–	0.20	3.70	–	–	nr	3.70
large	–	0.30	5.55	–	–	nr	5.55
extra large	–	0.40	7.40	–	–	nr	7.40
extra large; including forming ring arch cover	–	0.50	9.25	–	–	nr	9.25
Inspection chambers; polypropylene; Hepworth Plc							
Mini access chamber; up to 600 mm deep; including cover and frame							
300 mm diameter × 600 mm deep; three 100/110 mm inlets	81.59	3.00	55.50	–	83.92	nr	139.42
Up to 1200 mm deep; including polymer cover and frame with screw down lid							
475 mm diameter × 940 mm deep; five 100/110mm inlets; supplied with four stoppers in inlets	120.62	4.00	74.00	–	162.07	nr	236.07
Extra for square ductile iron cover and frame to 1 tonne load; screw down lid	55.95	–	–	–	16.83	nr	16.83
Step irons; drainage systems; malleable cast iron; galvanized; building into joints							
General purpose pattern; for one brick walls	4.14	0.17	3.15	–	4.14	nr	7.29

R DISPOSAL SYSTEMS

Item Excluding site overheads and profit	PC £	Labour hours	Labour £	Plant £	Material £	Unit	Total rate £
R12 DRAINAGE BELOW GROUND – cont							
Best quality vitrified clay half section channels; Hepworth Plc; bedding and jointing in cement: mortar (1:2)							
Channels; straight							
100 mm	4.72	0.80	14.80	–	7.28	m	**22.08**
150 mm	7.86	1.00	18.50	–	10.42	m	**28.92**
225 mm	17.65	1.35	24.98	–	20.21	m	**45.19**
300 mm	36.23	1.80	33.30	–	38.79	m	**72.09**
Bends; 15, 30, 45 or 90°							
100 mm bends	4.25	0.75	13.88	–	5.53	nr	**19.41**
150 mm bends	7.34	0.90	16.65	–	9.27	nr	**25.92**
225 mm bends	28.48	1.20	22.20	–	31.04	nr	**53.24**
300 mm bends	58.07	1.10	20.35	–	61.28	nr	**81.63**
Best quality vitrified clay channels; Hepworth Plc; bedding and jointing in cement: mortar (1:2)							
Branch bends; 15, 30, 45 or 90°; left or right hand							
100 mm	4.25	0.75	13.88	–	5.53	nr	**19.41**
150 mm	7.34	0.90	16.65	–	9.27	nr	**25.92**
Intercepting traps; Hepworth Plc							
Vitrified clay; inspection arms; brass stoppers; iron levers; chains and staples; galvanized; staples cut and pinned to brickwork; cement: mortar (1:2) joints to vitrified clay pipes and channels; bedding and surrounding in concrete; 11.50 N/mm² – 40 mm aggregate; cutting and fitting brickwork; making good facings							
100 mm inlet; 100 mm outlet	44.93	3.00	55.50	–	52.00	nr	**107.50**
150 mm inlet; 150 mm outlet	64.78	2.00	37.00	–	76.34	nr	**113.34**
Excavating trenches; using 3 tonne tracked excavator; to receive pipes; grading bottoms; earthwork support; filling with excavated material to within 150 mm of finished surfaces and compacting; completing fill with topsoil; disposal of surplus soil							
Services not exceeding 200 mm nominal size							
average depth of run not exceeding 0.50 m	1.19	0.12	2.22	1.01	1.19	m	**4.42**
average depth of run not exceeding 0.75 m	1.19	0.16	3.01	1.39	1.19	m	**5.59**
average depth of run not exceeding 1.00 m	1.19	0.28	5.24	2.42	1.19	m	**8.85**
average depth of run not exceeding 1.25 m	0.99	0.38	7.09	3.27	0.99	m	**11.35**
Granular beds to trenches; lay granular material to trenches excavated separately; to receive pipes (not included)							
300 mm wide × 100 mm thick							
reject sand	–	0.05	0.93	0.22	1.78	m	**2.93**
reject gravel	–	0.05	0.93	0.22	1.70	m	**2.85**
shingle 40 mm aggregate	–	0.05	0.93	0.22	3.31	m	**4.46**
sharp sand	3.37	0.05	0.93	0.22	3.37	m	**4.52**

R DISPOSAL SYSTEMS

Item Excluding site overheads and profit	PC £	Labour hours	Labour £	Plant £	Material £	Unit	Total rate £
300 mm wide × 150 mm thick							
reject sand	–	0.08	1.39	0.33	2.67	m	**4.39**
reject gravel	–	0.08	1.39	0.33	2.55	m	**4.27**
shingle 40 mm aggregate	–	0.08	1.39	0.33	4.96	m	**6.68**
sharp sand	5.05	0.08	1.39	0.33	5.05	m	**6.77**
Excavating trenches; using 3 tonne tracked excavator; to receive pipes; grading bottoms; earthwork support; filling with imported granular material type 2 and compacting; disposal of surplus soil							
Services not exceeding 200 mm nominal size							
average depth of run not exceeding 0.50 m	1.44	0.09	1.60	0.72	2.88	m	**5.20**
average depth of run not exceeding 0.75 m	2.16	0.11	2.00	0.90	4.32	m	**7.22**
average depth of run not exceeding 1.00 m	2.88	0.14	2.58	1.17	5.76	m	**9.51**
average depth of run not exceeding 1.25 m	3.60	0.23	4.23	1.95	7.20	m	**13.38**
Excavating trenches; using 3 tonne tracked excavator; to receive pipes; grading bottoms; earthwork support; filling with concrete, ready mixed ST2; disposal of surplus soil							
Services not exceeding 200 mm nominal size							
average depth of run not exceeding 0.50 m	9.86	0.11	1.97	0.35	11.30	m	**13.62**
average depth of run not exceeding 0.75 m	14.78	0.13	2.41	0.44	16.94	m	**19.79**
average depth of run not exceeding 1.00 m	19.71	0.17	3.08	0.59	22.59	m	**26.26**
average depth of run not exceeding 1.25 m	24.64	0.23	4.16	0.88	28.24	m	**33.28**
Earthwork support; providing support to opposing faces of excavation; moving along as work proceeds							
Maximum depth not exceeding 2.00 m							
distance between opposing faces not exceeding 2.00 m	–	0.80	14.80	17.96	–	m	**32.76**
Clay pipes and fittings; Hepworth Plc; Supersleve							
100 mm clay pipes; polypropylene slip coupling; in trenches (trenches not included)							
laid straight	2.33	0.25	4.63	–	4.06	m	**8.69**
short runs under 3.00 m	2.33	0.31	5.78	–	4.06	m	**9.84**
Extra over 100 mm clay pipes for							
bends; 15–90°; single socket	7.77	0.25	4.63	–	7.77	nr	**12.40**
junction; 45 or 90°; double socket	16.37	0.25	4.63	–	16.37	nr	**21.00**
slip couplings; polypropylene	2.76	0.08	1.54	–	2.76	nr	**4.30**
gully with P trap; 100 mm; 154 × 154 mm plastic grating	25.37	1.00	18.50	–	37.80	nr	**56.30**
150 mm clay pipes; polypropylene slip coupling; in trenches (trenches not included)							
laid straight	9.31	0.30	5.55	–	9.31	m	**14.86**
short runs under 3.00 m	7.17	0.33	6.17	–	7.17	m	**13.34**
Extra over 150 mm clay pipes for							
bends; 15–90°	10.37	0.28	5.18	–	15.37	nr	**20.55**
junction; 45 or 90°; 100 × 150 mm	13.88	0.40	7.40	–	23.88	nr	**31.28**
junction; 45 or 90°; 150 × 150 mm	15.23	0.40	7.40	–	25.23	nr	**32.63**

R DISPOSAL SYSTEMS

Item Excluding site overheads and profit	PC £	Labour hours	Labour £	Plant £	Material £	Unit	Total rate £
R12 DRAINAGE BELOW GROUND – cont							
Clay pipes and fittings – cont							
Extra over 150 mm clay pipes for – cont							
slip couplings; polypropylene	5.00	0.05	0.93	–	5.00	nr	**5.93**
tapered pipe; 100–150 mm	15.60	0.50	9.25	–	15.60	nr	**24.85**
tapered pipe; 150–225 mm	40.06	0.50	9.25	–	40.06	nr	**49.31**
socket adaptor; connection to traditional							
pipes and fittings	10.52	0.33	6.11	–	15.52	nr	**21.63**
Accessories in clay							
access pipe; 150 mm	39.12	–	–	–	49.12	nr	**49.12**
rodding eye; 150 mm	37.42	0.50	9.25	–	41.30	nr	**50.55**
gully with P traps; 150 mm; 154 × 154 mm							
plastic grating	50.81	0.80	14.80	–	57.37	nr	**72.17**
PVC-u pipes and fittings; Wavin Plastics Ltd; OsmaDrain							
110 mm PVC-u pipes; in trenches (trenches not included)							
laid straight	6.20	0.08	1.48	–	6.20	m	**7.68**
short runs under 3.00 m	6.98	0.12	2.22	–	6.98	m	**9.20**
Extra over 110 mm PVC-u pipes for							
bends; short radius	12.08	0.25	4.63	–	12.08	nr	**16.71**
bends; long radius	22.63	0.25	4.63	–	22.63	nr	**27.26**
junctions; equal; double socket	14.41	0.25	4.63	–	14.41	nr	**19.04**
slip couplings	6.99	0.25	4.63	–	6.99	nr	**11.62**
adaptors to clay	13.61	0.50	9.25	–	13.61	nr	**22.86**
160 mm PVC-u pipes; in trenches (trenches not included)							
laid straight	14.55	0.08	1.48	–	14.55	m	**16.03**
short runs under 3.00 m	27.64	0.12	2.22	–	27.64	m	**29.86**
Extra over 160 mm PVC-u pipes for							
socket bend; double; 90 or 45°	45.69	0.20	3.70	–	45.69	nr	**49.39**
socket bend; double; 15 or 30°	42.71	0.20	3.70	–	42.71	nr	**46.41**
socket bend; single; 87.5 or 45°	26.04	0.20	3.70	–	26.04	nr	**29.74**
socket bend; single; 15 or 30°	23.08	0.20	3.70	–	23.08	nr	**26.78**
bends; short radius	33.94	0.25	4.63	–	46.01	nr	**50.64**
bends; long radius	88.49	0.25	4.63	–	88.49	nr	**93.12**
junctions; single	83.64	0.33	6.17	–	83.64	nr	**89.81**
pipe coupler	16.97	0.05	0.93	–	16.97	nr	**17.90**
slip couplings PVC-u	8.45	0.05	0.93	–	8.45	nr	**9.38**
adaptors to clay	37.01	0.50	9.25	–	37.01	nr	**46.26**
level invert reducer	13.93	0.50	9.25	–	13.93	nr	**23.18**
spiggot	30.90	0.20	3.70	–	30.90	nr	**34.60**
Accessories in PVC-u							
110 mm screwed access cover	12.75	–	–	–	12.75	nr	**12.75**
110 mm rodding eye	24.68	0.50	9.25	–	28.56	nr	**37.81**
gully with P traps; 110 mm; 154 × 154 mm grating	33.47	1.00	18.50	–	35.02	nr	**53.52**
Kerbs; to gullies; in one course Class B engineering bricks; to four sides; rendering in cement: mortar (1:3); dished to gully gratings	2.01	1.00	18.50	–	2.76	nr	**21.26**
Gullies; concrete; FP McCann							
Concrete road gullies; trapped with rodding eye and stoppers; 450 mm diameter × 1.07 m deep	46.00	6.00	111.00	–	67.46	nr	**178.46**

R DISPOSAL SYSTEMS

Item Excluding site overheads and profit	PC £	Labour hours	Labour £	Plant £	Material £	Unit	Total rate £
Gullies; vitrified clay; Hepworth Plc; bedding in concrete; 11.50 N/mm² – 40 mm aggregate							
Yard gullies (mud); trapped; domestic duty (up to 1 tonne)							
100 mm outlet; 100 mm diameter; 225 mm internal width; 585 mm internal depth	58.43	3.50	64.75	–	58.97	nr	123.72
150 mm outlet; 100 mm diameter; 225 mm internal width; 585 mm internal depth	58.43	3.50	64.75	–	58.97	nr	123.72
Yard gullies (mud); trapped; medium duty (up to 5 tonnes)							
100 mm outlet; 100 mm diameter; 225 mm internal width; 585 mm internal depth	82.60	3.50	64.75	–	83.15	nr	147.90
150 mm outlet; 100 mm diameter; 225 mm internal width; 585 mm internal depth	90.50	3.50	64.75	–	91.05	nr	155.80
Combined filter and silt bucket for yard gullies							
225 mm diameter	21.13	–	–	–	21.13	nr	21.13
Road gullies; trapped with rodding eye							
100 mm outlet; 300 mm internal diameter; 600 mm internal depth	56.20	3.50	64.75	–	56.75	nr	121.50
150 mm outlet; 300 mm internal diameter; 600 mm internal depth	57.55	3.50	64.75	–	58.10	nr	122.85
150 mm outlet; 400 mm internal diameter; 750 mm internal depth	66.74	3.50	64.75	–	67.29	nr	132.04
150 mm outlet; 450 mm internal diameter; 900 mm internal depth	90.30	3.50	64.75	–	90.85	nr	155.60
Hinged gratings and frames for gullies; alloy							
193 mm for 150 mm diameter gully	–	–	–	–	14.00	nr	14.00
120 × 120 mm	–	–	–	–	4.98	nr	4.98
150 × 150 mm	–	–	–	–	9.01	nr	9.01
230 × 230 mm	–	–	–	–	16.47	nr	16.47
316 × 316 mm	–	–	–	–	43.66	nr	43.66
Hinged gratings and frames for gullies; cast iron							
265 mm for 225 mm diameter gully	–	–	–	–	27.98	nr	27.98
150 × 150 mm	–	–	–	–	9.01	nr	9.01
230 × 230 mm	–	–	–	–	16.47	nr	16.47
316 × 316 mm	–	–	–	–	43.66	nr	43.66
Universal gully trap PVC-u; Wavin Plastics Ltd; OsmaDrain system; bedding in concrete; 11.50 N/mm² – 40 mm aggregate							
Universal gully fitting; comprising gully trap only							
110 mm outlet; 110 mm diameter; 205 mm internal depth	10.24	3.50	64.75	–	10.79	nr	75.54
Vertical inlet hopper; c\w plastic grate							
272 × 183 mm	13.76	0.25	4.63	–	13.76	nr	18.39
Sealed access hopper							
110 × 110 mm	32.37	0.25	4.63	–	32.37	nr	37.00

R DISPOSAL SYSTEMS

Item Excluding site overheads and profit	PC £	Labour hours	Labour £	Plant £	Material £	Unit	Total rate £
R12 DRAINAGE BELOW GROUND – cont							
Universal gully PVC-u; Wavin Plastics Ltd; **OsmaDrain system; accessories to** **universal gully trap**							
Hoppers; backfilling with clean granular material; tamping; surrounding in lean mix concrete							
plain hopper; with 110 mm spigot; 150 mm long	11.25	0.40	7.40	–	11.58	nr	18.98
vertical inlet hopper; with 110 mm spigot; 150 mm long	13.76	0.40	7.40	–	13.76	nr	21.16
sealed access hopper; with 110 mm spigot; 150 mm long	32.37	0.40	7.40	–	32.37	nr	39.77
plain hopper; solvent weld to trap	7.68	0.40	7.40	–	7.68	nr	15.08
vertical inlet hopper; solvent wed to trap	13.20	0.40	7.40	–	13.20	nr	20.60
sealed access cover; PVC-u	16.48	0.10	1.85	–	16.48	nr	18.33
Gullies PVC-u; Wavin Plastics Ltd; **OsmaDrain system; bedding in concrete;** **11.50 N/mm² – 40 mm aggregate**							
Bottle gully; providing access to the drainage system for cleaning							
bottle gully; 228 × 228 × 317 mm deep	27.33	0.50	9.25	–	27.79	nr	37.04
sealed access cover; PVC-u; 217 × 217 mm	21.24	0.10	1.85	–	21.24	nr	23.09
grating; ductile iron; 215 × 215 mm	16.38	0.10	1.85	–	16.38	nr	18.23
bottle gully riser; 325 mm	3.52	0.50	9.25	–	4.53	nr	13.78
Yard gully; trapped; 300 mm diameter × 600 mm deep; including catchment bucket and ductile iron cover and frame; medium duty loading							
305 mm diameter × 600 mm deep	165.72	2.50	46.25	–	168.65	nr	214.90
Kerbs to gullies							
One course Class B engineering bricks to four sides; rendering in cement: mortar (1:3); dished to gully gratings							
150 × 150 mm	1.20	0.33	6.17	–	1.51	nr	7.68
Linear drainage							
Marshalls Plc Mini Beany combined kerb **and channel drainage system; to trenches** **(not included)**							
Precast concrete drainage channel base; 185–385 mm deep; bedding, jointing and pointing in cement mortar (1:3); on 150 mm deep concrete (ready mixed) foundation; including haunching with in situ concrete; 11.50 N/mm² – 40 mm aggregate one side; channels 250 mm wide × 1.00 m long							
straight; 1.00 m long	21.59	1.00	18.50	–	24.57	m	43.07
straight; 500 mm long	22.61	1.05	19.43	–	25.59	m	45.02
radial; 30/10 or 9/6 internal or external	20.89	1.33	24.67	–	23.87	m	48.54
angles 45 or 90°	62.33	1.00	18.50	–	65.31	nr	83.81

R DISPOSAL SYSTEMS

Item Excluding site overheads and profit	PC £	Labour hours	Labour £	Plant £	Material £	Unit	Total rate £
Mini Beany Top Block; perforated kerb unit to drainage channel above; natural grey							
straight	11.44	0.33	6.17	–	12.60	m	**18.77**
radial; 30/10 or 9/6 internal or external	14.94	0.50	9.25	–	16.10	m	**25.35**
angles 45 or 90°	34.80	0.50	9.25	–	35.96	nr	**45.21**
Mini Beany; outfalls; two section concrete trapped outfall with Mini Beany cast iron access cover and frame; to concrete foundation							
high capacity outfalls; silt box 150/225 mm outlet; two section trapped outfall silt box and cast iron access cover	42.23	1.00	18.50	–	260.56	nr	**279.06**
inline side or end outlet; outfall 150 mm; 2 section concrete trapped outfall; cast iron Mini Beany access cover and frame	226.08	1.00	18.50	–	226.71	nr	**245.21**
Ancillaries to Mini Beany							
end cap	12.66	0.25	4.63	–	12.66	nr	**17.29**
end cap outlets	32.90	0.25	4.63	–	32.90	nr	**37.53**
Precast concrete channels; Charcon Hard Landscaping; on 150 mm deep concrete foundations; including haunching with in situ concrete; 21.00 N/mm² – 20 mm aggregate; both sides							
Charcon Safeticurb; slotted safety channels; for pedestrians and light vehicles							
DBJ; 305 × 305 mm	69.58	0.67	12.33	–	84.78	m	**97.11**
DBA; 250 × 250 mm	35.82	0.67	12.33	–	51.02	m	**63.35**
Charcon Safeticurb; slotted safety channels; for heavy vehicles							
DBM; 248 × 248 mm	63.78	0.80	14.80	–	78.98	m	**93.78**
Clearway; 324 × 257 mm	73.15	0.80	14.80	–	88.35	m	**103.15**
Charcon Safeticurb; inspection units; ductile iron lids; including jointing to drainage channels							
248 × 248 × 914 mm	86.79	1.50	27.75	–	87.68	nr	**115.43**
Silt box tops; concrete frame; cast iron grid lids; type 1; set over gully							
457 × 610 mm	389.03	2.00	37.00	–	389.92	nr	**426.92**
Manhole covers; type K; cast iron; providing inspection to blocks and back gullies	494.27	2.00	37.00	–	496.36	nr	**533.36**

R DISPOSAL SYSTEMS

Item Excluding site overheads and profit	PC £	Labour hours	Labour £	Plant £	Material £	Unit	Total rate £
R12 DRAINAGE BELOW GROUND – cont							
Linear drainage							
Loadings for slot drains							
A15; 1.5 tonne; pedestrian							
B125; 12.5 tonne; domestic use							
C250; 25 tonne; car parks, supermarkets, industrial units							
D400; 40 tonne; highways							
E600; 60 tonne; forklifts							
Slot drains; ACO Technologies; laid to concrete bed C25 on compacted granular base on 200 mm deep concrete bed; haunched with 200 mm concrete surround; all in 750 × 430 mm wide trench with compacted 200 mm granular base surround (excavation and subbase not included)							
ACO MultiDrain M100PPD; recycled polypropylene drainage channel; range of gratings to complement installations which require discreet slot drainage							
142 mm wide × 150 mm deep	57.90	1.20	22.20	–	67.63	m	**89.83**
Accessories for M100PPD							
connectors; vertical outlet 110 mm	12.63	–	–	–	12.63	nr	**12.63**
connectors; vertical outlet 160 mm	12.63	–	–	–	12.63	nr	**12.63**
sump units; 110 mm	105.25	1.50	27.75	–	123.13	nr	**150.88**
universal gully and bucket; 440 × 440 × 1315 mm deep	611.80	3.00	55.50	–	645.33	nr	**700.83**
Channel drains; ACO Technologies; ACO MultiDrain MD polymer concrete channel drainage system; traditional channel and grate drainage solution							
ACO MultiDrain MD Brickslot; offset galvanized slot drain grating for M100PPD; load class C250							
brickslot; galvanized steel; 1.00 mm	62.80	–	–	–	62.80	m	**62.80**
ACO MultiDrain MD Brickslot; offset slot drain grating; load class C250 – 400							
brickslot; galvanized steel; 1000 mm	120.70	1.20	22.20	–	120.70	m	**142.90**
brickslot; stainless steel; 1000 mm	125.95	1.20	22.20	–	251.90	m	**274.10**
Accessories for MultiDrain MD							
sump unit; complete with sediment bucket and access unit	105.25	1.00	18.50	–	206.70	nr	**225.20**
end cap; closing piece	8.50	–	–	–	8.50	nr	**8.50**
end cap; inlet/outlet	17.20	–	–	–	17.20	nr	**17.20**
ACO Drainlock Gratings for M100PPD and M100D system							
A15 loading							
slotted galvanized steel	17.00	0.15	2.77	–	17.00	m	**19.77**
perforated galvanized steel	23.40	0.15	2.77	–	23.40	m	**26.17**

R DISPOSAL SYSTEMS

Item Excluding site overheads and profit	PC £	Labour hours	Labour £	Plant £	Material £	Unit	Total rate £
C250 loading							
Heelguard composite black; 500 mm long							
with security locking	42.40	–	–	–	42.40	m	**42.40**
Intercept; ductile iron; 500 mm long	93.40	–	–	–	93.40	m	**93.40**
slotted galvanized steel; 1.00 mm long	44.00	–	–	–	44.00	m	**44.00**
perforated galvanized steel; 1.00 mm long	48.50	–	–	–	48.50	m	**48.50**
mesh galvanized steel; 1.00 mm long	32.95	–	–	–	32.95	m	**32.95**
Wade Ltd; stainless steel channel drains;							
specialized applications; bespoke							
manufacture							
Drain in stainless steel; c/w tie in lugs and							
inbuilt falls to 100 mm spigot outlet; stainless							
steel gratings							
NE channel; Ref 12430; secured gratings	380.00	1.20	22.20	–	389.73	m	**411.93**
Fin drains; Cooper Clarke Civils and							
Lintels; Geofin Shallow Drain; drainage of							
sports fields and grassed landscaped							
areas							
Geofin Shallow Drain; to trenches; excavation							
by trenching machine; backfilled with single							
size aggregate 20 mm and covered with sharp							
sand rootzone 200 mm thick							
Geofin; 25 mm thick × 150 mm deep	4.42	0.05	0.93	0.72	7.72	m	**9.37**
Geofin; 25 mm thick × 450 mm deep	3.27	0.07	1.23	0.96	9.89	m	**12.08**
Geofin; 25 mm thick × 900 mm deep	7.18	0.10	1.85	0.96	19.87	m	**22.68**
Fin drains; Cooper Clarke Civils and							
Lintels; Geofin Geocomposite Finn Drain;							
laid to slabs							
Geofin fin drain laid horizontally to slab or							
blinded ground; covered with 20 mm shingle							
200 mm thick							
Geofin; 25 mm thick × 900 mm wide; laid							
flat to falls	7.97	0.02	0.37	0.14	14.27	m²	**14.78**
Maxit Ltd; Leca (light expanded clay							
aggregate); drainage aggregate to							
roofdecks and planters							
Placed mechanically to planters; average							
100 mm thick; by mechanical plant tipped into							
planters							
aggregate size 10–20 mm; delivered in							
30 m³ loads	44.31	0.40	7.40	1.57	44.31	m³	**53.28**
aggregate size 10–20 mm; delivered in							
70 m³ loads	39.24	0.20	3.70	5.27	39.24	m³	**48.21**
Placed by light aggregate blower (maximum							
40 m)							
aggregate size 10–20 mm; delivered in							
30 m³ loads	44.31	0.14	2.64	–	48.48	m³	**51.12**
aggregate size 10–20 mm; delivered in							
55 m³ loads on blower vehicle	62.41	0.14	2.64	–	62.41	m³	**65.05**
aggregate size 10–20 mm; delivered in							
70 m³ loads	39.24	0.14	2.64	–	43.41	m³	**46.05**

R DISPOSAL SYSTEMS

Item Excluding site overheads and profit	PC £	Labour hours	Labour £	Plant £	Material £	Unit	Total rate £
R12 DRAINAGE BELOW GROUND – cont							
Maxit Ltd – cont							
By hand							
aggregate size 10–20 mm; delivered in 30 m³ loads	44.31	1.33	24.67	–	44.31	m³	68.98
aggregate size 10–20 mm; delivered in 70 m³ loads	39.24	1.33	24.67	–	39.24	m³	63.91
Drainage boards laid to insulated slabs on roof decks; boards laid below growing medium and granulated drainage layer and geofabric (all not included) to collect and channel water to drainage outlets (not included)							
Alumasc Floradrain; polyethylene irrigation/ drainage layer; inclusive of geofabric laid over the surface of the drainage board							
Floradrain FD40; 0.96 × 2.08 m panels	12.05	0.05	0.93	–	13.20	m²	14.13
Floradrain FD60; 1.00 × 2.00 m panels	21.42	0.07	1.23	–	22.57	m²	23.80
Access covers and frames							
FACTA (Fabricated Access Cover Trade Association) class:							
A – 0.5 tonne maximum slow moving wheel load							
AA – 1.5 tonne maximum slow moving wheel load							
AAA – 2.5 tonne maximum slow moving wheel load							
B – 5 tonne maximum slow moving wheel load							
C – 6.5 tonne maximum slow moving wheel load							
D – 11 tonne maximum slow moving wheel load							
Access covers and frames; solid top; galvanized; Steelway Brickhouse; Bristeel; bedding frame in cement mortar (1:3); cover in grease and sand; clear opening sizes; base size shown in brackets (50 mm depth)							
FACTA AA; single seal							
450 × 450 mm (520 × 520 mm)	69.71	1.50	27.75	–	78.69	nr	106.44
600 × 450 mm (670 × 520 mm)	78.38	1.80	33.30	–	88.36	nr	121.66
600 × 600 mm (670 × 670 mm)	84.48	2.00	37.00	–	96.45	nr	133.45
FACTA AA; double seal							
450 × 450 mm (560 × 560 mm)	119.99	1.50	27.75	–	128.96	nr	156.71
600 × 450 mm (710 × 560 mm)	132.96	1.80	33.30	–	142.93	nr	176.23
600 × 600 mm (710 × 710 mm)	146.60	2.00	37.00	–	158.57	nr	195.57
FACTA AAA; single seal							
450 × 450 mm (520 × 520 mm)	99.97	1.50	27.75	–	108.95	nr	136.70
600 × 450 mm (670 × 520 mm)	108.68	1.80	33.30	–	118.66	nr	151.96
600 × 600 mm (670 × 670 mm)	116.45	2.00	37.00	–	128.42	nr	165.42
FACTA AAA; double seal							
450 mm × 450 mm (560 × 560)	147.17	1.50	27.75	–	156.14	nr	183.89
600 mm × 450 mm (710 × 560)	161.82	1.80	33.30	–	171.79	nr	205.09
600 mm × 600 mm (710 × 710)	181.60	2.00	37.00	–	193.57	nr	230.57

R DISPOSAL SYSTEMS

Item Excluding site overheads and profit	PC £	Labour hours	Labour £	Plant £	Material £	Unit	Total rate £
FACTA B; single seal							
450 × 450 mm (520 × 520 mm)	87.30	1.50	27.75	–	96.28	nr	124.03
600 × 450 mm (670 × 520 mm)	105.24	1.80	33.30	–	115.21	nr	148.51
600 × 600 mm (670 × 670 mm)	123.95	2.00	37.00	–	135.92	nr	172.92
FACTA B; double seal							
450 × 450 mm (560 × 560 mm)	139.46	1.50	27.75	–	148.43	nr	176.18
600 × 450 mm (710 × 560 mm)	156.54	1.80	33.30	–	166.51	nr	199.81
600 × 600 mm (710 × 710 mm)	187.24	2.00	37.00	–	199.21	nr	236.21
Access covers and frames; recessed; galvanized; Steelway Brickhouse; Bripave; bedding frame in cement mortar (1:3); cover in grease and sand; clear opening sizes; base size shown in brackets (for block depths 50 mm, 65 mm, 80 mm or 100 mm)							
FACTA AA							
450 × 450 mm (562 × 562 mm)	188.99	1.50	27.75	–	197.97	nr	225.72
600 × 450 mm (712 × 562 mm)	211.69	1.80	33.30	–	221.66	nr	254.96
600 × 600 mm (712 × 712 mm)	221.09	2.00	37.00	–	232.59	nr	269.59
750 × 600 mm (862 × 712 mm)	251.95	2.40	44.40	–	264.91	nr	309.31
FACTA B							
450 × 450 mm (562 × 562 mm)	189.40	1.50	27.75	–	198.37	nr	226.12
600 × 450 mm (712 × 562 mm)	227.54	1.80	33.30	–	237.51	nr	270.81
600 × 600 mm (712 × 712 mm)	231.83	2.00	37.00	–	243.80	nr	280.80
750 × 600 mm (862 × 712 mm)	280.17	2.40	44.40	–	293.14	nr	337.54
FACTA D							
450 × 450 mm (562 × 562 mm)	217.75	1.50	27.75	–	226.72	nr	254.47
600 × 450 mm (712 × 562 mm)	234.17	1.80	33.30	–	244.15	nr	277.45
600 × 600 mm (712 × 712 mm)	243.44	2.00	37.00	–	255.41	nr	292.41
750 × 600 mm (862 × 712 mm)	306.10	2.40	44.40	–	319.07	nr	363.47
Access covers and frames; Jones of Oswestry; bedding frame in cement mortar (1:3); cover in grease and sand; clear opening sizes							
Access covers and frames; Suprabloc; to paved areas; filling with blocks cut and fitted to match surrounding paving							
pedestrian weight; 300 × 300 mm	121.47	2.50	46.25	–	127.48	nr	173.73
pedestrian weight; 450 × 450 mm	207.27	2.50	46.25	–	207.27	nr	253.52
pedestrian weight; 450 × 600 mm	221.01	2.50	46.25	–	227.27	nr	273.52
light vehicular weight; 300 × 300 mm	155.63	2.40	44.40	–	161.64	nr	206.04
light vehicular weight; 450 × 450 mm	119.98	3.00	55.50	–	126.87	nr	182.37
light vehicular weight; 450 × 600 mm	233.57	4.00	74.00	–	239.58	nr	313.58
heavy vehicular weight; 300 × 300 mm	155.97	3.00	55.50	–	162.23	nr	217.73
heavy vehicular weight; 450 × 600 mm	271.60	3.00	55.50	–	278.49	nr	333.99
heavy vehicular weight; 600 × 600 mm	290.00	4.00	74.00	–	300.02	nr	374.02
Extra over manhole frames and covers for							
filling recessed manhole covers with brick paviors; PC £305.00/1000	13.42	1.00	18.50	–	13.96	m²	32.46
filling recessed manhole covers with vehicular paving blocks; PC £7.63/m²	7.63	0.75	13.88	–	8.16	m²	22.04
filling recessed manhole covers with concrete paving flags; PC £12.54/m²	12.54	0.35	6.47	–	12.86	m²	19.33

R DISPOSAL SYSTEMS

Item Excluding site overheads and profit	PC £	Labour hours	Labour £	Plant £	Material £	Unit	Total rate £
R13 LAND DRAINAGE							
Market prices of backfilling materials							
Sand	19.44	–	–	–	19.44	m³	19.44
Gravel rejects	17.82	–	–	–	17.82	m³	17.82
Topsoil; allowing for 20% settlement	26.40	–	–	–	26.40	m³	26.40
Soakaways							
Excavating mechanical; to reduce levels							
maximum depth not exceeding 1.00 m;							
JCB sitemaster	–	0.05	0.93	1.68	–	m³	2.61
maximum depth not exceeding 1.00 m; 360							
tracked excavator	–	0.04	0.74	1.68	–	m³	2.42
maximum depth not exceeding 2.00 m; 360							
tracked excavator	–	0.06	1.11	2.52	–	m³	3.63
Disposal							
Excavated material; off site; to tip;							
mechanically loaded (JCB)							
inert	–	–	–	–	–	m³	17.12
In situ concrete ring beam foundations to							
base of soakaway; 300 mm wide × 250 mm							
deep; poured on or against earth or							
unblinded hardcore							
Internal diameters of rings							
900 mm	18.10	4.00	74.00	–	18.10	nr	92.10
1200 mm	24.13	4.50	83.25	–	24.13	nr	107.38
1500 mm	30.16	5.00	92.50	–	30.16	nr	122.66
2400 mm	48.27	5.50	101.75	–	48.27	nr	150.02
Concrete soakaway rings; Milton Pipes Ltd;							
perforations and step irons to concrete							
rings at manufacturers recommended							
centres; placing of concrete ring							
soakaways to in situ concrete ring beams							
(1:3:6) (not included); filling and							
surrounding base with gravel 225 mm deep							
(not included)							
Ring diameter 900 mm							
1.00 m deep; volume 636 litres	64.50	1.25	23.13	18.41	167.31	nr	208.85
1.50 m deep; volume 954 litres	96.75	1.65	30.52	48.60	248.46	nr	327.58
2.00 m deep; volume 1272 litres	129.00	1.65	30.52	48.60	329.61	nr	408.73
Ring diameter 1200 mm							
1.00 m deep; volume 1131 litres	81.00	3.00	55.50	44.18	199.50	nr	299.18
1.50 m deep; volume 1696 litres	121.50	4.50	83.25	66.27	294.80	nr	444.32
2.00 m deep; volume 2261 litres	162.00	4.50	83.25	66.27	390.10	nr	539.62
2.50 m deep; volume 2827 litres	202.50	6.00	111.00	88.36	444.10	nr	643.46
Ring diameter 1500 mm							
1.00 m deep; volume 1767 litres	135.00	4.50	83.25	44.18	282.12	nr	409.55
1.50 m deep; volume 2651 litres	202.50	6.00	111.00	61.85	416.22	nr	589.07
2.00 m deep; volume 3534 litres	270.00	6.00	111.00	61.85	550.32	nr	723.17
2.50 m deep; volume 4418 litres	202.50	7.50	138.75	110.45	549.42	nr	798.62

R DISPOSAL SYSTEMS

Item Excluding site overheads and profit	PC £	Labour hours	Labour £	Plant £	Material £	Unit	Total rate £
Ring diameter 2400 mm							
1.00 m deep; volume 4524 litres	500.00	6.00	111.00	44.18	700.22	nr	855.40
1.50 m deep; volume 6786 litres	750.00	7.50	138.75	61.85	1040.42	nr	1241.02
2.00 m deep; volume 9048 litres	1000.00	7.50	138.75	61.85	1386.52	nr	1587.12
2.50 m deep; volume 11310 litres	1250.00	9.00	166.50	110.45	1726.72	nr	2003.67
Extra over for							
250 mm depth chamber ring	–	–	–	–	–	100%	–
500 mm depth chamber ring	–	–	–	–	–	50%	–
Cover slabs to soakaways							
Heavy duty precast concrete							
900 mm diameter	102.50	1.00	18.50	22.09	102.50	nr	143.09
1200 mm diameter	124.50	1.00	18.50	22.09	124.50	nr	165.09
1500 mm diameter	188.50	1.00	18.50	22.09	188.50	nr	229.09
2400 mm diameter	720.50	1.00	18.50	22.09	720.50	nr	761.09
Step irons to concrete chamber rings	27.00	–	–	–	27.00	m	27.00
Extra over soakaways for filter wrapping with a proprietary filter membrane							
900 mm diameter × 1.00 m deep	1.52	1.00	18.50	–	1.52	nr	20.02
900 mm diameter × 2.00 m deep	3.05	1.50	27.75	–	3.05	nr	30.80
1050 mm diameter × 1.00 m deep	1.78	1.50	27.75	–	1.78	nr	29.53
1050 mm diameter × 2.00 m deep	3.56	2.00	37.00	–	3.56	nr	40.56
1200 mm diameter × 1.00 m deep	2.03	2.00	37.00	–	2.03	nr	39.03
1200 mm diameter × 2.00 m deep	4.06	2.50	46.25	–	4.06	nr	50.31
1500 mm diameter × 1.00 m deep	2.54	2.50	46.25	–	2.54	nr	48.79
1500 mm diameter × 2.00 m deep	5.07	2.50	46.25	–	5.07	nr	51.32
1800 mm diameter × 1.00 m deep	3.04	3.00	55.50	–	3.04	nr	58.54
1800 mm diameter × 2.00 m deep	6.09	3.25	60.13	–	6.09	nr	66.22
Gravel surrounding to concrete ring soakaway							
40 mm aggregate backfilled to vertical face of soakaway wrapped with geofabric (not included)							
250 mm thick	31.50	0.20	3.70	8.84	31.50	m³	44.04
Backfilling to face of soakaway; carefully compacting as work proceeds							
Arising from the excavations							
average thickness exceeding 0.25 m; depositing in layers 150 mm maximum thickness	–	0.03	0.62	4.42	–	m³	5.04
Aquacell soakaway; Wavin Plastics Ltd; preformed polypropylene soakaway infiltration crate units; to trenches; surrounded by geotextile and 40 mm aggregate laid 100 mm thick in trenches (excavation, disposal and backfilling not included)							
1.00 m × 500 × 400 mm; internal volume 190 litres							
4 crates; 2.00 × 1.00x 400 mm; 760 litres	146.56	1.60	29.60	14.14	189.82	nr	233.56
8 crates; 2.00 × 1.00x 800 mm; 1520 litres	293.12	3.20	59.20	18.18	346.70	nr	424.08
12 crates; 6.00 m × 500 × 800 mm; 2280 litres	439.68	4.80	88.80	37.70	543.10	nr	669.60

R DISPOSAL SYSTEMS

Item Excluding site overheads and profit	PC £	Labour hours	Labour £	Plant £	Material £	Unit	Total rate £
R13 LAND DRAINAGE – cont							
Aquacell soakaway – cont							
1.00 m × 500 × 400 mm – cont							
16 crates; 4.00 × 1.00 × 800 mm; 3040 litres	586.24	6.40	118.40	33.66	679.35	nr	**831.41**
20 crates; 5.00 × 1.00 × 800 mm; 3800 litres	732.80	8.00	148.00	33.32	832.21	nr	**1013.53**
30 crates; 15.00 × 1.00 × 400 mm; 5700 litres	1099.20	12.00	222.00	90.88	1345.58	nr	**1658.46**
60 crates; 15.00 × 1.00 × 800 mm; 11400 litres	2198.40	20.00	370.00	118.82	2530.45	nr	**3019.27**
Geofabric surround to Aquacell units;							
Terram Ltd							
Terram synthetic fibre filter fabric; to face of							
concrete rings (not included); anchoring whilst							
backfilling (not included)							
Terram 1000; 0.70 mm thick; mean water							
flow 50 l/m²/s	0.52	0.05	0.93	–	0.52	m²	**1.45**
Sand slitting; Agripower Ltd							
Drainage slits; at 1.00 m centres; using							
spinning disc trenching machine; backfilling to							
100 mm of surface with pea gravel; finished							
surface with medium grade sand 100; arisings							
to be loaded, hauled and tipped onsite;							
minimum pitch size 6000 m²							
250 mm depth	–	–	–	–	–	m	**1.80**
300 mm depth	–	–	–	–	–	m	**1.97**
400 mm depth	–	–	–	–	–	m	**3.01**
Sand grooving or banding; Agripower Ltd							
Drainage bands at 260 mm m centres; using							
Blec Sandmaster; to existing grass at 260 mm							
centres; minimum pitch size 4000 m²							
sand banding or sand grooving	–	–	–	–	–	m²	**1.17**
Land drainage; calculation table							
For calculation of drainage per hectare the							
following table can be used; rates show the							
lengths of drains per unit and not the value							
Lateral drains							
10.00 m centres	–	–	–	–	–	m/ha	**1000.00**
15.00 m centres	–	–	–	–	–	m/ha	**650.00**
25.00 m centres	–	–	–	–	–	m/ha	**400.00**
30.00 m centres	–	–	–	–	–	m/ha	**330.00**
Main drains							
1 nr (at 100 m centres)	–	–	–	–	–	m/ha	**100.00**
2 nr (at 50 m centres)	–	–	–	–	–	m/ha	**200.00**
3 nr (at 33.333 m centres)	–	–	–	–	–	m/ha	**300.00**
4 nr (at 25 m centres)	–	–	–	–	–	m/ha	**400.00**
Land drainage; excavating							
Removing 150 mm depth of topsoil; 300 mm							
wide; depositing beside trench; by machine	–	1.50	27.75	13.20	–	100 m	**40.95**
Removing 150 mm depth of topsoil; 300 mm							
wide; depositing beside trench; by hand	–	8.00	148.00	–	–	100 m	**148.00**

R DISPOSAL SYSTEMS

Item Excluding site overheads and profit	PC £	Labour hours	Labour £	Plant £	Material £	Unit	Total rate £
Disposing on site; to spoil heaps; by machine							
not exceeding 100 m distance	–	0.07	1.22	2.32	–	m³	3.54
average 100–150 m distance	–	0.08	1.47	2.78	–	m³	4.25
average 150–200 m distance	–	0.09	1.71	3.25	–	m³	4.96
Removing excavated material from site to tip							
not exceeding 13 km; mechanically loaded							
excavated material and clean hardcore							
rubble	–	–	–	1.12	8.00	m³	9.12
Land drainage; Agripower Ltd; excavating trenches (with minimum project size of 1000 m) by trenching machine; spreading arisings on site; laying perforated pipe; as shown below; backfilling with shingle to 350 mm from surface							
Width 150 mm; 80 mm perforated pipe							
depth 750 mm	–	–	–	–	–	m	5.50
Width 175 mm; 100 mm perforated pipe							
depth 800 mm	–	–	–	–	–	m	6.65
Width 250 mm; 150 mm perforated pipe							
depth 900 mm	–	–	–	–	–	m	9.70
Sports or amenity drainage; Agripower Ltd Ltd; excavating trenches (with minimum project size of 1000 m) by trenching machine; arisings to spoil heap on site maximum 100 m; backfill with shingle to within 125 mm from surface and with rootzone to ground level							
Width 145 mm; 80 mm perforated pipe							
depth 550 mm	–	–	–	–	–	m	7.94
Width 145 mm; 100 mm perforated pipe							
depth 650 mm	–	–	–	–	–	m	9.12
Width 145 mm; 150 mm perforated pipe							
depth 700 mm	–	–	–	–	–	100 m	11.75
Land drainage; excavating for drains; by backacter excavator JCB C3X Sitemaster; including disposing spoil to spoil heaps not exceeding 100 m; boning to levels by laser							
Width 150–225 mm							
depth 450 mm	–	9.67	178.83	257.55	–	100 m	436.38
depth 600 mm	–	13.00	240.50	386.32	–	100 m	626.82
depth 700 mm	–	15.50	286.75	482.91	–	100 m	769.66
depth 900 mm	–	23.00	425.50	772.65	–	100 m	1198.15
Width 300 mm							
depth 450 mm	–	9.67	178.83	257.55	–	100 m	436.38
depth 700 mm	–	14.00	259.00	424.96	–	100 m	683.96
depth 900 mm	–	23.00	425.50	772.65	–	100 m	1198.15
depth 1000 mm	–	25.00	462.50	849.91	–	100 m	1312.41
depth 1200 mm	–	25.00	462.50	965.81	–	100 m	1428.31
depth 1500 mm	–	31.00	573.50	1081.71	–	100 m	1655.21

R DISPOSAL SYSTEMS

Item Excluding site overheads and profit	PC £	Labour hours	Labour £	Plant £	Material £	Unit	Total rate £
R13 LAND DRAINAGE – cont							
Land drainage; excavating for drains; by 7							
tonne tracked excavator; including							
disposing spoil to spoil heaps not							
exceeding 100 m							
Width 225 mm							
depth 450 mm	–	7.00	129.50	73.18	–	100 m	**202.68**
depth 600 mm	–	7.17	132.58	76.23	–	100 m	**208.81**
depth 700 mm	–	7.35	135.93	79.55	–	100 m	**215.48**
depth 900 mm	–	7.76	143.60	87.12	–	100 m	**230.72**
depth 1000 mm	–	8.00	148.00	91.48	–	100 m	**239.48**
depth 1200 mm	–	8.26	152.87	96.30	–	100 m	**249.17**
Width 300 mm							
depth 450 mm	–	7.35	135.93	79.55	–	100 m	**215.48**
depth 600 mm	–	7.76	143.60	87.12	–	100 m	**230.72**
depth 700 mm	–	8.26	152.87	96.30	–	100 m	**249.17**
depth 900 mm	–	9.25	171.13	114.35	–	100 m	**285.48**
depth 1000 mm	–	10.14	187.64	130.69	–	100 m	**318.33**
depth 1200 mm	–	11.00	203.50	146.37	–	100 m	**349.87**
Width 600 mm							
depth 450 mm	–	13.00	240.50	182.96	–	100 m	**423.46**
depth 600 mm	–	14.11	261.06	203.29	–	100 m	**464.35**
depth 700 mm	–	16.33	302.17	243.95	–	100 m	**546.12**
depth 900 mm	–	17.29	319.79	261.37	–	100 m	**581.16**
depth 1000 mm	–	18.38	340.12	281.48	–	100 m	**621.60**
depth 1200 mm	–	21.18	391.86	332.66	–	100 m	**724.52**
Land drainage; excavating for drains; by							
hand, including disposing spoil to spoil							
heaps not exceeding 100 m							
Width 150 mm							
depth 450 mm	–	22.03	407.56	–	–	100 m	**407.56**
depth 600 mm	–	29.38	543.53	–	–	100 m	**543.53**
depth 700 mm	–	34.27	634.00	–	–	100 m	**634.00**
depth 900 mm	–	44.06	815.11	–	–	100 m	**815.11**
Width 225 mm							
depth 450 mm	–	33.05	611.42	–	–	100 m	**611.42**
depth 600 mm	–	44.06	815.11	–	–	100 m	**815.11**
depth 700 mm	–	51.41	951.09	–	–	100 m	**951.09**
depth 900 mm	–	66.10	1222.85	–	–	100 m	**1222.85**
depth 1000 mm	–	73.44	1358.64	–	–	100 m	**1358.64**
Width 300 mm							
depth 450 mm	–	44.06	815.11	–	–	100 m	**815.11**
depth 600 mm	–	58.75	1086.88	–	–	100 m	**1086.88**
depth 700 mm	–	68.54	1267.99	–	–	100 m	**1267.99**
depth 900 mm	–	88.13	1630.40	–	–	100 m	**1630.40**
depth 1000 mm	–	97.92	1811.52	–	–	100 m	**1811.52**
Width 375 mm							
depth 450 mm	–	55.08	1018.98	–	–	100 m	**1018.98**
depth 600 mm	–	73.44	1358.64	–	–	100 m	**1358.64**
depth 700 mm	–	85.68	1585.08	–	–	100 m	**1585.08**
depth 900 mm	–	110.16	2037.96	–	–	100 m	**2037.96**
depth 1000 mm	–	122.40	2264.40	–	–	100 m	**2264.40**

R DISPOSAL SYSTEMS

Item Excluding site overheads and profit	PC £	Labour hours	Labour £	Plant £	Material £	Unit	Total rate £
Width 450 mm							
depth 450 mm	–	66.10	1222.85	–	–	100 m	1222.85
depth 600 mm	–	88.13	1630.40	–	–	100 m	1630.40
depth 700 mm	–	102.82	1902.17	–	–	100 m	1902.17
depth 900 mm	–	132.19	2445.51	–	–	100 m	2445.51
depth 1000 mm	–	146.88	2717.28	–	–	100 m	2717.28
Width 600 mm							
depth 450 mm	–	88.13	1630.40	–	–	100 m	1630.40
depth 600 mm	–	117.50	2173.75	–	–	100 m	2173.75
depth 700 mm	–	137.09	2536.16	–	–	100 m	2536.16
depth 900 mm	–	176.26	3260.81	–	–	100 m	3260.81
depth 1000 mm	–	195.84	3623.04	–	–	100 m	3623.04
Width 900 mm							
depth 450 mm	–	132.19	2445.51	–	–	100 m	2445.51
depth 600 mm	–	176.26	3260.81	–	–	100 m	3260.81
depth 700 mm	–	205.63	3804.16	–	–	100 m	3804.16
depth 900 mm	–	264.38	4891.03	–	–	100 m	4891.03
depth 1000 mm	–	293.76	5434.56	–	–	100 m	5434.56
Earthwork support; moving along as work proceeds							
Maximum depth not exceeding 2.00 m							
distance between opposing faces not exceeding 2.00 m	–	0.80	14.80	17.96	–	m	32.76
Land drainage; pipe laying							
Hepworth; agricultural clay drain pipes; 300 mm length; butt joints; in straight runs							
75 mm diameter	450.55	8.00	148.00	–	450.55	100 m	598.55
100 mm diameter	772.76	9.00	166.50	–	772.76	100 m	939.26
150 mm diameter	1581.02	10.00	185.00	–	1581.02	100 m	1766.02
Extra over clay drain pipes for filter-wrapping pipes with Terram or similar filter fabric							
Terram 700	0.23	0.04	0.74	–	0.23	m²	0.97
Terram 1000	0.22	0.04	0.74	–	0.22	m²	0.96
Junctions between drains in clay pipes							
75 × 75 mm	14.97	0.25	4.63	–	15.29	nr	19.92
100 × 100 mm	19.87	0.25	4.63	–	20.38	nr	25.01
150 × 150 mm	24.46	0.25	4.63	–	25.23	nr	29.86
Wavin Plastics Ltd; flexible plastic perforated pipes in trenches (not included); to a minimum depth of 450 mm (couplings not included)							
OsmaDrain; flexible plastic perforated pipes in trenches (not included); to a minimum depth of 450 mm (couplings not included)							
80 mm diameter; available in 100 m coil	58.63	2.00	37.00	–	58.63	100 m	95.63
100 mm diameter; available in 100 m coil	95.12	2.00	37.00	–	95.12	100 m	132.12
160 mm diameter; available in 35 m coil	230.42	2.00	37.00	–	230.42	100 m	267.42
WavinCoil; plastic pipe junctions							
80 × 80 mm	2.59	0.05	0.93	–	2.59	nr	3.52
100 × 100 mm	2.89	0.05	0.93	–	2.89	nr	3.82
100 × 60 mm	2.75	0.05	0.93	–	2.75	nr	3.68
100 × 80 mm	2.75	0.05	0.93	–	2.75	nr	3.68
160 × 160 mm	6.89	0.05	0.93	–	6.89	nr	7.82

R DISPOSAL SYSTEMS

Item Excluding site overheads and profit	PC £	Labour hours	Labour £	Plant £	Material £	Unit	Total rate £
R13 LAND DRAINAGE – cont							
Wavin Plastics Ltd – cont							
WavinCoil; couplings for flexible pipes							
80 mm diameter	0.91	0.03	0.62	–	0.91	nr	**1.53**
100 mm diameter	1.01	0.03	0.62	–	1.01	nr	**1.63**
160 mm diameter	1.36	0.03	0.62	–	1.36	nr	**1.98**
Land drainage; backfilling trench after							
laying pipes with gravel rejects or similar;							
blind filling with ash or sand; topping with							
150 mm topsoil from dumps not exceeding							
100 m; by machine							
Width 150 mm							
depth 450 mm	–	3.30	61.05	54.35	129.38	100 m	**244.78**
depth 600 mm	–	4.30	79.55	71.18	168.75	100 m	**319.48**
depth 750 mm	–	4.96	91.76	82.29	190.13	100 m	**364.18**
depth 900 mm	–	6.30	116.55	104.84	238.73	100 m	**460.12**
Width 225 mm							
depth 450 mm	–	4.95	91.58	81.53	198.61	100 m	**371.72**
depth 600 mm	–	6.45	119.33	106.77	253.21	100 m	**479.31**
depth 750 mm	–	7.95	147.08	132.02	307.96	100 m	**587.06**
depth 900 mm	–	9.45	174.83	157.26	362.56	100 m	**694.65**
Width 375 mm							
depth 450 mm	–	8.25	152.63	135.88	330.91	100 m	**619.42**
depth 600 mm	–	10.75	198.88	177.96	421.96	100 m	**798.80**
depth 750 mm	–	13.25	245.13	220.03	513.16	100 m	**978.32**
depth 900 mm	–	15.75	291.38	262.11	604.21	100 m	**1157.70**
Land drainage; backfilling trench after							
laying pipes with gravel rejects or similar,							
blind filling with ash or sand; topping with							
150 mm topsoil from dumps not exceeding							
100 m; by hand							
Width 150 mm							
depth 450 mm	–	18.63	344.65	–	119.48	100 m	**464.13**
depth 600 mm	–	24.84	459.54	–	168.75	100 m	**628.29**
depth 750 mm	–	31.05	574.42	–	140.63	100 m	**715.05**
depth 900 mm	–	37.26	689.31	–	238.73	100 m	**928.04**
Width 225 mm							
depth 450 mm	–	27.94	516.89	–	198.61	100 m	**715.50**
depth 600 mm	–	37.26	689.31	–	253.21	100 m	**942.52**
depth 750 mm	–	46.57	861.54	–	307.96	100 m	**1169.50**
depth 900 mm	–	55.89	1033.96	–	362.56	100 m	**1396.52**
Width 375 mm							
depth 450 mm	–	46.57	861.54	–	330.91	100 m	**1192.45**
depth 600 mm	–	61.10	1130.35	–	421.96	100 m	**1552.31**
depth 750 mm	–	77.63	1436.15	–	513.16	100 m	**1949.31**
depth 900 mm	–	93.15	1723.28	–	604.21	100 m	**2327.49**

R DISPOSAL SYSTEMS

Item Excluding site overheads and profit	PC £	Labour hours	Labour £	Plant £	Material £	Unit	Total rate £
Catchwater or french drains; laying pipes to drain excavated separately; 100 mm diameter non-coilable perforated plastic pipes; including straight jointing; pipes laid with perforations uppermost; lining trench; wrapping pipes with filter fabric; backfilling with shingle							
Width 300 mm							
depth 450 mm	772.80	9.10	168.35	30.60	1168.12	100 m	1367.07
depth 600 mm	777.16	9.84	182.04	47.80	1314.24	100 m	1544.08
depth 750 mm	781.72	10.60	196.10	60.59	1460.55	100 m	1717.24
depth 900 mm	786.19	11.34	209.79	73.04	1606.76	100 m	1889.59
depth 1000 mm	789.16	11.84	219.04	81.46	1704.24	100 m	2004.74
depth 1200 mm	795.11	12.84	237.54	98.29	1899.18	100 m	2235.01
Width 450 mm							
depth 450 mm	779.49	10.44	193.14	88.19	1387.44	100 m	1668.77
depth 600 mm	786.19	11.34	209.79	73.04	1606.76	100 m	1889.59
depth 750 mm	792.88	12.46	230.51	91.89	1826.08	100 m	2148.48
depth 900 mm	799.57	13.60	251.60	111.08	2045.40	100 m	2408.08
depth 1000 mm	804.03	14.34	265.29	123.53	2191.61	100 m	2580.43
depth 1200 mm	812.96	15.84	293.04	148.78	2484.03	100 m	2925.85
depth 1500 mm	826.34	18.10	334.85	186.81	2922.67	100 m	3444.33
depth 2000 mm	848.65	21.84	404.04	249.76	3653.73	100 m	4307.53
Width 600 mm							
depth 450 mm	786.19	11.24	207.94	73.04	1606.76	100 m	1887.74
depth 600 mm	795.11	12.84	237.54	98.29	1899.18	100 m	2235.01
depth 750 mm	804.03	14.34	265.29	123.53	2191.61	100 m	2580.43
depth 900 mm	812.96	15.84	293.04	148.78	2484.03	100 m	2925.85
depth 1000 mm	810.01	16.84	311.54	165.61	2670.08	100 m	3147.23
depth 1200 mm	830.80	18.84	348.54	199.27	3068.88	100 m	3616.69
depth 1500 mm	848.65	21.84	404.04	519.76	3653.73	100 m	4577.53
depth 2000 mm	878.40	19.84	367.04	451.72	4628.47	100 m	5447.23
Width 900 mm							
depth 450 mm	799.57	13.60	251.60	111.08	2045.40	100 m	2408.08
depth 600 mm	812.96	7.92	146.52	148.78	2484.03	100 m	2779.33
depth 750 mm	826.34	9.05	167.43	186.81	2922.67	100 m	3276.91
depth 900 mm	839.73	10.17	188.15	224.51	3361.30	100 m	3773.96
depth 1000 mm	848.65	10.92	202.02	249.76	3653.73	100 m	4105.51
depth 1200 mm	866.50	12.42	229.77	300.25	4238.57	100 m	4768.59
depth 1500 mm	893.27	14.67	271.39	375.98	5115.85	100 m	5763.22
depth 2000 mm	937.89	18.92	350.02	592.21	6577.96	100 m	7520.19
depth 2500 mm	982.51	22.17	410.14	718.43	8040.08	100 m	9168.65
depth 3000 mm	1027.13	25.92	479.52	754.66	9502.20	100 m	10736.38
Catchwater or french drains; laying pipes to drains excavated separately; 160 mm diameter non-coilable perforated plastic pipes; including straight jointing; pipes laid with perforations uppermost; lining trench; wrapping pipes with filter fabric; backfilling with shingle							
Width 600 mm							
depth 450 mm	1373.50	11.20	207.20	70.69	2168.25	100 m	2446.14
depth 600 mm	1382.42	12.70	234.95	95.93	2460.67	100 m	2791.55
depth 750 mm	1391.35	14.20	262.70	121.18	2753.09	100 m	3136.97

R DISPOSAL SYSTEMS

Item Excluding site overheads and profit	PC £	Labour hours	Labour £	Plant £	Material £	Unit	Total rate £
R13 LAND DRAINAGE – cont							
Catchwater or french drains – cont							
Width 600 mm – cont							
depth 900 mm	1400.27	15.70	290.45	146.42	3045.52	100 m	**3482.39**
depth 1000 mm	1406.22	16.70	308.95	163.25	3240.47	100 m	**3712.67**
depth 1200 mm	1418.12	18.70	345.95	197.01	3630.36	100 m	**4173.32**
depth 1500 mm	1435.97	21.70	401.45	247.40	4215.21	100 m	**4864.06**
depth 2000 mm	1465.71	26.70	493.95	331.55	5189.96	100 m	**6015.46**
depth 2500 mm	1495.46	31.70	586.45	415.70	6164.70	100 m	**7166.85**
depth 3000 mm	1525.20	36.70	678.95	499.85	7139.45	100 m	**8318.25**
Width 900 mm							
depth 450 mm	799.57	13.60	251.60	111.08	2045.40	100 m	**2408.08**
depth 600 mm	812.96	15.84	293.04	148.78	2484.03	100 m	**2925.85**
depth 750 mm	826.34	18.10	334.85	186.81	2922.67	100 m	**3444.33**
depth 900 mm	839.73	20.34	376.29	224.51	3361.30	100 m	**3962.10**
depth 1000 mm	848.65	21.84	404.04	249.76	3653.73	100 m	**4307.53**
depth 1200 mm	866.50	24.84	459.54	300.25	4238.57	100 m	**4998.36**
depth 1500 mm	893.27	29.34	542.79	375.98	5115.85	100 m	**6034.62**
depth 2000 mm	937.89	36.84	681.54	502.21	6577.96	100 m	**7761.71**
depth 2500 mm	982.51	44.34	820.29	628.43	8040.08	100 m	**9488.80**
depth 3000 mm	1027.13	51.84	959.04	754.66	9502.20	100 m	**11215.90**
Catchwater or french drains; Exxon **Chemical Geopolymers Ltd; Filtram filter** **drain; in trenches (trenches not included);** **comprising filter fabric, liquid conducting** **core and 110 mm uPVC slitpipes; all in** **accordance with manufacturer's** **instructions; backfilling with shingle**							
Width 600 mm							
depth 1000 mm	818.91	16.84	311.54	165.61	2678.98	100 m	**3156.13**
depth 1200 mm	830.80	18.84	348.54	199.27	3068.88	100 m	**3616.69**
depth 1500 mm	848.65	21.84	404.04	519.76	3653.73	100 m	**4577.53**
depth 2000 mm	878.40	19.84	367.04	451.72	4628.47	100 m	**5447.23**
Width 900 mm							
depth 450 mm	799.57	13.60	251.60	111.08	2045.40	100 m	**2408.08**
depth 600 mm	812.96	15.84	293.04	148.78	2484.03	100 m	**2925.85**
depth 750 mm	826.34	18.10	334.85	186.81	2922.67	100 m	**3444.33**
depth 900 mm	839.73	20.34	376.29	224.51	3361.30	100 m	**3962.10**
depth 1000 mm	848.65	21.84	404.04	249.76	3653.73	100 m	**4307.53**
depth 1200 mm	866.50	24.84	459.54	300.25	4238.57	100 m	**4998.36**
depth 1500 mm	893.27	29.34	542.79	375.98	5115.85	100 m	**6034.62**
depth 2000 mm	937.89	37.84	700.04	592.21	6577.96	100 m	**7870.21**
depth 2500 mm	982.51	44.34	820.29	718.43	8040.08	100 m	**9578.80**
depth 3000 mm	1027.13	51.84	959.04	844.66	9502.20	100 m	**11305.90**
Ditching; clear silt and bottom ditch not **exceeding 1.50 m deep; strim back** **vegetation; disposing to spoil heaps; by** **machine**							
Up to 1.50 m wide at top	–	6.00	111.00	26.39	–	100 m	**137.39**
1.50–2.50 m wide at top	–	8.00	148.00	35.19	–	100 m	**183.19**
2.50–4.00 m wide at top	–	9.00	166.50	39.59	–	100 m	**206.09**

R DISPOSAL SYSTEMS

Item Excluding site overheads and profit	PC £	Labour hours	Labour £	Plant £	Material £	Unit	Total rate £
Ditching; clear only vegetation from ditch not exceeding 1.50 m deep; disposing to spoil heaps; by strimmer							
Up to 1.50 m wide at top	–	5.00	92.50	7.01	–	100 m	99.51
1.50–2.50 m wide at top	–	6.00	111.00	12.61	–	100 m	123.61
2.50–4.00 m wide at top	–	7.00	129.50	19.62	–	100 m	149.12
Ditching; clear silt from ditch not exceeding 1.50 m deep; trimming back vegetation; disposing to spoil heaps; by hand							
Up to 1.50 m wide at top	–	15.00	277.50	7.01	–	100 m	284.51
1.50–2.50 m wide at top	–	27.00	499.50	12.61	–	100 m	512.11
2.50–4.00 m wide at top	–	42.00	777.00	19.62	–	100 m	796.62
Ditching; excavating and forming ditch and bank to given profile (normally 45°); in loam or sandy loam; by machine							
Width 300 mm							
depth 600 mm	–	3.70	68.45	32.55	–	100 m	101.00
depth 900 mm	–	5.20	96.20	45.75	–	100 m	141.95
depth 1200 mm	–	7.20	133.20	63.34	–	100 m	196.54
depth 1500 mm	–	9.40	173.90	72.36	–	100 m	246.26
Width 600 mm							
depth 600 mm	–	–	–	82.61	–	100 m	82.61
depth 900 mm	–	–	–	122.58	–	100 m	122.58
depth 1200 mm	–	–	–	166.55	–	100 m	166.55
depth 1500 mm	–	–	–	239.83	–	100 m	239.83
Width 900 mm							
depth 600 mm	–	–	–	321.30	–	100 m	321.30
depth 900 mm	–	–	–	566.10	–	100 m	566.10
depth 1200 mm	–	–	–	760.72	–	100 m	760.72
depth 1500 mm	–	–	–	942.48	–	100 m	942.48
Width 1200 mm							
depth 600 mm	–	–	–	504.90	–	100 m	504.90
depth 900 mm	–	–	–	749.70	–	100 m	749.70
depth 1200 mm	–	–	–	992.97	–	100 m	992.97
depth 1500 mm	–	–	–	1272.35	–	100 m	1272.35
Width 1500 mm							
depth 600 mm	–	–	–	622.71	–	100 m	622.71
depth 900 mm	–	–	–	949.21	–	100 m	949.21
depth 1200 mm	–	–	–	1262.25	–	100 m	1262.25
depth 1500 mm	–	–	–	1582.02	–	100 m	1582.02
Extra for ditching in clay	–	–	–	–	–	20%	–
Ditching; excavating and forming ditch and bank to given profile (normal 45°); in loam or sandy loam; by hand							
Width 300 mm							
depth 600 mm	–	36.00	666.00	–	–	100 m	666.00
depth 900 mm	–	42.00	777.00	–	–	100 m	777.00
depth 1200 mm	–	56.00	1036.00	–	–	100 m	1036.00
depth 1500 mm	–	70.00	1295.00	–	–	100 m	1295.00

R DISPOSAL SYSTEMS

Item Excluding site overheads and profit	PC £	Labour hours	Labour £	Plant £	Material £	Unit	Total rate £
R13 LAND DRAINAGE – cont							
Ditching – cont							
Width 600 mm							
depth 600 mm	–	56.00	1036.00	–	–	100 m	**1036.00**
depth 900 mm	–	84.00	1554.00	–	–	100 m	**1554.00**
depth 1200 mm	–	112.00	2072.00	–	–	100 m	**2072.00**
depth 1500 mm	–	140.00	2590.00	–	–	100 m	**2590.00**
Width 900 mm							
depth 600 mm	–	84.00	1554.00	–	–	100 m	**1554.00**
depth 900 mm ·	–	126.00	2331.00	–	–	100 m	**2331.00**
depth 1200 mm	–	168.00	3108.00	–	–	100 m	**3108.00**
depth 1500 mm	–	210.00	3885.00	–	–	100 m	**3885.00**
Width 1200 mm ·₃							
depth 600 mm	–	112.00	2072.00	–	–	100 m	**2072.00**
depth 900 mm	–	168.00	3108.00	–	–	100 m	**3108.00**
depth 1200 mm	–	224.00	4144.00	–	–	100 m	**│ 4144.00**
depth 1500 mm	–	280.00	5180.00	–	–	100 m	**5180.00**
Extra for ditching in clay	–	–	–	–	–	50%	**–**
Extra for ditching in compacted soil	–	–	–	–	–	90%	**–**
Piped ditching							
Jointed concrete pipes; FP McCann Ltd; including bedding, haunching and topping with 150 mm concrete; 11.50 N/mm² – 40 mm aggregate; to existing ditch							
300 mm diameter	16.13	0.67	12.33	6.73	32.39	m	**51.45**
450 mm diameter	23.97	0.67	12.33	6.73	48.36	m	**67.42**
600 mm diameter	38.66	0.67	12.33	6.73	73.29	m	**92.35**
900 mm diameter	102.05	1.00	18.50	10.10	120.13	m	**148.73**
Jointed concrete pipes; FP McCann Ltd; including bedding, haunching and topping with 150 mm concrete; 11.50 N/mm² – 40 mm aggregate; to existing ditch							
1200 mm diameter	175.39	1.00	18.50	10.10	199.04	m	**227.64**
extra over jointed concrete pipes for bends to 45°	161.30	0.67	12.33	8.41	168.51	nr	**189.25**
extra over jointed concrete pipes for single junctions 300 mm diameter	110.22	0.67	12.33	8.41	117.43	nr	**138.17**
extra over jointed concrete pipes for single junctions 450 mm diameter	163.39	0.67	12.33	7.65	170.60	nr	**190.58**
extra over jointed concrete pipes for single junctions 600 mm diameter	264.66	0.67	12.33	7.65	271.87	nr	**291.85**
extra over jointed concrete pipes for single junctions 900 mm diameter	447.98	0.67	12.33	7.65	455.19	nr	**475.17**
extra over jointed concrete pipes for single junctions 1200 mm diameter	684.30	0.67	12.33	7.65	691.51	nr	**711.49**
Concrete road gullies; trapped; cement: mortar (1:2) joints to concrete pipes; bedding and surrounding in concrete; 11.50 N/mm² – 40 mm aggregate; 450 mm diameter × 1.07 m deep; rodding eye; stoppers	46.00	6.00	111.00	–	62.15	nr	**173.15**

R DISPOSAL SYSTEMS

Item excluding site overheads and profit	PC £	Labour hours	Labour £	Plant £	Material £	Unit	Total rate £
Preamble: Flat rate hourly rainfall = 50 mm/hr and assumes 100 impermeability of the run-off area. A storage capacity of the soakaway should be ⅓ of the hourly rainfall. Formulae for calculating soakaway depths are provided in the publications mentioned below and in the Tables and Memoranda section of this publication. The design of soakaways is dependent on, amongst other factors, soil conditions, permeability, groundwater level and runoff. The definitive documents for design of soakaways are CIRIA 156 and BRE Digest 365 dated September 1991. The suppliers of the systems below will assist through their technical divisions. Excavation earthwork support of pits and disposal not included.							
Outfalls Reinforced concrete outfalls to water course; tank walls; for 150 mm drain outlets; overall dimensions							
900 × 1050 × 900 mm high	–	–	–	–	–	m	487.60
GRC outfall headwalls	–	–	–	–	–	nr	105.00
JKH Unit; standard small	–	–	–	–	–	nr	85.00

S PIPED SUPPLY SYSTEMS

Item Excluding site overheads and profit	PC £	Labour hours	Labour £	Plant £	Material £	Unit	Total rate £
S10 COLD WATER							
Borehole drilling; Agripower Ltd							
Drilling operations to average 100 m depth;							
lining with 150 mm diameter sieve							
drilling to 100 m	–	–	–	–	–	nr	12000.00
rate per m over 100 m	–	–	–	–	–	m	120.00
pump, rising main, cable and kiosk	–	–	–	–	–	nr	5000.00
Blue MDPE polythene pipes; type 50; for							
cold water services; with compression							
fittings; bedding on 100 mm DOT type							
1 granular fill material							
Pipes							
20 mm diameter	0.34	0.08	1.48	–	3.49	m	4.97
25 mm diameter	0.41	0.08	1.48	–	3.56	m	5.04
32 mm diameter	1.00	0.08	1.48	–	4.14	m	5.62
50 mm diameter	1.73	0.10	1.85	–	4.87	m	6.72
60 mm diameter	3.73	0.10	1.85	–	6.88	m	8.73
Hose union bib taps; including fixing to wall;							
making good surfaces							
15 mm	11.65	0.75	13.88	–	13.55	nr	27.4
22 mm	16.44	0.75	13.88	–	19.29	nr	33.17
Stopcocks; including fixing to wall; making							
good surfaces							
15 mm	6.92	0.75	13.88	–	8.82	nr	22.70
22 mm	10.77	0.75	13.88	–	12.67	nr	26.55
Standpipes; to existing 25 mm water mains							
1.00 m high	–	–	–	–	–	nr	225.00
Hose junction bib taps; to standpipes							
19 mm	–	–	–	–	–	nr	80.00
S14 IRRIGATION							
KAR UK; water recycling							
Fully automatic self cleaning water recycling							
system for use with domestic grey water;							
including subsurface tank, self-flushing filter							
pump and control system; excludes plumbing							
of the drainage system from the house into the							
tank	–	–	–	–	–	nr	2987.00
KAR UK; main or ring main supply							
Excavate and lay mains supply pipe to supply							
irrigated area							
PE 80 20 mm	–	–	–	–	–	m	4.87
PE 80 25 mm	–	–	–	–	–	m	5.00
Pro-Flow PEHD 63 mm PE	–	–	–	–	–	m	8.68
Pro-Flow PEHD 50 mm PE	–	–	–	–	–	m	7.09
Pro-Flow PEHD 40 mm PE	–	–	–	–	–	m	6.11
Pro-Flow PEHD 32 mm PE	–	–	–	–	–	m	5.49

S PIPED SUPPLY SYSTEMS

Item Excluding site overheads and profit	PC £	Labour hours	Labour £	Plant £	Material £	Unit	Total rate £
KAR UK; head control to sprinkler stations Valves installed at supply point on irrigation supply manifold; includes for pressure control and filtration							
2 valve unit	–	–	–	–	–	nr	132.03
4 valve unit	–	–	–	–	–	nr	206.11
6 valve unit	–	–	–	–	–	nr	291.49
12 valve unit	–	–	–	–	–	nr	558.92
Solenoid valve; 25 mm with chamber; extra over for each active station	–	–	–	–	–	nr	106.16
Cable to solenoid valves (alternative to valves on manifold as above)							
4 core	–	–	–	–	–	m	1.57
6 core	–	–	–	–	–	m	7.63
12 core	–	–	–	–	–	m	3.19
KAR UK; supply irrigation infrastructure for irrigation system Header tank and submersible pump and pressure stat							
1000 litre (500 gallon 25 mm/10 days to 1500 m²) 2 m³/hr pump	–	–	–	–	–	nr	489.67
4540 litre (1000 gallon 25 mm/10 days to 3500 m²)	–	–	–	–	–	nr	991.89
9080 litre (2000 gallon 25 mm/10 days to 7000 m²)	–	–	–	–	–	nr	1642.27
Electric multi-station controllers 240 V							
WeatherMatic Smartline; 4 station; standard	–	–	–	–	–	nr	149.21
WeatherMatic Smartline; 6 station; standard	–	–	–	–	–	nr	167.62
WeatherMatic Smartline; 12 station; standard	–	–	–	–	–	nr	284.18
WeatherMatic Smartline; 12 station; radio controlled	–	–	–	–	–	nr	583.80
WeatherMatic Smartline; 12 station; with moisture sensor	–	–	–	–	–	nr	981.43
WeatherMatic Smartline; 12 station; radio controlled; with moisture sensor	–	–	–	–	–	nr	1178.61
KAR UK; station consisting of multiple sprinklers; inclusive of all trenching, wiring and connections to ring main or main supply Sprayheads; WeatherMatic LX; including nozzle; 21 m² coverage							
100 mm (4')	–	–	–	–	–	nr	15.52
150 mm (6')	–	–	–	–	–	nr	21.92
300 mm (12')	–	–	–	–	–	nr	25.28
Matched precipitation sprinklers; WeatherMatic LX							
MP 1000; 16 m² coverage	–	–	–	–	–	nr	35.57
MP 2000; 30 m² coverage	–	–	–	–	–	nr	35.57
MP 3000; 81 m² coverage	–	–	–	–	–	nr	35.57

S PIPED SUPPLY SYSTEMS

Item Excluding site overheads and profit	PC £	Labour hours	Labour £	Plant £	Material £	Unit	Total rate £
S14 IRRIGATION – cont							
KAR UK – cont							
Gear drive sprinklers; WeatherMatic; placed to							
provide head to head (100%) overlap; average							
T3; 121 m² coverage	–	–	–	–	–	nr	46.72
CT70; 255 m² coverage	–	–	–	–	–	nr	82.36
CT70SS; 361 m² coverage	–	–	–	–	–	nr	91.46
KAR UK; driplines; inclusive of							
connections and draindown systems							
Metzerplas TechLand 30 cm drip emitter							
fixed on soil surface	–	–	–	–	–	m	1.07
subsurface	–	–	–	–	–	m	1.57
Metzerplas TechLand 50 cm drip emitter							
fixed on soil surface	–	–	–	–	–	m	0.90
subsurface	–	–	–	–	–	m	1.41
Commissioning and testing of irrigation							
system							
per station	–	–	–	–	–	nr	39.56
Annual maintenance costs of irrigation							
system							
Call out charge per visit	–	–	–	–	–	nr	329.59
Extra over per station	–	–	–	–	–	nr	13.19
KAR UK; KAR Moisture Control moisture							
sensing control system							
Large garden comprising 7000 m² and 24							
stations with four moisture sensors							
turf only	–	–	–	–	–	nr	13333.33
turf/shrub beds; 70/30	–	–	–	–	–	nr	7777.78
Medium garden comprising 3500 m² and 12							
stations with two moisture sensors							
turf only	–	–	–	–	–	nr	8888.89
turf/shrub beds; 70/30	–	–	–	–	–	nr	11555.56
Smaller gardens garden comprising 1000 m²							
and 6 stations with one moisture sensor							
turf only	–	–	–	–	–	nr	5444.44
turf/shrub beds; 70/30	–	–	–	–	–	nr	6000.00
turf/shrub beds; 50/50	–	–	–	–	–	nr	6444.44
Leaky Pipe Systems Ltd; Leaky Pipe;							
moisture leaking pipe irrigation system							
Main supply pipe inclusive of machine							
excavation; exclusive of connectors							
20 mm LDPE polytubing	1.12	0.05	0.93	0.61	1.12	m	2.66
16 mm LDPE polytubing	0.75	0.05	0.93	0.61	0.75	m	2.29
Water filters and cartridges							
No 10; 20 mm	–	–	–	–	54.47	nr	54.47
Big Blue and RR30 cartridge; 25 mm	–	–	–	–	139.36	nr	139.36
Water filters and pressure regulator sets;							
complete assemblies							
No 10; flow rate 3.1–82 litres per minute	–	–	–	–	92.95	nr	92.95

S PIPED SUPPLY SYSTEMS

Item Excluding site overheads and profit	PC £	Labour hours	Labour £	Plant £	Material £	Unit	Total rate £
Leaky pipe hose; placed 150 mm sub surface or turf irrigation; distance between laterals 450 mm; excavation and backfilling priced separately							
LP12L low leak	4.73	0.04	0.74	–	4.73	m²	5.47
LP12H high leak	4.10	0.04	0.74	–	4.10	m²	4.84
LP12UH ultra high leak	5.02	0.04	0.74	–	5.02	m²	5.76
Leaky pipe hose; laid to surface for landscape irrigation; distance between laterals 600 mm							
LP12L low leak	2.82	0.03	0.46	–	2.82	m²	3.28
LP12H high leak	2.45	0.03	0.46	–	2.45	m²	2.91
LP12UH ultra high leak	2.99	0.03	0.46	–	2.99	m²	3.45
Leaky pipe hose; laid to surface for landscape irrigation; distance between laterals 900 mm							
LP12L low leak	1.89	0.02	0.31	–	1.89	m²	2.20
LP12H high leak	1.64	0.02	0.31	–	1.64	m²	1.95
LP12UH ultra high leak	2.00	0.02	0.31	–	2.00	m²	2.31
Leaky pipe hose; laid to surface for tree irrigation							
laid around circumference of tree pit							
LP12L low leak	2.79	0.13	2.31	–	2.79	nr	5.10
LP12H high leak	2.42	0.13	2.31	–	2.42	nr	4.73
LP12UH ultra high leak	2.96	0.13	2.31	–	2.96	nr	5.27
Accessories							
automatic multi-station controller stations; inclusive of connections	375.70	2.00	41.00	–	375.70	nr	416.70
solenoid valves; inclusive of wiring and connections to a multi-station controller; nominal distance from controller 25 m	62.40	0.50	9.25	–	517.40	nr	526.65
Rainwater harvesting; Britannia Rainwater Recycling Systems; tanks for collection of water for use in landscape irrigation; inclusive of filtration of first stage particle and leaf matter; installed on surface; inclusive of base, gullies for water catchment pipework from catchment area and submersible pumps; excavated material to stockpile							
Tanks installed below ground; gullies for rainwater catchment and pipework not included							
23000 litre	6820.00	12.00	222.00	240.58	7281.05	nr	7743.63
13000 litre	5874.00	10.00	185.00	206.76	6124.00	nr	6515.76
6500 litre	4158.00	8.00	148.00	103.38	4619.05	nr	4870.43

S PIPED SUPPLY SYSTEMS

Item Excluding site overheads and profit	PC £	Labour hours	Labour £	Plant £	Material £	Unit	Total rate £
S15 FOUNTAINS/WATER FEATURES							
Lakes and ponds – General Preamble: The pressure of water against a retaining wall or dam is considerable, and where water retaining structures form part of the design of water features, the landscape architect is advised to consult a civil engineer. Artificially contained areas of water in raised reservoirs over 25,000 m³ have to be registered with the local authority and their dams will have to be covered by a civil engineer's certificate of safety.							
Typical linings – General Preamble: In addition to the traditional methods of forming the linings of lakes and ponds in puddled clay or concrete, there are a number of lining materials available. They are mainly used for reservoirs but can also help to form comparatively economic water features especially in soil which is not naturally water retentive. Information on the construction of traditional clay puddle ponds can be obtained from the British Trust for Conservation Volunteers, 36 St. Mary's Street, Wallingford, Oxfordshire, OX10 0EU. Tel: (01491) 39766. The cost of puddled clay ponds depends on the availability of suitable clay, the type of hand or machine labour that can be used and the use to which the pond is to be put.							
Lake liners; Fairwater Water Garden Design Consultants Ltd; to evenly graded surface of excavations (excavating not included); all stones over 75 mm; removing debris; including all welding and jointing of liner sheets Geotextile underlay; inclusive of spot welding to prevent dragging							
to water features	–	–	–	–	–	m²	2.27
to lakes or large features	–	–	–	–	–	1000 m²	2265.70
Butyl rubber liners; Varnamo; inclusive of site vulcanizing laid to geotextile above							
0.75 mm thick	–	–	–	–	–	m²	7.57
0.75 mm thick	–	–	–	–	–	1000 m²	5570.00
1.00 mm thick	–	–	–	–	–	m²	8.62
1.00 mm thick	–	–	–	–	–	1000 m²	8620.00

S PIPED SUPPLY SYSTEMS

Item Excluding site overheads and profit	PC £	Labour hours	Labour £	Plant £	Material £	Unit	Total rate £
Lake liners; Landline Ltd; Landflex or Alkorplan geomembranes; to prepared surfaces (surfaces not included); all joints fully welded; installation by Landline employees							
Landflex HC polyethylene geomembranes							
0.50 mm thick	–	–	–	–	–	1000 m²	3250.00
0.75 mm thick	–	–	–	–	–	1000 m²	3750.00
1.00 mm thick	–	–	–	–	–	1000 m²	4250.00
1.50 mm thick	–	–	–	–	–	1000 m²	5000.00
2.00 mm thick	–	–	–	–	–	1000 m²	5750.00
Alkorplan PVC geomembranes							
0.80 mm thick	–	–	–	–	–	1000 m	5950.00
1.20 mm thick	–	–	–	–	–	1000 m	8650.00
Lake liners; Monarflex							
Low density polyethylene (LDPE) lake and reservoir lining system; welding on site by Monarflex technicians (surface preparation and backfilling not included)							
Blackline 500	–	–	–	–	–	100 m²	847.40
Blackline 750	–	–	–	–	–	100 m²	930.73
Blackline 1000	–	–	–	–	–	100 m²	1149.95
Operations over surfaces of lake liners							
Dug ballast; evenly spread over excavation already brought to grade							
150 mm thick	534.60	2.00	37.00	26.65	534.60	100 m²	598.25
200 mm thick	712.80	3.00	55.50	39.97	712.80	100 m²	808.27
300 mm thick	1069.20	3.50	64.75	46.63	1069.20	100 m²	1180.58
Imported topsoil; evenly spread over excavation							
100 mm thick	264.00	1.50	27.75	19.99	264.00	100 m²	311.74
150 mm thick	396.00	2.00	37.00	26.65	396.00	100 m²	459.65
200 mm thick	528.00	3.00	55.50	39.97	528.00	100 m²	623.47
Blinding existing subsoil with 50 mm sand	168.30	1.00	18.50	26.65	168.30	100 m²	213.45
Topsoil from excavation; evenly spread over excavation							
100 mm thick	–	–	–	19.99	–	100 m²	19.99
200 mm thick	–	–	–	26.65	–	100 m²	26.65
300 mm thick	–	–	–	39.97	–	100 m²	39.97
Extra over for screening topsoil using a Powergrid screener; removing debris	–	–	–	5.19	0.40	m³	5.59
Lake construction; Fairwater Water Garden Design Consultants Ltd; lake construction at ground level; excavation; forming of lake; commissioning; excavated material spread on site							
Lined with existing site clay; natural edging with vegetation meeting the water							
500 m²	–	–	–	–	–	nr	9261.00
1000 m²	–	–	–	–	–	nr	16096.50
1500 m²	–	–	–	–	–	nr	22050.00
2000 m²	–	–	–	–	–	nr	27562.50

S PIPED SUPPLY SYSTEMS

Item Excluding site overheads and profit	PC £	Labour hours	Labour £	Plant £	Material £	Unit	Total rate £
S15 FOUNTAINS/WATER FEATURES – cont							
Lake construction – cont							
Lined with 0.75 mm butyl on geotextile							
underlay; natural edging with vegetation							
meeting the water							
500 m²	–	–	–	–	–	nr	**14883.75**
1000 m²	–	–	–	–	–	nr	**25798.50**
1500 m²	–	–	–	–	–	nr	**35280.00**
2000 m²	–	–	–	–	–	nr	**44100.00**
Extra over to the above for hard block							
edging to secure and protect liner;							
measured at lake perimeter	–	–	–	–	–	m	46.31
Lined with imported puddling clay; natural							
edging with vegetation meeting the water							
500 m²	–	–	–	–	–	nr	24255.00
1000 m²	–	–	–	–	–	nr	41895.00
1500 m²	–	–	–	–	–	nr	57330.00
2000 m²	–	–	–	–	–	nr	**71662.50**
Waterfall construction; Fairwater Water							
Garden Design Consultants Ltd							
Stone placed on top of butyl liner; securing							
with concrete and dressing to form natural							
rock pools and edgings							
Portland stone; m³ rate	–	–	–	–	–	m³	697.31
Portland stone; tonne rate	–	–	–	–	–	tonne	386.12
Balancing tank; blockwork construction;							
inclusive of recirculation pump and pond level							
control; 110 mm balancing pipe to pond;							
waterproofed with preformed polypropylene							
membrane; pipework mains water top-up and							
overflow							
450 × 600 × 1000 mm	–	–	–	–	–	nr	1450.00
Extra for pump							
2000 gallons per hour; submersible	–	–	–	–	–	nr	215.00
Water features; Fairwater Water Garden							
Design Consultants Ltd							
Natural stream; butyl lined level changes of							
1.00 m; water pumped from lower pond (not							
included) via balancing tank; level changes via							
Purbeck stone water falls							
20 m long stream	–	–	–	–	–	nr	20000.00
Bespoke freestanding water wall; steel or							
glass panel; self contained recirculation							
system and reservoir							
water wall; 2.00 m high × 850 mm wide	–	–	–	–	–	nr	9371.25

S PIPED SUPPLY SYSTEMS

Item Excluding site overheads and profit	PC £	Labour hours	Labour £	Plant £	Material £	Unit	Total rate £
Ornamental pools – General							
Preamble: Small pools may be lined with one of the materials mentioned under lakes and ponds, or may be in rendered brickwork, puddled clay or, for the smaller sizes, fibreglass. Most of these tend to be cheaper than waterproof concrete. Basic prices for various sizes of concrete pools are given in the Approximate Estimates section. Prices for excavation, grading, mass concrete, and precast concrete retaining walls are given in the relevant sections. The manufacturers should be consulted before specifying the type and thickness of pool liner, as this depends on the size, shape and proposed use of the pool. The manufacturer's recommendation on foundations and construction should be followed.							
Ornamental pools							
Pool liners; to 50 mm sand blinding to excavation (excavating not included); all stones over 50 mm; removing debris from surfaces of excavation; including underlay, all welding and jointing of liner sheets							
black polythene; 1000 gauge	–	–	–	–	–	m²	5.59
blue polythene; 1000 gauge	–	–	–	–	–	m²	7.38
coloured PVC; 1500 gauge	–	–	–	–	–	m²	8.46
black PVC; 1500 gauge	–	–	–	–	–	m²	6.38
black butyl; 0.75 mm thick	–	–	–	–	–	m²	8.41
black butyl; 1.00 mm thick	–	–	–	–	–	m²	8.82
black butyl; 1.50 mm thick	–	–	–	–	–	m²	30.82
Fine gravel; 100 mm; evenly spread over area of pool; by hand	2.45	0.13	2.31	–	2.45	m²	4.76
Selected topsoil from excavation; 100 mm; evenly spread over area of pool; by hand	–	0.13	2.31	–	–	m²	2.31
Extra over selected topsoil for spreading imported topsoil over area of pool; by hand	26.40	–	–	–	26.40	m³	26.40
Pool surrounds and ornament; **Haddonstone Ltd; Portland Bath or** **terracotta cast stone**							
Pool surrounds; installed to pools or water feature construction priced separately; surrounds and copings to 112.5 mm internal brickwork							
C4HSKVP half small pool surround; internal diameter 1780 mm; kerb features continuous moulding enriched with ovolvo and palmette designs; inclusive of plinth and integral conch shell vases flanked by dolphins	1162.00	16.00	296.00	–	1174.39	nr	1470.39
C4SKVP small pool surround as above but with full circular construction; internal diameter 1780 mm	2324.00	48.00	888.00	–	2338.92	nr	3226.92
C4MKVP medium pool surround; internal diameter 2705 mm; inclusive of plinth and integral vases	3486.00	48.00	888.00	–	3517.18	nr	4405.18
C4XLKVP extra large pool surround; internal diameter 5450 mm	4648.00	140.00	2590.00	–	4684.77	nr	7274.77

S PIPED SUPPLY SYSTEMS

Item Excluding site overheads and profit	PC £	Labour hours	Labour £	Plant £	Material £	Unit	Total rate £
S15 FOUNTAINS/WATER FEATURES – cont							
Pool surrounds and ornament – cont							
Pool centre pieces and fountains; inclusive of plumbing and pumps							
HC350 Lotus Bowl; 1830 mm wide with C1700 triple Dolphin Fountain; HD2900 Doric Pedestal	3582.00	8.00	148.00	–	3584.50	nr	**3732.50**
C251 Gothic Fountain and Gothic Upper Base A350; freestanding fountain	985.00	4.00	74.00	–	987.50	nr	**1061.50**
HC521 Romanesque Fountain; freestanding bowl with self-circulating fountain; filled with cobbles; 815 mm diameter × 348 mm high	453.00	2.00	37.00	–	544.12	nr	**581.12**
C300 Lion Fountain; 610 mm high on fountain base C305; 280 mm high	215.00	2.00	37.00	–	217.50	nr	**254.50**
Wall fountains, watertanks and fountains; inclusive of installation drainage, automatic top up, pump and balancing tank							
Capital Garden Products Ltd							
Lion Wall Fountain F010; 970 × 940 × 520 mm	–	4.00	74.00	–	1710.70	nr	**1784.70**
Dolphin Wall Fountain F001; 740 × 510 mm	–	4.00	74.00	–	1541.75	nr	**1615.75**
Dutch Master Wall Fountain F012; 740 × 410 mm	–	4.00	74.00	–	1568.75	nr	**1642.75**
James II Watertank 2801; 730 × 730 × 760 mm high; 405 litres	–	4.00	74.00	–	1722.80	nr	**1796.80**
James II Fountain 4901 bp; 710 × 1300 × 1440 mm high; 655 litres	–	4.00	74.00	–	1564.75	nr	**1638.75**
Marble wall fountains; Architectural Heritage Ltd; inclusive of installation, drainage, automatic top up, pump and balancing tank							
Breccia Pernice; marble wall fountain supported by two dolphins, 1770 mm high × 1000 mm wide × 640 mm deep	16400.00	4.00	74.00	–	17850.00	nr	**17924.00**
The River God; Verona marble wall mask fountain; 890 mm high × 810 mm wide × 230 mm deep	5400.00	4.00	74.00	–	6850.00	nr	**6924.00**
Fountain kits; typical prices of submersible units comprising fountain pumps, fountain nozzles, underwater spotlights, nozzle extension armatures, underwater terminal boxes and electrical control panels							
Single aerated white foamy water columns; ascending jet 70 mm diameter; descending water up to four times larger; jet height adjustable between 1.00–1.70 m	4095.00	–	–	–	4694.00	nr	**4694.00**
Single aerated white foamy water columns; ascending jet 110 mm diameter; descending water up to four times larger; jet height adjustable between 1.50–3.00 m	6825.00	–	–	–	7749.80	nr	**7749.80**

S PIPED SUPPLY SYSTEMS

Item Excluding site overheads and profit	PC £	Labour hours	Labour £	Plant £	Material £	Unit	Total rate £
SWIMMING POOLS							
Natural swimming pools; Fairwater Water Garden Design Consultants Ltd							
Natural swimming pool of 50 m² area; shingle regeneration zone 50 m² planted with marginal and aquatic plants	–	–	–	–	–	nr	60637.50
Swimming pools; Thompson Landscapes							
Conventional swimming pool construction; reinforced gunnite, heating, lighting, pumping, filtration and mosaic lined (excavation not included)							
rectangular swimming pool construction; 12 × 5 m	–	–	–	–	–	nr	65500.00
hydraulic undercoping safety cover; inclusive of all electrical works	–	–	–	–	–	nr	25300.00
automatic pool cleaning system with main drains	–	–	–	–	–	nr	14950.00

V ELECTRICAL SUPPLY/POWER/LIGHTING SYSTEMS

Item Excluding site overheads and profit	PC £	Labour hours	Labour £	Plant £	Material £	Unit	Total rate £
V ELECTRICAL INSTALLATION							
Trenching for electrical services							
3 tonne excavator (bucket volume 0.13 m³);							
arisings laid alongside							
600 mm deep	–	–	–	1.84	–	m	1.84
800 mm deep	–	–	–	2.15	–	m	2.15
1.00 m deep	–	–	–	2.57	–	m	2.57
Extra over any types of excavating irrespective							
of depth for breaking out existing materials;							
heavy duty 110 volt breaker tool							
hard rock	–	5.00	92.50	19.10	–	m³	111.60
concrete	–	3.00	55.50	11.46	–	m³	66.96
reinforced concrete	–	4.00	74.00	19.50	–	m³	93.50
brickwork, blockwork or stonework	–	1.50	27.75	5.73	–	m³	33.48
By hand							
600 mm deep	–	0.24	4.45	–	–	m	4.45
800 mm deep	–	0.43	7.91	–	–	m	7.91
1.00 m deep	–	0.67	12.33	–	–	m	12.33
Backfilling of trenches; including laying of							
electrical marker tape; compacting lightly							
as work proceeds							
To trenches containing armoured cable							
by machine	0.13	–	–	2.28	0.13	m	2.41
by hand	0.13	0.20	3.70	–	0.13	m	3.83
To trenches containing ducted cable; including							
150 mm sharp sand over the duct							
by machine	1.97	–	–	2.85	1.97	m	4.82
by hand	1.97	0.25	4.63	–	1.97	m	6.60
Cable to trenches (trenching operations							
not included); laying only cable							
Twin core steel wire armoured cable; 50 m							
drums							
1.5 mm core	0.80	–	–	–	0.80	m	0.80
2.5 mm core	0.96	–	–	–	0.96	m	0.96
Twin core steel wire armoured cable; lengths							
less than 50 m runs							
1.5 mm core	0.86	–	–	–	0.86	m	0.86
2.5 mm core	1.04	–	–	–	1.04	m	1.04
Three core steel wire armoured cable; 50 m							
drums							
1.5 mm core	0.90	–	–	–	0.90	m	0.90
2.5 mm core	1.23	–	–	–	1.23	m	1.23
4.0 mm core	1.48	–	–	–	1.48	m	1.48
6.0 mm core	1.92	–	–	–	1.92	m	1.92
10.0 mm core	3.39	–	–	–	3.39	m	3.39
Three core steel wire armoured cable; lengths							
less than 50 m runs							
1.5 mm core	0.93	–	–	–	0.93	m	0.93
2.5 mm core	1.33	–	–	–	1.33	m	1.33

V ELECTRICAL SUPPLY/POWER/LIGHTING SYSTEMS

Item Excluding site overheads and profit	PC £	Labour hours	Labour £	Plant £	Material £	Unit	Total rate £
Ducts to trenches; twin wall flexible cable ducts with drawstrings; laid on 150 mm clean sharp sand							
Twin wall duct; laying to trenches							
63 mm × 50 m coils	1.00	0.02	0.37	–	1.00	m	1.37
110 mm × 50 m coils	1.50	0.02	0.37	–	1.50	m	1.87
Cable drawing through ducts							
straight runs up to 50 m lengths	–	1.00	18.50	–	–	nr	18.50
Terminations to armoured cables; cutting coiling and taping length of cable to receive connection to light fitting, transformer and the like; fixing to temporary post							
Armoured cable	–	0.33	6.16	–	–	m	6.16
Plain ducted cable	–	0.13	2.31	–	–	m	2.31
Junction boxes; fixing to cables to receive connections to light fittings or transformers; to IP68							
Underground jointing box							
2 way	29.00	0.33	6.83	–	29.00	m	35.83
3 way	34.00	0.57	11.71	–	34.00	m	45.71
Underground jointing box							
2 way	3.50	0.33	6.83	–	3.50	m	10.33
3 way	4.90	0.40	8.20	–	4.90	m	13.10
Electrical connections							
Connections to existing distribution boards per switching circuit located within 1 m of the distribuition board; chasing of cables to walls not included	–	–	–	–	–	nr	41.67
Connections to switches; external power cable circuits; connections to internal wall switches; drilling through walls and making good; chasing of cables to walls not included							
price for the first circuit	4.00	1.00	18.50	–	45.67	nr	64.17
additional circuits	4.00	–	–	–	14.00	nr	14.00

V ELECTRICAL SUPPLY/POWER/LIGHTING SYSTEMS

Item Excluding site overheads and profit	PC £	Labour hours	Labour £	Plant £	Material £	Unit	Total rate £
V41 STREET/AREA/FLOOD LIGHTING							
Street area floodlighting – General Preamble: There are an enormous number of luminaires available which are designed for small scale urban and garden projects. The designs are continually changing and the landscape designer is advised to consult the manufacturer's latest catalogue. Most manufacturers supply light fittings suitable for column, bracket, bulkhead, wall or soffit mounting. Highway lamps and columns for trafficked roads are not included in this section as the design of highway lighting is a very specialized subject outside the scope of most landscape contracts. The IP reference number refers to the waterproof properties of the fitting; the higher the number the more waterproof the fitting. Most items can be fitted with time clocks or PIR controls.							
Market prices of lamps							
Lamps							
70 w HQIT-S	–	–	–	–	8.08	nr	8.08
70 w HQIT	–	–	–	–	17.50	nr	17.50
70 w SON	–	–	–	–	6.80	nr	6.80
70 w SONT	–	–	–	–	10.20	nr	10.20
100 w SONT	–	–	–	–	12.05	nr	12.05
150 w SONT	–	–	–	–	14.55	nr	14.55
28 w 2D	–	–	–	–	9.80	nr	9.80
100 w GLS\E27	–	–	–	–	1.90	nr	1.90
Bulkhead and canopy fittings; including fixing to wall and light fitting (lamp, final painting, electric wiring, connections or related fixtures such as switch gear and time clock mechanisms not included unless otherwise indicated)							
Bulkhead and canopy fittings; Targetti Poulsen Nyhavn Wall small domed top conical shade with rings; copper wall lantern; finished untreated copper to achieve verdigris finish; also available in white aluminium; 310 mm diameter shade; with wall mounting arm; to IP 44	627.00	0.50	9.25	–	629.85	nr	639.10
Bulkhead and canopy fittings; Sugg Lighting Princess IP54 backlamp; 375 × 229 mm; copper frame; stove painted in black; with chimney and lampholder	380.00	0.50	9.25	–	382.85	nr	392.10
Victoria IP54 backlamp; 502 × 323 mm; copper frame; polished copper finish; with chimney, door and lampholder	478.00	0.50	9.25	–	480.85	nr	490.10
Palace IP54 backlamp; 457 × 321 mm; copper frame; stove painted black finish; with chimney, door and lampholder	425.00	0.50	9.25	–	427.85	nr	437.10

V ELECTRICAL SUPPLY/POWER/LIGHTING SYSTEMS

Item Excluding site overheads and profit	PC £	Labour hours	Labour £	Plant £	Material £	Unit	Total rate £
Windsor IP54 backlamp; 650 × 306 mm; copper frame; polished and lacquered finish; with door and lampholder	468.00	0.50	9.25	–	470.85	nr	**480.10**
Windsor IP54 gas backlamp; 650 × 306 mm; copper frame; polished and lacquered finish; hinged door; double inverted cluster mantle with permanent pilot and mains solenoid	836.00	0.50	9.25	–	838.85	nr	**848.10**
Floodlighting; ground, wall or pole mounted; including fixing (lamp, final painting, electric wiring, connections or related fixtures such as switch gear and time clock mechanisms not included unless otherwise indicated)							
Floodlighting; Targetti Poulsen – Landscape Division							
SPR-12; multi-purpose ground/spike mounted wide angle floodlight; 70 w HIT or SON; to IP65	319.50	1.00	18.50	–	325.17	nr	**343.67**
Floodlight accessories; Targetti Poulsen – Landscape Division							
earth spike for SPR-12	35.50	–	–	–	35.50	nr	**35.50**
louvre for SPR-12	48.00	–	–	–	48.00	nr	**48.00**
cowl for SPR-12	48.00	–	–	–	48.00	nr	**48.00**
barn doors for SPR-12	112.00	–	–	–	112.00	nr	**112.00**
Large area/pitch floodlighting; CU Phosco							
FL444 1000 w SON-T; floodlights with lamp and loose gear; narrow asymmetric beam	266.27	1.00	20.50	–	266.27	nr	**286.77**
FL444 2.0 kw MBIOS; floodlight with lamp and loose gear; projector beam	508.09	1.00	20.50	–	508.09	nr	**528.59**
Large area floodlighting; CU Phosco							
FL345/G/250S; floodlight with lamp and integral gear	328.70	1.00	20.50	–	328.70	nr	**349.20**
FL345/G/400 MBI; floodlight with lamp and integral gear	253.16	1.00	20.50	–	253.16	nr	**273.66**
Small area floodlighting; Sugg Lighting; floodlight for feature lighting; clear or toughened glass							
Scenario lamp; 150 w HPS-T	650.00	1.00	20.50	–	650.00	nr	**670.50**
Scenario lamp; 150 w HQI-T	650.00	1.00	20.50	–	650.00	nr	**670.50**
Spotlights for uplighting and for illuminating signs and notice boards, statuary and other features) ground, wall or pole mounted; including fixing, light fitting and priming (lamp, final painting, electric wiring, connections or related fixtures such as switch gear and time clock mechanisms not included); all mains voltage (240 v) unless otherwise stated							
Spotlights; Targetti Poulsen – Landscape Division							
WeeBee Spot SP-05 for low voltage halogen reflector; 20/35/50 w; c/w integral transformer and wall mounting box; to IP65	160.00	1.00	18.50	–	165.67	nr	**184.17**

V ELECTRICAL SUPPLY/POWER/LIGHTING SYSTEMS

Item Excluding site overheads and profit	PC £	Labour hours	Labour £	Plant £	Material £	Unit	Total rate £
V41 STREET/AREA/FLOOD LIGHTING – cont							
Spotlights for uplighting and for illuminating signs and notice boards, statuary and other features) ground, wall or pole mounted – cont							
Spotlighters, uplighters and cowl lighting; Havells Sylvania; TECHNO – SHORT ARM; head adjustable 130°; rotation 350°; projection 215 mm on 210 mm base plate; 355 mm high; integral gear; PG16 cable gland; black or aluminium							
Ref S.3517.09/14; 70 w CDMT/HQIT	463.03	1.00	20.50	–	463.03	nr	**483.53**
Ref S.3518.09/14; 150 w CDMT/HQIT	486.72	1.00	20.50	–	486.72	nr	**507.22**
Recessed uplighting; including walk/drive over fully recessed uplighting; excavating, ground fixing, concreting in and making good surfaces (electric wiring, connections or related fixtures such as switch gear and time-clock mechanisms not included unless otherwise stated) (Note: transformers will power multiple lights dependent on the distance between the light units); all mains voltage (240 v) unless otherwise stated							
Recessed uplighting; Targetti Poulsen – Landscape Division; 266 mm diameter; diecast aluminium with stainless steel top plate 10 mm toughened safety glass; 2000 kg drive over; integral control gear; to IP67							
IPR-14 HIT metal halide; white light; spot or flood or wall wash distribution	275.00	1.50	27.75	–	280.67	nr	**308.42**

IRR-14 HIT compact fluorescent; white
    light; low power consumption; flood
    distribution
	275.00	1.50	27.75	–	280.67	nr	**308.42**

Accessories for IPR-14 uplighters; Targetti Poulsen – Landscape Division							
rockguard	154.50	–	–	–	154.50	nr	**154.50**
stainless steel installation sleeve	49.50	–	–	–	49.50	nr	**49.50**
anti glare louvre (internal tilt)	98.50	–	–	–	98.50	nr	**98.50**
Accessories for Nimbus uplighters; Targetti Poulsen – Landscape Division							
IS installation sleeve	57.00	–	–	–	57.00	nr	**57.00**
Wall recessed; Havells Sylvania; EOS range; integral gear; asymmetric reflector; toughened reeded glass							
MINI EOS; 145 × 90 mm; black or aluminium; 20 W QT9 lamp	63.87	1.25	25.63	–	63.87	nr	**89.50**
RECTANGULAR EOS; 270 × 145 mm; black or aluminium; 20 W QT9 lamp	123.24	1.25	25.63	–	139.34	nr	**164.97**

V ELECTRICAL SUPPLY/POWER/LIGHTING SYSTEMS

Item Excluding site overheads and profit	PC £	Labour hours	Labour £	Plant £	Material £	Unit	Total rate £
Underwater lighting							
Lighted bollards; including excavating,							
ground fixing, concreting in and making							
good surfaces (lamp, final painting, electric							
wiring, connections or related fixtures such							
as switch gear and time-clock mechanisms							
not included) (Note: all illuminated bollards							
must be earthed); heights given are from							
ground level to top of bollards							
Lighted bollards; Targetti Poulsen							
Orbiter; vandal-resistant; head of cast							
aluminium; domed top; anti-glare rings;							
pole extruded aluminium; diffuser clear UV							
stabilized polycarbonate; powder coated;							
1040 mm high × 255 mm diameter; with							
root or base plate; IP44	654.00	2.50	46.25	–	659.72	nr	**705.97**
Waterfront; solidly proportioned; head of							
cast silumin; domed top; symmetrical							
distribution; pole extruded aluminium							
sandblasted or painted white; internal							
diffuser clear UV stabilized polycarbonate;							
865 mm high × diameter 260 mm; IP55	736.00	2.50	46.25	–	741.72	nr	**787.97**
Bysted; concentric louvred bollard; head							
cast iron; post COR-TEN steel; externally							
untreated to provide natural aging effect of							
uniform oxidized red surface finish; internal							
painted white; lamp diffuser rings of clear							
polycarbonate; 1130 mm high × 280 mm							
diameter; to IP44	1133.00	2.50	46.25	–	1138.72	nr	**1184.97**
Lighted bollards; Woodscape Ltd							
Illuminated bollard; in 'very durable							
hardwood'; integral die cast aluminium							
lighting unit; 165 × 165 mm × 1.00 m high	246.30	2.50	46.25	–	257.48	nr	**303.73**
Reproduction street lanterns and columns;							
including excavating, concreting in,							
backfilling and disposing of spoil, or fixing							
to ground or wall, making good surfaces,							
light fitting and priming (final painting,							
electric wiring, connections or related							
fixtures such as switch gear and time-clock							
mechanisms not included) (Note: lanterns							
up to 14 inches and columns up to 7 ft are							
suitable for residential lighting)							
Reproduction street lanterns; Sugg							
Lighting; reproduction lanterns handmade							
to original designs; all with ES lamp holder							
Westminster IP54 hexagonal hinged top							
lantern with door; two piece folded							
polycarbonate glazing; lamp 100 w HQI T;							
integral photo electric cell							
small; 1016 mm high × 356 mm wide	815.00	0.50	9.25	–	815.00	nr	**824.25**
large; 1124 mm high × 760 mm wide	1280.00	0.50	9.25	–	1280.00	nr	**1289.25**

V ELECTRICAL SUPPLY/POWER/LIGHTING SYSTEMS

Item Excluding site overheads and profit	PC £	Labour hours	Labour £	Plant £	Material £	Unit	Total rate £
V41 STREET/AREA/FLOOD LIGHTING – cont							
Reproduction street lanterns – cont							
Guildhall IP54; handcrafted copper frame;							
clear polycarbonate glazing circular tapered							
lantern with hemispherical top; integral photo							
electric cell; with door							
small; 711 mm high × 330 mm wide	697.00	0.50	9.25	–	699.85	nr	709.10
medium; 1150 mm high × 432 mm wide	735.00	0.50	9.25	–	737.85	nr	747.10
large; 1370 mm high × 550 mm wide	1020.00	0.50	9.25	–	1022.85	nr	1032.10
Grosvenor circular lantern with door; copper							
frame; polished copper finish; polycarbonate							
glazing							
small; 790 × 330 mm; IP54	710.00	0.50	9.25	–	712.85	nr	722.10
medium; 1080 × 435 mm; IP65	735.00	0.50	9.25	–	737.85	nr	747.10
Classic Globe IP54 lantern with hinged outer							
frame; cast aluminium frame; black polyester							
powder coating							
medium; 965 × 483 mm	650.00	0.50	9.25	–	652.85	nr	662.10
Windsor lantern; copper frame; polished							
copper finish							
small; 905 × 356 mm; IP54; with door	568.00	0.50	9.25	–	570.85	nr	580.10
small; 905 × 356 mm; IP65; without door	530.00	0.50	9.25	–	532.85	nr	542.10
medium; 1124 × 420 mm; IP54; with door	620.00	0.50	9.25	–	622.85	nr	632.10
large; 1124 × 470 mm; IP54; with door	720.00	0.50	9.25	–	722.85	nr	732.10
medium; gas lantern; 1124 × 420 mm;							
IP54; with door; integral solenoid and pilot	1025.00	0.50	9.25	–	1027.85	nr	1037.10
Reproduction brackets and suspensions;							
Sugg Lighting							
Iron brackets							
Bow bracket; 6'0"	996.80	2.00	37.00	–	999.65	nr	1036.65
Ornate iron bracket; large	445.00	2.00	37.00	–	447.85	nr	484.85
Ornate iron bracket; medium	430.00	2.00	37.00	–	432.85	nr	469.85
Swan neck iron bracket; large	420.00	2.00	37.00	–	422.85	nr	459.85
Swan neck iron bracket; medium	238.00	2.00	37.00	–	240.85	nr	277.85
Cast brackets							
Universal cast bracket	318.00	2.00	37.00	–	320.85	nr	357.85
Abbey bracket	188.00	2.00	37.00	–	190.85	nr	227.85
Short Abbey bracket	186.00	2.00	37.00	–	188.85	nr	225.85
Plaza cast bracket	155.50	2.00	37.00	–	158.35	nr	195.35
Base mountings							
Universal pedestal	316.00	2.00	37.00	–	318.85	nr	355.85
Universal plinth	180.00	2.00	37.00	–	182.85	nr	219.85
Reproduction lighting columns							
Reproduction lighting columns; Sugg Lighting							
Harborne C11; fabricated iron/steel heavy							
duty post with integral cast root; 3–5 m	1380.00	8.00	148.00	–	1387.15	nr	1535.15
Aylesbury C12; base fabricated heavy							
gauge 89 mm aluminium post; 3–5 m	1968.00	8.00	148.00	–	1975.15	nr	2123.15
Cannonbury C13; fabricated heavy gauge							
89 mm aluminium post; 3–5 m	1450.00	8.00	148.00	–	1457.15	nr	1605.15

V ELECTRICAL SUPPLY/POWER/LIGHTING SYSTEMS

Item Excluding site overheads and profit	PC £	Labour hours	Labour £	Plant £	Material £	Unit	Total rate £
standard column C14; rooted British Steel 168/89 mm embellished post; 5–8 m	653.12	8.00	148.00	–	660.27	nr	**808.27**
Large Constitution Hill C22X; cast aluminium, steel cored post with extended spigot; 4.3 m	5230.00	6.00	111.00	–	5237.15	nr	**5348.15**
Cardiff C29; cast aluminium post; 3.5 m	2567.00	6.00	111.00	–	2574.15	nr	**2685.15**
Seven Dials C36; cast aluminium rooted post; 3.8 m	1896.00	8.00	148.00	–	1903.15	nr	**2051.15**
Royal Exchange C42; traditional cast aluminium post; welded multi-arm construction; 2.1 m	2896.00	6.00	111.00	–	2903.15	nr	**3014.15**
Precinct lighting lanterns; ground, wall or pole mounted; including fixing and light fitting (lamps, poles, brackets, final painting, electric wiring, connections or related fixtures such as switch gear and time clock mechanisms not included)							
Precinct lighting lanterns; Targetti Poulsen Nyhavn Park side entry 90° curved arm mounted; steel rings over domed top; housing cast silumin sandblasted with integral gear; protected by UV stabilized clear polycarbonate diffuser; IP55	650.00	1.00	18.50	–	650.00	nr	**668.50**
Kipp; bottom entry pole-top; shot blasted or lacquered Hanover design award; hinged diffuser for simple maintenance; indirect lighting technique ensures low glare; IP55	638.00	1.00	18.50	–	638.00	nr	**656.50**
Precinct lighting lanterns; Sugg Lighting Juno Dome; IP66; molded GRP body; graphite finish	560.00	1.00	18.50	–	560.00	nr	**578.50**
Juno Cone; IP66; molded GRP body; graphite finish	560.00	1.00	18.50	–	560.00	nr	**578.50**
Sharkon; IP65; cast aluminium body; silver and blue finish	560.00	1.00	18.50	–	560.00	nr	**578.50**

Spon's First Stage Estimating Handbook

3rd Edition

By **Bryan Spain**

Have you ever had to provide accurate costs for a new supermarket or a pub "just an idea…a ballpark figure…" ?

The earlier a pricing decision has to be made, the more difficult it is to estimate the cost and the more likely the design and the specs are to change. And yet a rough-and-ready estimate is more likely to get set in stone.

Spon's First Stage Estimating Handbook is the only comprehensive and reliable source of first stage estimating costs. Covering the whole spectrum of building costs and a wide range of related M&E work and landscaping work, vital cost data is presented as:

- Costs per square metre
- Principal rates
- Elemental cost analyses
- Composite rates

Compact and clear, *Spon's First Stage Estimating Handbook* is ideal for those key early meetings with clients. And with additional sections on whole life costing and general information, this is an essential reference for all construction professionals and clients making early judgements on the viability of new projects.

January 2010: 216x138: 244pp
Pb: 978-0-415-54715-4: **£45.00**

To Order: Tel: +44 (0) 1235 400524 **Fax:** +44 (0) 1235 400525
or Post: Taylor and Francis Customer Services,
Bookpoint Ltd, Unit T1, 200 Milton Park, Abingdon, Oxon, OX14 4TA UK
Email: book.orders@tandf.co.uk

For a complete listing of all our titles visit:
www.tandf.co.uk

PART 4

Approximate Estimating Rates – Minor Works

APPROXIMATE ESTIMATES

Prices in this section are based upon the Prices for Measured Works, but allow for incidentals which would normally be measured separately in a Bill of Quantities. They do not include for Preliminaries which are priced elsewhere in this book.

Items shown as subcontract or specialist rates would normally include the specialist's overhead and profit. All other items which could fall within the scope of works of general landscape and external works contractors would not include profit.

Based on current commercial rates, profits of 15% to 35% may be added to these rates to indicate the likely 'with profit' values of the tasks below. The variation quoted above is dependent on the sector in which the works are taking place – domestic, public or commercial.

DEMOLITION AND SITE CLEARANCE

Item Excluding site overheads and profit	Unit	Total rate £
DEMOLITION AND SITE CLEARANCE		
Demolish existing surfaces by hand held electric breaker; removal by grab		
Break up plain concrete slab; remove to licensed tip;		
150 mm thick	m²	11.10
200 mm thick	m²	13.40
Break up reinforced concrete slab and remove to licensed tip		
150 mm thick	m²	23.00
200 mm thick	m²	29.50
300 mm thick	m²	56.00
Break out existing surface and associated 150 mm thick granular base load to remove off site by grab		
macadam 70 mm thick	m²	11.80
block paving 50 mm thick	m²	17.30
block paving 80 mm thick	m²	19.50
Demolish existing free standing walls; grub out foundations; remove arisings to skip; backfill with imported topsoil; works by excavator, dumper and diesel breaker		
Brick wall; 112 mm thick		
300 mm high	m	23.00
500 mm high	m	31.00
Brick wall; 225 mm thick		
300 mm high	m	32.00
500 mm high	m	34.00
1.00 m high	m	42.50
1.20 m high	m	49.50
1.50 m high	m	52.00
1.80 m high	m	60.00
Demolish existing free standing walls; grub out foundations; remove arisings to skip; backfill with imported topsoil; works by hand and diesel breaker		
Brick wall; 112 mm thick		
300 mm high	m	24.50
500 mm high	m	27.00
Brick wall; 225 mm thick		
300 mm high	m	34.50
500 mm high	m	37.50
1.00 m high	m	49.00
1.20 m high	m	59.00
1.50 m high	m	62.00
1.80 m high	m	71.00
Break out existing free standing building; break out plain concrete base 150 mm thick and remove to skip distance 50 m; backfill with imported topsoil; all works by hand		
Timber buildings		
shed 6.0 m²	nr	310.00
shed 10.0 m²	nr	680.00
shed 15.0 m²	nr	950.00

DEMOLITION AND SITE CLEARANCE

Item Excluding site overheads and profit	Unit	Total rate £
Site clearance – General		
Clear away light fencing and gates (chain link, chestnut paling, light boarded fence or similar) and remove to licensed tip	100 m	280.00
Strip turf; strip topsoil 250 mm thick move to stockpile 25 m		
all by machine; disposal of turf of to skip	100 m	710.00
by machine; disposal of turf by grab	100 m	650.00
strip and stack turf for preservation by hand; strip soil by machine	100 m	890.00
by hand; disposal to skip	100 m	2175.00
Clear mixed shrub area; dig out roots		
Groundcovers and small shrubs 20%; shrubs 1.0–2.0 m high 40%; shrubs 2.0-3.0 m high 20%; shrubs over 3.0 m high 20%		
clearance only	m²	30.50
disposal to skip; chipped	m²	17.70
disposal to skip; unchipped	m²	34.50
disposal on site; unchipped	m²	4.70

GROUNDWORK

Item Excluding site overheads and profit	Unit	Total rate £
EXCAVATION AND FILLING		
Cut and strip by machine turves 50 mm thick		
Load to barrows and stack on site not exceeding 25 m travel to stack	100 m²	510.00
Load to barrows and disposal off site by skip; distance 25 m	100 m²	1125.00
Excavate to reduce levels; mechanical		
Removal to spoil heaps		
excavated directly to loading position	m³	7.70
transporting to loading position 25 m distance	m³	13.30
Excavate to reduce levels; mechanical		
Removal off site by grab		
excavated directly to loading position	m³	34.00
transporting to loading position 25 m distance	m³	40.00
Excavate to reduce levels; mechanical		
Removal off site by skip		
excavated directly to loading position	m³	55.00
transporting to loading position 25 m distance	m³	60.00
Excavate to reduce levels; hand		
Removal off site by skip		
excavated directly to loading position	m³	175.00
excavated directly to loading position	m³	225.00
filled to bags and transporting to loading position 25 m distant	m³	350.00
Excavation and filling; mechanical		
Excavate existing soil on proposed turf or planting area to reduce levels; grade to levels; fill excavated area with topsoil from spoil heaps; removal of and excavated material by skip		
100 mm deep	m²	7.05
200 mm deep	m²	14.10
300 mm deep	m²	21.00
Spread excavated material to levels in layers not exceeding 150 mm; grade to finished levels to receive surface treatments		
By machine		
average thickness 100 mm	m²	1.10
average thickness 100 mm but with imported topsoil	m²	4.65
average thickness 200 mm	m²	1.65
average thickness 200 mm but with imported topsoil	m²	8.80
average thickness 250 mm	m²	1.95
average thickness 250 mm but with imported topsoil	m²	10.80
extra for work to banks exceeding 30° slope	30%	–
Filling by hand		
Excavate existing soil on proposed turf or planting area to reduce levels; grade to levels; fill excavated area with topsoil from spoil heaps; removal of excavated material by skip		
100 mm deep	m²	8.75
200 mm deep	m²	17.50
300 mm deep	m²	26.00
500 mm deep	m²	43.50

GROUNDWORK

Item Excluding site overheads and profit	Unit	Total rate £
TRENCHES		
Excavate trenches; remove excavated material off site by grab 25 m distance; fill trench to ground level with site mixed concrete 1:3:6; allow for movement of concrete and excavated material		
By machine		
300 mm wide × 250 mm deep	m	13.80
500 mm wide × 250 mm deep	m	20.00
750 mm wide × 350 mm deep	m	40.00
1200 mm wide × 600 mm deep	m	110.00
By hand		
300 mm wide × 250 mm deep	m	23.00
500 mm wide × 250 mm deep	m	36.50
750 mm wide × 350 mm deep	m	81.00
1200 mm wide × 600 mm deep	m	220.00

IN SITU CONCRETE

Item Excluding site overheads and profit	Unit	Total rate £
IN SITU CONCRETE		
Mix concrete on site; aggregates delivered in 10 tonne loads; deliver mixed concrete to location by mechanical dumper distance 25 m		
1:3:6	m³	110.00
1:2:4	m³	120.00
As above but ready mixed concrete		
10 N/mm²	m³	100.00
15 N/mm²	m³	100.00
Mix concrete on site; aggregates delivered in 10 tonne loads; deliver mixed concrete to location by barrow distance 25 m		
1:3:6	m³	150.00
1:2:4	m³	165.00
As above but aggregates delivered in 850 kg bulk bags		
1:3:6	m³	195.00
1:2:4	m³	210.00
As above but concrete discharged directly from ready mix lorry to required location		
10 N/mm²	m³	95.00
15 N/mm²	m³	96.00
Excavate foundation trench mechanically; remove spoil off site by grab; lay 1:3:6 site mixed concrete foundations; distance from mixer 25 m; depth of trench to be 225 mm deeper than foundation to allow for three underground brick courses priced separately		
Foundation size		
200 mm deep × 400 mm wide	m	16.90
300 mm deep × 500 mm wide	m	29.00
400 mm deep × 400 mm wide	m	30.00
400 mm deep × 600 mm wide	m	45.00
600 mm deep × 600 mm wide	m	64.00
As above but hand excavation and disposal to spoil heap 25 m by barrow; disposal off site by grab		
200 mm deep × 400 mm wide	m	39.00
300 mm deep × 500 mm wide	m	46.00
400 mm deep × 400 mm wide	m	44.00
400 mm deep × 600 mm wide	m	94.00
600 mm deep × 600 mm wide	m	130.00

BRICK/BLOCK WALLING

Item Excluding site overheads and profit	Unit	Total rate £
BRICK WALLING		
Excavate foundation trench 500 mm deep; remove spoil to dump off site; (all by machine) lay site mixed concrete foundations 1:3:6 350 × 150 mm thick; construct half brick wall with one brick piers at 2.0 m centres; laid in cement:lime:sand (1:1:6) mortar with flush joints; fair face one side; DPC two courses underground; engineering brick in cement:sand (1:3) mortar; coping of headers on end		
Wall 900 mm high above DPC		
in engineering brick (class B) PC £290.00/1000	m	230.00
in sandfaced facings PC £300.00/1000	m	240.00
in reclaimed bricks PC £1,000.00/1000	m	375.00
Excavate foundation trench 400 mm deep; remove spoil to dump off site; lay GEN 1 concrete foundations 450 mm wide × 250 mm thick; construct one brick wall with one and a half brick piers at 3.0 m centres; all in English Garden Wall bond; laid in cement:lime:sand (1:1:6) mortar with flush joints, fair face one side; DPC two courses engineering brick in cement:sand (1:3) mortar; engineering brick coping		
Wall 900 mm high above DPC		
in engineering brick (class B) PC £290.00/1000	m	355.00
in sandfaced facings PC £300.00/1000	m	360.00
in reclaimed bricks PC £1,000.00/1000	m	440.00
Wall 1200 mm high above DPC		
in engineering brick (class B) PC £290.00/1000	m	415.00
in sandfaced facings PC £300.00/1000	m	415.00
in reclaimed bricks PC £1,000.00/1000	m	520.00
Wall 1800 mm high above DPC		
in engineering brick (class B) PC £290.00/1000	m	610.00
in sandfaced facings PC £300.00/1000	m	620.00
in reclaimed bricks PC £1,000.00/1000	m	770.00
BLOCK WALLING		
Excavate foundation trench 450 mm deep; remove spoil to dump off site; lay GEN 1 concrete foundations 600 × 300 mm thick; construct wall of concrete block; two courses below ground		
Solid blocks 7 N/mm^2; wall 1.00 m high		
100 mm thick	m	94.00
Hollow blocks filled with concrete; wall 1.00 m high		
215 mm thick	m	150.00
Hollow blocks but with steel bar cast into the foundation; wall 1.00 m high		
215 mm thick	m	150.00
Solid blocks 7 N/mm^2; wall 1.80 m high		
100 mm thick	m	140.00
Hollow blocks filled with concrete; wall 1.80 m high		
215 mm thick	m	235.00
Hollow blocks but with steel bar cast into the foundation wall 1.80 m high		
215 mm thick	m	235.00

ROADS AND PAVINGS

Item Excluding site overheads and profit	Unit	Total rate £
BASES FOR PAVING		
Excavate ground and reduce levels to receive 38 mm thick slab and 25 mm mortar bed; dispose of excavated material off site; treat substrate with total herbicide		
Lay granular fill Type 1 150 mm thick laid to falls and compacted		
all by machine	m²	17.30
all by hand except disposal by grab	m²	34.00
Excavate ground and reduce levels to receive 65 mm thick surface and bed (not included); dispose of excavated material off site; treat substrate with total herbicide		
Lay 100 mm compacted hardcore; lay 1:2:4 concrete base 150 mm thick laid to falls		
all by machine	m²	38.00
all by hand except disposal by grab	m²	79.00
As above but inclusive of reinforcement fabric A142		
all by machine	m²	43.50
all by hand except disposal by grab	m²	85.00
Lay 100 mm compacted hardcore; lay 1:3:6 concrete base 150 mm thick laid to falls		
all by machine	m²	31.00
all by hand except disposal by grab	m²	73.00
As above but inclusive of reinforcement fabric A142		
all by machine	m²	37.00
all by hand except disposal by grab	m²	78.00
KERBS AND EDGINGS		
Note: excavation is by machine unless otherwise mentioned		
Excavate trench and construct concrete foundation 150 mm wide × 150 mm deep; lay precast concrete kerb units bedded in semi-dry concrete; slump 35 mm maximum; haunching one side; disposal of arisings off site		
Edgings laid straight		
50 × 150 mm	m	29.00
125 mm high × 150 mm wide; bullnosed	m	29.00
50 × 200 mm	m	30.00
Second hand granite setts		
100 × 100 mm	m	61.00
Single course; brick or block edgings laid to stretcher		
concrete blocks 200 × 100 mm	m	23.00
engineering bricks	m	29.00
paving bricks; PC £500.00/1000	m	29.50
Double course; brick or block edgings laid to stretcher		
concrete blocks 200 × 100 mm	m	26.00
engineering bricks	m	34.00
paving bricks; PC £500.00/1000	m	36.00
Single course brick or block edgings laid to header course (soldier course)		
blocks 200 × 100 × 60 mm; PC £8.03/m²; butt jointed	m	26.00
bricks 200 × 100 × 50 mm; PC £300.00/1000; butt jointed	m	29.00
bricks 200 × 100 × 50 mm; PC £300.00/1000; mortar joints	m	33.50

ROADS AND PAVINGS

Item Excluding site overheads and profit	Unit	Total rate £
Sawn yorkstone edgings; excavate for groundbeam; lay concrete 1:2:4 150mm deep × 33.3% wider than the edging; on 35mm thick mortar bed; inclusive of haunching one side Yorkstone 50mm thick		
100mm wide × random lengths	m	**34.50**
100 × 100mm	m	**38.50**
200mm wide × 100mm long	m	**47.00**
250mm wide × random lengths	m	**45.00**
500mm wide × random lengths	m	**79.00**
Timber edgings; softwood 150 × 38mm		
laid straight	m	**4.60**
laid to curves	m	**6.65**
INTERLOCKING BLOCK PAVING		
Excavate ground; supply and lay granular fill Type 1 150mm thick laid to falls and compacted; supply and lay block pavers; laid on 50mm compacted sharp sand; vibrated; joints filled with loose sand excluding edgings or kerbs measured separately Concrete blocks		
200 × 100 × 60mm	m²	**59.00**
200 × 100 × 80mm	m²	**61.00**
Reduce levels; lay 150mm granular material Type 1; lay precast concrete edging 50 × 150mm; on concrete foundation 1:2:4; lay 200 × 100 × 60mm vehicular block paving to 90° herringbone pattern; on 50mm compacted sand bed; vibrated; jointed in sand and vibrated		
1.0m wide clear width between edgings	m	**97.00**
1.5m wide clear width between edgings	m	**125.00**
2.0m wide clear width between edgings	m	**155.00**
3.0m wide clear width between edgings	m	**215.00**
Works by machine; reduce levels; lay 150mm granular material Type 1; lay edge restraint of block paving 200mm wide on 150mm thick concrete foundation 1:2:4 haunched; lay 200 × 100 × 60mm vehicular block paving to 90° herringbone pattern; on 50mm compacted sand bed; vibrated; jointed in sand and vibrated		
1.0m wide clear width between edgings	m	**91.00**
1.5m wide clear width between edgings	m	**120.00**
2.0m wide clear width between edgings	m	**150.00**
3.0m wide clear width between edgings	m	**205.00**

ROADS AND PAVINGS

Item Excluding site overheads and profit	Unit	Total rate £
INTERLOCKING BLOCK PAVING – cont		
Works by hand; reduce levels; lay 150 mm granular material Type 1; lay edge restraint of block paving 200 mm wide on 150 mm thick concrete foundation 1:2:4 haunched; lay 200 × 100 × 60 mm vehicular block paving to 90° herringbone pattern; on 50 mm compacted sand bed; vibrated; jointed in sand and vibrated		
1.0 m wide clear width between edgings	m	**110.00**
1.5 m wide clear width between edgings	m	**140.00**
2.0 m wide clear width between edgings	m	**180.00**
3.0 m wide clear width between edgings	m	**250.00**
BRICK PAVING		
WORKS BY MACHINE		
Excavate and lay base Type 1 150 mm thick remove arisings; all by machine; lay clay brick paving		
200 × 100 × 50 mm thick; butt jointed on 50 mm sharp sand bed		
PC £300.00/1000	m²	**69.00**
PC £600.00/1000	m²	**85.00**
200 × 100 × 50 mm thick; 10 mm mortar joints on 35 mm mortar bed		
PC £300.00/1000	m²	**84.00**
PC £600.00/1000	m²	**97.00**
Excavate and lay 150 mm site mixed concrete base 1:3:6 reinforced with A142 mesh; all by machine; remove arisings; lay clay brick paving		
200 × 100 × 50 mm thick; 10 mm mortar joints on 35 mm mortar bed; running or stretcher bond		
PC £300.00/1000	m²	**110.00**
PC £600.00/1000	m²	**120.00**
200 × 100 × 50 mm thick; 10 mm mortar joints on 35 mm mortar bed; butt jointed; herringbone bond		
PC £300.00/1000	m²	**92.00**
PC £600.00/1000	m²	**110.00**
Excavate and lay readymix concrete base 150 mm thick reinforced with A393 mesh; all by machine; remove arisings; lay clay brick paving		
215 × 102.5 × 50 mm thick; 10 mm mortar joints on 35 mm mortar bed		
PC £300.00/1000	m²	**80.00**
PC £600.00/1000	m²	**96.00**
WORKS BY HAND		
Excavate and lay base Type 1 150 mm thick by hand; arisings barrowed to spoil heap maximum distance 25 m and removal off site by grab; lay clay brick paving		
200 × 100 × 50 mm thick; butt jointed on 50 mm sharp sand bed		
PC £300.00/1000	m²	**82.00**
PC £600.00/1000	m²	**97.00**
Excavate and lay 150 mm concrete base; 1:3:6: site mixed concrete reinforced with A142 mesh; remove arisings to stockpile and then off site by grab; lay clay brick paving		
215 × 102.5 × 50 mm thick; 10 mm mortar joints on 35 mm mortar bed		
PC £300.00/1000	m²	**100.00**
PC £600.00/1000	m²	**115.00**

ROADS AND PAVINGS

Item Excluding site overheads and profit	Unit	Total rate £
NATURAL STONE/SLAB PAVING		
WORKS BY MACHINE		
Yorkstone slabs; excavate ground by machine and reduce levels to receive 65 mm thick slab and 35 mm mortar bed; dispose of excavated material off site; treat substrate with total herbicide; lay granular fill Type 1 150 mm thick laid to falls and compacted; lay to random rectangular pattern on 35 mm mortar bed; Yorkstone		
New riven slabs		
laid random rectangular	m²	150.00
New riven slabs; but to 150 mm plain concrete base		
laid random rectangular	m²	160.00
Reclaimed Cathedral grade riven slabs		
laid random rectangular	m²	190.00
Reclaimed Cathedral grade riven slabs; but to 150 mm plain concrete base		
laid random rectangular	m²	200.00
New slabs sawn 6 sides		
laid random rectangular	m²	120.00
3 sizes; laid to coursed pattern	m²	130.00
New slabs sawn 6 sides; but to 150 mm plain concrete base		
laid random rectangular	m²	135.00
3 sizes; laid to coursed pattern	m²	140.00
Indian sandstone slabs; excavate ground by machine and reduce levels to receive 65 mm thick slab and 35 mm mortar bed; dispose of excavated material off site; treat substrate with total herbicide; lay granular fill Type 1 150 mm thick laid to falls and compacted; lay to random rectangular pattern on 35 mm mortar bed		
New riven slabs		
laid random rectangular	m²	98.00
WORKS BY HAND		
Yorkstone slabs; excavate ground by hand and reduce levels to receive 65 mm thick slab and 35 mm mortar bed; barrow all materials and arisings 25 m; dispose of excavated material off site by grab; treat substrate with total herbicide; lay granular fill Type 1 150 mm thick laid to falls and compacted; lay to random rectangular pattern on 35 mm mortar bed		
New riven slabs		
laid random rectangular	m²	160.00
New riven slabs laid random rectangular; but to 150 mm plain concrete base		
laid random rectangular	m²	180.00
Reclaimed Cathedral grade riven slabs		
laid random rectangular	m²	205.00
Reclaimed Cathedral grade riven slabs; but to 150 mm plain concrete base		
laid random rectangular	m²	220.00
New slabs sawn 6 sides		
laid random rectangular	m²	135.00
3 sizes; laid to coursed pattern	m²	140.00

ROADS AND PAVINGS

Item Excluding site overheads and profit	Unit	Total rate £
NATURAL STONE/SLAB PAVING – cont		
Yorkstone slabs – cont		
New slabs sawn 6 sides; but to 150 mm plain concrete base		
laid random rectangular	m²	**150.00**
3 sizes; laid to coursed pattern	m²	**150.00**
Indian sandstone slabs; excavate ground by hand and reduce levels to receive 65 mm thick slab and 35 mm mortar bed; barrow all materials and arisings 25 m; dispose of excavated material off site by grab; treat substrate with total herbicide; lay granular fill Type 1 150 mm thick laid to falls and compacted; lay to random rectangular pattern on 35 mm mortar bed		
New riven slabs		
laid random rectangular	m²	**115.00**

PREPARATION FOR SEEDING/TURFING

em xcluding site overheads and profit	Unit	Total rate £
URFACE PREPARATIONS		
ultivate existing ground by pedestrian operated rotavator; spread and lightly consolidate		
ppsoil brought from spoil heap 25 m distance in layers not exceeding 150 mm; grade		
o specified levels; remove stones over 25 mm; rake and grade to a fine tilth		
y machine		
100 mm thick	100 m²	160.00
200 mm thick	100 m²	255.00
300 mm thick	100 m²	350.00
500 mm thick	100 m²	540.00
y hand		
100 mm thick	100 m²	480.00
150 mm thick	100 m²	690.00
300 mm thick	100 m²	1275.00
500 mm thick	100 m²	2350.00
ift and remove existing turf to skip 25 m distance; cultivate surface to receive new		
urf; rake to a fine tilth		
Works by hand		
normal turfed area	100 m²	600.00
compacted turfed area	100 m²	610.00
ift and remove existing turf to skip 25 m distance; cultivate surface to receive new		
urf; rake to a fine tilth; fill area to receive turf with 50 mm imported topsoil		
Works by hand		
normal turfed area	100 m²	840.00
compacted turfed area	100 m²	850.00
EEDING AND TURFING		
Domestic lawn areas		
ultivate recently filled topsoil area; grade to levels and falls and rake to remove		
tones and debris; add fertilizers; add surface treatment as specified		
urf areas		
Rolawn medallion	m²	5.70
eeded areas		
grass seed PC £4.50/kg; application rate 35 g/m²	m²	1.15
grass seed PC £4.50/kg; application rate 50 g/m²	m²	1.20
grass seed PC £6.00/kg; application rate 35 g/m²	m²	1.20
grass seed PC £6.00/kg; application rate 50 g/m²	m²	1.35
ultivate recently filled topsoil area; grade to levels and falls and rake to remove		
tones and debris; add fertilizers; add surface treatment as specified; maintain for		
one year watering and cutting 26 times during the summer; pedestrian mower with		
grass box; arisings removed off site		
urf areas		
Rolawn medallion	m²	7.90
eeded areas		
grass seed PC £4.50/kg; application rate 35 g/m²	m²	3.35
grass seed PC £4.50/kg; application rate 50 g/m²	m²	3.40
grass seed PC £6.00/kg; application rate 35 g/m²	m²	3.45
grass seed PC £6.00/kg; application rate 50 g/m²	m²	3.55

PLANTING

Item Excluding site overheads and profit	Unit	Total rate £
TREE PLANTING		
Excavate tree pit by hand; fork over bottom of pit; plant tree with roots well spread out; backfill with excavated material, incorporating treeplanting compost at 1 m³ per 3 m³ of soil, one tree stake and two ties; tree pits square in sizes shown		
Light standard bare root tree in pit; PC £9.45		
600 × 600 mm deep	each	37.50
900 × 900 mm deep	each	49.00
Standard tree bare root tree in pit; PC £11.81		
600 × 600 mm deep	each	42.00
900 × 600 mm deep	each	53.00
Standard root balled tree in pit; PC £19.69		
600 × 600 mm deep	each	51.00
900 × 600 mm deep	each	67.00
Selected standard bare root tree in pit; PC £19.16		
900 × 900 mm deep	each	69.00
1.00 × 1.00 m × 600 mm deep	each	96.00
Selected standard root ball tree in pit; PC £29.66		
900 × 600 mm deep	each	79.00
1.00 × 1.00 m × 600 mm deep	each	105.00
Heavy standard bare root tree in pit; PC £35.17		
900 × 900 mm deep	each	98.00
1.00 × 1.00 m × 600 mm deep	each	105.00
Heavy standard root ball tree in pit; PC £50.92		
900 × 600 mm deep	each	110.00
1.00 × 1.00 m × 600 mm deep	each	140.00
Extra heavy standard bare root tree in pit; PC £44.10		
1.00 × 1.00 m deep	each	140.00
Extra heavy standard root ball tree in pit; PC £59.85		
1.00 × 1.00 m deep	each	160.00
1.50 m × 750 mm deep	each	190.00
SHRUB PLANTING		
Treat recently filled ground with systemic weedkiller; cultivate ground and clear arisings; add mushroom composts and fertilizers		
Cultivation by rotavator; all other works by hand		
compost 50 mm; general purpose fertilizer 35 g/m²	m²	4.20
compost 100 mm; general purpose fertilizer 35 g/m²	m²	6.95
compost 100 mm; Enmag 35 g/m²	m²	7.00
All works by hand		
compacted ground; compost 50 mm; general purpose fertilizer 35 g/m²	m²	4.90
ground previously planted but cleared of vegetable matter; compost 50 mm; general purpose fertilizer 35 g/m²	m²	4.75
ground previously planted but cleared of vegetable matter; compost 50 mm; general purpose fertilizer 35 g/m²	m²	4.60

PLANTING

Item excluding site overheads and profit	Unit	Total rate £
Excavate planting holes on 300 × 300 × 300 mm deep to area previously prepared; plant shrubs PC £3.00 each in groups of 3 to 5 inclusive of transport from holding area setting out and final mulching 50 mm thick		
by hand		
300 mm centres (11.11 plants per m²)	m²	65.00
400 mm centres (6.26 plants per m²)	m²	40.50
500 mm centres (4 plants per m²)	m²	26.00
750 mm centres (1.78 plants per m²)	m²	12.50
Cultivate and grade shrub bed; bring top 300 mm of topsoil to a fine tilth, incorporating mushroom compost at 50 mm and Enmag slow release fertilizer; rake and bring to given levels; remove all stones and debris over 50 mm; dig planting holes average 300 × 300 × 300 mm deep; supply and plant specified shrubs in quantities as shown below; backfill with excavated material as above; water to field capacity and mulch 50 mm bark chips 20–40 mm size; water and weed regularly for 12 months; shrubs 2 litre PC £2.80; ground covers 9 cm PC £1.50		
Shrubs at centres shown below		
300 mm centres	m²	62.00
400 mm centres	m²	41.00
500 mm centres	m²	27.50
600 mm centres	m²	21.50
750 mm centres	m²	16.50
900 mm centres	m²	13.90
Groundcover 30%/shrubs 70% at the distances shown below		
200/300 mm	m²	67.00
300/400 mm	m²	56.00
300/500 mm	m²	33.00
Groundcover 50%/shrubs 50% at the distances shown below		
200/300 mm	m²	77.00
300/400 mm	m²	42.00
300/500 mm	m²	36.50
400/500 mm	m²	28.00
Cultivate ground by machine and rake to level; plant bulbs as shown; bulbs PC £25.00/100		
15 bulbs per m²	m²	6.85
25 bulbs per m²	m²	11.20
50 bulbs per m²	m²	22.00

PLANTING

Item Excluding site overheads and profit	Unit	Total rate £
BEDDING		
Spray surface with glyphosate; lift and dispose of turf when herbicide action is complete; cultivate new area for bedding plants to 400 mm deep; spread compost 100 mm deep and chemical fertilizer Enmag and rake to fine tilth to receive new bedding plants; remove all arisings to skip		
Existing turf area		
disposal to skip	100 m²	840.00
disposal to compost area on site; distance 25 m	100 m²	500.00
Plant bedding to existing planting area; bedding planting PC £0.25 each		
Clear existing bedding; cultivate soil to 230 mm deep; incorporate compost 75 mm and rake to fine tilth; collect bedding from nursery and plant at 100 mm ccs; irrigate on completion; maintain weekly for 12 weeks		
mass planted; 100 mm centres	m²	32.50
to patterns; 100 mm centres	m²	35.00
mass planted; 150 mm centres	m²	18.60
to patterns; 150 mm centres	m²	21.00
mass planted; 200 mm centres	m²	12.70
to patterns; 200 mm centres	m²	15.10
Extra for watering by hand held hose pipe		
Flow rate 25 litres/minute		
10 litres/m²	100 m²	15.10
15 litres/m²	100 m²	22.50
20 litres/m²	100 m²	30.00
25 litres/m²	100 m²	37.50
Flow rate 40 litres/minute		
10 litres/m²	100 m²	9.50
15 litres/m²	100 m²	14.20
20 litres/m²	100 m²	18.70
25 litres/m²	100 m²	23.50

MAINTENANCE

Item Excluding site overheads and profit	Unit	Total rate £
GRASS CUTTING		
Cut grass with pedestrian mower; strim edges of shrub beds and edges to pavings;		
length of edges is 10% of the grass area; remove arisings to compost heap on site		
Lawn cut 22 times per year; cost per occasion		
pedestrian mower 400 mm wide	100 m²	4.45
pedestrian mower 450 mm wide	100 m²	3.75
pedestrian mower 500 mm wide	100 m²	3.60
Lawn cut 18 times per year; cost per occasion		
pedestrian mower 400 mm wide	100 m²	5.00
pedestrian mower 450 mm wide	100 m²	4.30
pedestrian mower 500 mm wide	100 m²	4.10
SHRUB BED MAINTENANCE		
Prune shrubs; cut back to soft growth; shrubs which have previously been pruned in		
the last 3 years		
Shrubs 1.00–2.00 m high		
1.00 m centres	10 m²	11.40
750 mm centres	10 m²	20.00
Shrubs		
1.00 m centres	10 m²	8.20
750 mm centres	10 m²	14.60
500 mm centres	10 m²	33.00
400 mm centres	10 m²	51.00
Weed newly planted shrub bed; irrigate to field capacity by hand; fork over surface of		
plant bed every four visits		
Regular visits		
newly planted areas; per visit	m²	0.80
established areas; per visit	m²	0.50
HEDGE MAINTENANCE		
Cut hedge by hand operated machinery; remove arising to compost heap on site; 2		
sides and top		
Hedge up to 2.00 m high; depth of cut 300 mm		
hedge up to 1.0 m high	m	0.95
hedge up to 1.5 m high	m	1.70
hedge up to 2.0 m high	m	3.40
Reduce height of ornamental hedge using hand tools or mechanical hand tools;		
thickness of branches average 20–50 mm diameter; collect arisings and dispose on		
site; hedge measured 2 sides and top; assumed thickness 600 mm		
Hedge up to 2.00 m high		
reduce by 600 mm	m	2.30
reduce by 900 mm	m	7.10
Hedge 2.00–4.00 m high		
reduce by 600 mm	m	3.40
reduce by 900 mm	m	8.55

DRAINAGE

Item Excluding site overheads and profit	Unit	Total rate £
PIPE LAYING		
Pipe laying		
Excavate trench 600 mm deep by excavator; lay bedding and backfill as per material specification below; lay non woven geofabric and fill with topsoil to ground level		
100 mm vitrified clay; laid on earth with excavated backfill	m	**14.90**
110 mm PVC-u drainpipe; laid on sand bed with gravel backfill	m	**21.00**
Excavate trench 600 mm deep by hand; lay bedding and backfill as per material specification below; lay non woven geofabric and fill with topsoil to ground level		
100 mm vitrified clay; laid on earth with excavated backfill	m	**19.20**
110 mm PVC-u drainpipe; laid on sand bed with gravel backfill	m	**25.50**
LINEAR DRAINAGE		
Linear drainage to design sensitive areas		
Excavate trench by machine; lay Aco Brickslot channel drain on concrete base and surround to falls; all to manufacturers specifications		
paving surround to both sides of channel	m	**28.50**
Linear drainage to vehicular area		
Excavate trench by machine; lay Aco MultiDrain MD Brickslot; offset galvanized slot drain grating for M100PPD; load class C250 channel drain on concrete base and surround to falls; all to manufacturers specifications; paving surround to channel with brick paving PC £300.00/1000		
brickslot galvanized grating; paving surround to one side of channel	m	**91.00**
slotted galvanized grating; paving surround to both sides of channel	m	**100.00**
stainless steel grating; paving surround to one side of channel	m	**170.00**
Linear drainage to pedestrian area		
Excavate trench by machine; lay Aco Hexdrain channel drain on concrete base and surround to falls; all to manufacturers specifications; paving surround to channel with brick paving PC £300.00/1000		
with black plastic; paving surround to one side of channel	m	**75.00**
slotted galvanized grating; paving surround to both sides of channel	m	**87.00**
with brickslot grating; paving surround to both side of channel	m	**140.00**
Heelguard ductile grating; paving surround to both sides of channel	m	**3800.00**
Accessories for channel drain		
Sump unit with sediment bucket	nr	**155.00**
End cap; inlet/outlet	nr	**3.30**

DRAINAGE

Item Excluding site overheads and profit	Unit	Total rate £
MANHOLES		
Inspection chambers; brick manhole; excavate pit for inspection chamber including earthwork support and disposal of spoil to dump on site not exceeding 100 m; lay concrete (1:2:4) base 1500 mm diameter × 200 mm thick; 110 mm vitrified clay channels; benching in concrete (1:3:6) allowing one outlet and two inlets for 110 mm diameter pipe; construct inspection chamber one brick thick walls of engineering brick Class B; backfill with excavated material; complete with two cast iron step irons; cover 600 × 450 mm ductile iron; light vehicle loading		
1200 × 1200 × 1200 mm		
excavation by machine	each	1025.00
excavation by hand	each	2050.00
1200 × 1200 × 1200 mm; by machine but with recessed cover		
600 × 450 mm; 5 tonne loading; filled with block pavers	each	1150.00
1200 × 1200 × 1500 mm		
excavation by machine	each	1175.00
excavation by hand	each	2350.00
Inspection chambers; polypropylene; excavate pit for inspection chamber including earthwork support and disposal of spoil to dump on site not exceeding 100 m; lay concrete (1:2:4) base 700 mm diameter × 200 mm thick		
Polypropylene 600 mm deep		
excavation by machine	each	325.00
excavation by hand	each	290.00
Polypropylene 1200 mm		
excavation by machine	each	395.00
excavation by hand	each	460.00
GULLIES		
Clay gully		
Excavate hole; supply and set in concrete vitrified clay trapped mud (dirt) gully complete with galvanized bucket and cast iron hinged locking grate and frame; lay kerb to gully		
1 tonne loading; RGP5; 100 mm outlet; 100 mm diameter; 225 mm internal width; 585 mm internal depth	each	300.00
Gullies PVC-u		
Excavate hole and lay 100 mm concrete (C20P) base 150 × 150 mm to suit given invert level of drain; connect to drainage system; backfill with Type 1 granular fill; install gully; complete with grate and frame; brick kerb to gully surround		
PVC-u universal gully with vertical hopper; plastic grid and frame included	each	195.00
PVC-u gully 100 mm with P trap; grid and frame included	each	200.00
yard gully trapped; 300 mm diameter; 600 mm deep; sediment bucket and ductile iron cover	each	425.00

DRAINAGE

Item Excluding site overheads and profit	Unit	Total rate £
SOAKAWAYS		
Soakaway Aquacell Wavin Plastics Ltd; preformed polypropylene soakaway infiltration crate units		
Excavate pits or trenches for soakaway units; place Aquacell polypropylene soakaway units in recommended configurations laid on 100 mm shingle; cover with terram and backfill to sides and top with 100 mm shingle; backfill with 150 mm topsoil		
4 crates; 760 litres	nr	**285.00**
8 crates; 1520 litres	nr	**510.00**
12 crates; 2280 litres	nr	**830.00**
16 crates; 3050 litres	nr	**1025.00**

GARDEN LIGHTING

Item Excluding site overheads and profit	Unit	Total rate £
TRENCHING AND CABLE INSTALLATION		
Small garden lighting; 12 × 12 m		
Lighting to domestic gardens; trenching to 600 mm deep and installation of cable; backfilling and terminating cable; connections to switches and to mains distribution board		
Works by machine; main supply of 50 m armoured cables to 4 lighting transformers; transformers at maximum 10 m distance from mains supply		
4 nr Hunza spike lights on one circuit	nr	1250.00
4 nr Hunza spike lights and 4 nr step lights on one circuit; 4 transformers	nr	3050.00
Works by hand; main supply of 50 m armoured cables to 4 lighting transformers; transformers at maximum 10 m distance from mains supply		
4 nr Hunza spike lights on one circuit	nr	1775.00
4 nr Hunza spike lights and 4 nr step lights on one circuit; 4 transformers	nr	3800.00
Medium garden lighting; 25 × 15 m		
Lighting to domestic gardens; trenching to 600 mm deep and installation of cable; backfilling and terminating cable; connections to switches and to mains distribution board		
Works by machine; main supply of 100 m armoured cables to 6 lighting transformers; transformers at maximum 10 m distance from mains supply		
8 nr Hunza spike lights on one circuit	nr	2075.00
8 nr Hunza spike lights and 4 nr step lights on one circuit; 6 transformers	nr	3800.00
Works by hand; main supply of 100 m armoured cables to 8 lighting transformers; transformers at maximum 10 m distance from mains supply		
8 nr Hunza spike lights on one circuit	nr	3100.00
8 nr Hunza spike lights and 4 nr step lights on one circuit; 4 transformers	nr	4400.00
Large garden lighting; 60 m × 30 m		
Lighting to domestic gardens; trenching to 600 mm deep and installation of cable; backfilling and terminating cable; connections to switches and to mains distribution board		
Works by machine and by hand 70/30%; main supply of 180 m armoured cables to 10 lighting transformers; transformers at maximum 10 m distance from mains supply		
8 nr Hunza spike lights, 4 Hunza wall lights and 4 Hunza step lights on 1 circuit.	nr	5600.00
Works by hand 70/30 %; main supply of 180 m armoured cables to 10 lighting transformers; transformers at maximum 10 m distance from mains supply		
8 nr Hunza spike lights, 4 Hunza wall lights and 4 Hunza step lights on 1 circuit.	nr	6600.00

Construction Contracts Questions and Answers

2nd Edition

By **David Chappell**

What they said about the first edition: "A fascinating concept, full of knowledgeable gems put in the most frank of styles... A book to sample when the time is right and to come back to when another time is right, maybe again and again." – *David A Simmonds,* Building Engineer *magazine*

- Is there a difference between inspecting and supervising?
- What does 'time-barred' mean?
- Is the contractor entitled to take possession of a section of the work even though it is the contractor's fault that possession is not practicable?

Construction law can be a minefield. Professionals need answers which are pithy and straightforward, as well as legally rigorous. The two hundred questions in the book are real questions, picked from the thousands of telephone enquiries David Chappell has received as a Specialist Adviser to the Royal Institute of British Architects. Although the enquiries were originally from architects, the answers to most of them are of interest to project managers, contractors, QSs, employers and others involved in construction.

The material is considerably updated from the first edition – weeded, extended and almost doubled in coverage. The questions range in content from extensions of time, liquidated damages and loss and/or expense to issues of warranties, bonds, novation, practical completion, defects, valuation, certificates and payment, architects' instructions, adjudication and fees. Brief footnotes and a table of cases will be retained for those who may wish to investigate further.

August 2010: 216x138: 352pp
Pb: 978-0-415-56650-6: **£34.99**

PART 5

Prices for Measured Works – Minor Works

INTRODUCTION
Typical Project Profile

Contract value	£10,000.00–£70,000.00
Labour rate (see pages 6–9)	£20.50 per hour
Labour rate for maintenance contracts	–
Number of site staff	6–9
Project area	1200 m^2
Project location	Outer London
Project components	20% hard landscape 80% soft landscape and planting
Access to works areas	Very good
Contract	Main contract
Delivery of materials	Part loads
Profit and site overheads	Excluded

NEW ITEMS FOR THIS EDITION

Item Excluding site overheads and profit	PC £	Labour hours	Labour £	Plant £	Material £	Unit	Total rate £
A PRELIMINARY COSTS FOR SMALL GARDENS							
Site accommodation							
Toilet							
weekly rate	–	–	–	27.00	–	week	**27.00**
delivery and collection charge	–	–	–	50.00	–	nr	**50.00**
Site office; 3.6 × 2.4 m							
weekly rate	–	–	–	36.75	–	week	**36.75**
delivery and collection charge	–	–	–	252.00	–	nr	**252.00**
Secure storage							
site storage; 3.0 × 2.4 m	–	–	–	12.00	–	week	**12.00**
delivery and collection charge	–	–	–	50.00	–	nr	**50.00**
Site compound; inclusive of small office, toilet, secure storage; all on 100 mm compacted crushed concrete hardcore with security fencing							
first week including setup costs	–	8.00	164.00	598.64	173.50	nr	**951.14**
weekly costs	–	–	–	255.75	–	nr	**255.75**
Site plant hire delivery costs							
Delivery and collection charges for small plant to 5 tonne							
per machine	–	–	–	80.00	–	nr	**80.00**
Non-productive time; unloading costs							
Costs for hand offloading on minor works items where materials are delivered in small loads and machine or crane offloads are impractical							
project value; £10,000	–	16.00	328.00	–	–	nr	**328.00**
project value; £20,000	–	20.00	410.00	–	–	nr	**410.00**
project value; £30,000	–	24.00	492.00	–	–	nr	**492.00**
project value; £40,000	–	28.00	574.00	–	–	nr	**574.00**
project value; £50,000	–	32.00	656.00	–	–	nr	**656.00**
Tendering costs							
Produce quotations for projects issued by landscape designer; plans and sections issued to contractors on paper copies; includes for initial site visit							
project value; £10,000	–	6.00	210.00	–	–	nr	**210.00**
project value; £20,000	–	8.00	280.00	–	–	nr	**280.00**
project value; £30,000	–	12.00	420.00	–	–	nr	**420.00**
project value; £40,000	–	14.00	490.00	–	–	nr	**490.00**
project value; £50,000	–	17.00	595.00	–	–	nr	**595.00**
F10 PIERS							
Isolated brick piers in English bond							
Engineering brick PC £239.00/1000 (not SMM)							
one brick thick (225 mm)	–	1.63	33.49	–	9.95	m	**43.44**
one and a half bricks thick (337.5 mm)	–	3.00	61.50	–	29.39	m	**90.89**
two bricks thick (450 mm)	–	4.50	92.25	–	47.01	m	**139.26**
Brick PC £300.00/1000 (not SMM)							
one brick thick (225 mm)	–	1.63	33.49	–	10.23	m	**43.72**
one and a half bricks thick (337.5 mm)	–	3.00	61.50	–	29.99	m	**91.49**
two bricks thick (450 mm)	–	4.50	92.25	–	48.09	m	**140.34**
three bricks thick (675 mm)	–	8.37	171.59	–	98.58	m	**270.17**

NEW ITEMS FOR THIS EDITION

Item Excluding site overheads and profit	PC £	Labour hours	Labour £	Plant £	Material £	Unit	Total rate £
Brick PC £800.00/1000 (not SMM)							
one brick thick (225 mm)	–	1.63	33.49	–	24.23	m	**57.72**
one and a half bricks thick (337.5 mm)	–	3.00	61.50	–	60.37	m	**121.87**
two bricks thick (450 mm)	–	4.50	92.25	–	102.09	m	**194.34**
three bricks thick (675 mm)	–	8.37	171.59	–	220.08	m	**391.67**
Q10 BESPOKE MILD STEEL EDGINGS							
Mild steel edgings; 6 mm galvanized steel; Gardenlink Ltd							
Bespoke edgings; complete with bolts for adjoining sections and mild steel pins at 1.50 m centres							
100 mm straight	–	–	–	–	–	m	**15.40**
150 mm straight	–	–	–	–	–	m	**17.90**
300 mm straight	–	–	–	–	–	m	**26.40**
Q31 TERRAVENT ROOT DECOMPACTION							
Root decompaction; Terravent; Goroots Ltd							
Decompaction and aeration of soil in the root zone of a tree or plant; followed by an infusion of beneficial mychorrhizal fungi and other liquid amendments into the freshly aerated soil; mulching the root zone	–	–	–	–	1.00	m²	**1.00**
R12 DRAINAGE BELOW GROUND – ACO CHANNELS FOR DOMESTIC USE							
Linear drainage; channels; Aco Building Products; laid to concrete bed C25 on compacted granular base on 200 mm deep concrete bed; haunched with 200 mm concrete surround; all in 750 × 430 mm wide trench with compacted 200 mm granular base surround (excavation and subbase not included)							
HexDrain channel drain; recycled polypropylene; load class A15; 1000 × 125 × 80 mm							
with black plastic grating	16.35	1.20	24.60	–	26.18	ea	**50.78**
with galvanized steel grating	19.55	1.20	24.60	–	29.38	ea	**53.98**
with black plastic brickslot grating	19.55	1.20	24.60	–	29.38	ea	**53.98**
Accessories for HexDrain							
endcap; black plastic	3.30	–	–	–	3.30	ea	**3.30**
corner unit; with metallic effect plastic grating	17.90	0.50	10.25	–	22.42	ea	**32.67**
sump unit; black plastic; 250 mm depth	28.80	1.00	20.50	–	33.32	ea	**53.82**
corner unit; with black plastic grating; 125 mm vertical outlet	18.70	0.75	15.38	–	19.23	ea	**34.61**
corner unit; brickslot; black plastic; 125 mm vertical outlet	20.20	0.50	10.25	–	24.72	ea	**34.97**
Universal 820 drain union; PVC-u; 110 mm diameter	3.35	–	–	–	3.35	ea	**3.35**

NEW ITEMS FOR THIS EDITION

Item Excluding site overheads and profit	PC £	Labour hours	Labour £	Plant £	Material £	Unit	Total rate £
R12 DRAINAGE BELOW GROUND – ACO CHANNELS FOR DOMESTIC USE – CONT							
Linear drainage – cont							
RainDrain lightweight polymer concrete channel; load class A15; 1000 × 118 × 97 mm							
with galvanized steel grating	23.60	1.20	24.60	–	28.12	ea	**52.72**
Accessories for RainDrain							
foul air trap; horizontal; 110 mm diameter	15.40	–	–	–	15.40	ea	**15.40**
closing endcap	3.95	–	–	–	3.95	ea	**3.95**
outlet endcap; 110 mm diameter	8.65	–	–	–	8.65	ea	**8.65**
sump; with galvanized steel grating and sediment bucket; 500 mm	59.80	0.50	10.25	–	64.32	ea	**74.57**
RainDrain Pro drainage channel							
with heelguard ductile iron grating	72.20	1.20	24.60	–	76.72	ea	**101.32**
Accessories for RainDrain Pro							
Universal silver closing end cap	3.55	0.13	2.56	–	3.55	ea	**6.11**
step connector; 50 mm	8.15	0.13	2.56	–	8.15	ea	**10.71**
vertical outlet connector; 110 mm or 160 mm	13.30	0.13	2.56	–	13.30	ea	**15.86**
roddable foul air trap; 110 mm	91.60	0.13	2.56	–	91.60	ea	**94.16**
roddable foul air trap; 160 mm	113.05	0.13	2.56	–	113.05	ea	**115.61**
V ELECTRICAL – TRENCHING FOR ELECTRICAL SERVICES							
Excavating; mechanical; trenches for electrical services							
3 tonne excavator (bucket volume 0.13 m³); arisings laid alongside							
600 mm deep	–	–	–	1.59	–	m	**1.59**
800 mm deep	–	–	–	1.86	–	m	**1.86**
1.00 m deep	–	–	–	2.23	–	m	**2.23**
Extra over any types of excavating irrespective of depth for breaking out existing materials; heavy duty 110 volt breaker tool							
hard rock	–	5.00	102.50	59.69	–	m³	**162.19**
concrete	–	3.00	61.50	35.81	–	m³	**97.31**
reinforced concrete	–	4.00	82.00	60.94	–	m³	**142.94**
brickwork, blockwork or stonework	–	1.50	30.75	17.91	–	m³	**48.66**
By hand							
600 mm deep	–	0.24	4.93	–	–	m	**4.93**
800 mm deep	–	0.43	8.76	–	–	m	**8.76**
1.00 mm deep	–	0.67	13.67	–	–	m	**13.67**
Backfilling of trenches; including laying of electrical marker tape; compacting lightly as work proceeds; spreading any remaining material							
To trenches containing armoured cable							
by machine	0.13	–	–	1.67	0.13	m	**1.80**
by hand	0.13	0.20	4.10	–	0.13	m	**4.23**
To trenches containing ducted cable including 150 mm sharp sand over the duct							
by machine	2.40	–	–	2.09	2.40	m	**4.49**
by hand	0.13	0.25	5.13	–	2.40	m	**7.53**

NEW ITEMS FOR THIS EDITION

Item Excluding site overheads and profit	PC £	Labour hours	Labour £	Plant £	Material £	Unit	Total rate £
V ELECTRICAL CABLES							
Cable to trenches (trenching operations **not included); laying only cable**							
Twin core steel wire armoured cable; 50 m drums							
1.5 mm core	0.80	0.02	0.41	–	0.80	m	**1.21**
2.5 mm core	0.96	0.02	0.41	–	0.96	m	**1.37**
Twin core steel wire armoured cable; lengths less than 50 m runs							
1.5 mm core	0.88	0.02	0.41	–	0.88	m	**1.29**
2.5 mm core	1.04	0.02	0.41	–	1.04	m	**1.45**
Three core steel wire armoured cable; 50 m drums							
1.5 mm core	0.90	0.02	0.41	–	0.90	m	**1.31**
2.5 mm core	1.23	0.02	0.41	–	1.23	m	**1.64**
4.0 mm core	1.48	0.02	0.41	–	1.48	m	**1.89**
6.0 mm core	1.92	0.02	0.46	–	1.92	m	**2.38**
10.0 mm core	3.39	0.03	0.51	–	3.39	m	**3.90**
Three core steel wire armoured cable; lengths less than 50 m runs							
1.5 mm core	0.93	0.02	0.41	–	0.93	m	**1.34**
2.5 mm core	1.33	0.02	0.41	–	1.33	m	**1.74**
Ducts to trenches; twin wall flexible cable **ducts with drawstrings; laid on 150 mm** **clean sharp sand**							
Twin wall duct; laying to trenches							
63 mm × 50 m coils	1.00	0.02	0.41	–	1.00	m	**1.41**
110 mm × 50 m coils	1.50	0.02	0.41	–	1.50	m	**1.91**
Cable drawing through ducts							
straight runs up to 50 m lengths	–	1.00	20.50	–	–	nr	**20.50**
Terminations to armoured cables; cutting, **coiling and taping length of cable to** **receive connection to light fitting,** **transformer and the like (all not included);** **fixing to temporary posts**							
Armoured cable	–	0.33	6.83	–	–	nr	**6.83**
Plain ducted cable	–	0.13	2.56	–	–	nr	**2.56**
Junction boxes; fixing to cables to receive **connections to light fittings or** **transformers; to IP68**							
Underground jointing box							
2 way	29.00	–	–	–	46.50	nr	**46.50**
3 way	34.00	–	–	–	54.00	nr	**54.00**
Above ground jointing box							
2 way	3.50	0.33	8.33	–	3.50	nr	**11.83**
3 way	4.90	–	–	–	18.90	nr	**18.90**

NEW ITEMS FOR THIS EDITION

Item Excluding site overheads and profit	PC £	Labour hours	Labour £	Plant £	Material £	Unit	Total rate £
V ELECTRICAL CABLES – CONT							
Power connections							
Internal connections and switches for control of external lighting; connections of switches or switch bank to distribution board and internal cables to switches							
per circuit located within 1.00 m of the distribution board; chasing of cables to walls not included	–	–	–	–	–	nr	**35.00**
Power connections to switches							
price per circuit for the first circuit; including builders work, drilling through walls and making good; chasing of cables to walls not included	–	1.00	20.50	–	74.00	nr	**94.50**
additional circuits at the same location	–	–	–	–	12.75	nr	**12.75**
V41 ELECTRICAL LIGHTS – LOW VOLTAGE							
Low voltage lights; connecting to transformers (transformers shown separately); fixing as described							
Hunza Spike spotlight; 63.5 mm diameter × 75 mm long; c/w 20/35/50 w lamp							
stainless steel	107.16	–	–	–	118.83	nr	**118.83**
copper	66.99	–	–	–	78.66	nr	**78.66**
black	43.32	–	–	–	54.99	nr	**54.99**
Hunza Spike spotlight; adjustable; 63.5 mm diameter × 75 mm long; c/w 20/35/50 w lamp							
stainless steel	134.68	–	–	–	146.35	nr	**146.35**
copper	77.25	–	–	–	88.92	nr	**88.92**
black	50.92	–	–	–	62.59	nr	**62.59**
Hunza pole lights; single; fixing to concrete base 150 × 150 × 150 mm							
stainless steel	166.94	1.00	20.50	–	178.61	nr	**199.11**
copper	113.96	1.00	20.50	–	126.13	nr	**146.63**
black	73.32	1.00	20.50	–	85.49	nr	**105.99**
Hunza pole lights; double; fixing to concrete base 150 × 150 × 150 mm							
stainless steel	268.28	1.00	20.50	–	280.45	nr	**300.95**
copper	113.96	1.00	20.50	–	126.13	nr	**146.63**
black	73.32	1.00	20.50	–	85.49	nr	**105.99**
Recessed wall lights; eyelid steplights; fixing to brick or stone walls; inclusive of core drilling							
stainless steel	268.28	1.50	30.75	–	279.95	nr	**310.70**
copper	136.29	1.50	30.75	–	147.96	nr	**178.71**
black	92.04	1.50	30.75	–	103.71	nr	**134.46**
Hunza; recessed deck path or lawn lights; stainless steel; inclusive of core drilling excavation and all making good							
installed to deck	250.00	0.66	16.50	–	250.00	nr	**266.50**
installed to path	250.00	1.08	23.63	–	250.00	nr	**273.63**
installed to lawn	197.00	0.20	4.10	–	208.55	nr	**212.65**
installed to driveway	250.00	1.08	23.63	–	250.00	nr	**273.63**

NEW ITEMS FOR THIS EDITION

Item Excluding site overheads and profit	PC £	Labour hours	Labour £	Plant £	Material £	Unit	Total rate £
V90 ELECTRICAL LIGHTING – LOW VOLTAGE TRANSFORMERS							
Outdoor transformers for 12 volt exterior lighting; connecting to junction box (not included); IP67							
Single light transformers; 100 × 68 × 72 mm; 50 vA							
fused	21.00	–	–	–	67.67	nr	67.67
unfused	20.34	–	–	–	20.34	nr	20.34
Two light transformer; 130 × 85 × 85 mm; 100 vA							
fused	34.81	–	–	–	34.81	nr	34.81
unfused	33.21	–	–	–	33.21	nr	33.21
Three light transformer; 130 × 85 × 85 mm; 150 vA							
fused	44.67	–	–	–	44.67	nr	44.67
unfused	42.21	–	–	–	42.21	nr	42.21
Four light transformer; 130 × 85 × 100 mm; 200 vA							
fused	52.56	–	–	–	52.56	nr	52.56
unfused	49.13	–	–	–	49.13	nr	49.13

A PRELIMINARIES

Item Excluding site overheads and profit	PC £	Labour hours	Labour £	Plant £	Material £	Unit	Total rate £
A PRELIMINARIES							
Site accommodation							
Toilet							
weekly rate	–	–	–	27.00	–	week	**27.00**
delivery and collection charge	–	–	–	50.00	–	nr	**50.00**
Site office; 3.6 × 2.4 m							
weekly rate	–	–	–	36.75	–	week	**36.75**
delivery and collection charge	–	–	–	252.00	–	nr	**252.00**
Secure storage							
site storage; 3.0 × 2.4 m	–	–	–	12.00	–	week	**12.00**
delivery and collection charge	–	–	–	50.00	–	nr	**50.00**
Site compound; inclusive of small office, toilet, secure storage; all on 100 mm compacted crushed concrete hardcore with security fencing							
first week including setup costs	–	8.00	164.00	598.64	173.50	nr	**951.14**
weekly costs	–	–	–	255.75	–	nr	**255.75**
Site plant hire delivery costs							
Delivery and collection charges for small plant to 5 tonne							
per machine	–	–	–	80.00	–	nr	**80.00**
Non-productive time; unloading costs							
Costs for hand offloading on minor works items where materials are delivered in small loads and machine or crane offloads are impractical							
project value; £10,000	–	16.00	328.00	–	–	nr	**328.00**
project value; £20,000	–	20.00	410.00	–	–	nr	**410.00**
project value; £30,000	–	24.00	492.00	–	–	nr	**492.00**
project value; £40,000	–	28.00	574.00	–	–	nr	**574.00**
project value; £50,000	–	32.00	656.00	–	–	nr	**656.00**
Tendering costs							
Produce quotations for projects issued by landscape designer; plans and sections issued to contractors on paper copies; includes for initial site visit							
project value; £10,000	–	6.00	210.00	–	–	nr	**210.00**
project value; £20,000	–	8.00	280.00	–	–	nr	**280.00**
project value; £30,000	–	12.00	420.00	–	–	nr	**420.00**
project value; £40,000	–	14.00	490.00	–	–	nr	**490.00**
project value; £50,000	–	17.00	595.00	–	–	nr	**595.00**

C EXISTING SITE/BUILDINGS/SERVICES

Item Excluding site overheads and profit	PC £	Labour hours	Labour £	Plant £	Material £	Unit	Total rate £
C20 DEMOLITION							
Demolish existing structures; removal to **skip maximum distance 50 m; mechanical** **demolition; with 3 tonne excavator and** **dumper**							
Brick wall							
112.5 mm thick	–	0.13	2.73	6.24	–	m²	**8.97**
225 mm thick	–	0.17	3.42	11.88	–	m²	**15.30**
337.5 mm thick	–	0.20	4.10	18.77	–	m²	**22.87**
450 mm thick	–	0.27	5.47	22.55	–	m²	**28.02**
Demolish existing structures; removal to **skip maximum distance 50 m; all works by** **hand**							
Brick wall							
112.5 mm thick	–	0.33	6.83	5.44	–	m²	**12.27**
225 mm thick	–	0.50	10.25	10.88	–	m²	**21.13**
337.5 mm thick	–	0.67	13.67	17.57	–	m²	**31.24**
450 mm thick	–	1.00	20.50	20.94	–	m²	**41.44**
Demolish existing structures; removal to **skip maximum distance 50 m; by diesel or** **electric breaker; all other works by hand**							
Brick wall							
112.5 mm thick	–	0.17	3.42	6.77	–	m²	**10.19**
225 mm thick	–	0.20	4.10	12.48	–	m²	**16.58**
337.5 mm thick	–	0.25	5.13	19.57	–	m²	**24.70**
450 mm thick	–	0.33	6.83	23.61	–	m²	**30.44**
Break out concrete footings associated **with free standing walls; inclusive of all** **excavation, removal to skip and backfilling** **with excavated material**							
By mechanical breaker; diesel or electric							
plain concrete	–	1.50	30.75	69.10	–	m³	**99.85**
as above but loading to skip by hand;							
backfilling by hand		5.50	112.75	82.35	–	m³	**195.10**
reinforced concrete	–	2.50	51.25	84.97	–	m³	**136.22**
Remove existing free standing buildings; **demolition by hand**							
Timber buildings with suspended timber floor; hardstanding or concrete base not included; loading to skip							
shed 6.0 m²	–	2.00	41.00	77.33	–	nr	**118.33**
shed 10.0 m²	–	3.00	61.50	108.74	–	nr	**170.24**
shed 15.0 m²	–	3.50	71.75	108.74	–	nr	**180.49**
Timber building; insulated; with timber or concrete posts set in concrete, felt covered timber or tiled roof; internal walls cladding with timber or plasterboard; load arisings to skip							
timber structure 6.0 m²	–	2.50	51.25	260.98	–	nr	**312.23**
timber structure 12.0 m²	–	3.50	71.75	365.37	–	nr	**437.12**
timber structure 20.0 m²	–	8.00	164.00	494.42	–	nr	**658.42**

C EXISTING SITE/BUILDINGS/SERVICES

Item Excluding site overheads and profit	PC £	Labour hours	Labour £	Plant £	Material £	Unit	Total rate £
C20 DEMOLITION – CONT							
Demolition of free standing brick buildings with tiled or sheet roof; concrete foundations measured separately; mechanical demolition maximum distance to stockpile 25 m; including for all access scaffolding and the like; maximum height of roof 4.00 m; including all doors, windows, guttering and down pipes; including disposal by grab							
Half brick thick							
10 m²	–	8.00	164.00	243.06	147.33	nr	**554.39**
20 m²	–	16.00	328.00	453.13	189.51	nr	**970.64**
1 brick thick							
10 m²	–	10.00	205.00	420.13	294.67	nr	**919.80**
20 m²	–	16.00	328.00	630.19	379.02	nr	**1337.21**
Cavity wall with blockwork inner skin and brick outer skin; insulated							
10 m²	–	12.00	246.00	597.19	409.50	nr	**1252.69**
20 m²	–	20.00	410.00	807.26	526.50	nr	**1743.76**
Extra over to the above for disconnection of services							
Electrical							
disconnection	–	–	–	–	–	nr	**70.00**
grub out cables and dispose; backfilling; by machine	–	–	–	4.77	–	m	**4.77**
grub out cables and dispose; backfilling; by hand	–	0.50	10.25	–	–	m	**10.25**
Water supply; foul or surface water drainage							
disconnection; capping off	–	1.00	20.50	–	70.00	nr	**90.50**
grub out pipes and dispose; backfilling; by machine	–	–	–	5.49	–	m	**5.49**
grub out pipes and dispose; backfilling; by hand	–	0.50	10.25	0.72	–	m	**10.97**

D GROUNDWORK

Item Excluding site overheads and profit	PC £	Labour hours	Labour £	Plant £	Material £	Unit	Total rate £
D11 SOIL STABILIZATION							
Timber log retaining walls; timber fencing; AVS Fencing Supplies Ltd; tanalized softwood posts; all timber posts are kiln dried redwood with 15 year guarantee; walls below which have cut posts would not be subject to the suppliers guarantee Machine rounded softwood logs to trenches priced separately; disposal of excavated material priced separately; inclusive of 75 mm hardcore blinding to trench and backfilling trench with site mixed concrete 1:3:6; geofabric pinned to rear of logs; heights of logs above ground							
500 mm (constructed from 1.80 m lengths)	31.15	1.50	30.75	–	41.22	m	71.97
1.20 m (constructed from 1.80 m lengths)	62.30	1.30	26.65	–	80.23	m	106.88
1.60 m (constructed from 2.40 m lengths)	83.20	2.50	51.25	–	107.08	m	158.33
2.00 m (constructed from 3.00 m lengths)	104.00	3.50	71.75	–	133.84	m	205.59
As above but with 150 mm machine rounded timbers							
500 mm	42.70	2.50	51.25	–	52.77	m	104.02
1.20 m	142.19	1.75	35.88	–	160.12	m	196.00
1.60 m	142.33	3.00	61.50	–	166.22	m	227.72
As above but with 200 mm machine rounded timbers							
1.80 m (constructed from 2.40 m lengths)	189.75	4.00	82.00	–	219.59	m	301.59
Railway sleeper walls; AVS Fencing Supplies Ltd							
Construct retaining wall from railway sleepers; fixed with steel galvanized pins 12 mm driven into the ground; sleepers laid flat Grade 1 softwood; 2590 × 250 × 125 mm							
150 mm; 1 sleeper high	8.05	0.50	10.25	–	13.65	m	23.90
300 mm; 2 sleepers high	16.10	1.00	20.50	–	17.22	m	37.72
450 mm; 3 sleepers high	23.96	1.50	30.75	–	25.43	m	56.18
600 mm; 4 sleepers high	32.28	2.00	41.00	–	33.75	m	74.75
Grade 1 softwood as above but with 2 nr galvanized angle iron stakes set into concrete internally and screwed to the inside face of the sleepers							
750 mm; 5 sleepers high	39.93	2.50	51.25	–	67.47	m²	118.72
900 mm; 6 sleepers high	47.79	2.75	56.38	–	78.29	m²	134.67
Grade 1 hardwood; 2590 × 250 × 150 mm							
150 mm; 1 sleeper high	7.72	0.50	10.25	–	8.28	m	18.53
300 mm; 2 sleepers high	15.43	1.00	20.50	–	16.55	m	37.05
450 mm; 3 sleepers high	22.96	1.50	30.75	–	24.43	m	55.18
600 mm; 4 sleepers high	30.93	2.00	41.00	–	32.40	m	73.40
Grade 1 hardwood as above but with 2 nr galvanized angle iron stakes set into concrete internally and screwed to the inside face of the sleepers							
750 mm; 5 sleepers high	38.27	2.50	51.25	–	65.81	m²	117.06
900 mm; 6 sleepers high	45.81	2.75	56.38	–	76.31	m²	132.69

D GROUNDWORK

Item Excluding site overheads and profit	PC £	Labour hours	Labour £	Plant £	Material £	Unit	Total rate £
D11 SOIL STABILIZATION – CONT							
Construct retaining wall from railway							
sleepers – cont							
New pine softwood; 2400 × 250 × 125 mm							
120 mm; 1 sleeper high	9.57	0.50	10.25	–	10.13	m	**20.38**
240 mm; 2 sleepers high	19.14	1.00	20.50	–	20.26	m	**40.76**
360 mm; 3 sleepers high	28.72	1.50	30.75	–	30.19	m	**60.94**
480 mm; 4 sleepers high	38.29	2.00	41.00	–	39.76	m	**80.76**
New pine softwood as above but with 2 nr							
galvanized angle iron stakes set into concrete							
internally and screwed to the inside face of the							
sleepers							
600 mm; 5 sleepers high	47.86	2.50	51.25	–	75.40	m²	**126.65**
720 mm; 6 sleepers high	57.43	2.75	56.38	–	87.93	m²	**144.31**
New oak hardwood; 2600 × 220 × 130 mm							
130 mm; 1 sleeper high	11.42	0.50	10.25	–	11.98	m	**22.23**
260 mm; 2 sleepers high	22.85	1.00	20.50	–	23.97	m	**44.47**
390 mm; 3 sleepers high	34.27	1.50	30.75	–	35.74	m	**66.49**
520 mm; 4 sleepers high	45.70	2.00	41.00	–	47.17	m	**88.17**
New oak hardwood as above but with 2 nr							
galvanized angle iron stakes set into concrete							
internally and screwed to the inside face of the							
sleepers							
640 mm; 5 sleepers high	57.12	2.50	51.25	–	84.66	m²	**135.91**
760 mm; 6 sleepers high	68.54	2.75	56.38	–	99.04	m²	**155.42**
Excavate foundation trench; set railway							
sleepers vertically on end in concrete 1:3:6							
continuous foundation to 33.3% of their							
length to form retaining wall							
Grade 1 hardwood; finished height above							
ground level							
300 mm	20.52	3.00	61.50	2.09	25.49	m	**89.08**
500 mm	29.75	3.00	61.50	2.09	34.72	m	**98.31**
600 mm	41.35	3.00	61.50	2.09	50.79	m	**114.38**
750 mm	44.62	3.50	71.75	2.09	54.06	m	**127.90**
1.00 m	50.81	3.75	76.88	2.09	61.15	m	**140.12**
Excavate and place vertical steel universal							
beams 165 wide in concrete base at 2.590							
centres; fix railway sleepers set							
horizontally between beams to form							
horizontal fence or retaining wall							
Grade 1 hardwood; bay length 2.590 m							
0.50 m high; 2 sleepers	16.24	2.50	51.25	–	59.15	bay	**110.40**
0.75 m high; 3 sleepers	23.86	2.60	53.30	–	87.87	bay	**141.17**
1.00 m high; 4 sleepers	31.98	3.00	61.50	–	113.74	bay	**175.24**
1.25 m high; 5 sleepers	40.84	3.50	71.75	–	146.31	bay	**218.06**
1.50 m high; 6 sleepers	47.97	3.00	61.50	–	174.17	bay	**235.67**
1.75 m high; 7 sleepers	55.84	3.00	61.50	–	182.05	bay	**243.55**

D GROUNDWORK

Item Excluding site overheads and profit	PC £	Labour hours	Labour £	Plant £	Material £	Unit	Total rate £
Retaining walls; Maccaferri Ltd Wire mesh gabions; galvanized mesh 80 mm × 100 mm; filling with broken stones 125–200 mm size; wire down securely to manufacturer's instructions; filling front face by hand							
2 × 1 × 0.50 m	22.16	2.00	41.00	7.01	91.46	nr	**139.47**
2 × 1 × 1.00 m	31.04	4.00	82.00	14.03	169.64	nr	**265.67**
PVC coated gabions							
2 × 1 × 0.5 m	28.67	2.00	41.00	7.01	97.97	nr	**145.98**
2 × 1 × 1.00 m	40.36	4.00	82.00	14.03	178.96	nr	**274.99**
D20 EXCAVATION AND FILLING							
MACHINE SELECTION TABLE							
Road Equipment Ltd; machine volumes for **excavating/filling only and placing** **excavated material alongside or to a** **dumper; no bulkages are allowed for in the** **material volumes; these rates should be** **increased by user-preferred percentages to** **suit prevailing site conditions; the figures** **in the next section for 'Excavation** **mechanical' and filling allow for the use of** **banksmen within the rates shown below** 1.5 tonne excavators; digging volume							
1 cycle/minute; 0.04 m³	–	0.42	8.54	2.74	–	m³	**11.28**
2 cycles/minute; 0.08 m³	–	0.21	4.27	1.69	–	m³	**5.96**
3 cycles/minute; 0.12 m³	–	0.14	2.85	1.36	–	m³	**4.21**
3 tonne excavators; digging volume							
1 cycle/minute; 0.13 m³	–	0.13	2.63	1.54	–	m³	**4.17**
2 cycles/minute; 0.26 m³	–	0.06	1.31	1.04	–	m³	**2.35**
3 cycles/minute; 0.39 m³	–	0.04	0.88	1.00	–	m³	**1.88**
5 tonne excavators; digging volume							
1 cycle/minute; 0.28 m³	–	0.06	1.22	1.02	–	m³	**2.24**
2 cycles/minute; 0.56 m³	–	0.03	0.61	0.54	–	m³	**1.15**
3 cycles/minute; 0.84 m³	–	0.02	0.41	0.49	–	m³	**0.90**
Dumpers; Road Equipment Ltd 1 tonne high tip skip loader; volume 0.485 m³ (775 kg)							
5 loads per hour	–	0.41	8.45	1.48	–	m³	**9.93**
7 loads per hour	–	0.29	6.04	1.10	–	m³	**7.14**
10 loads per hour	–	0.21	4.23	0.83	–	m³	**5.06**
3 tonne dumper; maximum volume 2.40 m³ (3.38 t); available volume 1.9 m³							
4 loads per hour	–	0.14	2.85	0.54	–	m³	**3.39**
5 loads per hour	–	0.11	2.28	0.44	–	m³	**2.72**
7 loads per hour	–	0.08	1.63	0.33	–	m³	**1.96**
10 loads per hour	–	0.06	1.14	0.25	–	m³	**1.39**
Lifting turf for preservation							
hand lift and stack	–	0.08	1.71	–	–	m²	**1.71**
Load to wheelbarrows and dispose off site by skip							
25 m distance	–	0.02	0.43	2.98	–	m²	**3.41**

D GROUNDWORK

Item Excluding site overheads and profit	PC £	Labour hours	Labour £	Plant £	Material £	Unit	Total rate £
D20 EXCAVATION AND FILLING – CONT							
Lifting turf for preservation – cont							
Note: The figures in this section relate to the machine capacities shown in the Major Works section of this book. The figures below allow for dig efficiency based on depth, bulkages of 25% on loamy soils (adjustments should be made for different soil types) and for a banksman.							
Site clearance; by machine; clear site of mature shrubs from existing cultivated beds; dig out roots by machine							
Mixed shrubs in beds; planting centres 500 mm average							
height less than 1 m	–	0.03	0.51	0.83	–	m²	**1.34**
1.00–1.50 m	–	0.04	0.82	1.34	–	m²	**2.16**
1.50–2.00 m; pruning to ground level by hand	–	0.10	2.05	3.34	–	m²	**5.39**
2.00–3.00 m; pruning to ground level by hand	–	0.10	2.05	6.68	–	m²	**8.73**
3.00–4.00 m; pruning to ground level by hand	–	0.20	4.10	11.13	–	m²	**15.23**
Site clearance; by hand; clear site of mature shrubs from existing cultivated beds; dig out roots							
Mixed shrubs in beds; planting centres 500 mm average							
height less than 1 m	–	0.33	6.83	–	–	m²	**6.83**
1.00–1.50 m	–	0.50	10.25	–	–	m²	**10.25**
1.50–2.00 m	–	1.00	20.50	–	–	m²	**20.50**
2.00–3.00 m	–	2.00	41.00	–	–	m²	**41.00**
3.00–4.00 m	–	3.00	61.50	–	–	m²	**61.50**
Disposal of material from site clearance operations							
Shrubs and groundcovers less than 1.00 m height; disposal to skip							
deciduous shrubs; not chipped; winter	–	0.05	1.02	11.93	–	m²	**12.95**
deciduous shrubs; chipped; winter	–	0.05	1.02	2.72	–	m²	**3.74**
evergreen or deciduous shrubs; chipped; summer	–	0.08	1.54	8.09	–	m²	**9.63**
Shrubs 1.00–2.00 m height; disposal to skip							
deciduous shrubs; not chipped; winter	–	0.17	3.42	35.80	–	m²	**39.22**
deciduous shrubs; chipped; winter	–	0.25	5.13	8.09	–	m²	**13.22**
evergreen or deciduous shrubs; chipped; summer	–	0.30	6.15	15.56	–	m²	**21.71**
Shrubs or hedges 2.00–3.00 m height; disposal to skip							
deciduous plants non woody growth; not chipped; winter	–	0.25	5.13	35.80	–	m²	**40.93**
deciduous shrubs; chipped; winter	–	0.50	10.25	18.83	–	m²	**29.08**
evergreen or deciduous shrubs non woody growth; chipped; summer	–	0.67	13.67	23.34	–	m²	**37.01**
evergreen or deciduous shrubs woody growth; chipped; summer	–	1.50	30.75	25.20	–	m²	**55.95**

D GROUNDWORK

Item Excluding site overheads and profit	PC £	Labour hours	Labour £	Plant £	Material £	Unit	Total rate £
Shrubs and groundcovers less than 1.00 m height; disposal to spoil heap							
deciduous shrubs; not chipped; winter	–	0.05	1.02	–	–	m²	**1.02**
deciduous shrubs; chipped; winter	–	0.05	1.02	0.93	–	m²	**1.95**
evergreen or deciduous shrubs; chipped; summer	–	0.08	1.54	0.93	–	m²	**2.47**
Shrubs 1.00–2.00 m height; disposal to spoil heap							
deciduous shrubs; chipped; winter	–	0.25	5.13	0.93	–	m²	**6.06**
deciduous shrubs; not chipped; winter	–	0.17	3.42	–	–	m²	**3.42**
evergreen or deciduous shrubs; chipped; summer	–	0.30	6.15	1.24	–	m²	**7.39**
Shrubs or hedges 2.00–3.00 m height; disposal to spoil heaps							
deciduous plants non woody growth; not chipped; winter	–	0.25	5.13	–	–	m²	**5.13**
deciduous shrubs; chipped; winter	–	0.50	10.25	0.93	–	m²	**11.18**
evergreen or deciduous shrubs non woody growth; chipped; summer	–	0.67	13.67	1.86	–	m²	**15.53**
evergreen or deciduous shrubs woody growth; chipped; summer	–	1.50	30.75	3.72	–	m²	**34.47**
Excavating; mechanical; topsoil for preservation; depositing alongside							
3 tonne tracked excavator (bucket volume 0.13 m³)							
average depth 100 mm	–	0.02	0.49	0.80	–	m²	**1.29**
average depth 150 mm	–	0.03	0.69	1.12	–	m²	**1.81**
average depth 200 mm	–	0.04	0.82	1.34	–	m²	**2.16**
average depth 250 mm	–	0.04	0.90	1.47	–	m²	**2.37**
average depth 300 mm	–	0.05	0.98	1.60	–	m²	**2.58**
Excavating; mechanical; topsoil for preservation; depositing to spoil heaps by wheeled dumper; spoil heaps maximum 50 m distance							
3 tonne tracked excavator (bucket volume 0.13 m³)							
average depth 100 mm	–	0.02	0.49	1.22	–	m²	**1.71**
average depth 150 mm	–	0.03	0.69	1.76	–	m²	**2.45**
average depth 200 mm	–	0.04	0.82	2.18	–	m²	**3.00**
average depth 250 mm	–	0.04	0.90	2.53	–	m²	**3.43**
average depth 300 mm	–	0.05	0.98	2.87	–	m²	**3.85**
Excavating; mechanical; to reduce levels							
3 tonne excavator (bucket volume 0.13 m³)							
maximum depth not exceeding 0.25 m	–	0.13	2.63	5.10	–	m³	**7.73**
maximum depth not exceeding 1.00 m	–	0.13	2.63	4.84	–	m³	**7.47**
Pits; 3 tonne tracked excavator							
maximum depth not exceeding 0.25 m	–	0.58	11.89	1.02	–	m³	**12.91**
maximum depth not exceeding 1.00 m	–	0.50	10.25	5.03	–	m³	**15.28**
maximum depth not exceeding 2.00 m	–	0.60	12.30	6.03	–	m³	**18.33**
Trenches; width not exceeding 0.30 m; 3 tonne excavator							
maximum depth not exceeding 0.25 m	–	1.33	27.33	5.59	–	m³	**32.92**
maximum depth not exceeding 1.00 m	–	0.69	14.04	2.87	–	m³	**16.91**
maximum depth not exceeding 2.00 m	–	0.60	12.30	2.51	–	m³	**14.81**

D GROUNDWORK

Item Excluding site overheads and profit	PC £	Labour hours	Labour £	Plant £	Material £	Unit	Total rate £
D20 EXCAVATION AND FILLING – CONT							
Excavating – cont							
Trenches; width exceeding 0.30 m; 3 tonne excavator							
maximum depth not exceeding 0.25 m	–	0.60	12.30	2.51	–	m³	14.81
maximum depth not exceeding 1.00 m	–	0.50	10.25	2.09	–	m³	12.34
maximum depth not exceeding 2.00 m	–	0.38	7.88	1.61	–	m³	9.49
Extra over any types of excavating irrespective of depth for breaking out existing materials; heavy duty 110 volt breaker tool							
hard rock	–	5.00	102.50	59.69	–	m³	162.19
concrete	–	3.00	61.50	35.81	–	m³	97.31
reinforced concrete	–	4.00	82.00	60.94	–	m³	142.94
brickwork, blockwork or stonework	–	1.50	30.75	17.91	–	m³	48.66
Extra over any types of excavating irrespective of depth for breaking out existing hard pavings; 1600 watt, 110 volt breaker							
concrete; 100 mm thick	–	0.20	4.10	1.60	–	m²	5.70
concrete; 150 mm thick	–	0.25	5.13	2.00	–	m²	7.13
concrete; 200 mm thick	–	0.29	5.86	2.29	–	m²	8.15
concrete; 300 mm thick	–	0.40	8.20	3.20	–	m²	11.40
reinforced concrete; 100 mm thick	–	0.33	6.83	9.26	–	m²	16.09
reinforced concrete; 150 mm thick	–	0.40	8.20	11.11	–	m²	19.31
reinforced concrete; 200 mm thick	–	0.50	10.25	13.89	–	m²	24.14
reinforced concrete; 300 mm thick	–	1.00	20.50	27.78	–	m²	48.28
tarmacadam; 75 mm thick	–	0.14	2.93	1.14	–	m²	4.07
tarmacadam and hardcore; 150 mm thick	–	0.20	4.10	1.60	–	m²	5.70
Extra over any types of excavating irrespective of depth for taking up							
precast concrete paving slabs	–	0.08	1.71	0.67	–	m²	2.38
natural stone paving	–	0.13	2.56	1.00	–	m²	3.56
cobbles	–	0.17	3.42	1.33	–	m²	4.75
brick paviors	–	0.17	3.42	1.33	–	m²	4.75
Excavating; hand							
Topsoil for preservation; loading to barrows							
average depth 100 mm	–	0.30	6.15	–	–	m²	6.15
average depth 200 mm	–	0.60	12.30	–	–	m²	12.30
average depth 300 mm	–	0.90	18.45	–	–	m²	18.45
over 300 mm	–	3.00	61.50	–	–	m³	61.50
Excavating; hand							
Topsoil to reduce levels							
maximum depth not exceeding 0.25 m	–	2.52	51.66	–	–	m³	51.66
maximum depth not exceeding 1.00 m	–	3.12	63.96	–	–	m³	63.96
Pits							
maximum depth not exceeding 0.25 m	–	2.67	54.67	–	–	m³	54.67
maximum depth not exceeding 1.00 m	–	3.47	71.07	–	–	m³	71.07
maximum depth not exceeding 2.00 m (includes earthwork support)	–	6.93	142.13	–	–	m³	204.50
Trenches; width not exceeding 0.30 m							
maximum depth not exceeding 0.25 m	–	2.86	58.57	–	–	m³	58.57
maximum depth not exceeding 1.00 m	–	3.72	76.28	–	–	m³	76.28
maximum depth not exceeding 1.00 m	–	3.72	76.28	–	–	m³	107.47

D GROUNDWORK

Item Excluding site overheads and profit	PC £	Labour hours	Labour £	Plant £	Material £	Unit	Total rate £
Trenches; width exceeding 0.30 m wide							
maximum depth not exceeding 0.25 m	–	2.86	58.57	–	–	m³	**58.57**
maximum depth not exceeding 1.00 m	–	4.00	82.00	–	–	m³	**82.00**
maximum depth not exceeding 2.00 m							
(includes earthwork support)	–	6.00	123.00	–	–	m³	**185.37**
Excavating; mechanical; trenches for							
electrical services							
3 tonne excavator (bucket volume 0.13 m³);							
arisings laid alongside							
600 mm deep	–	–	–	1.59	–	m	**1.59**
800 mm deep	–	–	–	1.86	–	m	**1.86**
1.00 m deep	–	–	–	2.23	–	m	**2.23**
By hand							
600 mm deep	–	0.24	4.93	–	–	m	**4.93**
800 mm deep	–	0.43	8.76	–	–	m	**8.76**
1.00 mm deep	–	0.67	13.67	–	–	m	**13.67**
Filling to make up levels; mechanical (3							
tonne tracked excavator); depositing in							
layers 150 mm maximum thickness							
Arising from the excavations							
maximum thickness less than 0.25 m	–	0.10	2.13	3.47	–	m³	**5.60**
average thickness exceeding 0.25 m	–	0.08	1.64	2.67	–	m³	**4.31**
Obtained from on site spoil heaps; average							
25 m distance; multiple handling							
maximum thickness less than 0.25 m		0.16	3.29	6.26	–	m³	**9.55**
average thickness exceeding 0.25 m	–	0.13	2.67	5.25	–	m³	**7.92**
Obtained off site; planting quality topsoil PC							
£26.00/m³							
maximum thickness less than 0.25 m	32.50	0.14	2.94	5.69	32.50	m³	**41.13**
average thickness exceeding 0.25 m	32.50	0.13	2.67	5.25	32.50	m³	**40.42**
Obtained off site; crushed concrete hardcore;							
PC £13.00/m³							
maximum thickness less than 0.25 m	15.86	0.14	2.87	13.34	15.86	m³	**32.07**
maximum thickness exceeding 0.25 m	15.86	0.13	2.61	12.80	15.86	m³	**31.27**
Backfilling of trenches; including laying of							
electrical marker tape; compacting lightly							
as work proceeds							
To trenches containing armoured cable							
by machine	–	–	–	–	–	m	**1.80**
by hand	–	–	–	–	–	m	**4.23**
To trenches containing ducted cable including							
150 mm sharp sand over the duct							
by machine	–	–	–	–	–	m	**4.49**
by hand	–	–	–	–	–	m	**7.53**
Filling to make up levels; hand; depositing							
in layers 150 mm maximum thickness							
Arising from the excavations							
average thickness exceeding 0.25 m	–	0.75	15.38	–	–	m³	**15.38**
Obtained from on site spoil heaps; average							
25 m distance; multiple handling							
average thickness exceeding 0.25 m thick	–	1.25	25.62	–	–	m³	**25.62**
Obtained off site; planting quality topsoil PC							
£43.50/m³ (10 tonne loads)							
average thickness exceeding 0.25 m thick	93.45	1.25	25.62	–	93.45	m³	**119.07**

D GROUNDWORK

Item Excluding site overheads and profit	PC £	Labour hours	Labour £	Plant £	Material £	Unit	Total rate £
D20 EXCAVATION AND FILLING – CONT							
Note: Bulkage factors have been applied to the figures below for disposal							
Disposal; mechanical;							
Light soils and loams (bulking factor of 1.25)							
Excavated material; off site; to tip;							
mechanically loaded by grab; capacity of load							
10–12 m³ (18 tonne)							
inert material (mixed building waste and							
landfill	–	–	–	–	–	m³	26.40
soil (sandy and loam) dry	–	–	–	–	–	m³	26.40
soil (sandy and loam) wet	–	–	–	–	–	m³	28.00
broken out compacted materials such as							
road bases and the like	–	–	–	–	–	m³	29.33
soil (clay) dry	–	–	–	–	–	m³	37.12
soil (clay) wet	–	–	–	–	–	m³	44.50
Disposal by skip; 6 yd³ (4.6 m³)							
Excavated material; off site; to tip							
by machine	–	–	–	47.13	–	m³	47.13
by hand	–	4.00	82.00	42.96	–	m³	124.96
Disposal on site							
Excavated material to spoil heaps							
average 25 m distance	–	–	–	5.57	–	m³	5.57
average 50 m distance	–	–	–	6.13	–	m³	6.13
average 100 m distance	–	–	–	6.68	–	m³	6.68
average 200 m distance	–	–	–	8.35	–	m³	8.35
Excavated material; spreading on site							
average 25 m distance	–	–	–	6.74	–	m³	6.74
average 50 m distance	–	–	–	7.38	–	m³	7.38
average 100 m distance	–	–	–	8.45	–	m³	8.45
average 200 m distance	–	–	–	9.30	–	m³	9.30
Disposal; hand							
Excavated material; on site; in spoil heaps							
average 25 m distance	–	2.40	49.20	–	–	m³	49.20
average 50 m distance	–	2.64	54.12	–	–	m³	54.12
average 100 m distance	–	3.00	61.50	–	–	m³	61.50
average 200 m distance	–	3.60	73.80	–	–	m³	73.80
Disposal hand; excavated material in bags							
Filling bags 25 kg and barrowing through							
buildings							
average 25 m distance	–	8.40	172.20	–	–	m³	172.20
average 50 m distance	–	9.20	188.60	–	–	m³	188.60
average 100 m distance	–	10.10	207.05	–	–	m³	207.05
average 200 m distance	–	11.60	237.80	–	–	m³	237.80
Excavated material; spreading on site							
average 25 m distance	–	2.64	54.12	–	–	m³	54.12
average 50 m distance	–	3.00	61.50	–	–	m³	61.50
average 100 m distance	–	3.60	73.80	–	–	m³	73.80
average 200 m distance	–	4.20	86.10	–	–	m³	86.10

D GROUNDWORK

Item Excluding site overheads and profit	PC £	Labour hours	Labour £	Plant £	Material £	Unit	Total rate £
Surface treatments							
Compacting							
bottoms of excavations	–	0.01	0.10	0.10	–	m²	0.20
Grading operations; surface previously							
excavated to reduce levels to prepare to							
receive subsequent treatments; grading to							
accurate levels and falls 20 mm tolerances							
Clay or heavy soils or hardcore							
3 tonne excavator	–	0.02	0.41	0.67	–	m²	1.08
5 tonne excavator	–	0.04	0.82	0.26	–	m²	1.08
by hand	–	0.10	2.05	–	–	m²	2.05
Loamy topsoils							
3 tonne excavator	–	0.01	0.27	0.44	–	m²	0.71
5 tonne excavator	–	0.03	0.55	0.17	–	m²	0.72
by hand	–	0.05	1.02	–	–	m²	1.02
Sand or graded granular materials							
3 tonne excavator	–	0.01	0.20	0.33	–	m²	0.53
5 tonne excavator	–	0.02	0.41	0.13	–	m²	0.54
by hand	–	0.03	0.68	–	–	m²	0.68
Excavating; by hand							
Topsoil for preservation; loading to barrows							
average depth 100 mm	–	0.24	4.92	–	–	m²	4.92
average depth 150 mm	–	0.36	7.38	–	–	m²	7.38
average depth 200 mm	–	0.58	11.81	–	–	m²	11.81
average depth 250 mm	–	0.72	14.76	–	–	m²	14.76
average depth 300 mm	–	0.86	17.71	–	–	m²	17.71
Grading operations; surface recently filled							
to raise levels to prepare to receive							
subsequent treatments; grading to							
accurate levels and falls 20 mm tolerances							
Clay or heavy soils or hardcore							
3 tonne excavator	–	0.02	0.31	0.51	–	m²	0.82
by hand	–	0.08	1.71	–	–	m²	1.71
Loamy topsoils							
3 tonne excavator	–	0.01	0.21	0.34	–	m²	0.55
by hand	–	0.04	0.82	–	–	m²	0.82
Sand or graded granular materials							
3 tonne excavator	–	0.01	0.15	0.25	–	m²	0.40
by hand	–	0.03	0.59	–	–	m²	0.59
Surface preparation							
Trimming surfaces of cultivated ground to							
final levels, removing roots stones and							
debris exceeding 50 mm in any direction to							
tip off site; slopes less than 15°							
clean ground with minimal stone							
content	–	0.25	5.13	–	–	100 m²	5.13
slightly stony; 0.5 kg stones per m²	–	0.33	6.83	0.02	–	100 m²	6.85
very stony; 1.0-3.0 kg stones per m²	–	0.50	10.25	0.04	–	100 m²	10.29
clearing mixed slightly contaminated							
rubble inclusive of roots and vegetation	–	0.50	10.25	8.95	–	100 m²	19.20
clearing brick-bats, stones and clean							
rubble	–	0.60	12.30	0.16	–	100 m²	12.46

D GROUNDWORK

Item Excluding site overheads and profit	PC £	Labour hours	Labour £	Plant £	Material £	Unit	Total rate £
D20 EXCAVATION AND FILLING – CONT							
Hand cultivation; preparation for turfing/ seeding areas							
Cultivate surface to receive landscape surface treatments using hand fork or implement; 1 spit (300 mm deep), break up ground and bring to a fine tilth by rake							
compacted ground	–	0.07	1.37	–	–	m²	**1.37**
ground previously supporting turf; lifting of turf not included	–	0.06	1.28	–	–	m²	**1.28**
ground previously planted with small to medium shrubs; removal of planting not included	–	0.06	1.14	–	–	m²	**1.14**
soft ground	–	0.04	0.82	–	–	m²	**0.82**
Hand cultivation; preparation for planting							
Cultivate surface to receive landscape surface treatments using hand fork or implement; 1 spit (300 mm deep), break up ground and bring to a fine tilth by rake							
compacted ground	–	0.05	1.02	–	–	m²	**1.02**
ground previously supporting turf; lifting of turf not included	–	0.04	0.85	–	–	m²	**0.85**
ground previously planted with small to medium shrubs; removal of planting not included	–	0.04	0.73	–	–	m²	**0.73**
soft ground	–	0.03	0.59	–	–	m²	**0.59**

E IN SITU CONCRETE/LARGE PRECAST CONCRETE

Item Excluding site overheads and profit	PC £	Labour hours	Labour £	Plant £	Material £	Unit	Total rate £
E10 MIXING/CASTING/CURING IN SITU CONCRETE							
General							
Please see the notes on concrete mix designations in the Major Works section of this book under E10 In Situ Concrete.							
Concrete mixes; mixed on site; costs for producing concrete; prices for commonly used mixes for various types of work; based on an 85 litre concrete mixer							
Roughest type mass concrete such as footings, road haunchings 300 mm thick; aggregates delivered in 10 tonne loads							
1:3:6	102.61	–	–	–	102.61	m³	**102.61**
1:3:6 sulphate-resisting	117.74	–	–	–	117.74	m³	**117.74**
As above but aggregates delivered in 850 kg bulk bags							
1:3:6	146.20	–	–	–	146.20	m³	**146.20**
1:3:6 sulphate-resisting	160.48	–	–	–	160.48	m³	**160.48**
Most ordinary use of concrete such as mass walls above ground, road slabs, etc. and general reinforced concrete work; aggregates delivered in 10 tonne loads							
1:2:4	85.95	1.51	31.02	–	85.95	m³	**116.97**
1:2:4 sulphate-resisting	138.11	–	–	–	138.11	m³	**138.11**
As above but aggregates delivered in 850 kg bulk bags							
1:2:4	48.15	1.51	31.02	–	128.69	m³	**159.71**
1:2:4 sulphate-resisting	–	–	–	–	180.84	m³	**180.84**
Watertight floors, pavements and walls, tanks, pits, steps, paths, surface of two course roads; reinforced concrete where extra strength is required (aggregates delivered in 10 tonne loads)							
1:1.5:3	99.42	1.51	31.02	–	99.42	m³	**130.44**
As above but aggregates delivered in 850 kg bulk bags							
1:1.5:3	61.62	1.51	31.02	–	142.16	m³	**173.18**
Plain in situ concrete; site mixed; 10 N/ mm² – 40 mm aggregate (1:3:6) (aggregate delivery indicated)							
Foundations							
ordinary Portland cement; 10 tonne ballast loads	90.31	1.00	20.50	–	90.31	m³	**110.81**
ordinary Portland cement; 850 kg bulk bags	146.20	1.00	20.50	–	146.20	m³	**166.70**
sulphate-resistant cement; 10 tonne ballast loads	117.74	1.00	20.50	–	117.74	m³	**138.24**
sulphate-resistant cement; 850 kg bulk bags	160.48	1.00	20.50	–	160.48	m³	**180.98**

E IN SITU CONCRETE/LARGE PRECAST CONCRETE

Item Excluding site overheads and profit	PC £	Labour hours	Labour £	Plant £	Material £	Unit	Total rate £
E10 MIXING/CASTING/CURING IN SITU CONCRETE – CONT							
Plain in situ concrete – cont							
Foundations; poured on or against earth or unblinded hardcore							
ordinary Portland cement; 10 tonne ballast loads	94.82	1.00	20.50	–	94.82	m³	**115.32**
ordinary Portland cement; 850 kg bulk bags	153.51	1.00	20.50	–	153.51	m³	**174.01**
sulphate-resistant cement; 10 tonne ballast loads	123.63	1.00	20.50	–	123.63	m³	**144.13**
sulphate-resistant cement; 850 kg bulk bags	168.50	1.00	20.50	–	168.50	m³	**189.00**
Isolated foundations							
ordinary Portland cement; 10 tonne ballast loads	92.56	2.00	41.00	–	92.56	m³	**133.56**
ordinary Portland cement; 850 kg bulk bags	149.85	2.00	41.00	–	149.85	m³	**190.85**
sulphate-resistant cement; 10 tonne ballast loads	120.69	2.00	41.00	–	120.69	m³	**161.69**
sulphate-resistant cement; 850 kg bulk bags	164.49	2.00	41.00	–	164.49	m³	**205.49**
Plain in situ concrete; site mixed; 21 N/mm² – 20 mm aggregate (1:2:4)							
Foundations							
ordinary Portland cement; 10 tonne ballast loads	121.02	1.00	20.50	–	121.02	m³	**141.52**
ordinary Portland cement; 850 kg bulk bags	–	1.00	20.50	–	164.83	m³	**185.33**
sulphate-resistant cement; 10 tonne ballast loads	141.56	1.00	20.50	–	141.56	m³	**162.06**
sulphate-resistant cement; 850 kg bulk bags	185.37	1.00	20.50	–	185.37	m³	**205.87**
Foundations; poured on or against earth or unblinded hardcore							
ordinary Portland cement; 10 tonne ballast loads	123.97	1.00	20.50	–	123.97	m³	**144.47**
ordinary Portland cement; 850 kg bulk bags	123.97	1.50	30.75	–	123.97	m³	**154.72**
sulphate-resistant cement; 10 tonne ballast loads	145.01	1.10	22.55	–	145.01	m³	**167.56**
sulphate-resistant cement; 850 kg bulk bags	189.89	1.50	30.75	–	189.89	m³	**220.64**
Isolated foundations							
ordinary Portland cement	138.11	2.00	41.00	–	138.11	m³	**179.11**
sulphate-resistant cement	141.56	2.00	41.00	–	141.56	m³	**182.56**
Reinforced in situ concrete; site mixed; 21 N/mm² – 20 mm aggregate (1:2:4); aggregates delivered in 10 tonne loads							
Foundations							
ordinary Portland cement	121.02	1.10	22.55	–	121.02	m³	**143.57**
sulphate-resistant cement	141.56	1.10	22.55	–	141.56	m³	**164.11**
Foundations; poured on or against earth or unblinded hardcore							
ordinary Portland cement	145.01	1.10	22.55	–	145.01	m³	**167.56**
sulphate-resistant cement	145.01	1.10	22.55	–	145.01	m³	**167.56**
Isolated foundations							
ordinary Portland cement	121.02	2.20	45.10	–	121.02	m³	**166.12**
sulphate-resistant cement	145.01	2.20	45.10	–	145.01	m³	**190.11**

E IN SITU CONCRETE/LARGE PRECAST CONCRETE

Item Excluding site overheads and profit	PC £	Labour hours	Labour £	Plant £	Material £	Unit	Total rate £
Plain in situ concrete; ready mixed; Tarmac Southern; 10 N/mm mixes; suitable for mass concrete fill and blinding							
Foundations							
GEN1; designated mix	64.37	1.50	30.75	–	64.37	m³	**95.12**
ST2; standard mix	65.70	1.50	30.75	–	65.70	m³	**96.45**
Foundations; poured on or against earth or unblinded hardcore							
GEN1; designated mix	65.94	1.57	32.29	–	65.94	m³	**98.23**
ST2; standard mix	67.31	1.57	32.29	–	67.31	m³	**99.60**
Isolated foundations							
GEN1; designated mix	64.37	2.00	41.00	–	64.37	m³	**105.37**
ST2; standard mix	65.70	2.00	41.00	–	65.70	m³	**106.70**
Plain in situ concrete; ready mixed; Tarmac Southern; 15 N/mm mixes; suitable for oversite below suspended slabs and strip footings in non aggressive soils							
Foundations							
GEN 2; designated mix	66.01	1.50	30.75	–	66.01	m³	**96.76**
ST3; standard mix	70.03	1.50	30.75	–	70.03	m³	**100.78**
Foundations; poured on or against earth or unblinded hardcore							
GEN2; designated mix	66.01	1.57	32.29	–	66.01	m³	**98.30**
ST3; standard mix	68.37	1.57	32.29	–	68.37	m³	**100.66**
Isolated foundations							
GEN2; designated mix	66.01	2.00	41.00	–	66.01	m³	**107.01**
ST3; standard mix	68.37	2.00	41.00	–	68.37	m³	**109.37**
Concrete mixed on site; Easymix Concrete Ltd							
Concrete mixer on lorry; maximum standing time 3 m³/hr							
C20 concrete	95.00	1.00	20.50	–	95.00	m³	**115.50**
C25 concrete	95.00	1.00	20.50	–	95.00	m³	**115.50**
C30 concrete	105.00	1.00	20.50	–	105.00	m³	**125.50**
Concrete mixed on site; Easymix Concrete Ltd; concrete delivered by pump							
Concrete mixer on lorry; maximum standing time 3 m³/hr							
C20 concrete	95.00	0.05	1.02	–	112.50	m³	**113.52**
C25 concrete	95.00	1.00	20.50	–	112.50	m³	**133.00**
C30 concrete	105.00	0.05	1.02	–	122.50	m³	**123.52**
E20 FORMWORK FOR IN SITU CONCRETE							
Plain vertical formwork; basic finish							
Sides of foundations							
height exceeding 1.00 m	–	2.00	41.00	–	5.34	m²	**46.34**
height not exceeding 250 mm	–	1.00	20.50	–	1.52	m	**22.02**
height 250–500 mm	–	1.00	20.50	–	2.67	m	**23.17**
height 500 mm–1.00 m	–	1.50	30.75	–	5.34	m	**36.09**

E IN SITU CONCRETE/LARGE PRECAST CONCRETE

Item Excluding site overheads and profit	PC £	Labour hours	Labour £	Plant £	Material £	Unit	Total rate £
E20 FORMWORK FOR IN SITU CONCRETE – CONT							
Plain vertical formwork – cont							
Sides of foundations; left in							
height over 1.00 m	–	2.00	41.00	–	15.10	m²	56.10
height not exceeding 250 mm	–	1.00	20.50	–	5.61	m	26.11
height 250–500 mm	–	1.00	20.50	–	10.85	m	31.35
height 500 mm–1.00 m	–	1.50	30.75	–	21.70	m	52.45
E30 REINFORCEMENT FOR IN SITU CONCRETE							
For reinforcement bar prices please see the Major Works section of this book.							
Reinforcement fabric; lapped; in beds or suspended slabs							
Fabric 3.6 × 2.0 m²							
ref A142 (2.22 kg/m²)	4.08	0.22	4.51	–	4.08	m²	8.59
ref A393 (6.16 kg/m²)	11.32	0.28	5.74	–	11.32	m²	17.06
Fabric 2.4 × 4.8 m²							
ref A98 (1.54 kg/m²)	1.96	0.22	4.51	–	1.96	m²	6.47
ref A142 (2.22 kg/m²)	2.82	0.22	4.51	–	2.82	m²	7.33
ref A193 (3.02 kg/m²)	3.84	0.22	4.51	–	3.84	m²	8.35
ref A252 (3.95 kg/m²)	4.54	0.24	4.92	–	4.54	m²	9.46
ref A393 (6.16 kg/m²)	7.08	0.28	5.74	–	7.08	m²	12.82

F MASONRY

em Excluding site overheads and profit	PC £	Labour hours	Labour £	Plant £	Material £	Unit	Total rate £
F10 BRICK/BLOCK WALLING							
Batching quantities for these mortar mixes may be found in the Memoranda section of this book.							
Mortar mixes; common mixes for various types of work; mortar mixed on site; prices based on builders merchant rates for cement; mechanically mixed							
Aggregates delivered in 850 kg bulk bags							
1:3	–	0.75	15.38	–	169.51	m³	**184.89**
1:4	–	0.75	15.38	–	133.59	m³	**148.97**
1:1:6	–	0.75	15.38	–	162.19	m³	**177.57**
1:1:6 sulphate-resisting	–	0.75	15.38	–	181.20	m³	**196.58**
Aggregates delivered in 10 tonne loads							
1:2	–	0.75	15.38	–	153.25	m³	**168.63**
1:3	–	0.75	15.38	–	125.14	m³	**140.52**
1:4	–	0.75	15.38	–	102.96	m³	**118.34**
1:1:6	–	0.75	15.38	–	127.16	m³	**142.54**
1:1:6 sulphate-resisting	–	0.75	15.38	–	143.97	m³	**159.35**
Mortar mixes; common mixes for various types of work; mortar mixed on site; prices based on builders merchant rates for cement; hand mixed							
Aggregates delivered in 850 kg bulk bags							
1:3	–	2.00	41.00	–	171.61	m³	**212.61**
1:4	–	2.00	41.00	–	135.35	m³	**176.35**
1:1:6	–	2.00	41.00	–	163.95	m³	**204.95**
1:1:6 sulphate-resisting	–	2.00	41.00	–	182.96	m³	**223.96**
Variation in brick prices Add or subtract the following amounts for every £1.00/1000 difference in the PC price of the measured items below							
half brick thick	–	–	–	–	0.06	m²	**0.06**
one brick thick	–	–	–	–	0.13	m²	**0.13**
one and a half brick thick	–	–	–	–	0.19	m²	**0.19**
two brick thick	–	–	–	–	0.25	m²	**0.25**
Movement of materials Loading to wheelbarrows and transporting to location; per 215 mm thick walls; maximum 25 m distance	–	0.42	8.54	–	–	m²	**8.54**
Class B engineering bricks; PC £290.00/ 1000; double Flemish bond in cement mortar (1:3) Mechanically offloading; maximum 25 m distance; loading to wheelbarrows; transporting to location; per 215 mm thick walls	–	0.42	8.54	–	–	m²	**8.54**
Walls							
half brick thick	–	3.19	65.44	–	19.54	m²	**84.98**
one brick thick	–	6.38	130.87	–	39.08	m²	**169.95**
one and a half brick thick	–	9.58	196.34	–	58.62	m²	**254.96**
two brick thick	–	12.77	261.74	–	78.16	m²	**339.90**

F MASONRY

Item Excluding site overheads and profit	PC £	Labour hours	Labour £	Plant £	Material £	Unit	Total rate £
F10 BRICK/BLOCK WALLING – CONT							
Class B engineering bricks – cont							
Walls; curved; mean radius 6 m							
half brick thick	–	4.81	98.51	–	20.25	m²	118.7
one brick thick	–	9.61	196.93	–	39.08	m²	236.0
Walls; curved; mean radius 1.5 m							
half brick thick	–	6.40	131.16	–	20.25	m²	151.4
one brick thick	–	12.80	262.48	–	39.08	m²	301.5
Walls; tapering; one face battering; average							
one and a half brick thick	–	10.34	212.07	–	58.62	m²	270.6
two brick thick	–	15.52	318.10	–	78.16	m²	396.2
Walls; battering (retaining)							
one and a half brick thick	–	11.49	235.63	–	58.62	m²	294.2
two brick thick	–	17.24	353.45	–	78.16	m²	431.6
Isolated piers							
one brick thick	–	9.31	190.85	–	43.36	m²	234.2
one and a half brick thick	–	11.97	245.38	–	65.04	m²	310.4
two brick thick	–	13.30	272.65	–	88.14	m²	360.7
three brick thick	–	16.49	338.09	–	130.08	m²	468.1
Projections; vertical							
one brick × half brick	–	0.70	14.35	–	4.36	m	18.7
one brick × one brick	–	1.40	28.70	–	8.73	m	37.4
one and a half brick × one brick	–	2.10	43.05	–	13.10	m	56.1
two brick × one brick	–	2.30	47.15	–	17.46	m	64.6
Walls; half brick thick							
in honeycomb bond	–	1.80	36.90	–	14.45	m²	51.3
in quarter bond	–	1.67	34.17	–	18.83	m²	53.0
Facing bricks; PC £300.00/1000; English **garden wall bond; in gauged mortar (1:1:6);** **facework one side**							
Mechanically offloading; maximum 25 m distance; loading to wheelbarrows and transporting to location; per 215 mm thick walls	–	0.42	8.54	–	–	m²	8.5
Walls							
half brick thick	–	3.19	65.44	–	21.05	m²	86.4
half brick thick (using site cut snap headers to form bond)	–	4.99	102.24	–	21.05	m²	123.2
one brick thick	–	6.38	130.87	–	39.95	m²	170.8
one and a half brick thick	–	9.58	196.31	–	63.16	m²	259.4
two brick thick	–	12.77	261.74	–	84.22	m²	345.9
Walls; curved; mean radius 6 m							
half brick thick	–	4.81	98.51	–	22.67	m²	121.1
one brick thick	–	9.61	196.93	–	43.91	m²	240.8
Walls; curved; mean radius 1.5 m							
half brick thick	–	6.40	131.16	–	22.22	m²	153.3
one brick thick	–	12.80	262.48	–	43.01	m²	305.4
Walls; tapering; one face battering; average							
one and a half brick thick	–	11.49	235.59	–	65.86	m²	301.4
two brick thick	–	17.24	353.45	–	87.82	m²	441.2
Walls; battering (retaining)							
one and a half brick thick	–	10.34	212.03	–	65.86	m²	277.8
two brick thick	–	15.52	318.10	–	87.82	m²	405.9
Isolated piers; English bond; facework all round							
one brick thick	–	9.31	190.85	–	46.42	m²	237.2
one and a half brick thick	–	11.97	245.38	–	69.62	m²	315.0
two brick thick	–	13.30	272.65	–	94.27	m²	366.9
three brick thick	–	16.49	338.09	–	139.25	m²	477.3

F MASONRY

Item Excluding site overheads and profit	PC £	Labour hours	Labour £	Plant £	Material £	Unit	Total rate £
Projections; vertical							
one brick × half brick	–	0.93	19.09	–	4.70	m	**23.79**
one brick × one brick	–	1.86	38.17	–	9.40	m	**47.57**
one and a half brick × one brick	–	2.79	57.26	–	14.11	m	**71.37**
two brick × one brick	–	3.06	62.71	–	18.81	m	**81.52**
Brickwork fair faced both sides; facing bricks in gauged mortar (1:1:6)							
Extra for fair face both sides; flush, struck, weathered or bucket-handle pointing		0.89	18.18	–	–	m²	**18.18**
Reclaimed bricks; PC £1000.00/1000; English garden wall bond; in gauged mortar (1:1:6)							
Walls							
half brick thick (stretcher bond)	–	3.19	65.44	–	65.15	m²	**130.59**
half brick thick (using site cut snap headers to form bond)	–	4.99	102.24	–	65.15	m²	**167.39**
one brick thick	–	6.38	130.87	–	127.91	m²	**258.78**
one and a half brick thick	–	9.58	196.34	–	191.86	m²	**388.20**
two brick thick	–	12.77	261.74	–	255.82	m²	**517.56**
Walls; curved; mean radius 6 m							
half brick thick	–	4.81	98.51	–	68.87	m²	**167.38**
one brick thick	–	9.61	196.93	–	136.31	m²	**333.24**
Walls; curved; mean radius 1.50 m							
half brick thick	–	6.40	131.16	–	71.87	m²	**203.03**
one brick thick	–	12.77	261.74	–	142.31	m²	**404.05**
Walls; stretcher bond; wall ties at 450 centres vertically and horizontally							
one brick thick	–	6.96	142.77	–	128.43	m²	**271.20**
two brick thick	–	13.93	285.50	–	257.37	m²	**542.87**
Brickwork fair faced both sides; facing bricks in gauged mortar (1:1:6)							
extra for fair face both sides; flush, struck, weathered or bucket-handle pointing	–	0.67	13.67	–	–	m²	**13.67**
Brick copings							
Copings; all brick headers-on-edge; two angles rounded 53 mm radius; flush pointing top and both sides as work proceeds; one brick wide; horizontal							
machine-made specials	51.49	0.71	14.57	–	52.34	m	**66.91**
handmade specials	65.23	0.71	14.57	–	66.09	m	**80.66**
Extra over copings for two courses machine-made tile creasings, projecting 25 mm each side; 260 mm wide copings; horizontal	6.47	0.50	10.25	–	6.90	m	**17.15**
Copings; all brick headers-on-edge; flush pointing top and both sides as work proceeds; one brick wide; horizontal							
bricks PC £300.00/1000	4.30	0.71	14.57	–	4.58	m	**19.15**
bricks PC £500.00/1000	6.83	0.71	14.57	–	7.13	m	**21.70**

F MASONRY

Item Excluding site overheads and profit	PC £	Labour hours	Labour £	Plant £	Material £	Unit	Total rate £
F10 BRICK/BLOCK WALLING – CONT							
Dense aggregate concrete blocks; Tarmac							
Topblock or other equal and approved; in							
gauged mortar (1:2:9)							
Solid blocks 7 N/mm²							
440 × 215 × 100 mm thick	8.83	1.20	24.60	–	9.97	m²	34.57
440 × 215 × 140 mm thick	14.19	1.30	26.65	–	15.47	m²	42.12
Solid blocks 7 N/mm² laid flat							
440 × 100 × 215 mm thick	18.02	3.22	66.01	–	21.09	m²	87.10
Hollow concrete blocks							
440 × 215 × 215 mm thick	20.78	1.30	26.65	–	22.06	m²	48.71
Filling of hollow concrete blocks with concrete							
as work proceeds; tamping and compacting							
440 × 215 × 215 mm thick	21.58	0.20	4.10	–	21.58	m²	25.68
Isolated brick piers in English bond							
Engineering brick PC £239.00/1000 (not							
SMM)							
one brick thick (225 mm)	–	1.63	33.49	–	9.95	m	43.44
one and a half bricks thick (337.5 mm)	–	3.00	61.50	–	29.39	m	90.89
two bricks thick (450 mm)	–	4.50	92.25	–	47.01	m	139.26
Brick PC £300.00/1000 (not SMM)							
one brick thick (225 mm)	–	1.63	33.49	–	10.23	m	43.72
one and a half bricks thick (337.5 mm)	–	3.00	61.50	–	29.99	m	91.49
two bricks thick (450 mm)	–	4.50	92.25	–	48.09	m	140.34
three bricks thick (675 mm)	–	8.37	171.59	–	98.58	m	270.17
Brick PC £800.00/1000 (not SMM)							
one brick thick (225 mm)	–	1.63	33.49	–	24.23	m	57.72
one and a half bricks thick (337.5 mm)	–	3.00	61.50	–	60.37	m	121.87
two bricks thick (450 mm)	–	4.50	92.25	–	102.09	m	194.34
three bricks thick (675 mm)	–	8.37	171.59	–	220.08	m	391.67

G STRUCTURAL/CARCASSING METAL/TIMBER

Item Excluding site overheads and profit	PC £	Labour hours	Labour £	Plant £	Material £	Unit	Total rate £
G31 TIMBER DECKING							
Timber decking							
Supports for timber decking; softwood joists to receive decking boards; joists at 400 mm centres; Southern yellow pine							
38 × 88 mm	18.20	1.00	20.50	–	20.19	m²	**40.69**
50 × 150 mm	12.54	1.00	20.50	–	14.53	m²	**35.03**
50 × 125 mm	10.45	1.00	20.50	–	12.44	m²	**32.94**
Hardwood decking; yellow Balau; grooved or smooth; 6 mm joints							
deck boards; 90 mm wide × 19 mm thick	19.99	1.00	20.50	–	22.26	m²	**42.76**
deck boards; 145 mm wide × 21 mm thick	32.26	1.00	20.50	–	34.54	m²	**55.04**
deck boards; 145 mm wide × 28 mm thick	33.21	1.00	20.50	–	35.48	m²	**55.98**
Hardwood decking; Ipe; smooth; 6 mm joints							
deck boards; 90 mm wide × 19 mm thick	45.18	1.00	20.50	–	47.46	m²	**67.96**
deck boards; 145 mm wide × 19 mm thick	46.34	1.00	20.50	–	48.62	m²	**69.12**
Western Red Cedar; 6 mm joints							
Prime deck grade; 90 mm wide × 40 mm thick	38.13	1.00	20.50	–	40.41	m²	**60.91**
Prime deck grade; 142 mm wide × 40 mm thick	40.54	1.00	20.50	–	42.82	m²	**63.32**
Handrails and base rail; fixed to posts at 2.00 m centres							
posts 100 × 100 × 1370 mm high	9.19	1.00	20.50	–	9.73	m	**30.23**
posts turned 1220 mm high	17.83	1.00	20.50	–	18.36	m	**38.86**
Handrails; balusters							
square balusters at 100 mm centres	39.90	0.50	10.25	–	40.43	m	**50.68**
square balusters at 300 mm centres	13.29	0.33	6.83	–	13.71	m	**20.54**
turned balusters at 100 mm centres	63.00	0.50	10.25	–	63.53	m	**73.78**
turned balusters at 300 mm centres	20.98	0.33	6.83	–	21.41	m	**28.24**

J WATERPROOFING

Item Excluding site overheads and profit	PC £	Labour hours	Labour £	Plant £	Material £	Unit	Total rate £
J30 LIQUID APPLIED TANKING/DAMP-PROOFING							
Tanking and damp-proofing; Ruberoid Building Products; Synthaprufe cold applied bituminous emulsion waterproof coating							
Synthaprufe; to smooth finished concrete or screeded slabs; flat; blinding with sand							
two coats	2.75	0.22	4.56	–	2.91	m²	**7.47**
three coats	4.12	0.31	6.36	–	4.29	m²	**10.65**
Synthaprufe; to fair faced brickwork with flush joints, rendered brickwork or smooth finished concrete walls; vertical							
two coats	3.09	0.29	5.86	–	3.26	m²	**9.12**
three coats	4.53	0.40	8.20	–	4.70	m²	**12.90**
Tanking and damp-proofing; RIW Ltd							
Liquid asphaltic composition; to smooth finished concrete screeded slabs or screeded slabs; flat							
two coats	6.24	0.33	6.83	–	6.24	m²	**13.07**
Liquid asphaltic composition; fair faced brickwork with flush joints, rendered brickwork or smooth finished concrete walls; vertical							
two coats	6.24	0.50	10.25	–	6.24	m²	**16.49**
Heviseal; to smooth finished concrete or screeded slabs; to surfaces of ponds, tanks, planters; flat							
two coats	9.46	0.33	6.83	–	9.46	m²	**16.29**
Heviseal; to fair faced brickwork with flush joints, rendered brickwork or smooth finished concrete walls; to surfaces of retaining walls, ponds, tanks, planters; vertical							
two coats	9.46	0.50	10.25	–	9.46	m²	**19.71**

M SURFACE FINISHES

Item Excluding site overheads and profit	PC £	Labour hours	Labour £	Plant £	Material £	Unit	Total rate £
M60 PAINTING/CLEAR FINISHING							
Prepare; touch up primer; two undercoats and one finishing coat of gloss oil paint; on metal surfaces							
General surfaces							
girth exceeding 300 mm	1.38	0.67	13.67	–	1.38	m²	**15.05**
isolated surfaces; girth not exceeding 300 mm	0.46	0.40	8.20	–	0.46	m	**8.66**
isolated areas not exceeding 0.50 m²							
irrespective of girth	0.69	0.67	13.67	–	0.69	nr	**14.36**
Ornamental railings; each side measured overall							
girth exceeding 300 mm	1.38	0.75	15.38	–	1.38	m²	**16.76**
Prepare; one coat primer; two undercoats and one finishing coat of gloss oil paint; on wood surfaces							
General surfaces							
girth exceeding 300 mm	1.29	0.40	8.20	–	1.29	m²	**9.49**
isolated surfaces; girth not exceeding 300 mm	0.39	0.20	4.10	–	0.39	m	**4.49**
isolated areas not exceeding 0.50 m²							
irrespective of girth	0.64	0.36	7.45	–	0.64	nr	**8.09**
Prepare; two coats of Protek wood preservative; on wood surfaces							
Planed surfaces							
girth exceeding 300 mm	0.33	0.07	1.40	–	0.33	m²	**1.73**
isolated surfaces; girth not exceeding 300 mm	1.10	0.17	3.50	–	1.10	m	**4.60**
Sawn surfaces							
girth exceeding 300 mm	0.22	0.10	2.10	–	0.22	m²	**2.32**
isolated surfaces; girth not exceeding 300 mm	0.07	0.17	3.50	–	0.07	m	**3.57**
Prepare; two coats of Sadolin wood preservative; on wood surfaces							
Planed surfaces							
girth exceeding 300 mm	2.32	0.07	1.40	–	2.32	m²	**3.72**
isolated surfaces; girth not exceeding 300 mm	0.77	0.14	2.80	–	0.77	m	**3.57**
Sawn surfaces							
girth exceeding 300 mm	3.72	0.10	2.10	–	3.72	m²	**5.82**
isolated surfaces; girth not exceeding 300 mm	0.62	0.17	3.50	–	0.62	m	**4.12**
Prepare; proprietary solution primer; two coats of dark stain; on wood surfaces							
General surfaces							
girth exceeding 300 mm	1.33	0.17	3.42	–	1.33	m²	**4.75**
isolated surfaces; girth not exceeding 300 mm	0.44	0.13	2.56	–	0.44	m	**3.00**
Two coats Weathershield exterior paint; to clean, dry surfaces; in accordance with manufacturer's instructions							
Brick or block walls							
girth exceeding 300 mm	–	0.33	6.83	–	1.09	m²	**7.92**
Cement render or concrete walls							
girth exceeding 300 mm	–	0.28	5.74	–	1.09	m²	**6.83**

Q PAVING/PLANTING/FENCING/SITE FURNITURE

Item Excluding site overheads and profit	PC £	Labour hours	Labour £	Plant £	Material £	Unit	Total rate £
Q10 KERBS/EDGINGS/CHANNELS/PAVING ACCESSORIES							
Foundations to kerbs							
Excavating trenches; width 300 mm; 3 tonne excavator							
depth 300 mm	–	0.10	2.05	0.42	1.64	m	**4.11**
depth 400 mm	–	0.11	2.28	0.47	2.20	m	**4.95**
By hand	–	0.60	12.35	–	2.74	m	**15.09**
Excavating trenches; width 450 mm; 3 tonne excavator; disposal off site							
depth 300 mm	–	0.13	2.56	0.63	2.47	m	**5.66**
depth 400 mm	–	0.14	2.93	0.72	3.30	m	**6.95**
Disposal							
Excavated material; off site; to tip							
grab loaded 7.25 m³	–	–	–	–	–	m³	**1.64**
Disposal by skip; 6 yd³ (4.6 m³)							
by machine	–	–	–	40.09	–	m³	**40.09**
by hand	–	4.00	82.00	42.96	–	m³	**124.96**
Foundations; to kerbs, edgings or channels; in situ concrete; 21 N/mm² – 20 mm aggregate (1:2:4) site mixed; one side against earth face, other against formwork (not included); site mixed concrete							
Site mixed concrete							
150 wide × 100 mm deep	–	0.13	2.73	–	1.77	m	**4.50**
150 wide × 150 mm deep	–	0.17	3.42	–	2.66	m	**6.08**
200 wide × 150 mm deep	–	0.20	4.10	–	3.54	m	**7.64**
300 wide × 150 mm deep	–	0.23	4.78	–	5.31	m	**10.09**
600 wide × 200 mm deep	–	0.29	5.86	–	14.17	m	**20.03**
Ready mixed concrete							
150 wide × 100 mm deep	–	0.13	2.73	–	0.99	m	**3.72**
150 wide × 150 mm deep	–	0.17	3.42	–	1.48	m	**4.90**
200 wide × 150 mm deep	–	0.20	4.10	–	1.97	m	**6.07**
300 wide × 150 mm deep	–	0.23	4.78	–	2.96	m	**7.74**
600 wide × 200 mm deep	–	0.29	5.86	–	7.88	m	**13.74**
Formwork; sides of foundations (this will usually be required to one side of each kerb foundation adjacent to road subbases)							
100 mm deep	–	0.04	0.85	–	0.32	m	**1.17**
150 mm deep	–	0.04	0.85	–	0.47	m	**1.32**
Precast concrete edging units; including haunching with in situ concrete; 11.50 N/mm² – 40 mm aggregate both sides							
Edgings; rectangular, bullnosed or chamfered							
50 × 150 mm	1.57	0.33	6.83	–	6.08	m	**12.91**
125 × 150 mm bullnosed	1.71	0.33	6.83	–	6.22	m	**13.05**
50 × 200 mm	2.34	0.33	6.83	–	6.86	m	**13.69**
50 × 250 mm	2.72	0.33	6.83	–	7.23	m	**14.06**
50 × 250 mm flat top	3.35	0.33	6.83	–	7.87	m	**14.70**

Q PAVING/PLANTING/FENCING/SITE FURNITURE

Item Excluding site overheads and profit	PC £	Labour hours	Labour £	Plant £	Material £	Unit	Total rate £
Marshalls Mono; small element precast concrete kerb system; Keykerb Large (KL) upstand of 100–125 mm; on 150 mm concrete foundation including haunching with in situ concrete 1:3:6 1 side							
Bullnosed or half battered; 100 × 127 × 200 mm							
laid straight	12.79	0.80	16.40	–	13.74	m	**30.14**
radial blocks laid to curve; 8 blocks/1/4 circle – 500 mm radius	9.85	0.80	16.40	–	10.80	m	**27.20**
radial blocks laid to curve; 8 radial blocks; alternating 8 standard blocks/1/4 circle – 1000 mm radius	16.24	1.25	25.63	–	17.19	m	**42.82**
radial blocks; alternating 16 standard blocks/1/4 circle – 1500 mm radius	14.94	1.50	30.75	–	15.89	m	**46.64**
internal angle 90°	5.40	0.20	4.10	–	5.40	nr	**9.50**
external angle	5.40	0.20	4.10	–	5.40	nr	**9.50**
drop crossing kerbs; KL half battered to KL splay; LH and RH	11.30	1.00	20.50	–	12.25	pair	**32.75**
drop crossing kerbs; KL half battered to KS bullnosed	11.00	1.00	20.50	–	11.95	pair	**32.45**
Marshalls Mono; small element precast concrete kerb system; Keykerb Small (KS) upstand of 25–50 mm; on 150 mm concrete foundation including haunching with in situ concrete 1:3:6 1 side							
Half battered							
laid straight	8.99	0.80	16.40	–	9.94	m	**26.34**
radial blocks laid to curve; 8 blocks/1/4 circle – 500 mm radius	7.39	1.00	20.50	–	8.34	m	**28.84**
radial blocks laid to curve; 8 radial blocks; alternating 8 standard blocks/1/4 circle – 1000 mm radius	11.88	1.25	25.63	–	12.83	m	**38.46**
radial blocks; alternating 16 standard blocks/1/4 circle – 1500 mm radius	12.72	1.50	30.75	–	13.67	m	**44.42**
internal angle 90°	5.40	0.20	4.10	–	5.40	nr	**9.50**
external angle	5.40	0.20	4.10	–	5.40	nr	**9.50**
Second-hand granite setts; 100 × 100 mm; bedding in cement mortar (1:4); on 150 mm deep concrete foundations; including haunching with in situ concrete; 11.50 N/mm² – 40 mm aggregate one side							
Edgings							
300 mm wide	13.24	1.33	27.33	–	17.79	m	**45.12**

Q PAVING/PLANTING/FENCING/SITE FURNITURE

Item Excluding site overheads and profit	PC £	Labour hours	Labour £	Plant £	Material £	Unit	Total rate £
Q10 KERBS/EDGINGS/CHANNELS/PAVING ACCESSORIES – CONT							
Brick or block stretchers; bedding in cement mortar (1:4); on 150 mm deep concrete foundations, including haunching with in situ concrete; 11.50 N/mm² – 40 mm aggregate one side							
Single course							
concrete paving blocks; PC £8.72/m²; 200 × 100 × 60 mm	0.89	0.31	6.31	–	3.92	m	**10.23**
engineering bricks; PC £300.00/1000; 215 × 102.5 × 65 mm	1.35	0.40	8.20	–	4.39	m	**12.59**
paving bricks; PC £500.00/1000; 215 × 102.5 × 65 mm	2.22	0.40	8.20	–	5.26	m	**13.46**
Two courses							
concrete paving blocks; PC £8.72/m²; 200 × 100 × 60 mm	1.78	0.40	8.20	–	5.35	m	**13.55**
engineering bricks; PC £300.00/1000; 215 × 102.5 × 65 mm	2.71	0.57	11.71	–	6.28	m	**17.99**
paving bricks; PC £500.00/1000; 215 × 102.5 × 65 mm	4.44	0.57	11.71	–	8.02	m	**19.73**
Three courses							
concrete paving blocks; PC £8.72/m²; 200 × 100 × 60 mm	2.67	0.44	9.11	–	6.24	m	**15.35**
engineering bricks; PC £300.00/1000; 215 × 102.5 × 65 mm	4.06	0.67	13.67	–	7.63	m	**21.30**
paving bricks; PC £500.00/1000; 215 × 102.5 × 65 mm	6.67	0.67	13.67	–	10.24	m	**23.91**
Brick or block headers; brick on flat; bedding in cement mortar (1:4); on 150 mm deep concrete foundations; including haunching with in situ concrete; 11.50 N/mm² – 40 mm aggregate one side							
Single course							
concrete paving blocks; PC £8.72/m²; 200 × 100 × 60 mm	1.78	0.62	12.61	–	4.81	m	**17.42**
engineering bricks; PC £300.00/1000; 215 × 102.5 × 65 mm	2.71	0.80	16.40	–	5.74	m	**22.14**
paving bricks; PC £500.00/1000; 215 × 102.5 × 65 mm	4.44	0.80	16.40	–	7.48	m	**23.88**
Bricks on edge; bedding in cement mortar (1:4); on 150 mm deep concrete foundations; including haunching with in situ concrete; 11.50 N/mm² – 40 mm aggregate one side							
One brick wide							
engineering bricks; 215 × 102.5 × 65 mm	4.06	0.57	11.71	–	7.76	m	**19.47**
paving bricks; 215 × 102.5 × 65 mm	6.67	0.57	11.71	–	10.37	m	**22.08**
Two courses; stretchers laid on edge; 225 mm wide							
engineering bricks; 215 × 102.5 × 65 mm	8.12	1.20	24.60	–	13.02	m	**37.62**
paving bricks; 215 × 102.5 × 65 mm	13.33	1.20	24.60	–	18.23	m	**42.83**
Extra over bricks on edge for standard kerbs to one side; haunching in concrete							
125 × 255 mm; ref HB2; SP	4.01	0.44	9.11	–	6.72	m	**15.83**

Q PAVING/PLANTING/FENCING/SITE FURNITURE

Item Excluding site overheads and profit	PC £	Labour hours	Labour £	Plant £	Material £	Unit	Total rate £
New granite sett edgings; CED Ltd; 100 × **100 × 100 mm; bedding in cement mortar** **(1:4); including haunching with in situ** **concrete; 11.50 N/mm² – 40 mm aggregate** **one side; concrete foundations (not** **included)**							
Edgings 100 × 100 × 100 mm; ref 1R 'better quality' with 20 mm pointing gaps							
100 mm wide	2.91	1.33	27.27	–	5.70	m	32.97
2 rows; 220 mm wide	5.83	2.22	45.51	–	9.43	m	54.94
3 rows; 340 mm wide	8.74	3.00	61.50	–	11.57	m	73.07
Edgings 100 × 100 × 200 mm; ref 1R 'better quality' with 20 mm pointing gaps							
100 mm wide	2.58	0.75	15.38	–	5.36	m	20.74
2 rows; 220 mm wide	5.15	1.50	30.75	–	8.75	m	39.50
3 rows; 340 mm wide	7.73	2.25	46.13	–	10.56	m	56.69
Edgings 100 × 100 × 100 mm; ref 8R 'standard quality' with 20 mm pointing gaps							
100 mm wide	2.89	1.33	27.27	–	5.67	m	32.94
2 rows; 220 mm wide	5.77	2.22	45.51	–	9.37	m	54.88
3 rows; 340 mm wide	8.66	3.00	61.50	–	11.49	m	72.99
Edgings 100 × 100 × 200 mm; ref 8R 'standard quality' with 20 mm pointing gaps							
100 mm wide	4.04	0.75	15.38	–	6.83	m	22.21
2 rows; 220 mm wide	8.09	1.50	30.75	–	11.69	m	42.44
3 rows; 340 mm wide	12.13	2.25	46.13	–	14.97	m	61.10
Reclaimed granite setts edgings; CED Ltd; **100 × 100 mm; bedding in cement mortar** **(1:4); including haunching with in situ** **concrete; 11.50 N/mm² – 40 mm aggregate** **one side; on concrete foundations (not** **included)**							
Edgings 100 × 100 × 100 mm; with 20 mm pointing gaps							
100 mm wide	3.94	1.33	27.27	–	6.72	m	33.99
2 rows; 220 mm wide	7.88	2.22	45.51	–	11.48	m	56.99
3 rows; 340 mm wide	11.83	3.00	61.50	–	14.66	m	76.16
Channels; bedding in cement mortar (1:3); **joints pointed flush; on concrete** **foundations (not included)**							
Three courses stretchers; 350 mm wide; quarter bond to form dished channels							
engineering bricks; PC £300.00/1000; 215 × 102.5 × 65 mm	4.06	1.00	20.50	–	5.27	m	25.77
paving bricks; PC £500.00/1000; 215 × 102.5 × 65 mm	6.67	1.00	20.50	–	7.88	m	28.38
Three courses granite setts; 340 mm wide; to form dished channels							
340 mm wide	13.24	2.00	41.00	–	16.02	m	57.02

Q PAVING/PLANTING/FENCING/SITE FURNITURE

Item Excluding site overheads and profit	PC £	Labour hours	Labour £	Plant £	Material £	Unit	Total rate £
Q10 KERBS/EDGINGS/CHANNELS/PAVING ACCESSORIES – CONT							
Mild steel edgings; 6 mm galvanized steel; Gardenlink Ltd							
Bespoke edgings; complete with bolts for adjoining sections and mild steel pins at 1.50 m centres							
100 mm straight	–	–	–	–	–	m	**15.40**
150 mm straight	–	–	–	–	–	m	**17.90**
300 mm straight	–	–	–	–	–	m	**26.40**
Permaloc AshphaltEdge; RTS Ltd; extruded aluminium alloy L shaped edging with 5.33 mm exposed upper lip; edging fixed to roadway base and edge profile with 250 mm steel fixing spike; laid to straight or curvilinear road edge; subsequently filled with macadam (not included)							
Depth of macadam							
38 mm	7.52	0.02	0.34	–	7.52	m	**7.86**
51 mm	8.48	0.02	0.35	–	8.48	m	**8.83**
64 mm	9.37	0.02	0.37	–	9.37	m	**9.74**
76 mm	11.33	0.02	0.41	–	11.33	m	**11.74**
102 mm	13.18	0.02	0.43	–	13.18	m	**13.61**
Permaloc Cleanline; RTS Ltd; heavy duty straight profile edging; for edgings to soft landscape beds or turf areas; 3.2 mm × 102 high; 3.2 mm thick with 4.75 mm exposed upper lip; fixed to form straight or curvilinear edge with 305 mm fixing spike							
Milled aluminium							
100 mm deep	8.62	0.02	0.37	–	8.62	m	**8.99**
Black							
100 mm deep	8.62	0.02	0.37	–	8.62	m	**8.99**
Permaloc Permastrip; RTS Ltd; heavy duty L shaped profile maintenance strip; 3.2 mm × 89 mm high with 5.2 mm exposed top lip; for straight or gentle curves on paths or bed turf interfaces; fixed to form straight or curvilinear edge with standard 305 mm stake; other stake lengths available							
Milled aluminium							
89 mm deep	8.62	0.02	0.37	–	8.62	m	**8.99**
Black							
89 mm deep	8.62	0.02	0.37	–	8.62	m	**8.99**

Q PAVING/PLANTING/FENCING/SITE FURNITURE

Item Excluding site overheads and profit	PC £	Labour hours	Labour £	Plant £	Material £	Unit	Total rate £
Permaloc Proline; RTS Ltd; medium duty straight profiled maintenance strip; 3.2 mm × 102 mm high with 3.18 mm exposed top lip; for straight or gentle curves on paths or bed turf interfaces; fixed to form straight or curvilinear edge with standard 305 mm stake; other stake lengths available							
Milled aluminium							
89 mm deep	7.50	0.02	0.37	–	7.50	m	7.87
Black							
89 mm deep	7.50	0.02	0.37	–	7.50	m	7.87
Exceledge SuperEdge; recycled vinyl edging; black; 3 × 150 mm high with 5 mm diameter bead on each edge; for edging to soft landscape turf or bedding areas; formed by heat to required angle, or fixed to form straight or curvilinear lines using 350 mm galvanized fixing stakes							
150 mm deep	5.65	0.02	0.37	–	5.65	m	6.02
Exceledge BrickEdge; recycled vinyl hard landscape edging former; black; 50 mm high with 76 mm return; 3 mm thick wall for the retention of bricks, blocks and loose or bonded gravels; fixed with steel fixing spikes at 300 mm centres							
Flexible; 50 mm deep	6.20	0.02	0.37	–	6.20	m	6.57
Rigid; 50 mm deep	6.20	0.02	0.37	–	6.20	m	6.57
Exceledge EdgeKing; polyethylene flexible edging; black; used for turfed areas and retention of loose fill materials; with 25 mm diameter rounded hose top for bold demarcation and safety in play areas; fixed with steel vee-stakes at 45° angle into adjacent material or to prefixed timber supports using screws and washers							
125 mm deep	3.96	0.02	0.37	–	3.96	m	4.33
Corner pieces	9.52	–	–	–	9.52	ea	9.52
Joiner plugs	0.75	–	–	–	0.75	ea	0.75
Exceledge Borderline; tree rings; preformed unit from 100 mm high rolled steel in galvanized or powder coated (black or brown) finish; for circular traditional lawn edging around trees, bushes, flag poles or part of a feature; fixed with flat steel matching stakes							
700 mm	18.20	0.50	10.25	–	18.20	ea	28.45
900 mm	24.85	0.55	11.28	–	24.85	ea	36.13
1200 mm	29.65	0.60	12.30	–	29.65	ea	41.95
1500 mm	35.78	0.65	13.32	–	35.78	ea	49.10
1800 mm	39.98	0.70	14.35	–	39.98	ea	54.33

Q PAVING/PLANTING/FENCING/SITE FURNITURE

Item Excluding site overheads and profit	PC £	Labour hours	Labour £	Plant £	Material £	Unit	Total rate £
Q20 GRANULAR SUBBASES TO ROADS/ PAVINGS							
Hardcore bases; obtained off site; PC £8.00/m³							
By machine							
100 mm thick	2.07	0.05	1.02	2.19	2.07	m²	**5.28**
150 mm thick	3.10	0.07	1.37	2.75	3.10	m²	**7.22**
over 150 thick	20.70	0.50	10.25	21.90	20.70	m³	**52.85**
By hand							
100 mm thick	2.07	0.20	4.10	0.52	2.07	m²	**6.69**
150 mm thick	3.73	0.30	6.16	0.52	3.73	m²	**10.41**
Hardcore; difference for each £1.00 increase/ decrease in PC price per m³; price will vary with type and source of hardcore							
average 75 mm thick	–	–	–	–	0.08	m²	**0.08**
average 100 mm thick	–	–	–	–	0.10	m²	**0.10**
average 150 mm thick	–	–	–	–	0.15	m²	**0.15**
average 200 mm thick	–	–	–	–	0.20	m²	**0.20**
average 250 mm thick	–	–	–	–	0.25	m²	**0.25**
average 300 mm thick	–	–	–	–	0.30	m²	**0.30**
exceeding 300 mm thick	–	–	–	–	1.10	m³	**1.10**
Type 1 granular fill base; PC £23.00/tonne (£50.60/m³ compacted)							
By machine							
100 mm thick	5.06	0.03	0.57	0.59	5.06	m²	**6.22**
150 mm thick	7.59	0.03	0.51	0.82	7.59	m²	**8.92**
By hand (mechanical compaction)							
100 mm thick	5.06	0.17	3.42	0.17	5.06	m²	**8.65**
150 mm thick	7.59	0.25	5.13	0.26	7.59	m²	**12.98**
Surface treatments							
Sand blinding; to hardcore base (not included); 25 mm thick	0.94	0.03	0.68	–	0.94	m²	**1.62**
Sand blinding; to hardcore base (not included); 50 mm thick	1.89	0.05	1.02	–	1.89	m²	**2.91**
Filter fabrics; to hardcore base (not included)	0.51	0.01	0.20	–	0.51	m²	**0.71**
Q21 IN SITU CONCRETE ROADS/PAVINGS							
Unreinforced concrete; on prepared subbase (not included)							
Paths and roads; 21.00 N/mm² – 20 mm aggregate (1:2:4) mechanically mixed on site							
100 mm thick	12.40	0.13	2.56	–	12.40	m²	**14.96**
150 mm thick	18.60	0.17	3.42	–	18.60	m²	**22.02**
Formwork for in situ concrete							
Sides of foundations							
height not exceeding 250 mm	0.79	0.05	1.02	–	1.08	m	**2.10**
Extra over formwork for curved work 6 m radius	–	0.25	5.13	–	–	m	**5.13**
Steel road forms; to edges of beds or faces of foundations							
150 mm wide	–	0.20	4.10	1.40	–	m	**5.50**

Q PAVING/PLANTING/FENCING/SITE FURNITURE

Item Excluding site overheads and profit	PC £	Labour hours	Labour £	Plant £	Material £	Unit	Total rate £
Reinforcement; fabric; side laps 150 mm; **head laps 300 mm; mesh 200 × 200 mm; in** **roads, footpaths or pavings**							
Fabric							
A142 (2.22 kg/m²)	4.08	0.08	1.70	–	4.08	m²	**5.78**
A193 (3.02 kg/m²)	5.55	0.08	1.70	–	5.55	m²	**7.25**
Reinforced in situ concrete; mechanically **mixed on site; normal Portland cement; on** **hardcore base (not included);** **reinforcement (not included)**							
Roads; 11.50 N/mm² – 40 mm aggregate (1:3:6)							
100 mm thick	9.26	0.40	8.20	–	9.26	m²	**17.46**
150 mm thick	13.89	0.60	12.30	–	13.89	m²	**26.19**
200 mm thick	18.51	0.80	16.40	–	18.51	m²	**34.91**
Roads; 21.00 N/mm² – 20 mm aggregate (1:2:4)							
100 mm thick	12.10	0.40	8.20	–	12.10	m²	**20.30**
150 mm thick	18.16	0.60	12.30	–	18.16	m²	**30.46**
200 mm thick	24.20	0.80	16.40	–	24.20	m²	**40.60**
Roads; 25.00 N/mm² – 20 mm aggregate GEN 4 ready mixed							
100 mm thick	6.83	0.40	8.20	–	6.83	m²	**15.03**
150 mm thick	10.24	0.60	12.30	–	10.24	m²	**22.54**
200 mm thick	13.65	0.80	16.40	–	13.65	m²	**30.05**
250 mm thick	17.06	1.00	20.50	0.70	17.06	m²	**38.26**
Reinforced in situ concrete; ready mixed; **discharged directly into location from** **supply lorry; normal Portland cement; on** **hardcore base (not included);** **reinforcement (not included)**							
Roads; 11.50 N/mm² – 40 mm aggregate (1:3:6)							
100 mm thick	6.41	0.16	3.28	–	6.41	m²	**9.69**
150 mm thick	9.62	0.24	4.92	–	9.62	m²	**14.54**
200 mm thick	12.82	0.36	7.38	–	12.82	m²	**20.20**
250 mm thick	16.02	0.54	11.07	0.70	16.02	m²	**27.79**
300 mm thick	19.23	0.66	13.53	0.70	19.23	m²	**33.46**
Roads; 21.00 N/mm² – 20 mm aggregate (1:2:4)							
100 mm thick	6.92	0.16	3.28	–	6.92	m²	**10.20**
150 mm thick	10.38	0.24	4.92	–	10.38	m²	**15.30**
200 mm thick	13.84	0.36	7.38	–	13.84	m²	**21.22**
250 mm thick	17.30	0.54	11.07	0.70	17.30	m²	**29.07**
300 mm thick	20.76	0.66	13.53	0.70	20.76	m²	**34.99**
Roads; 26.00 N/mm² – 20 mm aggregate (1:1:5:3)							
100 mm thick	7.06	0.16	3.28	–	7.06	m²	**10.34**
150 mm thick	10.59	0.24	4.92	–	10.59	m²	**15.51**
200 mm thick	14.12	0.36	7.38	–	14.12	m²	**21.50**
250 mm thick	17.65	0.54	11.07	0.70	17.65	m²	**29.42**
300 mm thick	21.18	0.66	13.53	0.70	21.18	m²	**35.41**

Q PAVING/PLANTING/FENCING/SITE FURNITURE

Item Excluding site overheads and profit	PC £	Labour hours	Labour £	Plant £	Material £	Unit	Total rate £
Q21 IN SITU CONCRETE ROADS/ PAVINGS – CONT							
Concrete sundries							
Treating surfaces of unset concrete; grading to cambers; tamping with 75 mm thick steel shod tamper or similar	–	0.13	2.73	–	–	m²	2.73
Expansion joints							
13 mm thick joint filler; formwork							
width or depth not exceeding 150 mm	1.82	0.20	4.10	–	2.90	m	7.00
width or depth 150–300 mm	1.82	0.25	5.13	–	3.98	m	9.11
width or depth 300–450 mm	0.82	0.30	6.15	–	4.06	m	10.21
25 mm thick joint filler; formwork							
width or depth not exceeding 150 mm	4.95	0.20	4.10	–	6.03	m	10.13
width or depth 150–300 mm	4.95	0.25	5.13	–	7.11	m	12.24
width or depth 300–450 mm	4.95	0.30	6.15	–	8.19	m	14.34
Sealants; sealing top 25 mm of joint with rubberized bituminous compound	1.28	0.25	5.13	–	1.28	m	6.41
Q22 COATED MACADAM/ASPHALT ROADS/PAVINGS							
Coated macadam/asphalt roads/pavings – General							
Preamble: The prices for all in situ finishings to roads and footpaths include for work to falls, crossfalls or slopes not exceeding 15° from horizontal; for laying on prepared bases (not included) and for rolling with an appropriate roller.							
Users should note the new terminology for the surfaces described below which is to European standard descriptions. The now redundant descriptions for each course are shown in brackets.							
Note: The rates below are for smaller areas up to 200 m².							
Macadam surfacing; Spadeoak Construction Co Ltd; surface (wearing) course; 20 mm of 6 mm dense bitumen macadam							
Hand lay							
limestone aggregate	–	–	–	–	–	m²	9.04
granite aggregate	–	–	–	–	–	m²	9.02
red	–	–	–	–	–	m²	11.12
Macadam surfacing; Spadeoak Construction Co Ltd; surface (wearing) course; 30 mm of 10 mm dense bitumen macadam							
Hand lay							
limestone aggregate	–	–	–	–	–	m²	10.47
granite aggregate	–	–	–	–	–	m²	10.60
red	–	–	–	–	–	m²	–

Q PAVING/PLANTING/FENCING/SITE FURNITURE

Item Excluding site overheads and profit	PC £	Labour hours	Labour £	Plant £	Material £	Unit	Total rate £
Macadam surfacing; Spadeoak **Construction Co Ltd; surface (wearing)** **course; 40 mm of 10 mm dense bitumen** **macadam** Hand lay							
limestone aggregate	–	–	–	–	–	m²	12.95
granite aggregate	–	–	–	–	–	m²	13.11
red	–	–	–	–	–	m²	22.89
Macadam surfacing; Spadeoak **Construction Co Ltd; binder (base) course;** **50 mm of 20 mm dense bitumen macadam** Hand lay							
limestone aggregate	–	–	–	–	–	m²	12.98
granite aggregate	–	–	–	–	–	m²	13.16
Macadam surfacing; Spadeoak **Construction Co Ltd; binder (base) course;** **60 mm of 20 mm dense bitumen macadam** Hand lay							
limestone aggregate	–	–	–	–	–	m²	14.12
granite aggregate	–	–	–	–	–	m²	14.33
Macadam surfacing; Spadeoak **Construction Co Ltd; base (roadbase)** **course; 75 mm of 28 mm dense bitumen** **macadam** Hand lay							
limestone aggregate	–	–	–	–	–	m²	17.01
granite aggregate	–	–	–	–	–	m²	17.29
Areas up to 300 m²							
Macadam surfacing; Spadeoak **Construction Co Ltd; base (roadbase)** **course; 100 mm of 28 mm dense bitumen** **macadam** Hand lay							
limestone aggregate	–	–	–	–	–	m²	72.46
granite aggregate	–	–	–	–	–	m²	21.10
Macadam surfacing; Spadeoak **Construction Co Ltd; base (roadbase)** **course; 150 mm of 28 mm dense bitumen** **macadam in two layers** Hand lay							
limestone aggregate	–	–	–	–	–	m²	33.29
granite aggregate	–	–	–	–	–	m²	33.84
Base (roadbase) course; 200 mm of 28 mm **dense bitumen macadam in two layers** Hand lay							
limestone aggregate	–	–	–	–	–	m²	28.16
granite aggregate	–	–	–	–	–	m²	42.25

Q PAVING/PLANTING/FENCING/SITE FURNITURE

Item Excluding site overheads and profit	PC £	Labour hours	Labour £	Plant £	Material £	Unit	Total rate £
Q22 COATED MACADAM/ASPHALT ROADS/PAVINGS – CONT							
Resin bound macadam pavings; hand laid; Bituchem; to pedestrian or vehicular hard landscape areas; laid to base course (not included)							
Natratex clear resin bound macadam							
25 mm thick to pedestrian areas	22.50	0.13	2.56	–	22.50	m²	**25.06**
30 mm thick to vehicular areas	26.25	0.17	3.42	–	26.25	m²	**29.67**
Colourtex coloured resin bound macadam to pedestrian or vehicular hard landscape areas							
25 mm thick	22.50	0.13	2.56	–	22.50	m²	**25.06**
Bound aggregates; Addagrip Surface Treatments UK Ltd; natural decorative resin bonded surface dressing laid to concrete, macadam or to plywood panels priced separately							
Primer coat to macadam or concrete base	–	–	–	–	–	m²	**5.00**
Golden pea gravel 1–3 mm							
buff adhesive	–	–	–	–	–	m²	**25.00**
red adhesive	–	–	–	–	–	m²	**25.00**
green adhesive	–	–	–	–	–	m²	**25.00**
Golden pea gravel 2–5 mm							
buff adhesive	–	–	–	–	–	m²	**28.00**
Chinese bauxite 1–3 mm							
buff adhesive	–	–	–	–	–	m²	**25.00**
Cobalt Blue Glass; 6 mm							
15 mm depth	–	–	–	–	–	m²	**85.00**
Midnight Grey, Chocolate, Derbyshire Cream, All Gold; 6 mm							
15 mm depth	–	–	–	–	–	m²	**50.00**
18 mm depth	–	–	–	–	–	m²	**55.00**
Q23 GRAVEL/HOGGIN/WOODCHIP ROADS/ PAVINGS							
Excavation and path preparation; excavating; 300 mm deep; to width of path; depositing excavated material at sides of excavation							
By machine							
width 1.00 m	–	–	–	4.45	–	m²	**4.45**
width 1.50 m	–	–	–	3.71	–	m²	**3.71**
width 2.00 m	–	–	–	3.18	–	m²	**3.18**
width 3.00 m	–	–	–	2.67	–	m²	**2.67**
By hand							
width 1.00 m	–	0.70	14.44	–	–	m²	**14.44**
width 1.50 m	–	1.05	21.53	–	–	m²	**21.53**
width 2.00 m	–	1.40	28.70	–	–	m²	**28.70**
width 3.00 m	–	2.10	43.05	–	–	m²	**43.05**

Q PAVING/PLANTING/FENCING/SITE FURNITURE

Item Excluding site overheads and profit	PC £	Labour hours	Labour £	Plant £	Material £	Unit	Total rate £
Excavating trenches; in centre of pathways; 100 mm flexible drain pipes; filling with clean broken stone or gravel rejects							
300 × 450 mm deep							
by machine	5.73	0.10	2.05	1.67	5.73	m	**9.45**
by hand	5.73	0.40	8.20	–	5.73	m	**13.93**
Hand trimming and compacting reduced surface of pathway; by vibrating roller							
width 1.00 m	–	0.05	1.02	0.52	–	m	**1.54**
width 1.50 m	–	0.04	0.91	0.46	–	m	**1.37**
width 2.00 m	–	0.04	0.82	0.42	–	m	**1.24**
width 3.00 m	–	0.04	0.82	0.42	–	m	**1.24**
Permeable membranes; to trimmed and compacted surface of pathway							
Terram 1000	0.51	0.02	0.41	–	0.51	m²	**0.92**
Permaloc AshphaltEdge; RTS Ltd; extruded aluminium alloy L shaped edging with 5.33 mm exposed upper lip; edging fixed to roadway base and edge profile with 250 mm steel fixing spike; laid to straight or curvilinear road edge; subsequently filled with macadam (not included)							
Depth of macadam							
38 mm	7.52	0.02	0.34	–	7.52	m	**7.86**
51 mm	8.48	0.02	0.35	–	8.48	m	**8.83**
64 mm	9.37	0.02	0.37	–	9.37	m	**9.74**
76 mm	11.33	0.02	0.41	–	11.33	m	**11.74**
102 mm	13.18	0.02	0.43	–	13.18	m	**13.61**
Permaloc Cleanline; RTS Ltd; heavy duty straight profile edging; for edgings to soft landscape beds or turf areas; 3.2 mm × 102 mm high; 3.2 mm thick with 4.75 mm exposed upper lip; fixed to form straight or curvilinear edge with 305 mm fixing spike							
Milled aluminium							
100 mm deep	8.62	0.02	0.37	–	8.62	m	**8.99**
Black							
100 mm deep	8.62	0.02	0.37	–	8.62	m	**8.99**
Permaloc Permastrip; RTS Ltd; heavy duty L shaped profile maintenance strip; 3.2 mm × 89 mm high with 5.2 mm exposed top lip; for straight or gentle curves on paths or bed turf interfaces; fixed to form straight or curvilinear edge with standard 305 mm stake; other stake lengths available							
Milled aluminium							
89 mm deep	8.62	0.02	0.37	–	8.62	m	**8.99**
Black							
89 mm deep	8.62	0.02	0.37	–	8.62	m	**8.99**

Q PAVING/PLANTING/FENCING/SITE FURNITURE

Item Excluding site overheads and profit	PC £	Labour hours	Labour £	Plant £	Material £	Unit	Total rate £
Q23 GRAVEL/HOGGIN/WOODCHIP ROADS/ PAVINGS – CONT							
Permaloc Proline; RTS Ltd; medium duty straight profiled maintenance strip; 3.2 mm × 102 mm high with 3.18 mm exposed top lip; for straight or gentle curves on paths or bed turf interfaces; fixed to form straight or curvilinear edge with standard 305 mm stake; other stake lengths available							
Milled aluminium							
89 mm deep	7.50	0.02	0.37	–	7.50	m	**7.87**
Black							
89 mm deep	7.50	0.02	0.37	–	7.50	m	**7.87**
Timber edging boards							
Boards; 50 × 50 × 750 mm timber pegs at 1000 mm centres (excavations and hardcore under edgings not included)							
Straight							
38 × 150 mm treated softwood edge boards	2.54	0.10	2.05	–	2.54	m	**4.59**
50 × 150 mm treated softwood edge boards	2.75	0.10	2.05	–	2.75	m	**4.80**
38 × 150 mm hardwood (iroko) edge boards	9.26	0.10	2.05	–	9.26	m	**11.31**
50 × 150 mm hardwood (iroko) edge boards	10.69	0.10	2.05	–	10.69	m	**12.74**
Curved							
38 × 150 mm treated softwood edge boards	2.54	0.20	4.10	–	2.54	m	**6.64**
50 × 150 mm treated softwood edge boards	2.75	0.25	5.13	–	2.75	m	**7.88**
38 × 150 mm hardwood (iroko) edge boards	9.26	0.20	4.10	–	9.26	m	**13.36**
50 × 150 mm hardwood (iroko) edge boards	10.69	0.25	5.13	–	10.69	m	**15.82**
Filling to make up levels							
Obtained off site; hardcore; PC £13.00/m³							
150 mm thick	3.10	0.07	1.37	2.75	3.10	m²	**7.22**
Obtained off site; granular fill type 1; PC £23.00/tonne (£50.60/m³ compacted)							
100 mm thick	5.06	0.03	0.57	0.59	5.06	m²	**6.22**
150 mm thick	7.59	0.03	0.51	0.82	7.59	m²	**8.92**
Surface treatments							
Sand blinding; to hardcore (not included)							
50 mm thick	2.08	0.04	0.82	–	2.08	m²	**2.90**
Filter fabric; to hardcore (not included)	0.51	0.01	0.20	–	0.51	m²	**0.71**
Granular Pavings							
Footpath gravels; porous self binding gravel							
Breedon Special Aggregates; Golden Gravel or equivalent; rolling wet; on hardcore base (not included); for pavements; to falls and crossfalls and to slopes not exceeding 15° from horizontal; over 300 mm wide							
50 mm thick	12.15	0.05	1.02	1.19	12.15	m²	**14.36**
by hand; 50 mm thick	12.15	0.20	4.10	1.04	12.15	m²	**17.29**

Q PAVING/PLANTING/FENCING/SITE FURNITURE

Item Excluding site overheads and profit	PC £	Labour hours	Labour £	Plant £	Material £	Unit	Total rate £
Breedon Special Aggregates; Wayfarer specially formulated fine gravel for use on golf course pathways							
50 mm thick	11.36	0.03	0.59	0.74	11.36	m²	**12.69**
by hand; 50 mm thick	11.36	0.10	2.05	0.60	11.36	m²	**14.01**
Hoggin (stabilized); PC £54.58/m³ on hardcore base (not included); to falls and crossfalls and to slopes not exceeding 15° from horizontal; over 300 mm wide							
100 mm thick	8.19	0.02	0.41	0.58	8.19	m²	**9.18**
by hand; 100 mm thick	8.19	0.10	2.05	0.42	8.19	m²	**10.66**
150 mm thick	12.28	0.03	0.51	0.73	12.28	m²	**13.52**
Ballast; as dug; watering; rolling; on hardcore base (not included)							
100 mm thick	4.72	0.13	2.73	1.95	4.72	m²	**9.40**
150 mm thick	7.09	0.03	0.51	0.63	7.09	m²	**8.23**
CED Ltd; Cedec gravel; self-binding; laid to inert (non-limestone) base measured separately; compacting							
red, gold or silver; 50 mm thick	13.81	0.03	0.59	0.45	13.81	m²	**14.85**
by hand; red, gold or silver; 50 mm thick	13.81	0.07	1.37	0.21	13.81	m²	**15.39**
Grundon Ltd; Coxwell self-binding path gravels laid and compacted to excavation or base measured separately							
50 mm thick	3.27	0.08	1.71	0.42	3.27	m²	**5.40**
Footpath gravels; porous loose gravels							
Washed shingle; on prepared base (not included)							
25–50 mm size; 25 mm thick	1.07	0.02	0.36	0.07	1.07	m²	**1.50**
25–50 mm size; 75 mm thick	3.20	0.05	1.08	0.22	3.20	m²	**4.50**
50–75 mm size; 25 mm thick	1.07	0.02	0.41	0.08	1.07	m²	**1.56**
50–75 mm size; 75 mm thick	3.20	0.07	1.37	0.34	3.20	m²	**4.91**
Pea shingle; on prepared base (not included)							
10–15 mm size; 25 mm thick	0.99	0.02	0.36	0.07	0.99	m²	**1.42**
5–10 mm size; 75 mm thick	2.98	0.05	1.08	0.22	2.98	m²	**4.28**
Breedon Special Aggregates; Breedon Buff decorative limestone chippings							
50 mm thick	5.64	0.01	0.20	0.08	5.64	m²	**5.92**
by hand; 50 mm thick	5.64	0.03	0.68	–	5.64	m²	**6.32**
Breedon Special Aggregates; Brindle or Moorland Black chippings							
50 mm thick	7.31	0.01	0.20	0.08	7.31	m²	**7.59**
by hand; 50 mm thick	7.31	0.03	0.68	–	7.31	m²	**7.99**
Breedon Special Aggregates; slate chippings; plum/blue/green							
50 mm thick	10.50	0.01	0.20	0.08	10.50	m²	**10.78**
by hand; 50 mm thick	10.50	0.03	0.68	–	10.50	m²	**11.18**
Wood chip surfaces; Melcourt Industries Ltd							
Wood chips; to surface of pathways by hand; levelling and spreading by hand (excavation and preparation not included) (items labelled FSC are Forest Stewardship Council certified)							
Walk Chips; FSC; 100 mm thick (25 m³ loads)	3.85	0.02	0.34	–	3.85	m²	**4.19**
Woodfibre; FSC; 100 mm thick	2.49	0.05	1.03	–	2.49	m²	**3.52**

Q PAVING/PLANTING/FENCING/SITE FURNITURE

Item Excluding site overheads and profit	PC £	Labour hours	Labour £	Plant £	Material £	Unit	Total rate £
Q24 INTERLOCKING BRICK/BLOCK ROADS/PAVINGS							
Precast concrete block edgings; PC £8.03/ m²; 200 × 100 × 60 mm; on prepared base (not included); haunching one side							
Edgings; butt joints							
stretcher course	0.82	0.17	3.42	–	3.36	m	**6.78**
header course	1.64	0.27	5.47	–	4.37	m	**9.84**
Precast concrete vehicular paving blocks; Marshalls Plc; on prepared base (not included); on 50 mm compacted sharp sand bed; blocks laid in 7 mm loose sand and vibrated; joints filled with sharp sand and vibrated; level and to falls only							
Trafica paving blocks; 450 × 450 × 70 mm							
Perfecta finish; colour natural	29.97	0.50	10.25	0.32	32.04	m²	**42.61**
Perfecta finish; colour buff	34.61	0.50	10.25	0.32	36.69	m²	**47.26**
Saxon finish; colour natural	26.28	0.50	10.25	0.32	28.36	m²	**38.93**
Saxon finish; colour buff	30.27	0.50	10.25	0.32	32.35	m²	**42.92**
Precast concrete vehicular paving blocks; Keyblock; Marshalls Plc; on prepared base (not included); on 50 mm compacted sharp sand bed; blocks laid in 7 mm loose sand and vibrated; joints filled with sharp sand and vibrated; level and to falls only							
Herringbone bond							
200 × 100 × 60 mm; natural grey	8.60	1.50	30.75	0.32	10.73	m²	**41.80**
200 × 100 × 60 mm; colours	9.33	1.50	30.75	0.32	11.47	m²	**42.54**
200 × 100 × 80 mm; natural grey	9.58	1.50	30.75	0.32	11.71	m²	**42.78**
200 × 100 × 80 mm; colours	10.81	1.50	30.75	0.32	12.95	m²	**44.02**
Basketweave bond							
200 × 100 × 60 mm; natural grey	8.60	1.20	24.60	0.32	10.73	m²	**35.65**
200 × 100 × 60 mm; colours	9.33	1.20	24.60	0.32	11.47	m²	**36.39**
200 × 100 × 80 mm; natural grey	9.58	1.20	24.60	0.32	11.71	m²	**36.63**
200 × 100 × 80 mm; colours	10.81	1.20	24.60	0.32	12.95	m²	**37.87**
Precast concrete vehicular paving blocks; Charcon Hard Landscaping; on prepared base (not included); on 50 mm compacted sharp sand bed; blocks laid in 7 mm loose sand and vibrated; joints filled with sharp sand and vibrated; level and to falls only							
Europa concrete blocks							
200 × 100 × 60 mm; natural grey	12.72	1.50	30.75	0.32	14.85	m²	**45.92**
200 × 100 × 60 mm; colours	13.61	1.50	30.75	0.32	15.74	m²	**46.81**
Parliament concrete blocks							
200 × 100 × 65 mm; natural grey	30.03	1.50	30.75	0.32	32.17	m²	**63.24**
200 × 100 × 65 mm; colours	30.03	1.50	30.75	0.32	32.17	m²	**63.24**

Q PAVING/PLANTING/FENCING/SITE FURNITURE

Item Excluding site overheads and profit	PC £	Labour hours	Labour £	Plant £	Material £	Unit	Total rate £
Recycled polyethylene grassblocks; **Boddingtons Grass Reinforcement** **Solutions; Boddingtons Ltd; interlocking** **units laid to prepared base or rootzone (not** **included)** BodPave85; load bearing <400 tonnes per m²; 500 × 500 × 50 mm deep; 35 mm ground spike							
1–50 m²	16.50	0.20	4.10	0.20	19.70	m²	**24.00**
51–500 m²	13.75	0.20	4.10	0.20	16.95	m²	**21.25**
GrassProtecta; extruded expanded polyethylene flexible mesh laid to existing grass surface or newly seeded areas to provide heavy surface protection from traffic and pedestrians							
Standard; 1.2 kg/m²; 20 × 2 m; up to 320 m²	7.50	0.01	0.17	–	7.50	m²	**7.67**
Standard; 1.2 kg/m²; 20 × 2 m; 321 m² to 3000 m²	6.50	0.01	0.17	–	6.50	m²	**6.67**
Heavy; 2 kg/m²; 20 × 2 m; up to 320 m²	8.50	0.01	0.17	–	8.50	m²	**8.67**
Heavy; 2 kg/m²; 20 × 2 m; 321 m² to 3000 m²	8.00	0.01	0.17	–	8.00	m²	**8.17**
TurfProtecta; extruded polyethylene flexible mesh laid to existing grass surface or newly seeded areas to provide surface protection from traffic including vehicle or animal wear and tear							
Standard; 30 m × 2 m; up to 300 m²	2.90	0.01	0.17	–	2.90	m²	**3.07**
Standard; 30 m × 2 m; 301 m² to 600 m²	2.41	0.01	0.17	–	2.41	m²	**2.58**
Premium; 30 m × 2 m; up to 300 m²	2.99	0.01	0.17	–	2.99	m²	**3.16**
Premium; 30 m × 2 m; 301 m² to 600 m²	2.57	0.01	0.17	–	2.57	m²	**2.74**
Q25 SLAB/BRICK/SETT/COBBLE PAVINGS							
Bricks – General Preamble: Bricks shall be hard, well burnt, non-dusting, resistant to frost and sulphate attack and true to shape, size and sample.							
Movement of materials Mechanically offloading bricks; loading wheelbarrows; transporting maximum 25 m distance	–	0.20	4.10	–	–	m²	**4.10**
Edge restraints; to brick paving; on **prepared base (not included); 65 mm thick** **bricks; PC £300.00/1000; haunching one** **side** Header course							
200 × 100 mm; butt joints	3.00	0.27	5.47	–	5.73	m	**11.20**
210 × 105 mm; mortar joints	2.67	0.50	10.25	–	5.66	m	**15.91**
Stretcher course							
200 × 100 mm; butt joints	1.50	0.17	3.42	–	4.04	m	**7.46**
210 × 105 mm; mortar joints	1.36	0.33	6.83	–	4.02	m	**10.85**

Q PAVING/PLANTING/FENCING/SITE FURNITURE

Item Excluding site overheads and profit	PC £	Labour hours	Labour £	Plant £	Material £	Unit	Total rate £
Q25 SLAB/BRICK/SETT/COBBLE PAVINGS – CONT							
Variation in brick prices; add or subtract the following amounts for every £1.00/1000 difference in the PC price							
Edgings							
100 mm wide	–	–	–	–	0.05	10 m	**0.05**
200 mm wide	–	–	–	–	0.11	10 m	**0.11**
102.5 mm wide	–	–	–	–	0.05	10 m	**0.05**
215 mm wide	–	–	–	–	0.09	10 m	**0.09**
Clay brick pavings; on prepared base (not included); bedding on 50 mm sharp sand; kiln dried sand joints							
Pavings; 200 × 100 × 65 mm wirecut chamfered paviors							
brick; PC £500.00/1000	25.63	1.97	40.37	0.25	27.95	m²	**68.57**
Clay brick pavings; 200 × 100 × 50 mm; laid to running stretcher or stack bond only; on prepared base (not included); bedding on cement: sand (1:4) pointing mortar as work proceeds							
PC £600.00/1000							
laid on edge	48.81	4.76	97.62	–	56.47	m²	**154.09**
laid on edge but pavior 65 mm thick	41.00	3.81	78.09	–	48.66	m²	**126.75**
laid flat	26.62	2.20	45.10	–	31.19	m²	**76.29**
PC £500.00/1000							
laid on edge	40.67	4.76	97.62	–	48.33	m²	**145.95**
laid on edge but pavior 65 mm thick	34.16	3.81	78.09	–	41.82	m²	**119.91**
laid flat	22.19	2.20	45.10	–	26.75	m²	**71.85**
PC £400.00/1000							
laid on edge	32.54	4.76	97.62	–	40.20	m²	**137.82**
laid on edge but pavior 65 mm thick	27.33	3.81	78.09	–	34.99	m²	**113.08**
laid flat	17.75	2.20	45.10	–	22.31	m²	**67.41**
PC £300.00/1000							
laid on edge	24.40	4.76	97.62	–	32.06	m²	**129.68**
laid on edge but pavior 65 mm thick	20.50	3.81	78.09	–	28.16	m²	**106.25**
laid flat	13.31	2.20	45.10	–	17.88	m²	**62.98**
Clay brick pavings; 200 × 100 × 50 mm; butt jointed laid herringbone or basketweave pattern only; on prepared base (not included); bedding on 50 mm sharp sand							
PC £600.00/1000							
laid on edge	61.50	2.26	46.30	0.47	64.09	m²	**110.86**
laid flat	30.75	1.44	29.59	0.47	33.10	m²	**63.16**
PC £500.00/1000							
laid flat	25.63	1.44	29.59	0.47	27.97	m²	**58.03**
PC £400.00/1000							
laid flat	20.50	1.44	29.59	0.47	22.85	m²	**52.91**
PC £300.00/1000							
laid flat	15.38	1.44	29.59	0.47	17.72	m²	**47.78**

Q PAVING/PLANTING/FENCING/SITE FURNITURE

Item Excluding site overheads and profit	PC £	Labour hours	Labour £	Plant £	Material £	Unit	Total rate £
Clay brick pavings; 215 × 102.5 × 65 mm; on prepared base (not included); bedding on cement: sand (1:4) pointing mortar as work proceeds							
Paving bricks; PC £600.00/1000; herringbone bond							
laid on edge	36.44	3.55	72.88	–	42.83	m²	115.71
laid flat	23.70	2.37	48.59	–	30.09	m²	78.68
Paving bricks; PC £600.00/1000; basketweave bond							
laid on edge	36.44	2.37	48.59	–	42.83	m²	91.42
laid flat	23.70	1.58	32.40	–	30.09	m²	62.49
Paving bricks; PC £600.00/1000; running or stack bond							
laid on edge	36.44	1.90	38.88	–	42.83	m²	81.71
laid flat	23.70	1.26	25.92	–	30.09	m²	56.01
Paving bricks; PC £500.00/1000; herringbone bond							
laid on edge	30.37	3.55	72.88	–	40.43	m²	113.31
laid flat	19.75	2.37	48.59	–	26.14	m²	74.73
Paving bricks; PC £500.00/1000; basketweave bond							
laid on edge	30.37	2.37	48.59	–	34.04	m²	82.63
laid flat	19.75	1.58	32.40	–	26.14	m²	58.54
Paving bricks; PC £500.00/1000; running or stack bond							
laid on edge	30.37	1.90	38.88	–	34.04	m²	72.92
laid flat	19.75	1.26	25.92	–	26.14	m²	52.06
Paving bricks; PC £400.00/1000; herringbone bond							
laid on edge	24.29	3.55	72.88	–	27.96	m²	100.84
laid flat	15.80	2.37	48.59	–	22.07	m²	70.66
Paving bricks; PC £400.00/1000; basketweave bond							
laid on edge	24.29	2.37	48.59	–	27.96	m²	76.55
laid flat	15.80	1.58	32.40	–	22.07	m²	54.47
Paving bricks; PC £400.00/1000; running or stack bond							
laid on edge	24.29	1.90	38.88	–	27.96	m²	66.84
laid flat	15.80	1.26	25.92	–	22.07	m²	47.99
Paving bricks; PC £300.00/1000; herringbone bond							
laid on edge	17.77	3.55	72.88	–	21.44	m²	94.32
laid flat	11.85	2.37	48.59	–	18.24	m²	66.83
Paving bricks; PC £300.00/1000; basketweave bond							
laid on edge	17.77	2.37	48.59	–	21.44	m²	70.03
laid flat	11.85	1.58	32.40	–	18.24	m²	50.64
Paving bricks; PC £300.00/1000; running or stack bond							
laid on edge	17.77	1.90	38.87	–	21.44	m²	60.31
laid flat	11.85	1.26	25.92	–	18.24	m²	44.16
Cutting							
curved cutting	–	0.44	9.11	15.03	–	m	24.14
raking cutting	–	0.33	6.83	11.46	–	m	18.29

Q PAVING/PLANTING/FENCING/SITE FURNITURE

Item Excluding site overheads and profit	PC £	Labour hours	Labour £	Plant £	Material £	Unit	Total rate £
Q25 SLAB/BRICK/SETT/COBBLE PAVINGS – CONT							
Add or subtract the following amounts for every £10.00/1000 difference in the prime cost of bricks							
Butt joints							
200 × 100 mm	–	–	–	–	0.50	m²	**0.50**
215 × 102.5 mm	–	–	–	–	0.45	m²	**0.45**
10 mm mortar joints							
200 × 100 mm	–	–	–	–	0.43	m²	**0.43**
215 × 102.5 mm	–	–	–	–	0.40	m²	**0.40**
Precast concrete pavings; Charcon Hard Landscaping; on prepared subbase (not included); bedding on 25 mm thick cement: sand mortar (1:4); butt joints; straight both ways; jointing in cement: sand (1:3) brushed in; on 50 mm thick sharp sand base							
Pavings; natural grey							
400 × 400 × 65 mm chamfered	27.31	0.40	8.20	–	32.20	m²	**40.40**
450 × 450 × 50 mm chamfered	14.72	0.44	9.11	–	19.60	m²	**28.71**
450 × 450 × 70 mm chamfered	16.94	0.44	9.11	–	21.83	m²	**30.94**
600 × 300 × 50 mm	13.33	0.44	9.11	–	18.22	m²	**27.33**
450 × 600 × 50 mm	11.78	0.44	9.11	–	16.67	m²	**25.78**
600 × 600 × 50 mm	9.47	0.40	8.20	–	14.36	m²	**22.56**
750 × 600 × 50 mm	8.96	0.40	8.20	–	13.84	m²	**22.04**
900 × 600 × 50 mm	8.19	0.40	8.20	–	13.07	m²	**21.27**
Pavings; coloured							
450 × 450 × 70 mm chamfered	19.90	0.44	9.11	–	24.79	m²	**33.90**
450 × 600 × 50 mm	17.78	0.44	9.11	–	22.67	m²	**31.78**
400 × 400 × 65 mm chamfered	27.31	0.40	8.20	–	32.20	m²	**40.40**
600 × 600 × 50 mm	14.81	0.40	8.20	–	19.69	m²	**27.89**
750 × 600 × 50 mm	13.67	0.40	8.20	–	18.55	m²	**26.75**
900 × 600 × 50 mm	12.14	0.40	8.20	–	17.02	m²	**25.22**
Precast concrete pavings; Charcon Hard Landscaping; on prepared subbase (not included); bedding on 25 mm thick cement: sand mortar (1:4); butt joints; straight both ways; jointing in cement: sand (1:3) brushed in; on 50 mm thick sharp sand base							
Appalacian rough textured exposed aggregate pebble paving							
600 × 600 × 65 mm	36.15	0.50	10.25	–	39.15	m²	**49.40**

Q PAVING/PLANTING/FENCING/SITE FURNITURE

Item Excluding site overheads and profit	PC £	Labour hours	Labour £	Plant £	Material £	Unit	Total rate £
Precast concrete pavings; Marshalls Plc; **Heritage imitation riven yorkstone paving;** **on prepared subbase measured separately;** **bedding on 25 mm thick cement: sand** **mortar (1:4); pointed straight both ways** **cement: sand (1:3)**							
Square and rectangular paving							
450 × 300 × 38 mm	40.77	1.00	20.50	–	44.82	m²	65.32
450 × 450 × 38 mm	24.55	0.75	15.38	–	28.60	m²	43.98
600 × 300 × 38 mm	26.82	0.80	16.40	–	30.87	m²	47.27
600 × 450 × 38 mm	27.20	0.75	15.38	–	31.25	m²	46.63
600 × 600 × 38 mm	24.81	0.50	10.25	–	28.93	m²	39.18
Extra labours for laying the a selection of the above sizes to random rectangular pattern	–	0.33	6.83	–	–	m²	6.83
Radial paving for circles							
circle with centre stone and first ring (8 slabs); 450 × 230/560 × 38 mm; diameter 1.54 m (total area 1.86 m²)	67.03	1.50	30.75	–	72.57	nr	103.32
circle with second ring (16 slabs); 450 × 300/460 × 38 mm; diameter 2.48 m (total area 4.83 m²)	172.79	4.00	82.00	–	187.02	nr	269.02
circle with third ring (16 slabs); 450 × 470/ 625 × 38 mm; diameter 3.42 m (total area 9.18 m²)	314.39	8.00	164.00	–	341.20	nr	505.20
Stepping stones							
380 diameter × 38 mm	5.05	0.20	4.10	–	11.05	nr	15.15
asymmetrical 560 × 420 × 38 mm	7.53	0.20	4.10	–	13.53	nr	17.63
Precast concrete pavings; Marshalls Plc; **Chancery imitation reclaimed riven** **yorkstone paving; on prepared subbase** **measured separately; bedding on 25 mm** **thick cement: sand mortar (1:4); pointed** **straight both ways cement: sand (1:3)**							
Square and rectangular paving							
300 × 300 × 45 mm	28.29	1.00	20.50	–	32.34	m²	52.84
450 × 300 × 45 mm	26.64	0.90	18.45	–	30.69	m²	49.14
600 × 300 × 45 mm	25.92	0.80	16.40	–	29.97	m²	46.37
600 × 450 × 45 mm	26.01	0.75	15.38	–	30.06	m²	45.44
450 × 450 × 45 mm	23.42	0.75	15.38	–	27.47	m²	42.85
600 × 600 × 45 mm	25.00	0.50	10.25	–	29.06	m²	39.31
Extra labours for laying the a selection of the above sizes to random rectangular pattern	–	0.33	6.83	–	–	m²	6.83
Radial paving for circles							
circle with centre stone and first ring (8 slabs); 450 × 230/560 × 38 mm; diameter 1.54 m (total area 1.86 m²)	85.65	1.50	30.75	–	91.19	nr	121.94
circle with second ring (16 slabs); 450 × 300/460 × 38 mm; diameter 2.48 m (total area 4.83 m²)	221.01	4.00	82.00	–	235.24	nr	317.24
circle with third ring (16 slabs); 450 × 470/ 625 × 38 mm; diameter 3.42 m (total area 9.18 m²)	402.29	8.00	164.00	–	429.10	nr	593.10

Q PAVING/PLANTING/FENCING/SITE FURNITURE

Item Excluding site overheads and profit	PC £	Labour hours	Labour £	Plant £	Material £	Unit	Total rate £
Q25 SLAB/BRICK/SETT/COBBLE PAVINGS – CONT							
Precast concrete pavings – cont							
Squaring off set for 2 ring circle							
16 slabs; 2.72 m^2	235.28	1.00	20.50	–	235.28	nr	**255.78**
Edge restraints; to block paving; on prepared base (not included); 200 × 100 × 80 mm; PC £8.94/m^2; haunching one side							
Header course							
200 × 100 mm; butt joints	1.82	0.27	5.47	–	4.55	m	**10.02**
Stretcher course							
200 × 100 mm; butt joints	0.91	0.17	3.42	–	3.45	m	**6.87**
Concrete paviors; Marshalls Plc; on prepared base (not included); bedding on 50 mm sand; kiln dried sand joints swept in							
Keyblock paviors							
200 × 100 × 60 mm; grey	8.19	0.40	8.20	0.32	10.32	m^2	**18.84**
200 × 100 × 60 mm; colours	8.89	0.40	8.20	0.32	11.02	m^2	**19.54**
200 × 100 × 80 mm; grey	9.12	0.44	9.11	0.32	11.25	m^2	**20.68**
200 × 100 × 80 mm; colours	10.30	0.44	9.11	0.32	12.43	m^2	**21.86**
Concrete cobble paviors; Charcon Hard Landscaping; Concrete Products; on prepared base (not included); bedding on 50 mm sand; kiln dried sand joints swept in							
Paviors							
Woburn blocks; 100–201 × 134 × 80 mm; random sizes	31.56	0.67	13.67	0.32	33.69	m^2	**47.68**
Woburn blocks; 100–201 × 134 × 80 mm; single size	31.56	0.50	10.25	0.32	33.69	m^2	**44.26**
Woburn blocks; 100–201 × 134 × 60 mm; random sizes	25.55	0.67	13.67	0.32	27.69	m^2	**41.68**
Woburn blocks; 100–201 × 134 × 60 mm; single size	25.53	0.50	10.25	0.32	27.66	m^2	**38.23**
Concrete setts; on 25 mm sand; compacted; vibrated; joints filled with sand; natural or coloured; well rammed hardcore base (not included)							
Marshalls Plc; Tegula cobble paving							
60 mm thick; random sizes	23.14	0.57	11.71	0.32	25.28	m^2	**37.31**
60 mm thick; single size	23.14	0.45	9.32	0.32	25.46	m^2	**35.10**
80 mm thick; random sizes	26.66	0.57	11.71	0.32	28.98	m^2	**41.01**
80 mm thick; single size	26.66	0.45	9.32	0.32	28.98	m^2	**38.62**
cobbles 80 × 80 × 60 mm thick; traditional	39.44	0.56	11.39	0.32	41.77	m^2	**53.48**
Cobbles							
Charcon Hard Landscaping; country setts							
100 mm thick; random sizes	38.03	1.00	20.50	0.32	40.35	m^2	**61.17**
100 mm thick; single size	38.98	0.67	13.67	0.32	41.30	m^2	**55.29**

Q PAVING/PLANTING/FENCING/SITE FURNITURE

Item Excluding site overheads and profit	PC £	Labour hours	Labour £	Plant £	Material £	Unit	Total rate £
Natural stone, slab or granite paving – **General** Preamble: Provide paving slabs of the specified thickness in random sizes but not less than 25 slabs per 10 m² of surface area, to be laid in parallel courses with joints alternately broken and laid to falls.							
Reconstituted yorkstone aggregate **pavings; Marshalls Plc; Saxon on prepared** **subbase measured separately; bedding on** **25 mm thick cement: sand mortar (1:4) ;on** **50 mm thick sharp sand base** Square and rectangular paving in various colours; butt joints straight both ways							
300 × 300 × 35 mm	37.52	0.80	16.40	–	43.08	m²	**59.48**
600 × 300 × 35 mm	23.24	0.65	13.32	–	28.81	m²	**42.13**
450 × 450 × 50 mm	25.38	0.75	15.38	–	30.95	m²	**46.33**
600 × 600 × 35 mm	17.42	0.50	10.25	–	22.98	m²	**33.23**
600 × 600 × 50 mm	17.42	0.55	11.28	–	22.98	m²	**34.26**
Square and rectangular paving in natural; butt joints straight both ways							
300 × 300 × 35 mm	31.30	0.80	16.40	–	36.87	m²	**53.27**
450 × 450 × 50 mm	21.66	0.70	14.35	–	27.22	m²	**41.57**
600 × 300 × 35 mm	20.44	0.75	15.38	–	26.01	m²	**41.39**
600 × 600 × 35 mm	15.04	0.50	10.25	–	20.60	m²	**30.85**
600 × 600 × 50 mm	18.23	0.60	12.30	–	23.79	m²	**36.09**
Radial paving for circles; 20 mm joints circle with centre stone and first ring (8 slabs); 450 × 230/560 × 35 mm; diameter 1.54 m (total area 1.86 m²)	69.31	1.50	30.75	–	75.16	nr	**105.91**
circle with second ring (16 slabs); 450 × 300/460 × 35 mm; diameter 2.48 m (total area 4.83 m²)	168.83	4.00	82.00	–	183.94	nr	**265.94**
circle with third ring (24 slabs); 450 × 310/ 430 × 35 mm; diameter 3.42 m (total area 9.18 m²)	318.11	8.00	164.00	–	347.01	nr	**511.01**
Granite setts; bedding on 25 mm cement: **sand (1:3)** Natural granite setts; 100 × 100 mm to 125 × 150 mm; × 150 to 250 mm length; riven surface; silver grey							
new; standard grade	40.12	2.00	41.00	–	48.68	m²	**89.68**
new; high grade	28.90	2.00	41.00	–	37.46	m²	**78.46**
reclaimed; cleaned	39.88	2.00	41.00	–	48.44	m²	**89.44**
Natural stone, slate or granite flag pavings; **CED Ltd; on prepared base (not included);** **bedding on 25 mm cement: sand (1:3);** **cement: sand (1:3) joints** Yorkstone; riven laid random rectangular							
new slabs; 40–60 mm thick	86.87	1.71	35.14	–	90.62	m²	**125.76**
reclaimed slabs; Cathedral grade; 50–75 mm thick	105.92	2.80	57.40	–	109.67	m²	**167.07**

Q PAVING/PLANTING/FENCING/SITE FURNITURE

Item Excluding site overheads and profit	PC £	Labour hours	Labour £	Plant £	Material £	Unit	Total rate £
Q25 SLAB/BRICK/SETT/COBBLE **PAVINGS – CONT**							
Natural stone, slate or granite flag **pavings – cont**							
Donegal quartzite slabs; standard tiles							
200 × random lengths × 15–25 mm	83.69	4.20	86.13	–	87.44	m²	173.57
250 × random lengths × 15–25 mm	83.69	3.82	78.28	–	87.44	m²	165.72
300 × random lengths × 15–25 mm	83.69	3.47	71.18	–	87.44	m²	158.62
350 × random lengths × 15–25 mm	83.69	3.16	64.69	–	87.44	m²	152.13
400 × random lengths × 15–25 mm	83.69	2.87	58.82	–	87.44	m²	146.26
450 × random lengths × 15–25 mm	87.87	2.61	53.47	–	91.63	m²	145.10
Natural yorkstone, pavings or edgings; **Johnsons Wellfield Quarries; sawn 6 sides;** **50 mm thick; on prepared base measured** **separately; bedding on 25 mm cement:** **sand (1:3); cement: sand (1:3) joints**							
Paving							
laid to random rectangular pattern	62.60	1.71	35.14	–	66.17	m²	101.31
laid to coursed laying pattern; 3 sizes	67.53	1.72	35.34	–	71.09	m²	106.43
Paving; single size							
600 × 600 mm	73.04	0.85	17.43	–	76.60	m²	94.03
600 × 400 mm	73.04	1.00	20.50	–	76.60	m²	97.10
300 × 200 mm	82.69	2.00	41.00	–	86.25	m²	127.25
215 × 102.5 mm	85.45	2.50	51.25	–	89.02	m²	140.27
Paving; cut to template off site; 600 × 600 mm; radius							
1.00 m	178.50	3.33	68.33	–	182.07	m²	250.40
2.50 m	178.50	2.00	41.00	–	182.07	m²	223.07
5.00 m	178.50	2.00	41.00	–	182.07	m²	223.07
Edgings							
100 mm wide × random lengths	7.72	0.50	10.25	–	8.07	m	18.32
100 mm × 100 mm	8.54	0.50	10.25	–	12.11	m	22.36
100 mm × 200 mm	17.09	0.50	10.25	–	20.66	m	30.91
250 mm wide × random lengths	16.88	0.40	8.20	–	20.45	m	28.65
500 mm wide × random lengths	33.77	0.33	6.83	–	37.33	m	44.16
Yorkstone edgings; 600 mm long × 250 mm wide; cut to radius							
1.00 m to 3.00 m	53.75	0.50	10.25	–	54.10	m	64.35
3.00 m to 5.00 m	53.75	0.44	9.11	–	54.10	m	63.21
exceeding 5.00 m	53.75	0.40	8.20	–	54.10	m	62.30
Natural yorkstone, pavings or edgings; **Johnsons Wellfield Quarries; sawn 6 sides;** **75 mm thick; on prepared base measured** **separately; bedding on 25 mm cement:** **sand (1:3); cement: sand (1:3) joints**							
Paving							
laid to random rectangular pattern	75.14	0.95	19.48	–	78.70	m²	98.18
laid to coursed laying pattern; 3 sizes	80.65	0.95	19.48	–	84.22	m²	103.70
Paving; single size							
600 × 600 mm	86.16	0.95	19.48	–	89.73	m²	109.21
600 × 400 mm	86.16	0.95	19.48	–	89.73	m²	109.21
300 × 200 mm	97.02	0.75	15.38	–	100.59	m²	115.97
215 × 102.5 mm	99.22	2.50	51.25	–	102.79	m²	154.04

Q PAVING/PLANTING/FENCING/SITE FURNITURE

Item Excluding site overheads and profit	PC £	Labour hours	Labour £	Plant £	Material £	Unit	Total rate £
Paving; cut to template off site; 600 × 600 mm; radius							
1.00 m	210.00	4.00	82.00	–	213.57	m²	295.57
2.50 m	210.00	2.50	51.25	–	213.57	m²	264.82
5.00 m	210.00	2.50	51.25	–	213.57	m²	264.82
Edgings							
100 mm wide × random lengths	9.65	0.60	12.30	–	10.01	m	22.31
100 mm × 100 mm	9.92	0.60	12.30	–	13.49	m	25.79
100 mm × 200 mm	1.98	0.50	10.25	–	5.55	m	15.80
250 mm wide × random lengths	20.16	0.50	10.25	–	23.73	m	33.98
500 mm wide × random lengths	40.33	0.40	8.20	–	43.90	m	52.10
Edgings; 600 mm long × 250 mm wide; cut to radius							
1.00 m to 3.00 m	62.02	0.60	12.30	–	62.37	m	74.67
3.00 m to 5.00 m	62.02	0.50	10.25	–	62.37	m	72.62
exceeding 5.00 m	62.02	0.44	9.11	–	62.37	m	71.48
CED Ltd; Indian sandstone, riven pavings or edgings; 25–35 mm thick; on prepared base measured separately; bedding on 25 mm cement: sand (1:3); cement: sand (1:3) joints							
Paving							
laid to random rectangular pattern	23.65	2.40	49.21	–	27.21	m²	76.42
laid to coursed laying pattern; 3 sizes	23.65	2.00	41.00	–	27.21	m²	68.21
Paving; single size							
600 × 600 mm	23.65	1.00	20.50	–	27.21	m²	47.71
600 × 400 mm	23.65	1.25	25.63	–	27.21	m²	52.84
400 × 400 mm	23.65	1.67	34.17	–	27.21	m²	61.38
Natural stone, slate or granite flag pavings; CED Ltd; on prepared base (not included); bedding on 25 mm cement: sand (1:3); cement: sand (1:3) joints							
Granite paving; sawn 6 sides; textured top							
new slabs; silver grey, blue grey or yellow; 50 mm thick	56.70	1.71	35.14	–	60.45	m²	95.59
new slabs; black; 50 mm thick	79.82	1.71	35.14	–	83.57	m²	118.71
Cobble pavings – General							
Cobbles should be embedded by hand, tight-butted, endwise to a depth of 60% of their length. A dry grout of rapid-hardening cement: sand (1:2) shall be brushed over the cobbles until the interstices are filled to the level of the adjoining paving. Surplus grout shall then be brushed off and a light, fine spray of water applied over the area.							
Cobble pavings							
Cobbles; to present a uniform colour in panels or varied in colour as required							
Scottish beach cobbles; 200–100 mm	57.42	2.00	41.00	–	65.12	m²	106.12
Scottish beach cobbles; 100–75 mm	40.62	2.50	51.25	–	48.32	m²	99.57
Scottish beach cobbles; 75–50 mm	23.65	3.33	68.33	–	31.35	m²	99.68

Q PAVING/PLANTING/FENCING/SITE FURNITURE

Item Excluding site overheads and profit	PC £	Labour hours	Labour £	Plant £	Material £	Unit	Total rate £
Q26 SPECIAL SURFACINGS/PAVINGS FOR SPORT/GENERAL AMENITY							
Market prices of surfacing materials							
Surfacings; Melcourt Industries Ltd (items labelled FSC are Forest Stewardship Council certified)							
Playbark® 10/50; per 25 m³ load	–	–	–	–	59.60	m³	59.60
Playbark® 8/25; per 25 m³ load	–	–	–	–	58.10	m³	58.10
Playchips®; per 25 m³ load; FSC	–	–	–	–	37.75	m³	37.75
Kushyfall; per 25 m³ load; FSC	–	–	–	–	36.20	m³	36.20
Softfall; per 25 m³ load	–	–	–	–	29.40	m³	29.40
Playsand; per 10 t load	–	–	–	–	94.86	m³	94.86
Woodfibre; per 25 m³ load; FSC	–	–	–	–	36.70	m³	36.70
Playgrounds; Wicksteed Leisure Ltd							
Safety tiles; on prepared base (not included)							
1000 × 1000 × 60 mm; red or green	75.00	0.20	4.10	–	75.00	m²	79.10
1000 × 1000 × 60 mm; black	72.00	0.20	4.10	–	72.00	m²	76.10
1000 × 1000 × 43 mm; red or green	72.00	0.20	4.10	–	72.00	m²	76.10
1000 × 1000 × 43 mm; black	66.00	0.20	4.10	–	66.00	m²	70.10
Playgrounds; SMP Playgrounds Ltd							
Tiles; on prepared base (not included)							
Premier 25; 1000 × 1000 × 25 mm; black; for general use; freefall height up to 0.8 m	48.00	0.20	4.10	–	57.00	m²	61.10
Premier 70; 1000 × 1000 × 70 mm; black; for higher equipment; freefall height up to 2.6 m	84.00	0.25	5.13	–	93.00	m²	98.13
Playgrounds; Melcourt Industries Ltd							
Playbark® ; on drainage layer (not included); minimum 300 mm settled depth							
Playbark® 8/25; 8–25 mm particles; red/brown	17.43	0.35	7.19	–	17.43	m²	24.62
Playbark® 10/50; 10–50 mm particles; red/brown	19.86	0.55	11.29	–	19.86	m²	31.15
Playgrounds; timber edgings							
Timber edging boards; 50 × 50 × 750 mm timber pegs at 1000 mm centres; excavations and hardcore under edgings (not included)							
50 × 150 mm; hardwood (iroko) edge boards	9.97	0.10	2.05	–	10.69	m	12.74
38 × 150 mm; hardwood (iroko) edge boards	8.55	0.10	2.05	–	9.26	m	11.31
50 × 150 mm; treated softwood edge boards	2.75	0.10	2.05	–	2.75	m	4.80
38 × 150 mm; treated softwood edge boards	1.82	0.10	2.05	–	2.54	m	4.59

Q PAVING/PLANTING/FENCING/SITE FURNITURE

Item Excluding site overheads and profit	PC £	Labour hours	Labour £	Plant £	Material £	Unit	Total rate £
Q30 SEEDING/TURFING							
Seeding/turfing – General							
Market prices of seeding materials							
Please see market prices in the Major Works section of this book.							
Market prices of chemicals and application rates							
Please see market prices in the Major Works section of this book.							
Cultivation							
Breaking up existing ground; using pedestrian operated tine cultivator or rotavator							
100 mm deep	–	0.50	10.25	16.58	–	100 m²	**26.83**
150 mm deep	–	0.57	11.71	18.95	–	100 m²	**30.66**
200 mm deep	–	0.67	13.67	22.11	–	100 m²	**35.78**
As above but in heavy clay or wet soils							
100 mm deep	–	0.67	13.67	22.11	–	100 m²	**35.78**
150 mm deep	–	0.80	16.40	26.53	–	100 m²	**42.93**
200 mm deep	–	1.00	20.50	33.17	–	100 m²	**53.67**
Rolling cultivated ground lightly; using self-propelled agricultural roller	–	0.06	1.14	0.84	–	100 m²	**1.98**
Importing and storing selected and approved topsoil; inclusive of settlement							
small quantities; less than 15 m³	31.20	–	–	–	31.20	m³	**31.20**
over 15 m³	31.20	–	–	–	31.20	m³	**31.20**
Spreading and lightly consolidating approved topsoil (imported or from spoil heaps); in layers not exceeding 150 mm; travel distance from spoil heaps not exceeding 25 m							
By machine (imported topsoil not included)							
minimum depth 100 mm	–	–	–	0.93	–	m²	**0.93**
minimum depth 200 mm	–	–	–	1.87	–	m²	**1.87**
minimum depth 300 mm	–	–	–	2.80	–	m²	**2.80**
minimum depth 500 mm	–	–	–	4.66	–	m²	**4.66**
over 500 mm	–	–	–	9.33	–	m³	**9.33**
By hand (imported topsoil not included)							
minimum depth 100 mm	–	0.20	4.10	–	–	m²	**4.10**
minimum depth 150 mm	–	0.30	6.16	–	–	m²	**6.16**
minimum depth 300 mm	–	0.60	12.28	–	–	m²	**12.28**
minimum depth 450 mm	–	0.90	18.47	–	–	m²	**18.47**
over 450 mm deep	–	2.22	45.51	–	–	m³	**45.51**
Extra over for spreading topsoil to slopes 15-30°; by machine or hand	–	–	–	–	–	10%	**–**
Extra over for spreading topsoil to slopes over 30°; by machine or hand	–	–	–	–	–	25%	**–**
Extra over for spreading topsoil from spoil heaps; travel exceeding 100 m; by machine							
100–150 m	–	0.01	0.25	0.07	–	m³	**0.32**
150–200 m	–	0.02	0.38	0.11	–	m³	**0.49**
200–300 m	–	0.03	0.57	0.16	–	m³	**0.73**

Q PAVING/PLANTING/FENCING/SITE FURNITURE

Item Excluding site overheads and profit	PC £	Labour hours	Labour £	Plant £	Material £	Unit	Total rate £
Q30 SEEDING/TURFING – CONT							
Spreading and lightly consolidating approved topsoil (imported or from spoil heaps) – cont							
Extra over spreading topsoil for travel exceeding 100 m; by hand							
100 m	–	0.83	17.08	–	–	m³	17.08
200 m	–	1.67	34.17	–	–	m³	34.17
300 m	–	2.50	51.25	–	–	m³	51.25
Evenly grading; to general surfaces to bring to finished levels							
by pedestrian operated rotavator	–	–	0.08	0.13	–	m²	0.21
by hand	–	0.01	0.20	–	–	m²	0.20
Extra over grading for slopes 15–30°; by machine or hand	–	–	–	–	–	10%	–
Extra over grading for slopes over 30°; by machine or hand	–	–	–	–	–	25%	–
Apply screened topdressing to grass surfaces; spread using Tru-Lute							
sand soil mixes 90/10 to 50/50	0.12	–	0.04	0.03	0.12	m²	0.19
Spread only existing cultivated soil to final levels using Tru-Lute							
cultivated soil	–	–	0.04	0.03	–	m²	0.07
Clearing stones; disposing off site; to distance not exceeding 13 km							
by hand; stones not exceeding 50 mm in any direction; loading to skip 4.6 m³	–	0.01	0.20	0.04	–	m²	0.24
Lightly cultivating; weeding; to fallow areas; disposing debris off site							
by hand	–	0.01	0.29	0.29	–	m²	0.58
Surface applications and soil additives; pre-seeding; material delivered to a maximum of 25 m from area of application; applied by hand							
Soil conditioners; to cultivated ground; medium bark; based on deliveries of 25 m³ loads; PC £35.65/m³; including turning in							
1 m³ per 40 m² = 25 mm thick	0.93	0.02	0.46	–	0.93	m²	1.39
1 m³ per 20 m² = 50 mm thick	1.86	0.04	0.91	–	1.86	m²	2.77
1 m³ per 13.33 m² = 75 mm thick	2.79	0.07	1.37	–	2.79	m²	4.16
1 m³ per 10 m² = 100 mm thick	3.71	0.08	1.64	–	3.71	m²	5.35
Soil conditioners; to cultivated ground; mushroom compost; delivered in 25 m³ loads; PC £21.20/m³; including turning in							
1 m³ per 40 m² = 25 mm thick	0.53	0.02	0.46	–	0.53	m²	0.99
1 m³ per 20 m² = 50 mm thick	1.06	0.04	0.91	–	1.06	m²	1.97
1 m³ per 13.33 m² = 75 mm thick	1.59	0.07	1.37	–	1.59	m²	2.96
1 m³ per 10 m² = 100 mm thick	2.12	0.08	1.64	–	2.12	m²	3.76
Soil conditioners; to cultivated ground; mushroom compost; delivered in 60 m³ loads; PC £9.50/m³; including turning in							
1 m³ per 40 m² = 25 mm thick	0.24	0.02	0.46	–	0.24	m²	0.70
1 m³ per 20 m² = 50 mm thick	0.42	0.04	0.91	–	0.42	m²	1.33
1 m³ per 13.33 m² = 75 mm thick	0.71	0.07	1.37	–	0.71	m²	2.08
1 m³ per 10 m² = 100 mm thick	0.95	0.08	1.64	–	0.95	m²	2.59

Q PAVING/PLANTING/FENCING/SITE FURNITURE

Item Excluding site overheads and profit	PC £	Labour hours	Labour £	Plant £	Material £	Unit	Total rate £
Preparation of seedbeds – General Preamble: for preliminary operations see 'Cultivation' section.							
Preparation of seedbeds; soil preparation Lifting selected and approved topsoil from spoil heaps; passing through 6 mm screen; removing debris	–	0.08	1.71	4.59	0.01	m³	6.31
Topsoil; supply only; PC £26.00/m³; allowing for 20% settlement; 20 tonne loads							
25 mm	0.78	–	–	–	0.78	m²	0.78
50 mm	1.56	–	–	–	1.56	m²	1.56
100 mm	3.12	–	–	–	3.12	m²	3.12
150 mm	4.68	–	–	–	4.68	m²	4.68
200 mm	6.24	–	–	–	6.24	m²	6.24
250 mm	7.80	–	–	–	7.80	m²	7.80
300 mm	9.36	–	–	–	9.36	m²	9.36
400 mm	12.48	–	–	–	12.48	m²	12.48
450 mm	14.04	–	–	–	14.04	m²	14.04
Topsoil; supply only; PC £45.31/m³; allowing for 20% settlement; 10 tonne loads							
25 mm	1.36	–	–	–	1.36	m²	1.36
50 mm	2.72	–	–	–	2.72	m²	2.72
100 mm	5.44	–	–	–	5.44	m²	5.44
150 mm	8.16	–	–	–	8.16	m²	8.16
200 mm	10.87	–	–	–	10.87	m²	10.87
250 mm	13.59	–	–	–	13.59	m²	13.59
300 mm	16.31	–	–	–	16.31	m²	16.31
400 mm	21.75	–	–	–	21.75	m²	21.75
450 mm	24.47	–	–	–	24.47	m²	24.47
Spreading topsoil to form seedbeds **average thickness not exceeding 200 mm** **(topsoil not included); by machine; grading** **and cultivation not included** Excavated material; from spoil heaps							
average 25 m distance	–	0.10	2.05	5.98	–	m³	8.03
average 50 m distance	–	0.10	2.05	6.60	–	m³	8.65
average 100 m distance	–	0.10	2.05	7.20	–	m³	9.25
average 200 m distance	–	0.10	2.05	9.00	–	m³	11.05
Excavated material; spreading on site							
25 mm deep	–	–	–	0.22	–	m²	0.22
50 mm deep	–	–	–	0.44	–	m²	0.44
75 mm deep	–	–	–	0.65	–	m²	0.65
100 mm deep	–	–	–	0.87	–	m²	0.87
150 mm deep	–	–	–	1.04	–	m²	1.04
Spreading only topsoil to form seedbeds **(topsoil not included); by hand**							
25 mm deep	–	0.03	0.51	–	–	m²	0.51
50 mm deep	–	0.03	0.68	–	–	m²	0.68
75 mm deep	–	0.04	0.88	–	–	m²	0.88
100 mm deep	–	0.05	1.02	–	–	m²	1.02
150 mm deep	–	0.08	1.54	–	–	m²	1.54

Q PAVING/PLANTING/FENCING/SITE FURNITURE

Item Excluding site overheads and profit	PC £	Labour hours	Labour £	Plant £	Material £	Unit	Total rate £
Q30 SEEDING/TURFING – CONT							
Spreading only topsoil to form seedbeds (topsoil not included) – cont							
Bringing existing topsoil to a fine tilth for seeding; by raking or harrowing; stones not to exceed 6 mm; by machine	–	–	0.08	0.04	–	m²	0.12
Bringing existing topsoil to a fine tilth for seeding; by raking or harrowing; stones not to exceed 6 mm; by hand	–	0.01	0.18	–	–	m²	0.18
Preparation of seedbeds; soil treatments For the following topsoil improvement and seeding operations add or subtrac#t the following amounts for every £0.10 difference in the material cost price							
35 g/m²	–	–	–	–	0.35	100 m²	0.35
50 g/m²	–	–	–	–	0.50	100 m²	0.50
70 g/m²	–	–	–	–	0.70	100 m²	0.70
100 g/m²	–	–	–	–	1.00	100 m²	1.00
Pre-seeding fertilizers (12+00+09); PC £0.85/ kg; to seedbeds; by hand							
35 g/m²	2.30	0.17	3.42	–	2.30	100 m²	5.72
50 g/m²	3.28	0.17	3.42	–	3.28	100 m²	6.70
70 g/m²	4.60	0.17	3.42	–	4.60	100 m²	8.02
100 g/m²	6.57	0.20	4.10	–	6.57	100 m²	10.67
125 g/m²	8.21	0.20	4.10	–	8.21	100 m²	12.31
Seeding Grass seed; spreading in two operations; PC £4.50/kg (for changes in material prices please refer to table above); by hand							
35 g/m²	15.75	0.17	3.42	–	15.75	100 m²	19.17
50 g/m²	22.50	0.17	3.42	–	22.50	100 m²	25.92
70 g/m²	31.50	0.17	3.42	–	31.50	100 m²	34.92
100 g/m²	45.00	0.17	3.42	–	45.00	100 m²	48.42
125 g/m²	56.25	0.20	4.10	–	56.25	100 m²	60.35
Grass seed; spreading in two operations; PC £6.00/kg (for changes in material prices please refer to table above); by hand							
35 g/m²	21.00	0.17	3.42	–	21.00	100 m²	24.42
50 g/m²	30.00	0.17	3.42	–	30.00	100 m²	33.42
70 g/m²	42.00	0.17	3.42	–	42.00	100 m²	45.42
100 g/m²	60.00	0.17	3.42	–	60.00	100 m²	63.42
125 g/m²	75.00	0.20	4.10	–	75.00	100 m²	79.10
Extra over seeding by hand for slopes over 30° (allowing for the actual area but measured in plan)							
35 g/m²	2.34	–	0.08	–	2.34	100 m²	2.42
50 g/m²	3.38	–	0.08	–	3.38	100 m²	3.46
70 g/m²	4.72	–	0.08	–	4.72	100 m²	4.80
100 g/m²	6.75	–	0.09	–	6.75	100 m²	6.84
125 g/m²	8.41	–	0.09	–	8.41	100 m²	8.50
Harrowing seeded areas; light chain harrow	–	–	–	0.12	–	100 m²	0.12
Raking over seeded areas							
by hand	–	0.80	16.40	–	–	100 m²	16.40

Q PAVING/PLANTING/FENCING/SITE FURNITURE

Item Excluding site overheads and profit	PC £	Labour hours	Labour £	Plant £	Material £	Unit	Total rate £
Rolling seeded areas; light roller							
by pedestrian operated mechanical roller	–	0.08	1.71	1.74	–	100 m²	**3.45**
by hand drawn roller	–	0.17	3.42	–	–	100 m²	**3.42**
Extra over harrowing, raking or rolling seeded areas for slopes over 30°; by machine or hand	–	–	–	–	–	25%	**–**
Turf edging; to seeded areas; 300 mm wide	1.99	0.13	2.73	–	1.99	m²	**4.72**
Preparation of turf beds							
Rolling turf to be lifted; lifting by hand or mechanical turf stripper; stacks to be not more than 1 m high							
cutting only preparing to lift; pedestrian turf cutter	–	0.75	15.38	9.01	–	100 m²	**24.39**
lifting and stacking; by hand	–	8.33	170.83	–	–	100 m²	**170.83**
Rolling up; moving to stacks							
distance not exceeding 100 m	–	2.50	51.25	–	–	100 m²	**51.25**
extra over rolling and moving turf to stacks to transport per additional 100 m	–	0.83	17.08	–	–	100 m²	**17.08**
Lifting selected and approved topsoil from spoil heaps							
passing through 6 mm screen; removing debris	–	0.17	3.42	9.18	–	m³	**12.60**
Extra over lifting topsoil and passing through screen for imported topsoil; plus 20% allowance for settlement	31.20	–	–	–	31.20	m³	**31.20**
Spreading topsoil to form turfbeds (topsoil not included); by machine							
25 mm deep	–	–	–	0.23	–	m²	**0.23**
50 mm deep	–	–	–	0.47	–	m²	**0.47**
75 mm deep	–	–	–	0.70	–	m²	**0.70**
100 mm deep	–	–	–	0.93	–	m²	**0.93**
150 mm deep	–	–	–	1.40	–	m²	**1.40**
Bringing existing topsoil to a fine tilth for turfing by raking or harrowing; stones not to exceed 6 mm; by hand	–	0.01	0.27	–	–	m²	**0.27**
Spreading topsoil to form turfbeds (topsoil not included); by hand							
25 mm deep	–	0.04	0.82	–	–	m²	**0.82**
50 mm deep	–	0.08	1.64	–	–	m²	**1.64**
75 mm deep	–	0.12	2.46	–	–	m²	**2.46**
100 mm deep	–	0.15	3.08	–	–	m²	**3.08**
150 mm deep	–	0.23	4.62	–	–	m²	**4.62**
Operations after spreading of topsoil; by hand							
Bringing existing topsoil to a fine tilth for turfing by raking or harrowing; stones not to exceed 6 mm							
topsoil spread by machine (soil compacted)	–	0.03	0.51	–	–	m²	**0.51**
topsoil spread by hand	–	0.02	0.34	–	–	m²	**0.34**

Q PAVING/PLANTING/FENCING/SITE FURNITURE

Item Excluding site overheads and profit	PC £	Labour hours	Labour £	Plant £	Material £	Unit	Total rate £
Q30 SEEDING/TURFING – CONT							
Turfing							
Turfing; laying only; to stretcher bond; butt							
joints; including providing and working from							
barrow plank runs where necessary to							
surfaces not exceeding 30° from horizontal							
specially selected lawn turves from							
previously lifted stockpile	–	0.20	4.10	–	–	m²	4.10
cultivated lawn turves; to larger open areas	–	0.11	2.19	–	–	m²	2.19
cultivated lawn turves; to domestic or							
garden areas	–	0.13	2.73	–	–	m²	2.73
Industrially grown turf; PC prices listed							
represent the general range of industrial							
turf prices for sportsfields and amenity							
purposes; prices will vary with quantity							
and site location							
Rolawn							
RB Medallion; sports fields, domestic							
lawns, general landscape	1.99	0.13	2.73	–	1.99	m²	4.72
Inturf							
Inturf Masters; formal lawns, golf greens,							
bowling greens and low maintenance areas	2.72	0.13	2.73	–	2.72	m²	5.45
Inturf Lawn; garden areas and open							
spaces, racecourses and hockey fields	2.02	0.13	2.73	–	2.02	m²	4.75
Inturf Classic; golf tees, surrounds, lawns,							
parks, general purpose landscaping areas							
and winter sports	2.37	0.13	2.73	–	2.37	m²	5.10
Inturf Ornamental; medium fine turf for							
ornamental lawns, fairways and green							
surrounds	2.37	0.13	2.73	–	2.37	m²	5.10
Firming turves with wooden beater	–	0.01	0.20	–	–	m²	0.20
Rolling turfed areas; light roller							
by pedestrian operated mechanical roller	–	0.08	1.71	1.74	–	100 m²	3.45
by hand drawn roller	–	0.17	3.42	–	–	100 m²	3.42
Dressing with finely sifted topsoil; brushing							
into joints	0.04	0.05	1.02	–	0.04	m²	1.06
Turfing; laying only							
to slopes over 30°; to diagonal bond							
(measured as plan area – add 15% to							
these rates for the incline area of 30°							
slopes)	–	0.12	2.46	–	–	m²	2.46
Extra over laying turfing for pegging down							
turves							
wooden or galvanized wire pegs; 200 mm							
long; 2 pegs per 0.50 m²	1.58	0.01	0.27	–	1.58	m²	1.85
Artificial grass; Artificial Lawn Company;							
laid to sharp sand bed priced separately							
15 kg kiln sand brushed in per m²							
Leisure Lawn; 24 mm thick artificial sports turf;							
sand filled	–	–	–	–	–	m²	27.80
Budget Grass; for general use; budget							
surface; sand filled	–	–	–	–	–	m²	21.95
Multi Grass; patios conservatories and pool							
surrounds; sand filled	–	–	–	–	–	m²	23.90

Q PAVING/PLANTING/FENCING/SITE FURNITURE

Item Excluding site overheads and profit	PC £	Labour hours	Labour £	Plant £	Material £	Unit	Total rate £
Premier; lawns and patios	–	–	–	–	–	m²	**30.00**
Play Lawn; grass/sand and rubber filled	–	–	–	–	–	m²	**33.85**
Grassflex; safety surfacing for play areas	–	–	–	–	–	m²	**49.50**
Maintenance operations (Note: the following rates apply to aftercare maintenance executed as part of a landscaping contract only)							
Initial cutting; to turfed areas							
20 mm high; using pedestrian guided power driven cylinder mower; including boxing off cuttings (stone picking and rolling not included)	–	0.18	3.69	0.29	–	100 m²	**3.98**
Repairing damaged grass areas							
scraping out; removing slurry; from ruts and holes; average 100 mm deep	–	0.13	2.73	–	–	m²	**2.73**
100 mm topsoil	–	0.13	2.73	–	3.12	m²	**5.85**
Repairing damaged grass areas; sowing grass seed to match existing or as specified; to individually prepared worn patches							
35 g/m²	0.18	0.01	0.20	–	0.18	m²	**0.38**
50 g/m²	0.26	0.01	0.20	–	0.26	m²	**0.46**
Using pedestrian operated mechanical equipment and blowers							
grassed areas with perimeters of mature trees such as sports fields and amenity areas	–	0.04	0.82	0.05	–	100 m²	**0.87**
grassed areas containing ornamental trees and shrub beds	–	0.10	2.05	0.12	–	100 m²	**2.17**
verges	–	0.07	1.37	0.08	–	100 m²	**1.45**
By hand							
grassed areas with perimeters of mature trees such as sports fields and amenity areas	–	0.05	1.02	0.06	–	100 m²	**1.08**
grassed areas containing ornamental trees and shrub beds	–	0.08	1.71	0.10	–	100 m²	**1.81**
verges	–	1.00	20.50	1.20	–	100 m²	**21.70**
Removal of arisings							
areas with perimeters of mature trees	–	0.01	0.14	0.10	1.28	100 m²	**1.52**
areas containing ornamental trees and shrub beds	–	0.02	0.41	0.37	3.20	100 m²	**3.98**
Cutting grass to specified height; per cut							
ride-on triple cylinder mower	–	0.01	0.29	0.14	–	100 m²	**0.43**
ride-on triple rotary mower	–	0.01	0.29	–	–	100 m²	**0.29**
pedestrian mower (open areas)	–	0.18	3.69	0.37	–	100 m²	**4.06**
pedestrian mower (small areas or areas with obstacles)	–	0.33	6.83	0.44	–	100 m²	**7.27**
Cutting rough grass; per cut							
power flail or scythe cutter	–	0.04	0.72	–	–	100 m²	**0.72**
Extra over cutting grass for slopes not exceeding 30°	–	–	–	–	–	10%	**–**
Extra over cutting grass for slopes exceeding 30°	–	–	–	–	–	40%	**–**
Cutting fine sward							
pedestrian operated seven-blade cylinder lawn mower	–	0.14	2.87	0.22	–	100 m²	**3.09**

Q PAVING/PLANTING/FENCING/SITE FURNITURE

Item Excluding site overheads and profit	PC £	Labour hours	Labour £	Plant £	Material £	Unit	Total rate £
Q30 SEEDING/TURFING – CONT							
Maintenance operations (Note: the following rates apply to aftercare maintenance executed as part of a landscaping contract only) – cont							
Extra over cutting fine sward for boxing off cuttings							
pedestrian mower	–	0.03	0.57	0.04	–	100 m²	**0.61**
Cutting areas of rough grass							
scythe	–	1.00	20.50	–	–	100 m²	**20.50**
sickle	–	2.00	41.00	–	–	100 m²	**41.00**
petrol operated strimmer	–	0.30	6.16	0.41	–	100 m²	**6.57**
Cutting areas of rough grass which contain trees or whips							
petrol operated strimmer	–	0.40	8.20	0.55	–	100 m²	**8.75**
Extra over cutting rough grass for on site raking up and dumping	–	0.33	6.83	–	–	100 m²	**6.83**
Trimming edge of grass areas; edging tool							
with petrol powered strimmer	–	0.13	2.73	0.18	–	100 m	**2.91**
by hand	–	0.67	13.67	–	–	100 m	**13.67**
Rolling grass areas; light roller							
by pedestrian operated mechanical roller	–	0.08	1.71	1.74	–	100 m²	**3.45**
by hand drawn roller	–	0.17	3.42	–	–	100 m²	**3.42**
Aerating grass areas; to a depth of 100 mm							
using pedestrian-guided motor powered solid or slitting tine turf aerator	–	0.18	3.59	2.68	–	100 m²	**6.27**
using hollow tine aerator; including sweeping up and dumping corings	–	0.50	10.25	5.36	–	100 m²	**15.61**
using hand aerator or fork	–	1.67	34.17	–	–	100 m²	**34.17**
Extra over aerating grass areas for on site sweeping up and dumping corings	–	0.17	3.42	–	–	100 m²	**3.42**
Switching off dew; from fine turf areas	–	0.20	4.10	–	–	100 m²	**4.10**
Scarifying grass areas to break up thatch; removing dead grass							
using self-propelled scarifier; including removing and disposing of grass on site	–	0.33	6.83	0.13	–	100 m²	**6.96**
by hand	–	3.03	62.12	–	–	100 m²	**62.12**
For the following topsoil improvement and seeding operations add or subtract the following amounts for every £0.10 difference in the material cost price							
35 g/m²	–	–	–	–	0.35	100 m²	**0.35**
50 g/m²	–	–	–	–	0.50	100 m²	**0.50**
70 g/m²	–	–	–	–	0.70	100 m²	**0.70**
100 g/m²	–	–	–	–	1.00	100 m²	**1.00**
Top dressing fertilizers (7+7+7); PC £1.13/kg; to seedbeds; by hand							
35 g/m²	2.86	0.17	3.42	–	2.86	100 m²	**6.28**
50 g/m²	4.08	0.17	3.42	–	4.08	100 m²	**7.50**
70 g/m²	5.71	0.17	3.42	–	5.71	100 m²	**9.13**
Watering turf; evenly; at a rate of 5 litre/m²							
using sprinkler equipment and with sufficient water pressure to run 1 nr 15 m radius sprinkler	–	0.02	0.41	–	–	100 m²	**0.41**
using hand-held watering equipment	–	0.25	5.13	–	–	100 m²	**5.13**

Q PAVING/PLANTING/FENCING/SITE FURNITURE

Item Excluding site overheads and profit	PC £	Labour hours	Labour £	Plant £	Material £	Unit	Total rate £
Q31 PLANTING							
Site protection; temporary protective fencing							
Cleft chestnut rolled fencing; to 100 mm diameter chestnut posts; driving into firm ground at 3 m centres; pales at 50 mm centres							
900 mm high	4.08	0.11	2.19	–	8.95	m	**11.14**
1100 mm high	4.61	0.11	2.19	–	9.48	m	**11.67**
1500 mm high	7.07	0.11	2.19	–	11.95	m	**14.14**
Extra over temporary protective fencing for removing and making good (no allowance for re-use of material)	–	0.07	1.37	0.22	–	m	**1.59**
Cultivation							
Breaking up existing ground; using pedestrian operated tine cultivator or rotavator							
100 mm deep	–	0.55	11.28	16.58	–	100 m²	**27.86**
150 mm deep	–	0.63	12.88	18.95	–	100 m²	**31.83**
200 mm deep	–	0.73	15.03	22.11	–	100 m²	**37.14**
As above but in heavy clay or wet soils							
100 mm deep	–	0.73	15.03	22.11	–	100 m²	**37.14**
150 mm deep	–	0.88	18.04	26.53	–	100 m²	**44.57**
200 mm deep	–	1.10	22.55	33.17	–	100 m²	**55.72**
Importing only selected and approved topsoil							
1–14 m³	78.00	–	–	–	78.00	m³	**78.00**
over 15 m³	26.00	–	–	–	26.00	m³	**26.00**
Imported topsoil; spreading and lightly consolidating approved topsoil in layers not exceeding 150 mm; travel distance from offloading point not exceeding 25 m							
By machine							
minimum depth 100 mm	–	–	–	0.93	3.25	m²	**4.18**
minimum depth 200 mm	–	–	–	1.87	6.50	m²	**8.37**
minimum depth 300 mm	–	–	–	2.80	9.75	m²	**12.55**
minimum depth 500 mm	–	–	–	4.66	16.25	m²	**20.91**
over 500 mm	–	–	–	9.33	32.50	m³	**41.83**
By hand							
minimum depth 100 mm	–	0.20	4.10	–	3.25	m²	**7.35**
minimum depth 150 mm	–	0.30	6.16	–	4.88	m²	**11.04**
minimum depth 300 mm	–	0.60	12.28	–	9.75	m²	**22.03**
minimum depth 500 mm	–	0.90	18.47	–	16.25	m²	**34.72**
over 500 mm deep	–	2.22	45.51	–	32.50	m³	**78.01**
Spreading and lightly consolidating approved topsoil (imported or from spoil heaps); in layers not exceeding 150 mm; travel distance from spoil heaps not exceeding 25 m							
By machine (imported topsoil not included)							
minimum depth 100 mm	–	–	–	0.93	–	m²	**0.93**
minimum depth 200 mm	–	–	–	1.87	–	m²	**1.87**
minimum depth 300 mm	–	–	–	2.80	–	m²	**2.80**
minimum depth 500 mm	–	–	–	4.66	–	m²	**4.66**
over 500 mm	–	–	–	9.33	–	m³	**9.33**

Q PAVING/PLANTING/FENCING/SITE FURNITURE

Item Excluding site overheads and profit	PC £	Labour hours	Labour £	Plant £	Material £	Unit	Total rate £
Q31 PLANTING – CONT							
Spreading and lightly consolidating approved topsoil (imported or from spoil heaps) – cont							
By hand (imported topsoil not included)							
minimum depth 100 mm	–	0.20	4.10	–	–	m²	**4.10**
minimum depth 150 mm	–	0.30	6.16	–	–	m²	**6.16**
minimum depth 300 mm	–	0.60	12.28	–	–	m²	**12.28**
minimum depth 500 mm	–	0.90	18.47	–	–	m²	**18.47**
over 500 mm deep	–	2.22	45.51	–	–	m³	**45.51**
Extra over for spreading topsoil to slopes 15–30° by machine or hand	–	–	–	–	–	10%	**–**
Extra over for spreading topsoil to slopes over 30° by machine or hand	–	–	–	–	–	25%	**–**
Extra over spreading topsoil for travel exceeding 100 m; by machine							
100–150 m	–	–	–	0.62	–	m³	**0.62**
150–200 m	–	–	–	0.82	–	m³	**0.82**
200–300 m	–	–	–	1.09	–	m³	**1.09**
Extra over spreading topsoil for travel exceeding 100 m; by hand							
100 m	–	2.50	51.25	–	–	m³	**51.25**
200 m	–	3.50	71.75	–	–	m³	**71.75**
300 m	–	4.50	92.25	–	–	m³	**92.25**
Evenly grading; to general surfaces to bring to finished levels							
by pedestrian operated rotavator	–	–	0.08	0.13	–	m²	**0.21**
by hand	–	0.01	0.20	–	–	m²	**0.20**
Extra over grading for slopes 15–30° by machine or hand	–	–	–	–	–	10%	**–**
Extra over grading for slopes over 30° by machine or hand	–	–	–	–	–	25%	**–**
Clearing stones; disposing off site							
by hand; stones not exceeding 50 mm in any direction; loading to skip 4.6 m³	–	0.01	0.20	0.04	–	m²	**0.24**
Lightly cultivating; weeding; to fallow areas; disposing debris off site							
by hand	–	0.01	0.29	0.29	–	m²	**0.58**
Preparation of planting operations; herbicides and granular additives							
For the following topsoil improvement and planting operations add or subtract the following amounts for every £0.10 difference in the material cost price							
35 g/m²	–	–	–	–	0.35	100 m²	**0.35**
50 g/m²	–	–	–	–	0.50	100 m²	**0.50**
70 g/m²	–	–	–	–	0.70	100 m²	**0.70**
100 g/m²	–	–	–	–	1.00	100 m²	**1.00**
General herbicides; in accordance with manufacturer's instructions; Knapsack spray application							
Roundup Pro Biactive 360 at 50 ml/100 m²	0.55	0.33	6.83	–	0.55	100 m²	**7.38**

Q PAVING/PLANTING/FENCING/SITE FURNITURE

Item Excluding site overheads and profit	PC £	Labour hours	Labour £	Plant £	Material £	Unit	Total rate £
Fertilizers; in top 150 mm of topsoil; at 35 g/m²							
Mascot Microfine; 8+0+6 + 2% Mg + 4% Fe	5.00	0.12	2.51	–	5.00	100 m²	7.51
Enmag; controlled release fertilizer							
(8–9 months); 11+22+09	8.69	0.12	2.51	–	8.69	100 m²	11.20
Super Phosphate Powder	4.48	0.12	2.51	–	4.48	100 m²	6.99
Bone meal	4.07	0.12	2.51	–	4.07	100 m²	6.58
Mascot Outfield; 9+5+5	3.01	0.12	2.51	–	3.01	100 m²	5.52
Mascot Outfield; 4+10+10	2.86	0.12	2.51	–	2.86	100 m²	5.37
Fertilizers; in top 150 mm of topsoil at 70 g/m²							
Mascot Microfine; 8+0+6 + 2% Mg + 4% Fe	10.01	0.12	2.51	–	10.01	100 m²	12.52
Enmag; controlled release fertilizer							
(8–9 months); 11+22+09	17.37	0.12	2.51	–	17.37	100 m²	19.88
Super Phosphate Powder	8.97	0.12	2.51	–	8.97	100 m²	11.48
Bone meal	8.14	0.12	2.51	–	8.14	100 m²	10.65
Mascot Oufield; 9+5+5	6.02	0.12	2.51	–	6.02	100 m²	8.53
Mascot Outfield; 4+10+10	5.71	0.12	2.51	–	5.71	100 m²	8.22
Preparation of planting areas; movement of materials to location maximum 25 m from offload location							
By machine	–	–	–	6.96	–	m³	6.96
By hand	–	1.00	20.50	–	–	m³	20.50
Preparation of planting operations; spreading only; movement of material to planting beds not included							
Composted bark and manure soil conditioner (20 m³ loads); from not further than 25 m from location; cultivating into topsoil by pedestrian operated cultivator							
50 mm thick	159.07	2.86	58.57	5.53	159.07	100 m²	223.17
100 mm thick	318.15	6.05	124.00	5.53	318.15	100 m²	447.68
150 mm thick	477.23	8.90	182.55	5.53	477.23	100 m²	665.31
200 mm thick	636.30	12.90	264.55	5.53	636.30	100 m²	906.38
Mushroom compost (20 m³ loads); from not further than 25 m from location; cultivating into topsoil by pedestrian operated cultivator							
50 mm thick	106.00	2.86	58.57	5.53	106.00	100 m²	170.10
100 mm thick	212.00	6.05	124.00	5.53	212.00	100 m²	341.53
150 mm thick	318.00	8.90	182.55	5.53	318.00	100 m²	506.08
Root decompaction; Terravent; Goroots Ltd							
200 mm thick	424.00	12.90	264.55	5.53	424.00	100 m²	694.08
Decompaction and aeration of soil in the root zone of a tree or plant; followed by an infusion of beneficial mychorrhizal fungi and other liquid amendments into the freshly aerated soil; mulching the root zone	–	–	–	–	1.00	m²	1.00
Tree planting; pre-planting operations							
Excavating tree pits; depositing soil alongside pits; by machine							
600 mm × 600 mm × 600 mm deep	–	0.15	3.02	0.62	–	nr	3.64
900 mm × 900 mm × 600 mm deep	–	0.33	6.76	1.38	–	nr	8.14
1.00 m × 1.00 m × 600 mm deep	–	0.61	12.58	1.71	–	nr	14.29

Q PAVING/PLANTING/FENCING/SITE FURNITURE

Item Excluding site overheads and profit	PC £	Labour hours	Labour £	Plant £	Material £	Unit	Total rate £
Q31 PLANTING – CONT							
Tree planting – cont							
Excavating tree pits – cont							
1.25 m × 1.25 m × 600 mm deep	–	0.96	19.69	2.68	–	nr	**22.37**
1.00 m × 1.00 m × 1.00 m deep	–	1.02	20.97	2.86	–	nr	**23.83**
1.50 m × 1.50 m × 750 mm deep	–	1.73	35.38	4.82	–	nr	**40.20**
1.50 m × 1.50 m × 1.00 m deep	–	2.30	47.05	6.41	–	nr	**53.46**
1.75 m × 1.75 m × 1.00 mm deep	–	3.13	64.21	8.75	–	nr	**72.96**
2.00 m × 2.00 m × 1.00 mm deep	–	4.09	83.86	11.43	–	nr	**95.29**
Excavating tree pits; depositing soil alongside pits; by hand							
600 mm × 600 mm × 600 mm deep	–	0.44	9.02	–	–	nr	**9.02**
900 mm × 900 mm × 600 mm deep	–	1.00	20.50	–	–	nr	**20.50**
1.00 m × 1.00 m × 600 mm deep	–	1.13	23.06	–	–	nr	**23.06**
1.25 m × 1.25 m × 600 mm deep	–	1.93	39.56	–	–	nr	**39.56**
1.00 m × 1.00 m × 1.00 m deep	–	2.06	42.23	–	–	nr	**42.23**
1.50 m × 1.50 m × 750 mm deep	–	3.47	71.14	–	–	nr	**71.14**
1.75 m × 1.50 m × 750 mm deep	–	4.05	83.03	–	–	nr	**83.03**
1.50 m × 1.50 m × 1.00 m deep	–	4.63	94.92	–	–	nr	**94.92**
2.00 m × 2.00 m × 750 mm deep	–	6.17	126.48	–	–	nr	**126.48**
2.00 m × 2.00 m × 1.00 m deep	–	8.23	168.73	–	–	nr	**168.73**
Breaking up subsoil in tree pits; to a depth of 200 mm	–	0.03	0.68	–	–	m²	**0.68**
Spreading and lightly consolidating approved topsoil (imported or from spoil heaps); in layers not exceeding 150 mm; distance from spoil heaps not exceeding 50 m (imported topsoil not included); by machine							
minimum depth 100 mm	–	–	–	1.32	–	m²	**1.32**
minimum depth 150 mm	–	–	–	1.98	–	m²	**1.98**
minimum depth 300 mm	–	–	–	3.63	–	m²	**3.63**
minimum depth 450 mm	–	–	–	5.22	–	m²	**5.22**
Spreading and lightly consolidating approved topsoil (imported or from spoil heaps); in layers not exceeding 150 mm; distance from spoil heaps not exceeding 100 m (imported topsoil not included); by hand							
minimum depth 100 mm	–	0.13	2.56	–	–	m²	**2.56**
minimum depth 150 mm	–	0.19	3.85	–	–	m²	**3.85**
minimum depth 300 mm	–	0.38	7.71	–	–	m²	**7.71**
minimum depth 450 mm	–	0.56	11.58	–	–	m²	**11.58**
Extra for filling tree pits with imported topsoil; PC £26.00/m³; plus allowance for 20% settlement							
depth 100 mm	–	–	–	–	3.12	m²	**3.12**
depth 150 mm	–	–	–	–	4.68	m²	**4.68**
depth 200 mm	–	–	–	–	6.24	m²	**6.24**
depth 300 mm	–	–	–	–	9.36	m²	**9.36**
depth 400 mm	–	–	–	–	12.48	m²	**12.48**
depth 450 mm	–	–	–	–	14.04	m²	**14.04**
depth 500 mm	–	–	–	–	15.60	m²	**15.60**
depth 600 mm	–	–	–	–	18.72	m²	**18.72**

Q PAVING/PLANTING/FENCING/SITE FURNITURE

Item Excluding site overheads and profit	PC £	Labour hours	Labour £	Plant £	Material £	Unit	Total rate £
Add or deduct the following amounts for every £0.50 change in the material price of topsoil							
depth 100 mm	–	–	–	–	0.06	m²	**0.06**
depth 150 mm	–	–	–	–	0.09	m²	**0.09**
depth 200 mm	–	–	–	–	0.12	m²	**0.12**
depth 300 mm	–	–	–	–	0.18	m²	**0.18**
depth 400 mm	–	–	–	–	0.24	m²	**0.24**
depth 450 mm	–	–	–	–	0.27	m²	**0.27**
depth 500 mm	–	–	–	–	0.30	m²	**0.30**
depth 600 mm	–	–	–	–	0.36	m²	**0.36**
Tree staking							
J Toms Ltd; extra over trees for tree stake(s); driving 500 mm into firm ground; trimming to approved height; including two tree ties to approved pattern							
one stake; 1.52 m long × 32 mm × 32 mm	1.47	0.20	4.10	–	1.47	nr	**5.57**
two stakes; 1.21 m long × 25 mm × 25 mm	1.62	0.30	6.15	–	1.62	nr	**7.77**
two stakes; 1.52 m long × 32 mm × 32 mm	2.16	0.30	6.15	–	2.16	nr	**8.31**
three stakes; 1.52 m long × 32 mm × 32 mm	3.24	0.36	7.38	–	3.24	nr	**10.62**
Tree anchors							
Platipus Anchors Ltd; extra over trees for tree anchors							
RF1 rootball kit; for 75 to 220 mm girth; 2 to 4.5 m high; inclusive of Plati-Mat PM1	34.99	1.00	20.50	–	34.99	nr	**55.49**
RF2; rootball kit; for 220 to 450 mm girth; 4.5 to 7.5 m high; inclusive of Plati-Mat PM2	58.48	1.33	27.27	–	58.48	nr	**85.75**
RF3; rootball kit; for 450 to 750 mm girth; 7.5 to 12 m high; inclusive of Plati-Mat PM3	116.90	1.50	30.75	–	116.90	nr	**147.65**
CG1; guy fixing kit; 75 to 220 mm girth; 2 to 4.5 m high	19.08	1.67	34.17	–	19.08	nr	**53.25**
CG2; guy fixing kit; 220 to 450 mm girth; 4.5 to 7.5 m high	38.59	2.00	41.00	–	38.59	nr	**79.59**
installation tools; drive rod for RF1/CG1 kits	–	–	–	–	62.79	nr	**62.79**
installation tools; drive rod for RF2/CG2 kits	–	–	–	–	93.20	nr	**93.20**
Extra over trees for land drain to tree pits; 100 mm diameter perforated flexible agricultural drain; including excavating drain trench; laying pipe; backfilling	1.79	1.00	20.50	–	1.79	m	**22.29**
Tree planting; tree pit additives							
Melcourt Industries Ltd; Topgrow; incorporating into topsoil at 1 part Topgrow to 3 parts excavated topsoil; supplied in 75 L bags; pit size							
600 × 600 × 600 mm	1.51	0.02	0.41	–	1.51	nr	**1.92**
900 × 900 × 900 mm	5.10	0.06	1.23	–	5.10	nr	**6.33**
1.00 × 1.00 × 1.00 m	6.99	0.24	4.92	–	6.99	nr	**11.91**
1.25 × 1.25 × 1.25 m	13.67	0.40	8.20	–	13.67	nr	**21.87**
1.50 × 1.50 × 1.50 m	23.63	0.90	18.45	–	23.63	nr	**42.08**

Q PAVING/PLANTING/FENCING/SITE FURNITURE

Item Excluding site overheads and profit	PC £	Labour hours	Labour £	Plant £	Material £	Unit	Total rate £
Q31 PLANTING – CONT							
Tree planting – cont							
Melcourt Industries Ltd; Topgrow;							
incorporating into topsoil at 1 part Topgrow to							
3 parts excavated topsoil; supplied in 65 m³							
loose loads; pit size							
600 × 600 × 600 mm	1.16	0.02	0.34	–	1.16	nr	**1.50**
900 × 900 × 900 mm	3.93	0.05	1.02	–	3.93	nr	**4.95**
1.00 × 1.00 × 1.00 m	5.39	0.20	4.10	–	5.39	nr	**9.49**
1.25 × 1.25 × 1.25 m	10.52	0.33	6.83	–	10.52	nr	**17.35**
1.50 × 1.50 × 1.50 m	18.18	0.75	15.38	–	18.18	nr	**33.56**
Mulching of tree pits; Melcourt Industries							
Ltd (items labelled FSC are Forest							
Stewardship Council certified)							
Spreading mulch; to individual trees;							
maximum distance 25 m (mulch not included)							
50 mm thick	–	0.05	1.00	–	–	m²	**1.00**
75 mm thick	–	0.07	1.49	–	–	m²	**1.49**
100 mm thick	–	0.10	1.99	–	–	m²	**1.99**
Mulch; Bark Nuggets®; to individual trees;							
delivered in 25 m³ loads; maximum distance							
25 m							
50 mm thick	2.71	0.05	1.00	–	2.71	m²	**3.71**
75 mm thick	4.07	0.05	1.02	–	4.07	m²	**5.09**
100 mm thick	5.43	0.07	1.37	–	5.43	m²	**6.80**
Mulch; Amenity Bark Mulch; FSC; to individual							
trees; delivered in 25 m³ loads; maximum							
distance 25 m							
50 mm thick	1.95	0.05	1.00	–	1.95	m²	**2.95**
75 mm thick	2.93	0.07	1.49	–	2.93	m²	**4.42**
100 mm thick	3.90	0.07	1.37	–	3.90	m²	**5.27**
Trees; planting labours only							
Bare root trees; including backfilling with							
previously excavated material (all other							
operations and materials not included)							
light standard; 6–8 cm girth	–	0.35	7.17	–	–	nr	**7.17**
standard; 8–10 cm girth	–	0.40	8.20	–	–	nr	**8.20**
selected standard; 10–12 cm girth	–	0.58	11.89	–	–	nr	**11.89**
heavy standard; 12–14 cm girth	–	0.83	17.08	–	–	nr	**17.08**
extra heavy standard; 14–16 cm girth	–	1.00	20.50	–	–	nr	**20.50**
Root balled trees; including backfilling with							
previously excavated material (all other							
operations and materials not included)							
standard; 8–10 cm girth	–	0.50	10.25	–	–	nr	**10.25**
selected standard; 10–12 cm girth	–	0.60	12.30	–	–	nr	**12.30**
heavy standard; 12–14 cm girth	–	0.80	16.40	–	–	nr	**16.40**
extra heavy standard; 14–16 cm girth	–	1.50	30.75	–	–	nr	**30.75**
16–18 cm girth	–	1.30	26.57	24.93	–	nr	**51.50**
18–20 cm girth	–	1.60	32.80	30.77	–	nr	**63.57**
20–25 cm girth	–	4.50	92.25	99.41	–	nr	**191.66**
25–30 cm girth	–	6.00	123.00	130.83	–	nr	**253.83**
30–35 cm girth	–	11.00	225.50	230.86	–	nr	**456.36**

Q PAVING/PLANTING/FENCING/SITE FURNITURE

Item Excluding site overheads and profit	PC £	Labour hours	Labour £	Plant £	Material £	Unit	Total rate £
Tree planting; root balled trees; advanced nursery stock and semi-mature – General Preamble: the cost of planting semi-mature trees will depend on the size and species, and on the access to the site for tree handling machines. Prices should be obtained for individual trees and planting.							
Tree planting; bare root trees; nursery stock; James Coles & Sons (Nurseries) Ltd Acer platanoides; including backfillling with excavated material (other operations not included)							
light standard; 6–8 cm girth	9.45	0.35	7.17	–	9.45	nr	**16.62**
standard; 8–10 cm girth	12.60	0.40	8.20	–	12.60	nr	**20.80**
selected standard; 10–12 cm girth	20.48	0.58	11.89	–	20.48	nr	**32.37**
heavy standard; 12–14 cm girth	44.10	0.83	17.08	–	44.10	nr	**61.18**
extra heavy standard; 14–16 cm girth	58.80	1.00	20.50	–	58.80	nr	**79.30**
Carpinus betulus; including backfillling with excavated material (other operations not included)							
light standard; 6–8 cm girth	13.91	0.35	7.17	–	13.91	nr	**21.08**
standard; 8–10 cm girth	27.82	0.40	8.20	–	27.82	nr	**36.02**
selected standard; 10–12 cm girth	39.64	0.58	11.89	–	39.64	nr	**51.53**
heavy standard; 12–14 cm girth	41.21	0.83	17.09	–	41.21	nr	**58.30**
extra heavy standard; 14–16 cm girth	48.56	1.00	20.50	–	48.56	nr	**69.06**
Fraxinus excelsior; including backfillling with excavated material (other operations not included)							
light standard; 6–8 cm girth	10.24	0.35	7.17	–	10.24	nr	**17.41**
standard; 8–10 cm girth	16.80	0.40	8.20	–	16.80	nr	**25.00**
selected standard; 10–12 cm girth	23.63	0.58	11.89	–	23.63	nr	**35.52**
heavy standard; 12–14 cm girth	41.21	0.83	17.02	–	41.21	nr	**58.23**
extra heavy standard; 14–16 cm girth	49.88	1.00	20.50	–	49.88	nr	**70.38**
Prunus avium 'Plena'; including backfillling with excavated material (other operations not included)							
light standard; 6–8 cm girth	10.24	0.36	7.46	–	10.24	nr	**17.70**
standard; 8–10 cm girth	16.80	0.40	8.20	–	16.80	nr	**25.00**
selected standard; 10–12 cm girth	30.98	0.58	11.89	–	30.98	nr	**42.87**
heavy standard; 12–14 cm girth	49.88	0.83	17.08	–	49.88	nr	**66.96**
extra heavy standard; 14–16 cm girth	46.99	1.00	20.50	–	46.99	nr	**67.49**
Quercus robur; including backfillling with excavated material (other operations not included)							
light standard; 6–8 cm girth	22.05	0.35	7.17	–	22.05	nr	**29.22**
standard; 8–10 cm girth	32.29	0.40	8.20	–	32.29	nr	**40.49**
selected standard; 10–12 cm girth	44.10	0.58	11.89	–	44.10	nr	**55.99**
heavy standard; 12–14 cm girth	61.69	0.83	17.08	–	61.69	nr	**78.77**
Robinia pseudoacacia 'Frisia'; including backfillling with excavated material (other operations not included)							
light standard; 6–8 cm girth	22.05	0.35	7.17	–	22.05	nr	**29.22**
standard; 8–10 cm girth	29.40	0.40	8.20	–	29.40	nr	**37.60**
selected standard; 10–12 cm girth	46.99	0.58	11.89	–	46.99	nr	**58.88**

Q PAVING/PLANTING/FENCING/SITE FURNITURE

Item Excluding site overheads and profit	PC £	Labour hours	Labour £	Plant £	Material £	Unit	Total rate £
Q31 PLANTING – CONT							
Tree planting; root balled trees; nursery stock; James Coles & Sons (Nurseries) Ltd							
Acer platanoides; including backfillling with excavated material (other operations not included)							
standard; 8–10 cm girth	20.48	0.48	9.84	–	20.48	nr	**30.32**
selected standard; 10–12 cm girth	30.98	0.56	11.49	–	30.98	nr	**42.47**
heavy standard; 12–14 cm girth	59.85	0.76	15.67	–	59.85	nr	**75.52**
extra heavy standard; 14–16 cm girth	74.55	1.20	24.60	–	74.55	nr	**99.15**
Carpinus betulus; including backfillling with excavated material (other operations not included)							
standard; 8–10 cm girth	35.70	0.48	9.84	–	35.70	nr	**45.54**
selected standard; 10–12 cm girth	50.14	0.56	11.49	–	50.14	nr	**61.63**
heavy standard; 12–14 cm girth	83.47	0.76	15.67	–	83.47	nr	**99.14**
extra heavy standard; 14–16 cm girth	111.30	1.20	24.60	–	111.30	nr	**135.90**
Tree planting; containerized trees; nursery stock; James Coles & Sons (Nurseries) Ltd							
Acer platanoides 'Emerald Queen'; including backfillling with excavated material (other operations not included)							
standard; 8–10 cm girth	46.99	0.48	9.84	–	46.99	ea	**56.83**
selected standard; 10–12 cm girth	73.50	0.56	11.49	–	73.50	ea	**84.99**
heavy standard; 12–14 cm girth	95.55	0.76	15.67	–	95.55	ea	**111.22**
extra heavy standard; 14–16 cm girth	110.25	1.20	24.60	–	110.25	ea	**134.85**
Carpinus betulus; including backfillling with excavated material (other operations not included)							
standard; 8–10 cm girth	46.99	0.48	9.84	–	46.99	ea	**56.83**
selected standard; 10–12 cm girth	73.50	0.56	11.49	–	73.50	ea	**84.99**
heavy standard; 12–14 cm girth	95.55	0.76	15.67	–	95.55	ea	**111.22**
extra heavy standard; 14–16 cm girth	117.60	1.20	24.60	–	117.60	ea	**142.20**
Fraxinus excelsior 'Altena'; including backfillling with excavated material (other operations not included)							
standard; 8–10 cm girth	44.10	0.48	9.84	–	44.10	ea	**53.94**
selected standard; 10–12 cm girth	70.61	0.56	11.49	–	70.61	ea	**82.10**
heavy standard; 12–14 cm girth	88.20	0.76	15.67	–	88.20	ea	**103.87**
extra heavy standard; 14–16 cm girth	117.60	1.20	24.60	–	117.60	ea	**142.20**
Prunus avium 'Plena'; including backfillling with excavated material (other operations not included)							
selected standard; 10–12 cm girth	73.50	0.56	11.49	–	73.50	ea	**84.99**
heavy standard; 12–14 cm girth	88.20	0.76	15.67	–	88.20	ea	**103.87**
extra heavy standard; 14–16 cm girth	102.90	1.20	24.60	–	102.90	ea	**127.50**
Quercus robur; including backfillling with excavated material (other operations not included)							
standard; 8–10 cm girth	51.45	0.48	9.84	–	51.45	ea	**61.29**
selected standard; 10–12 cm girth	80.85	0.56	11.49	–	80.85	ea	**92.34**
heavy standard; 12–14 cm girth	110.25	0.76	15.67	–	110.25	ea	**125.92**
extra heavy standard; 14–16 cm girth	124.95	1.20	24.60	–	124.95	ea	**149.55**

Q PAVING/PLANTING/FENCING/SITE FURNITURE

Item Excluding site overheads and profit	PC £	Labour hours	Labour £	Plant £	Material £	Unit	Total rate £
Betula utilis jaquemontii; multi-stemmed; including backfillling with excavated material (other operations not included)							
175/200 mm high	73.50	0.48	9.84	–	73.50	ea	**83.34**
200/250 mm high	102.90	0.56	11.49	–	102.90	ea	**114.39**
250/300 mm high	124.95	0.76	15.67	–	124.95	ea	**140.62**
300/350 mm high	205.80	1.20	24.60	–	205.80	ea	**230.40**
Tree planting; Airpot container grown **trees; advanced nursery stock and** **semi-mature; Deepdale Trees Ltd** Acer platanoides 'Emerald Queen'; including backfilling with excavated material (other operations not included)							
16–18 cm girth	95.00	1.98	40.59	33.04	95.00	nr	**168.63**
18–20 cm girth	130.00	2.18	44.65	36.35	130.00	nr	**211.00**
20–25 cm girth	190.00	2.38	48.71	39.65	190.00	nr	**278.36**
25–30 cm girth	250.00	2.97	60.88	66.09	250.00	nr	**376.97**
30–35 cm girth	450.00	3.96	81.18	91.80	450.00	nr	**622.98**
Aesculus briotti; including backfilling with excavated material (other operations not included)							
16–18 cm girth	110.00	1.98	40.59	33.04	110.00	nr	**183.63**
18–20 cm girth	130.00	1.60	32.80	36.35	130.00	nr	**199.15**
20–25 cm girth	200.00	2.38	48.71	39.65	200.00	nr	**288.36**
25–30 cm girth	300.00	2.97	60.88	83.62	300.00	nr	**444.50**
30–35 cm girth	450.00	3.96	81.18	91.80	450.00	nr	**622.98**
Tree planting; Airpot container grown **trees; semi-mature and mature trees;** **Deepdale Trees Ltd; planting and back** **filling; planted by telehandler or by crane;** **delivery included; all other operations** **priced separately** Semi mature trees indicative prices							
40–45 cm girth	550.00	4.00	82.00	53.38	550.00	nr	**685.38**
45–50 cm girth	750.00	4.00	82.00	53.38	750.00	nr	**885.38**
55–60 cm girth	1350.00	6.00	123.00	53.38	1350.00	nr	**1526.38**
60–70 cm girth	2500.00	7.00	143.50	71.72	2500.00	nr	**2715.22**
70–80 cm girth	3500.00	7.50	153.75	90.07	3500.00	nr	**3743.82**
80–90 cm girth	4500.00	8.00	164.00	106.76	4500.00	nr	**4770.76**
Tree planting; root balled trees; advanced **nursery stock and semi-mature; Lorenz** **von Ehren** Acer platanoides 'Emerald Queen'; including backfilling with excavated material (other operations not included)							
16–18 cm girth	93.50	1.30	26.57	4.19	93.50	nr	**124.26**
18–20 cm girth	115.50	1.60	32.80	4.19	115.50	nr	**152.49**
20–25 cm girth	143.00	4.50	92.25	19.14	143.00	nr	**254.39**
25–30 cm girth	187.00	6.00	123.00	23.80	187.00	nr	**333.80**
30–35 cm girth	335.50	11.00	225.50	31.85	335.50	nr	**592.85**

Q PAVING/PLANTING/FENCING/SITE FURNITURE

Item Excluding site overheads and profit	PC £	Labour hours	Labour £	Plant £	Material £	Unit	Total rate £
Q31 PLANTING – CONT							
Tree planting – cont							
Quercus palustris; including backfilling with excavated material (other operations not included)							
16–18 cm girth	137.50	1.30	26.57	4.19	137.50	nr	**168.26**
18–20 cm girth	165.00	1.60	32.80	4.19	165.00	nr	**201.99**
20–25 cm girth	209.00	4.50	92.25	19.14	209.00	nr	**320.39**
25–30 cm girth	247.50	6.00	123.00	23.80	247.50	nr	**394.30**
30–35 cm girth	374.00	11.00	225.50	31.85	374.00	nr	**631.35**
Tilia cordata 'Greenspire'; including backfilling with excavated material (other operations not included)							
16–18 cm girth	104.50	1.30	26.57	4.19	104.50	nr	**135.26**
18–20 cm girth	126.50	1.60	32.80	4.19	126.50	nr	**163.49**
20–25 cm girth	154.00	4.50	92.25	19.14	154.00	nr	**265.39**
25–30 cm girth; 5 × transplanted 4.0–5.0 m tall	165.00	6.00	123.00	23.80	165.00	nr	**311.80**
30–35 cm girth; 5 × transplanted 5.0–7.0 m tall	242.00	11.00	225.50	31.85	242.00	nr	**499.35**
Betula pendula; 3 stems; including backfilling with excavated material (other operations not included)							
3.0–3.5 m high	66.00	1.98	40.59	33.04	66.00	nr	**139.63**
3.5–4.0 m high	99.00	1.60	32.80	36.35	99.00	nr	**168.15**
4.0–4.5 m high	143.00	2.38	48.71	55.08	143.00	nr	**246.79**
4.5–5.0 m high	165.00	2.97	60.88	83.62	165.00	nr	**309.50**
5.0–6.0 m high	231.00	3.96	81.18	91.80	231.00	nr	**403.98**
6.0–7.0 m high	385.00	4.50	92.25	108.66	385.00	nr	**585.91**
Pinus sylvestris; including backfilling with excavated material (other operations not included)							
3.0–3.5 m high	440.00	1.98	40.59	33.04	440.00	nr	**513.63**
3.5–4.0 m high	550.00	1.60	32.80	36.35	550.00	nr	**619.15**
4.0–4.5 m high	759.00	2.38	48.71	39.65	759.00	nr	**847.36**
4.5–5.0 m high	1045.00	2.97	60.88	83.62	1045.00	nr	**1189.50**
5.0–6.0 m high	1650.00	3.96	81.18	91.80	1650.00	nr	**1822.98**
6.0–7.0 m high	2750.00	4.50	92.25	112.00	2750.00	nr	**2954.25**
Tree planting; containerized trees; Lorenz von Ehren; to the tree prices above add for Airpot containerization only							
Tree size							
20–25 cm	–	–	–	–	49.50	nr	**49.50**
25–30 cm	–	–	–	–	82.50	nr	**82.50**
30–35 cm	–	–	–	–	104.50	nr	**104.50**
35–40 cm	–	–	–	–	143.00	nr	**143.00**
40–45 cm	–	–	–	–	165.00	nr	**165.00**
45–50 cm	–	–	–	–	209.00	nr	**209.00**
50–60 cm	–	–	–	–	275.00	nr	**275.00**
60–70 cm	–	–	–	–	352.00	nr	**352.00**
70–80 cm	–	–	–	–	429.00	nr	**429.00**
80–90 cm	–	–	–	–	495.00	nr	**495.00**

Q PAVING/PLANTING/FENCING/SITE FURNITURE

Item Excluding site overheads and profit	PC £	Labour hours	Labour £	Plant £	Material £	Unit	Total rate £
Hedges							
Excavating trench for hedges; depositing soil alongside trench; by machine							
300 mm deep × 300 mm wide	–	0.03	0.61	0.30	–	m	0.91
300 mm deep × 450 mm wide	–	0.05	0.92	0.45	–	m	1.37
Excavating trench for hedges; depositing soil alongside trench; by hand							
300 mm deep × 300 mm wide	–	0.12	2.46	–	–	m	2.46
300 mm deep × 450 mm wide	–	0.23	4.62	–	–	m	4.62
Setting out; notching out; excavating trench; breaking up subsoil to minimum depth 300 mm							
minimum 400 mm deep	–	0.25	5.13	–	–	m	5.13
Hedge planting; including backfill with excavated topsoil; PC £0.36/nr							
single row; 200 mm centres	1.80	0.06	1.28	–	1.80	m	3.08
single row; 300 mm centres	1.20	0.06	1.14	–	1.20	m	2.34
single row; 400 mm centres	0.90	0.04	0.85	–	0.90	m	1.75
single row; 500 mm centres	0.72	0.03	0.68	–	0.72	m	1.40
double row; 200 mm centres	3.60	0.17	3.42	–	3.60	m	7.02
double row; 300 mm centres	2.40	0.13	2.73	–	2.40	m	5.13
double row; 400 mm centres	1.80	0.08	1.71	–	1.80	m	3.51
double row; 500 mm centres	1.44	0.07	1.37	–	1.44	m	2.81
Extra over hedges for incorporating manure; at 1 m³ per 30 m	0.80	0.03	0.51	–	0.80	m	1.31
Shrub planting – General							
Preamble: for preparation of planting areas see 'Cultivation' at the beginning of the planting section.							
Shrub planting							
Setting out; selecting planting from holding area; loading to wheelbarrows; planting as plan or as directed; distance from holding area maximum 50 m; plants 2–3 litre containers							
plants in groups of 100 nr minimum	–	0.01	0.24	–	–	nr	0.24
plants in groups of 10–100 nr	–	0.02	0.34	–	–	nr	0.34
plants in groups of 3–5 nr	–	0.03	0.51	–	–	nr	0.51
single plants not grouped	–	0.04	0.82	–	–	nr	0.82
Forming planting holes; in cultivated ground (cultivating not included); by mechanical auger; trimming holes by hand; depositing excavated material alongside holes							
250 mm diameter	–	0.03	0.68	0.14	–	nr	0.82
250 × 250 mm	–	0.04	0.82	0.25	–	nr	1.07
300 × 300 mm	–	0.08	1.54	0.31	–	nr	1.85
Hand excavation; forming planting holes; in cultivated ground (cultivating not included); depositing excavated material alongside holes							
100 × 100 × 100 mm deep; with mattock or hoe	–	0.01	0.14	–	–	nr	0.14
250 × 250 × 300 mm deep	–	0.04	0.82	–	–	nr	0.82
300 × 300 × 300 mm deep	–	0.06	1.14	–	–	nr	1.14
400 × 400 × 400 mm deep	–	0.13	2.56	–	–	nr	2.56
500 × 500 × 500 mm deep	–	0.25	5.13	–	–	nr	5.13
600 × 600 × 600 mm deep	–	0.43	8.87	–	–	nr	8.87
900 × 900 × 600 mm deep	–	1.00	20.50	–	–	nr	20.50
1.00 m × 1.00 m × 600 mm deep	–	1.23	25.21	–	–	nr	25.21
1.25 m × 1.25 m × 600 mm deep	–	1.93	39.56	–	–	nr	39.56

Q PAVING/PLANTING/FENCING/SITE FURNITURE

Item Excluding site overheads and profit	PC £	Labour hours	Labour £	Plant £	Material £	Unit	Total rate £
Q31 PLANTING – CONT							
Shrub planting – cont							
Hand excavation; forming planting holes; in uncultivated ground; depositing excavated material alongside holes							
100 × 100 × 100 mm deep; with mattock or hoe	–	0.03	0.51	–	–	nr	**0.51**
250 × 250 × 300 mm deep	–	0.06	1.14	–	–	nr	**1.14**
300 × 300 × 300 mm deep	–	0.06	1.28	–	–	nr	**1.28**
400 × 400 × 400 mm deep	–	0.25	5.13	–	–	nr	**5.13**
500 × 500 × 500 mm deep	–	0.33	6.67	–	–	nr	**6.67**
600 × 600 × 600 mm deep	–	0.55	11.28	–	–	nr	**11.28**
900 × 900 × 600 mm deep	–	1.25	25.63	–	–	nr	**25.63**
1.00 m × 1.00 m × 600 mm deep	–	1.54	31.52	–	–	nr	**31.52**
1.25 m × 1.25 m × 600 mm deep	–	2.41	49.46	–	–	nr	**49.46**
Bare root planting; to planting holes (forming holes not included); including backfilling with excavated material (bare root plants not included)							
bare root 1+1; 30–90 mm high	–	0.02	0.34	–	–	nr	**0.34**
bare root 1+2; 90–120 mm high	–	0.02	0.34	–	–	nr	**0.34**
Containerized planting; to planting holes (forming holes not included); including backfilling with excavated material (shrub or ground cover not included)							
9 cm pot	–	0.01	0.20	–	–	nr	**0.20**
2 litre container	–	0.02	0.41	–	–	nr	**0.41**
3 litre container	–	0.02	0.46	–	–	nr	**0.46**
5 litre container	–	0.03	0.68	–	–	nr	**0.68**
10 litre container	–	0.05	1.02	–	–	nr	**1.02**
15 litre container	–	0.07	1.37	–	–	nr	**1.37**
20 litre container	–	0.08	1.71	–	–	nr	**1.71**
Shrub planting; 2 litre containerized plants; in cultivated ground (cultivating not included); PC £3.00/nr							
average 2 plants per m²	–	0.06	1.15	–	6.00	m²	**7.15**
average 3 plants per m²	–	0.08	1.72	–	9.00	m²	**10.72**
average 4 plants per m²	–	0.11	2.30	–	12.00	m²	**14.30**
average 6 plants per m²	–	0.17	3.45	–	18.00	m²	**21.45**
Extra over shrubs for stakes	0.70	0.02	0.34	–	0.70	nr	**1.04**
Composted bark soil conditioners; 20 m³ loads; on beds by mechanical loader; spreading and rotavating into topsoil; by machine							
50 mm thick	151.50	–	–	6.94	151.50	100 m²	**158.44**
100 mm thick	318.15	–	–	11.89	318.15	100 m²	**330.04**
150 mm thick	477.23	–	–	17.04	477.23	100 m²	**494.27**
200 mm thick	636.30	–	–	22.09	636.30	100 m²	**658.39**
Mushroom compost; 25 m³ loads; delivered not further than 25 m from location; cultivating into topsoil by pedestrian operated machine							
50 mm thick	106.00	2.86	58.57	5.53	106.00	100 m²	**170.10**
100 mm thick	212.00	6.05	124.00	5.53	212.00	100 m²	**341.53**
150 mm thick	318.00	8.90	182.55	5.53	318.00	100 m²	**506.08**
200 mm thick	424.00	12.90	264.55	5.53	424.00	100 m²	**694.08**

Q PAVING/PLANTING/FENCING/SITE FURNITURE

Item Excluding site overheads and profit	PC £	Labour hours	Labour £	Plant £	Material £	Unit	Total rate £
Manure; 20 m³ loads; delivered not further than 25 m from location; cultivating into topsoil by pedestrian operated machine							
50 mm thick	236.25	2.86	58.57	5.53	236.25	100 m²	300.35
100 mm thick	496.13	6.05	124.00	5.53	496.13	100 m²	625.66
150 mm thick	744.19	8.90	182.55	5.53	744.19	100 m²	932.27
200 mm thick	992.25	12.90	264.55	5.53	992.25	100 m²	1262.33
Fertilizers (7+7+7); PC £1.13/kg; to beds; by hand							
35 g/m²	2.86	0.17	3.42	–	2.86	100 m²	6.28
50 g/m²	4.08	0.17	3.42	–	4.08	100 m²	7.50
70 g/m²	5.71	0.17	3.42	–	5.71	100 m²	9.13
Fertilizers; Enmag; PC £2.63/kg; controlled release fertilizer; to beds; by hand							
35 g/m²	8.27	0.17	3.42	–	8.27	100 m²	11.69
50 g/m²	11.82	0.17	3.42	–	11.82	100 m²	15.24
70 g/m²	17.73	0.17	3.42	–	17.73	100 m²	21.15
Note: for machine incorporation of fertilizers and soil conditioners see 'Cultivation'.							
Herbaceous and groundcover planting							
Herbaceous plants; PC £1.50/nr; including forming planting holes in cultivated ground (cultivating not included); backfilling with excavated material; 1 litre containers							
average 4 plants per m² – 500 mm centres	6.00	0.09	1.91	–	6.00	m²	7.91
average 6 plants per m² – 408 mm centres	–	0.14	2.87	–	9.00	m²	11.87
average 8 plants per m² – 354 mm centres	–	0.19	3.83	–	12.00	m²	15.83
Note: for machine incorporation of fertilizers and soil conditioners see 'Cultivation'.							
Bulb planting							
Bulbs; including forming planting holes in cultivated area (cultivating not included); backfilling with excavated material							
small	13.00	0.83	17.08	–	13.00	100 nr	30.08
medium	22.00	0.83	17.08	–	22.00	100 nr	39.08
large	25.00	0.91	18.64	–	25.00	100 nr	43.64
Bulbs; in grassed area; using bulb planter; including backfilling with screened topsoil or peat and cut turf plug							
small	13.00	1.67	34.17	–	13.00	100 nr	47.17
medium	22.00	1.67	34.17	–	22.00	100 nr	56.17
large	25.00	2.00	41.00	–	25.00	100 nr	66.00
Aquatic planting							
Aquatic plants; in prepared growing medium in pool; plant size 2–3 litre containerized (plants not included)	–	0.04	0.82	–	–	nr	0.82
Operations after planting							
Initial cutting back to shrubs and hedge plants; including disposal of all cuttings	–	1.00	20.50	–	–	100 m²	20.50

Q PAVING/PLANTING/FENCING/SITE FURNITURE

Item Excluding site overheads and profit	PC £	Labour hours	Labour £	Plant £	Material £	Unit	Total rate £
Q31 PLANTING – CONT							
Operations after planting – cont							
Mulch; Melcourt Industries Ltd; Bark							
Nuggets®; to plant beds; delivered in 25 m³							
loads; maximum distance 25 m							
50 mm thick	2.41	0.03	0.60	–	2.41	m²	**3.01**
75 mm thick	4.07	0.07	1.37	–	4.07	m²	**5.44**
100 mm thick	5.43	0.09	1.82	–	5.43	m²	**7.25**
Mulch; Melcourt Industries Ltd; Amenity Bark							
Mulch; FSC; to plant beds; delivered in 25 m³							
loads; maximum distance 25 m							
50 mm thick	1.95	0.04	0.91	–	1.95	m²	**2.86**
75 mm thick	2.93	0.07	1.37	–	2.93	m²	**4.30**
100 mm thick	3.90	0.09	1.82	–	3.90	m²	**5.72**
Maintenance operations (Note: the							
following rates apply to aftercare							
maintenance executed as part of a							
landscaping contract only)							
Weeding and hand forking planted areas;							
including disposing weeds and debris on site;							
areas maintained weekly	–	–	0.08	–	–	m²	**0.08**
Weeding and hand forking planted areas;							
including disposing weeds and debris on site;							
areas maintained monthly	–	0.01	0.20	–	–	m²	**0.20**
Mulch; Melcourt Industries Ltd; Bark							
Nuggets®; to plant beds; delivered in 25 m³							
loads; maximum distance 25 m							
50 mm thick	2.71	0.04	0.91	–	2.71	m²	**3.62**
75 mm thick	4.07	0.07	1.37	–	4.07	m²	**5.44**
Mulch; Melcourt Industries Ltd; Amenity Bark							
Mulch; FSC; to plant beds; delivered in 25 m³							
loads; maximum distance 25 m							
50 mm thick	1.95	0.04	0.91	–	1.95	m²	**2.86**
75 mm thick	2.93	0.07	1.37	–	2.93	m²	**4.30**
Watering planting; evenly; at a rate of 5 litre/m²							
using hand-held watering equipment	–	0.25	5.13	–	–	100 m²	**5.13**
using sprinkler equipment and with							
sufficient water pressure to run 1 nr 15 m							
radius sprinkler	–	0.14	2.85	–	–	100 m²	**2.85**
Work to existing planting							
Cutting and trimming ornamental hedges; to							
specified profiles; including cleaning out							
hedge bottoms; hedge cut 2 occasions per							
annum; by hand							
up to 2.00 m high	–	0.03	0.68	–	0.28	m	**0.96**
2.00–4.00 m high	–	0.05	1.02	1.88	0.56	m	**3.46**

Q PAVING/PLANTING/FENCING/SITE FURNITURE

Item Excluding site overheads and profit	PC £	Labour hours	Labour £	Plant £	Material £	Unit	Total rate £
Q35 LANDSCAPE MAINTENANCE							
Grass cutting – pedestrian operated equipment							
Using cylinder lawn mower fitted with not less than five cutting blades, front and rear rollers; on surface not exceeding 30° from horizontal; arisings let fly; width of cut							
51 cm	–	0.06	1.25	0.20	–	100 m	**1.45**
61 cm	–	0.05	1.04	0.18	–	100 m	**1.22**
Using rotary self-propelled mower; width of cut							
45 cm	–	0.08	1.58	0.06	–	100 m²	**1.64**
40 cm	–	0.11	2.28	0.06	–	100 m²	**2.34**
Add for using grass box for collecting and depositing arisings							
removing and depositing arisings	–	0.05	1.02	–	–	100 m²	**1.02**
Add for 30–50° from horizontal	–	–	–	–	–	33%	**–**
Add for slopes exceeding 50°	–	–	–	–	–	100%	**–**
Cutting grass or light woody undergrowth; using trimmer with nylon cord or metal disc cutter; on surface							
not exceeding 30° from horizontal	–	0.20	4.10	0.27	–	100 m²	**4.37**
30–50° from horizontal	–	0.40	8.20	0.55	–	100 m²	**8.75**
exceeding 50° from horizontal	–	0.50	10.25	0.69	–	100 m²	**10.94**
Grass cutting – collecting arisings							
Extra over for tractor drawn and self-propelled machinery using attached grass boxes; depositing arisings							
22 cuts per year	–	0.05	1.02	–	–	100 m²	**1.02**
18 cuts per year	–	0.08	1.54	–	–	100 m²	**1.54**
12 cuts per year	–	0.10	2.05	–	–	100 m²	**2.05**
4 cuts per year	–	0.25	5.13	–	–	100 m²	**5.13**
Disposing arisings							
22 cuts per year	–	0.01	0.16	0.03	0.04	100 m²	**0.23**
18 cuts per year	–	0.01	0.20	0.03	0.05	100 m²	**0.28**
12 cuts per year	–	0.01	0.30	0.04	0.08	100 m²	**0.42**
4 cuts per year	–	0.04	0.89	0.14	0.26	100 m²	**1.29**
Scarifying by hand							
hand implement	–	0.50	10.25	–	–	100 m²	**10.25**
add for disposal of arisings	–	0.03	0.51	1.10	16.00	100 m²	**17.61**
Rolling							
Rolling grassed area; equipment towed by tractor; once over; using							
smooth roller	–	0.01	0.29	0.22	–	100 m²	**0.51**
Turf aeration							
By hand							
hand fork; to effect a minimum penetration of 100 mm and spaced 150 mm apart	–	1.33	27.33	–	–	100 m²	**27.33**
hollow tine hand implement; to effect a minimum penetration of 100 mm and spaced 150 mm apart	–	2.00	41.00	–	–	100 m²	**41.00**
collection of arisings by hand	–	3.00	61.50	–	–	100 m²	**61.50**

Q PAVING/PLANTING/FENCING/SITE FURNITURE

Item Excluding site overheads and profit	PC £	Labour hours	Labour £	Plant £	Material £	Unit	Total rate £
Q35 LANDSCAPE MAINTENANCE – CONT							
Turf areas; surface treatments and top dressing; Boughton Loam Ltd							
Apply screened topdressing to grass surfaces; spread using Tru-Lute							
sand soil mixes 90/10 to 50/50	0.12	–	0.04	0.03	0.12	m²	**0.19**
Leaf clearance							
Using pedestrian operated mechanical equipment and blowers							
grassed areas with perimeters of mature trees such as sports fields and amenity areas	–	0.04	0.82	0.05	–	100 m²	**0.87**
grassed areas containing ornamental trees and shrub beds	–	0.10	2.05	0.12	–	100 m²	**2.17**
verges	–	0.07	1.37	0.08	–	100 m²	**1.45**
By hand							
grassed areas with perimeters of mature trees such as sports fields and amenity areas	–	0.05	1.02	0.06	–	100 m²	**1.08**
grassed areas containing ornamental trees and shrub beds	–	0.08	1.71	0.10	–	100 m²	**1.81**
verges	–	1.00	20.50	1.20	–	100 m²	**21.70**
Removal of arisings							
areas with perimeters of mature trees	–	0.01	0.14	0.10	1.28	100 m²	**1.52**
areas containing ornamental trees and shrub beds	–	0.02	0.41	0.37	3.20	100 m²	**3.98**
Litter clearance							
Collection and disposal of litter from grassed area							
areas exceeding 1000 m²	–	0.01	0.20	0.29	–	100 m²	**0.49**
area not exceeding 1000 m²	–	0.04	0.82	0.29	–	100 m²	**1.11**
Edge maintenance							
Maintain edges where lawn abuts pathway or hard surface using							
strimmer	–	0.01	0.10	0.01	–	m	**0.11**
shears	–	0.02	0.34	–	–	m	**0.34**
Maintain edges where lawn abuts plant bed using							
mechanical edging tool	–	0.01	0.14	0.03	–	m	**0.17**
shears	–	0.01	0.23	–	–	m	**0.23**
half moon edging tool	–	0.02	0.41	–	–	m	**0.41**
Tree guards, stakes and ties							
Adjusting existing tree tie	–	0.03	0.68	–	–	nr	**0.68**
Taking up single or double tree stake and ties; removing and disposing	–	0.05	1.02	–	–	nr	**1.02**
Pruning shrubs							
Trimming ground cover planting							
soft groundcover; vinca ivy and the like	–	1.00	20.50	–	–	100 m²	**20.50**
woody groundcover; cotoneaster and the like	–	1.50	30.75	–	–	100 m²	**30.75**

Q PAVING/PLANTING/FENCING/SITE FURNITURE

Item Excluding site overheads and profit	PC £	Labour hours	Labour £	Plant £	Material £	Unit	Total rate £
Pruning massed shrub border (measure ground area)							
shrub beds pruned annually	–	0.01	0.20	–	–	m²	0.20
shrub beds pruned hard every 3 years	–	0.03	0.57	–	–	m²	0.57
Cutting off dead heads							
bush or standard rose	–	0.05	1.02	–	–	nr	1.02
climbing rose	–	0.08	1.71	–	–	nr	1.71
Pruning roses							
bush or standard rose	–	0.05	1.02	–	–	nr	1.02
climbing or rambling rose; tying in as required	–	0.07	1.37	–	–	nr	1.37
Pruning ornamental shrubs; height before pruning (increase these rates by 50% if pruning work has not been executed during the previous two years)							
not exceeding 1m	–	0.04	0.82	–	–	nr	0.82
1–2 m	–	0.06	1.14	–	–	nr	1.14
exceeding 2 m	–	0.13	2.56	–	–	nr	2.56
Removing excess growth etc. from face of building etc.; height before pruning							
not exceeding 2 m	–	0.03	0.59	–	–	nr	0.59
2–4 m	–	0.05	1.02	–	–	nr	1.02
4–6 m	–	0.08	1.71	–	–	nr	1.71
6–8 m	–	0.13	2.56	–	–	nr	2.56
8–10 m	–	0.14	2.93	–	–	nr	2.93
Removing epicormic growth from base of shrub or trunk and base of tree; any height; any diameter; number of growths							
not exceeding 10	–	0.05	1.02	–	–	nr	1.02
10–20	–	0.07	1.37	–	–	nr	1.37
Beds, borders and planters							
Lifting							
bulbs	–	0.50	10.25	–	–	100 nr	10.25
tubers or corms	–	0.40	8.20	–	–	100 nr	8.20
established herbaceous plants; hoeing and depositing for replanting	–	2.00	41.00	–	–	100 nr	41.00
Temporary staking and tying in herbaceous plant	–	0.03	0.68	–	0.12	nr	0.80
Cutting down spent growth of herbaceous plant; clearing arisings							
unstaked	–	0.02	0.41	–	–	nr	0.41
staked; not exceeding 4 stakes per plant; removing stakes and putting into store	–	0.03	0.51	–	–	nr	0.51
Hand weeding							
newly planted areas	–	2.00	41.00	–	–	100 m²	41.00
established areas	–	0.50	10.25	–	–	100 m²	10.25
Removing grasses from groundcover areas	–	3.00	61.56	–	–	100 m²	61.56
Hand digging with fork; not exceeding 150 mm deep; breaking down lumps; leaving surface with a medium tilth	–	1.33	27.33	–	–	100 m²	27.33
Hand digging with fork or spade to an average depth of 230 mm; breaking down lumps; leaving surface with a medium tilth	–	2.00	41.00	–	–	100 m²	41.00
Hand hoeing; not exceeding 50 mm deep; leaving surface with a medium tilth	–	0.40	8.20	–	–	100 m²	8.20

Q PAVING/PLANTING/FENCING/SITE FURNITURE

Item Excluding site overheads and profit	PC £	Labour hours	Labour £	Plant £	Material £	Unit	Total rate £
Q35 LANDSCAPE MAINTENANCE – CONT							
Beds, borders and planters – cont							
Hand raking to remove stones etc.; breaking down lumps; leaving surface with a fine tilth prior to planting	–	0.67	13.67	–	–	100 m²	**13.67**
Hand weeding; planter or window box; not exceeding 1.00 m²							
ground level box	–	0.05	1.02	–	–	nr	**1.02**
box accessed by stepladder	–	0.08	1.71	–	–	nr	**1.71**
Spreading only compost, mulch or processed bark to a depth of 75 mm							
on shrub bed with existing mature planting	–	0.09	1.86	–	–	m²	**1.86**
recently planted areas	–	0.07	1.37	–	–	m²	**1.37**
groundcover and herbaceous areas	–	0.08	1.54	–	–	m²	**1.54**
Clearing cultivated area of leaves, litter and other extraneous debris; using hand implement							
weekly maintenance	–	0.13	2.56	–	–	100 m²	**2.56**
daily maintenance	–	0.02	0.34	–	–	100 m²	**0.34**
Bedding							
Lifting							
bedding plants; hoeing and depositing for disposal	–	3.00	61.50	–	–	100 m²	**61.50**
Hand digging with fork; not exceeding 150 mm deep; breaking down lumps; leaving surface with a medium tilth	–	0.75	15.38	–	–	100 m²	**15.38**
Hand weeding							
newly planted areas	–	2.00	41.00	–	–	100 m²	**41.00**
established areas	–	0.50	10.25	–	–	100 m²	**10.25**
Hand digging with fork or spade to an average depth of 230 mm; breaking down lumps; leaving surface with a medium tilth	–	0.50	10.25	–	–	100 m²	**10.25**
Hand hoeing: not exceeding 50 mm deep; leaving surface with a medium tilth	–	0.40	8.20	–	–	100 m²	**8.20**
Hand raking to remove stones etc.; breaking down lumps; leaving surface with a fine tilth prior to planting	–	0.67	13.67	–	–	100 m²	**13.67**
Hand weeding; planter or window box; not exceeding 1.00 m²							
ground level box	–	0.05	1.02	–	–	nr	**1.02**
box accessed by stepladder	–	0.08	1.71	–	–	nr	**1.71**
Spreading only; compost, mulch or processed bark to a depth of 75 mm							
on shrub bed with existing mature planting	–	0.09	1.86	–	–	m²	**1.86**
recently planted areas	–	0.07	1.37	–	–	m²	**1.37**
groundcover and herbaceous areas	–	0.08	1.54	–	–	m²	**1.54**
Collecting bedding from nursery	–	3.00	61.50	14.67	–	100 m²	**76.17**
Setting out							
mass planting single variety	–	0.13	2.56	–	–	m²	**2.56**
pattern	–	0.33	6.83	–	–	m²	**6.83**
Planting only							
massed bedding plants	–	0.20	4.10	–	–	m²	**4.10**

Q PAVING/PLANTING/FENCING/SITE FURNITURE

Item Excluding site overheads and profit	PC £	Labour hours	Labour £	Plant £	Material £	Unit	Total rate £
Clearing cultivated area of leaves, litter and other extraneous debris; using hand implement							
weekly maintenance	–	0.13	2.56	–	–	100 m²	**2.56**
daily maintenance	–	0.02	0.34	–	–	100 m²	**0.34**
Irrigation and watering							
Hand held hosepipe; flow rate 25 litres per minute; irrigation requirement							
10 litres/m²	–	0.74	15.11	–	–	100 m²	**15.11**
15 litres/m²	–	1.10	22.55	–	–	100 m²	**22.55**
20 litres/m²	–	1.46	29.99	–	–	100 m²	**29.99**
25 litres/m²	–	1.84	37.66	–	–	100 m²	**37.66**
Hand held hosepipe; flow rate 40 litres per minute; irrigation requirement							
10 litres/m²	–	0.46	9.47	–	–	100 m²	**9.47**
15 litres/m²	–	0.69	14.21	–	–	100 m²	**14.21**
20 litres/m²	–	0.91	18.72	–	–	100 m²	**18.72**
25 litres/m²	–	1.15	23.50	–	–	100 m²	**23.50**
Hedge cutting; field hedges cut once or twice annually							
Trimming sides and top using hand tool or hand held mechanical tools							
not exceeding 2 m high	–	0.10	2.05	0.15	–	10 m²	**2.20**
2 to 4 m high	–	0.33	6.83	0.49	–	10 m²	**7.32**
Hedge cutting; ornamental							
Trimming sides and top using hand tool or hand held mechanical tools							
not exceeding 2 m high	–	0.13	2.56	0.18	–	10 m²	**2.74**
2 to 4 m high	–	0.50	10.25	0.73	–	10 m²	**10.98**
Hedge cutting; reducing width; hand tool or hand held mechanical tools							
Not exceeding 2 m high							
average depth of cut not exceeding 300 mm	–	0.10	2.05	0.15	–	10 m²	**2.20**
average depth of cut 300 to 600 mm	–	0.83	17.08	1.23	–	10 m²	**18.31**
average depth of cut 600 to 900 mm	–	1.25	25.63	1.84	–	10 m²	**27.47**
2 to 4 m high							
average depth of cut not exceeding 300 mm	–	0.03	0.51	0.04	–	10 m²	**0.55**
average depth of cut 300 to 600 mm	–	0.13	2.56	0.18	–	10 m²	**2.74**
average depth of cut 600 to 900 mm	–	2.50	51.25	3.67	–	10 m²	**54.92**
4 to 6 m high							
average depth of cut not exceeding 300 mm	–	0.10	2.05	0.15	–	m²	**2.20**
average depth of cut 300 to 600 mm	–	0.17	3.42	0.25	–	m²	**3.67**
average depth of cut 600 to 900 mm	–	0.50	10.25	0.73	–	m²	**10.98**

Q PAVING/PLANTING/FENCING/SITE FURNITURE

Item Excluding site overheads and profit	PC £	Labour hours	Labour £	Plant £	Material £	Unit	Total rate £
Q35 LANDSCAPE MAINTENANCE – CONT							
Hedge cutting; reducing height; hand tool							
or hand held mechanical tools							
Not exceeding 2 m high							
average depth of cut not exceeding							
300 mm	–	0.03	0.51	0.04	–	m²	0.55
average depth of cut 300 to 600 mm	–	0.04	0.82	0.06	–	m²	0.88
average depth of cut 600 to 900 mm	–	0.10	2.05	0.15	–	m²	2.20
2 to 4 m high							
average depth of cut not exceeding							
300 mm	–	0.05	1.02	0.07	–	m²	1.09
average depth of cut 300 to 600 mm	–	0.07	1.37	0.10	–	m²	1.47
average depth of cut 600 to 900 mm	–	0.13	2.56	0.18	–	m²	2.74
4 to 6 m high							
average depth of cut not exceeding							
300 mm	–	0.11	2.28	0.16	–	m²	2.44
average depth of cut 300 to 600 mm	–	0.20	4.10	0.29	–	m²	4.39
average depth of cut 600 to 900 mm	–	0.50	10.25	0.73	–	m²	10.98
Hedge cutting; removal and disposal of							
arisings							
Sweeping up and depositing arisings							
300 mm cut	–	0.05	1.02	–	–	10 m²	1.02
600 mm cut	–	0.20	4.10	–	–	10 m²	4.10
900 mm cut	–	0.40	8.20	–	–	10 m²	8.20
Chipping arisings							
300 mm cut	–	0.02	0.41	0.37	–	10 m²	0.78
600 mm cut	–	0.08	1.71	1.55	–	10 m²	3.26
900 mm cut	–	0.20	4.10	3.72	–	10 m²	7.82
Disposal of unchipped arisings							
300 mm cut	–	0.02	0.34	0.55	2.13	10 m²	3.02
600 mm cut	–	0.03	0.68	1.10	3.20	10 m²	4.98
900 mm cut	–	0.08	1.71	2.75	8.00	10 m²	12.46
Disposal of chipped arisings							
300 mm cut	–	–	0.07	0.18	3.20	10 m²	3.45
600 mm cut	–	0.02	0.34	0.18	6.40	10 m²	6.92
900 mm cut	–	0.03	0.68	0.37	16.00	10 m²	17.05
Spraying; labour rates only; for chemical							
rates please use the tables at the beginning							
of the section Q35 Landscape Maintenance							
in the Major Works section of this book							
Herbicide applications; standard backpack							
spray applicators; application to maintain							
1.00 m diameter clear circles (0.79 m²) around							
new planting							
plants at 1.50 m centres; 4444 nr/ha	–	0.25	5.06	–	–	100 m²	5.06
plants at 1.75 m centres; 3265 nr/ha	–	0.18	3.72	–	–	100 m²	3.72
plants at 2.00 m centres; 2500 nr/ha	–	0.14	2.85	–	–	100 m²	2.85
understory of mature planting or trees	–	0.13	2.56	–	–	100 m²	2.56
mass spraying low vegetation or hard							
surfaces	–	0.10	2.05	–	–	100 m²	2.05

Q PAVING/PLANTING/FENCING/SITE FURNITURE

Item Excluding site overheads and profit	PC £	Labour hours	Labour £	Plant £	Material £	Unit	Total rate £
Q40 FENCING							
Temporary fencing; HSS Hire; mesh framed **mesh unclimbable fencing; including** **precast concrete supports and couplings** Weekly hire; 2.85 × 2 m high							
weekly hire rate	–	–	–	6.00	–	m	**6.00**
erection of fencing; labour only	–	0.13	2.56	–	–	m	**2.56**
removal of fencing; loading to collection vehicle	–	0.08	1.71	–	–	m	**1.71**
delivery charge	–	–	–	0.80	–	m	**0.80**
return haulage charge	–	–	–	0.60	–	m	**0.60**
Protective fencing Cleft chestnut rolled fencing; fixing to 100 mm diameter chestnut posts; driving into firm ground at 3 m centres							
900 mm high	4.08	0.11	2.19	–	8.95	m	**11.14**
1200 mm high	4.84	0.16	3.28	–	9.71	m	**12.99**
1500 mm high; 3 strand	7.07	0.21	4.38	–	11.95	m	**16.33**
Timber fencing; AVS Fencing Supplies Ltd; **tanalized softwood fencing** Timber lap panels; pressure treated; fixed to timber posts 75 × 75 mm in 1:3:6 concrete; at 1.90 centres							
900 mm high	7.87	0.67	13.67	–	18.84	m	**32.51**
1200 mm high	8.03	0.75	15.38	–	19.32	m	**34.70**
1500 mm high	8.17	0.80	16.40	–	19.27	m	**35.67**
1800 mm high	8.91	0.90	18.45	–	21.51	m	**39.96**
Timber lap panels; fixed to slotted concrete posts 100 × 100 mm in 1:3:6 concrete at 1.88 m centres							
900 mm high	7.95	0.75	15.38	–	23.80	m	**39.18**
1200 mm high	8.12	0.80	16.40	–	23.96	m	**40.36**
1500 mm high	14.46	0.85	17.43	–	28.31	m	**45.74**
1800 mm high	21.02	0.90	18.45	–	34.87	m	**53.32**
extra for corner posts	22.58	0.90	18.45	–	35.44	nr	**53.89**
Extra over panel fencing for 300 mm high trellis tops; slats at 100 mm centres; including additional length of posts	4.50	0.10	2.05	–	4.50	m	**6.55**
Close boarded fencing; to concrete posts 100 × 100 mm; 2 nr oak Arris rails; 100 × 22 mm softwood pales lapped 13 mm; including excavating and backfilling into firm ground at 3.00 m centres; setting in concrete 1:3:6; timber gravel board 150 × 50 mm							
900 mm high	16.10	1.00	20.50	–	22.05	m	**42.55**
1050 mm high	17.58	1.10	22.55	–	22.41	m	**44.96**
1500 mm high	19.62	1.15	23.57	–	24.45	m	**48.02**

Q PAVING/PLANTING/FENCING/SITE FURNITURE

Item Excluding site overheads and profit	PC £	Labour hours	Labour £	Plant £	Material £	Unit	Total rate £
Q40 FENCING – CONT							
Timber fencing – cont							
Close boarded fencing; to concrete posts							
100 × 100 mm; 3 nr oak Arris rails; 100 ×							
22 mm softwood pales lapped 13 mm;							
including excavating and backfilling into firm							
ground at 3.00 m centres; setting in concrete;							
timber gravel board 150 × 50 mm							
1650 mm high	24.47	1.25	25.63	–	29.29	m	**54.92**
1800 mm high	24.63	1.30	26.65	–	29.45	m	**56.10**
extra over to the above for concrete gravel							
board 150 × 50 mm in lieu of timber gravel							
board	4.41	0.11	2.28	–	4.47	m	**6.75**
Close boarded fencing; 2 nr oak Arris rails; to							
timber posts 100 × 100 mm; 100 × 22 mm							
softwood pales lapped 13 mm; including							
excavating and backfilling into firm ground at							
3.00 m centres; setting in concrete 1:3:6							
900 mm high	12.26	1.00	20.50	–	21.78	m	**42.28**
1050 mm high	17.32	1.10	22.55	–	22.14	m	**44.69**
1200 mm high	19.36	1.15	23.57	–	24.18	m	**47.75**
Close boarded fencing; 3 nr oak Arris rails; to							
timber posts 100 × 100 mm; 100 × 22 mm							
softwood pales lapped 13 mm; including							
excavating and backfilling into firm ground at							
3.00 m centres; setting in concrete 1:3:6							
1350 mm high	19.21	0.95	19.52	–	27.61	m	**47.13**
1650 mm high	23.75	1.30	26.65	–	28.57	m	**55.22**
1800 mm high	23.87	1.35	27.68	–	28.69	m	**56.37**
extra over for post 125 × 100 mm; for							
1800 mm high fencing	8.86	–	–	–	8.86	nr	**8.86**
extra over to the above for counter rail	0.68	0.10	2.05	–	0.68	m	**2.73**
extra over to the above for capping rail	1.29	0.10	2.05	–	1.29	m	**3.34**
Close boarded fencing; 2 nr cant rails; nailed							
to timber posts 100 × 100 mm; 125 × 22 mm							
softwood pales lapped 13 mm; including							
excavating and backfilling into firm ground at							
3.00 m centres; setting in concrete 1:3:6							
900 mm high	7.22	1.00	20.50	–	14.26	m	**34.76**
1050 mm high	9.79	1.10	22.55	–	14.62	m	**37.17**
1200 mm high	11.11	1.10	22.55	–	15.94	m	**38.49**
Close boarded fencing; 3 nr cant rails; nailed							
to timber posts 100 × 100 mm; 125 × 22 mm							
softwood pales lapped 13 mm; including							
excavating and backfilling into firm ground at							
3.00 m centres; setting in concrete 1:3:6							
1350 mm high	9.97	1.15	23.57	–	17.01	m	**40.58**
1800 mm high	14.00	1.20	24.60	–	18.83	m	**43.43**
extra over for post 125 × 100 mm; for							
1800 mm high fencing	8.86	–	–	–	8.86	nr	**8.86**

Q PAVING/PLANTING/FENCING/SITE FURNITURE

Item Excluding site overheads and profit	PC £	Labour hours	Labour £	Plant £	Material £	Unit	Total rate £
Palisade fencing; 19 × 75 mm softwood vertical palings with pointed tops at 150 mm centres; nailing to 2 nr horizontal softwood Arris rails; morticed to 100 × 100 mm softwood posts with weathered tops at 3.00 m centres; setting in concrete							
900 mm high	17.74	0.90	18.45	–	22.64	m	**41.09**
1050 mm high	18.88	0.90	18.45	–	23.70	m	**42.15**
1200 mm high	19.21	0.95	19.48	–	24.11	m	**43.59**
extra over for rounded tops	0.80	–	–	–	0.80	m	**0.80**
Post-and-rail fencing; 90 × 38 mm softwood horizontal rails; fixing with galvanized nails to 150 × 75 mm softwood posts; including excavating and backfilling into firm ground at 1.80 m centres; all treated timber							
1200 mm high; 3 horizontal rails	8.68	0.35	7.18	–	8.76	m	**15.94**
1200 mm high; 4 horizontal rails	9.96	0.35	7.18	–	10.04	m	**17.22**
Cleft rail fencing; oak or chestnut tapered rails 2.80 m long; morticed into joints; to 125 × 100 mm softwood posts 1.95 m long; including excavating and backfilling into firm ground at 2.80 m centres							
two rails	19.49	0.25	5.13	–	19.49	m	**24.62**
three rails	22.21	0.28	5.74	–	22.21	m	**27.95**
four rails	24.93	0.35	7.18	–	24.93	m	**32.11**
Hit and Miss horizontal rail fencing; treated softwood; 87 × 38 mm top and bottom rails; 100 × 22 mm vertical boards arranged alternately on opposite side of rails; to 100 × 100 mm posts; including excavating and backfilling into firm ground; setting in concrete at 1.8 m centres							
1600 mm high	28.94	1.20	24.60	–	31.78	m	**56.38**
1800 mm high	31.85	1.33	27.33	–	34.69	m	**62.02**
2000 mm high	34.77	1.40	28.70	–	37.62	m	**66.32**
Palisade fencing; 100 × 22 mm softwood vertical palings with flat tops; nailing to 3 nr 50 × 100 mm horizontal softwood rails; housing into 100 × 100 mm softwood posts with weathered tops at 3.00 m centres; setting in concrete; all treated timber							
1800 mm high	22.62	0.47	9.58	–	27.48	m	**37.06**
1800 mm high	30.71	0.47	9.57	–	35.57	m	**45.14**
Post-and-rail fencing; 2 nr 90 × 38 mm softwood horizontal rails; fixing with galvanized nails to 150 × 75 mm softwood posts; including excavating and backfilling into firm ground at 1.80 m centres; all treated timber							
1200 mm high	7.40	0.30	6.15	–	7.48	m	**13.63**
Post-and-rail fencing; 3 nr 90 × 38 mm softwood horizontal rails; fixing with galvanized nails to 150 × 75 mm softwood posts; including excavating and backfilling into firm ground at 1.80 m centres; all treated timber							
1200 mm high	8.68	0.35	7.18	–	8.76	m	**15.94**

Q PAVING/PLANTING/FENCING/SITE FURNITURE

Item Excluding site overheads and profit	PC £	Labour hours	Labour £	Plant £	Material £	Unit	Total rate £
Q40 FENCING – CONT							
Timber fencing – cont							
Morticed post-and-rail fencing; 3 nr horizontal							
90 × 38 mm softwood rails; fixing with							
galvanized nails; 90 × 38 mm softwood centre							
prick posts; to 150 × 75 mm softwood posts;							
including excavating and backfilling into firm							
ground at 2.85 m centres; all treated timber							
1200 mm high	8.18	0.35	7.18	–	8.26	m	**15.44**
1350 mm high five rails	10.42	0.40	8.20	–	10.50	m	**18.70**
Cleft rail fencing; chestnut tapered rails 2.85 m							
long; morticed into joints; to 125 × 100 mm							
softwood posts 2.1 m long; including							
excavating and backfilling into firm ground at							
2.75 m centres							
two rails	29.55	0.25	5.13	–	29.55	m	**34.68**
three rails	36.49	0.28	5.74	–	36.49	m	**42.23**
four rails	43.97	0.35	7.18	–	43.97	m	**51.15**
Boundary fencing; strained wire and wire							
mesh; AVS Fencing Supplies Ltd							
Strained wire fencing; concrete posts only at							
2750 mm centres; 610 mm below ground;							
excavating holes; filling with concrete;							
replacing topsoil; disposing surplus soil off site							
900 mm high	3.34	0.56	11.39	1.92	4.93	m	**18.24**
1200 mm high	4.18	0.56	11.39	1.92	5.77	m	**19.08**
1800 mm high	5.72	0.78	15.94	4.79	7.31	m	**28.04**
Extra over strained wire fencing for concrete							
straining posts with one strut; posts and struts							
610 mm below ground; struts, cleats,							
stretchers, winders, bolts and eye bolts;							
excavating holes; filling to within 150 mm of							
ground level with concrete (1:12) – 40 mm							
aggregate; replacing topsoil; disposing surplus							
soil off site							
900 mm high	22.50	0.67	13.73	4.31	54.39	nr	**72.43**
1200 mm high	24.69	0.67	13.67	4.32	57.03	nr	**75.02**
1800 mm high	34.69	0.67	13.67	4.32	69.49	nr	**87.48**
Extra over strained wire fencing for concrete							
straining posts with two struts; posts and struts							
610 mm below ground; excavating holes; filling							
to within 150 mm of ground level with concrete							
(1:12) – 40 mm aggregate; replacing topsoil;							
disposing surplus soil off site							
900 mm high	33.35	0.91	18.66	5.54	81.29	nr	**105.49**
1200 mm high	36.84	0.85	17.43	4.32	89.75	nr	**111.50**
1800 mm high	52.23	0.85	17.43	4.32	108.94	nr	**130.69**
Strained wire fencing; painted steel angle							
posts only at 2750 mm centres; 610 mm below							
ground; driving in							
900 mm high; 40 × 40 × 5 mm	2.78	0.03	0.62	–	2.78	m	**3.40**
1200 mm high; 40 × 40 × 5 mm	3.16	0.03	0.68	–	3.16	m	**3.84**
1400 mm high; 40 × 40 × 5 mm	4.16	0.06	1.24	–	4.16	m	**5.40**
1800 mm high; 40 × 40 × 5 mm	4.53	0.07	1.49	–	4.53	m	**6.02**
1800 mm high; 45 × 45 × 5 mm; with							
extension for three rows barbed wire	5.81	0.12	2.48	–	5.81	m	**8.29**

Q PAVING/PLANTING/FENCING/SITE FURNITURE

Item Excluding site overheads and profit	PC £	Labour hours	Labour £	Plant £	Material £	Unit	Total rate £
Painted steel straining posts with two struts for strained wire fencing; setting in concrete							
900 mm high; 50 × 50 × 6 mm	49.56	1.00	20.50	–	49.56	nr	**70.06**
1200 mm high; 50 × 50 × 6 mm	59.86	1.00	20.50	–	59.86	nr	**80.36**
1500 mm high; 50 × 50 × 6 mm	74.79	1.00	20.50	–	74.79	nr	**95.29**
1800 mm high; 50 × 50 × 6 mm	77.52	1.00	20.50	–	77.52	nr	**98.02**
Strained wire; to posts (posts not included); 3 mm galvanized wire; fixing with galvanized stirrups							
900 mm high; 2 wire	0.17	0.03	0.68	–	0.33	m	**1.01**
1200 mm high; 3 wire	0.25	0.05	0.96	–	0.41	m	**1.37**
1400 mm high; 3 wire	0.25	0.05	0.96	–	0.41	m	**1.37**
1800 mm high; 3 wire	0.25	0.05	0.96	–	0.41	m	**1.37**
Barbed wire; to posts (posts not included); 3 mm galvanized wire; fixing with galvanized stirrups							
900 mm high; 2 wire	0.26	0.07	1.37	–	0.42	m	**1.79**
1200 mm high; 3 wire	0.39	0.09	1.91	–	0.55	m	**2.46**
1400 mm high; 3 wire	0.39	0.09	1.91	–	0.55	m	**2.46**
1800 mm high; 3 wire	0.39	0.09	1.91	–	0.55	m	**2.46**
Chain link fencing; AVS Fencing Supplies Ltd; to strained wire and posts priced separately; 3 mm galvanized wire; 50 mm mesh; galvanized steel components; fixing to line wires threaded through posts and strained with eye-bolts; posts (not included)							
900 mm high	4.08	0.07	1.37	–	4.24	m	**5.61**
1200 mm high	5.71	0.07	1.37	–	5.87	m	**7.24**
1800 mm high	7.91	0.10	2.05	–	8.07	m	**10.12**
2400 mm high	10.71	0.10	2.05	–	10.87	m	**12.92**
Chain link fencing; to strained wire and posts priced separately; 3.15 mm plastic coated galvanized wire (wire only 2.50 mm); 50 mm mesh; galvanized steel components; fencing with line wires threaded through posts and strained with eye-bolts; posts (not included) (Note: plastic coated fencing can be cheaper than galvanized finish as wire of a smaller cross-sectional area can be used)							
900 mm high	2.89	0.07	1.37	–	3.21	m	**4.58**
1200 mm high	3.98	0.07	1.37	–	4.30	m	**5.67**
1800 mm high	5.67	0.13	2.56	–	6.95	m	**9.51**
Extra over strained wire fencing for cranked arms and galvanized barbed wire							
1 row	3.65	0.02	0.34	–	3.65	m	**3.99**
2 row	3.78	0.05	1.02	–	3.78	m	**4.80**
3 row	3.91	0.05	1.02	–	3.91	m	**4.93**
Field fencing; Jacksons Fencing; welded wire mesh; fixed to posts and straining wires measured separately							
cattle fence; 1100 m high; 114 × 300 mm at bottom to 230 × 300 mm at top	1.36	0.10	2.05	–	1.61	m	**3.66**
sheep fence; 900 mm high; 140 × 300 mm at bottom to 230 × 300 mm at top	1.12	0.10	2.05	–	1.37	m	**3.42**
deer fence; 1900 mm high; 89 × 150 mm at bottom to 267 × 300 mm at top	2.86	0.13	2.56	–	3.11	m	**5.67**
Extra for concreting in posts	–	0.50	10.25	–	4.33	nr	**14.58**
Extra for straining post	10.35	0.75	15.38	–	10.35	nr	**25.73**

Q PAVING/PLANTING/FENCING/SITE FURNITURE

Item Excluding site overheads and profit	PC £	Labour hours	Labour £	Plant £	Material £	Unit	Total rate £
Q40 FENCING – CONT							
Rabbit netting; AVS Fencing Supplies Ltd; **timber stakes; peeled kiln dried pressure** **treated; pointed; 1.8 m posts driven 900 mm** **into ground at 3 m centres (line wires and** **netting priced separately)**							
75–100 mm stakes	1.94	0.25	5.13	–	1.94	m	**7.07**
Corner posts or straining posts 150 mm **diameter 2.3 m high set in concrete;** **centres to suit local conditions or changes** **of direction**							
1 strut	12.69	1.00	20.50	12.31	20.25	each	**53.06**
2 strut	15.50	1.00	20.50	12.31	23.05	each	**55.86**
Strained wire; to posts (posts not **included); 3 mm galvanized wire; fixing** **with galvanized stirrups**							
900 mm high; 2 wire	0.17	0.03	0.68	–	0.33	m	**1.01**
1200 mm high; 3 wire	0.25	0.05	0.96	–	0.41	m	**1.37**
Rabbit netting; 31 mm; 19 gauge; 1050 mm **high netting fixed to posts; line wires and** **straining posts or corner posts all priced** **separately**							
900 high turned in	0.99	0.04	0.82	–	0.99	m	**1.81**
900 high buried 150 mm in trench	0.99	0.08	1.71	–	0.99	m	**2.70**
Boundary fencing; strained wire and wire **mesh; Jacksons Fencing**							
Tubular chain link fencing; galvanized; plastic coated; 60.3 mm diameter posts at 3.0 m centres; setting 700 mm into ground; choice of ten mesh colours; including excavating holes; backfilling and removing surplus soil; with top rail only							
900 mm high	–	–	–	–	–	m	**33.68**
1200 mm high	–	–	–	–	–	m	**35.78**
1800 mm high	–	–	–	–	–	m	**39.89**
2000 mm high	–	–	–	–	–	m	**42.10**
Tubular chain link fencing; galvanized; plastic coated; 60.3 mm diameter posts at 3.0 m centres; cranked arms and 3 lines barbed wire; setting 700 mm into ground; including excavating holes; backfilling and removing surplus soil; with top rail only							
1800 mm high	–	–	–	–	–	m	**41.47**
2000 mm high	–	–	–	–	–	m	**42.68**

Q PAVING/PLANTING/FENCING/SITE FURNITURE

Item Excluding site overheads and profit	PC £	Labour hours	Labour £	Plant £	Material £	Unit	Total rate £
Traditional trellis panels; The Garden Trellis Company; bespoke trellis panels for decorative, screening or security applications; timber planed all round; height of trellis 1800 mm Freestanding panels; posts 70 × 70 mm set in concrete; timber frames mitred and grooved 45 × 34 mm; heights and widths to suit; slats 32 × 10 mm joinery quality tanalized timber at 100 mm ccs; elements fixed by galvanized staples; capping rail 70 × 34 mm							
HV68; horizontal and vertical slats; softwood	79.20	1.00	20.50	–	84.98	m	**105.48**
D68; diagonal slats; softwood	93.60	1.00	20.50	–	99.38	m	**119.88**
HV68; horizontal and vertical slats; hardwood iroko	190.80	1.00	20.50	–	196.58	m	**217.08**
D68; diagonal slats; hardwood iroko or Western Red Cedar	216.00	1.00	20.50	–	221.78	m	**242.28**
Trellis panels fixed to face of existing wall or railings; timber frames mitred and grooved 45 × 34 mm; heights and widths to suit; slats 32 × 10 mm joinery quality tanalized timber at 100 mm ccs; elements fixed by galvanized staples; capping rail 70 × 34 mm							
HV68; horizontal and vertical slats; softwood	75.60	0.50	10.25	–	77.87	m	**88.12**
D68; diagonal slats; softwood	88.20	0.50	10.25	–	90.47	m	**100.72**
HV68; horizontal and vertical slats; iroko or Western Red Cedar	189.00	0.50	10.25	–	191.27	m	**201.52**
D68; diagonal slats; iroko or Western Red Cedar	198.00	0.50	10.25	–	200.27	m	**210.52**
Contemporary style trellis panels; The Garden Trellis Company; bespoke trellis panels for decorative, screening or security applications; timber planed all round Freestanding panels 30/15; posts 70 × 70 mm set in concrete with 90 × 30 mm top capping; slats 30 × 14 mm with 15 mm gaps; vertical support at 450 mm ccs							
joinery treated softwood	126.00	1.00	20.50	–	131.78	m	**152.28**
hardwood iroko or Western Red Cedar	207.00	1.00	20.50	–	212.78	m	**233.28**
Panels 30/15 face fixed to existing wall or fence; posts 70 × 70 mm set in concrete with 90 × 30 mm top capping; slats 30 × 14 mm with 15 mm gaps; vertical support at 450 mm ccs							
joinery treated softwood	117.00	0.50	10.25	–	119.27	m	**129.52**
hardwood iroko or Western Red Cedar	189.00	0.50	10.25	–	191.27	m	**201.52**
Integral arches to ornamental trellis panels; The Garden Trellis Company Arches to trellis panels in 45 × 34 mm grooved timbers to match framing; fixed to posts							
R450; 1/4 circle; joinery treated softwood	38.00	1.50	30.75	–	43.78	nr	**74.53**
R450; 1/4 circle; hardwood iroko or Western Red Cedar	49.00	1.50	30.75	–	54.78	nr	**85.53**

Q PAVING/PLANTING/FENCING/SITE FURNITURE

Item Excluding site overheads and profit	PC £	Labour hours	Labour £	Plant £	Material £	Unit	Total rate £
Q40 FENCING – CONT							
Integral arches to ornamental trellis panels – cont							
Arches to span 1800 mm wide							
joinery treated softwood	49.00	1.50	30.75	–	54.78	nr	85.53
hardwood iroko or Western Red Cedar	65.00	1.50	30.75	–	70.78	nr	101.53
Painting or staining of trellis panels; high quality coatings							
microporous opaque paint or spirit based stain	–	–	–	–	24.00	m²	24.00
Concrete fencing							
Panel fencing; to precast concrete posts; in 2 m bays; setting posts 600 mm into ground; sandfaced finish							
900 mm high	11.74	0.25	5.13	–	13.45	m	18.58
1200 mm high	15.68	0.25	5.13	–	17.39	m	22.52
Panel fencing; to precast concrete posts; in 2 m bays; setting posts 750 mm into ground; sandfaced finish							
1500 mm high	20.46	0.33	6.83	–	22.17	m	29.00
1800 mm high	24.52	0.36	7.45	–	26.22	m	33.67
2100 mm high	28.39	0.40	8.20	–	30.10	m	38.30
2400 mm high	32.63	0.40	8.20	–	34.34	m	42.54
extra over capping panel to all of the above	6.72	–	–	–	6.72	m	6.72
Windbreak fencing							
Fencing; English Woodlands; Shade and Shelter Netting windbreak fencing; green; to 100 mm diameter treated softwood posts; setting 450 mm into ground; fixing with 50 × 25 mm treated softwood battens nailed to posts; including excavating and backfilling into firm ground; setting in concrete at 3 m centres							
1200 mm high	1.24	0.16	3.19	–	5.37	m	8.56
1800 mm high	1.76	0.16	3.19	–	5.89	m	9.08
Gates – General							
Preamble: gates in fences; see specification for fencing, as gates in traditional or proprietary fencing systems are usually constructed of the same materials and finished as the fencing itself.							
Gates; hardwood; AVS Fencing Supplies Ltd							
Hardwood entrance gate; five bar diamond braced; curved hanging stile; planed iroko; fixed to 150 × 150 mm softwood posts; inclusive of hinges and furniture							
ref 1100 040; 0.9 m wide	211.08	5.00	102.50	–	269.69	nr	372.19
ref 1100 041; 1.2 m wide	225.06	5.00	102.50	–	283.67	nr	386.17
ref 1100 042; 1.5 m wide	293.35	5.00	102.50	–	351.96	nr	454.46
ref 1100 043; 1.8 m wide	310.63	5.00	102.50	–	369.24	nr	471.74
ref 1100 044; 2.1 m wide	340.12	5.00	102.50	–	398.73	nr	501.23
ref 1100 045; 2.4 m wide	320.10	5.00	102.50	–	378.71	nr	481.21
ref 1100 047; 3.0 m wide	390.95	5.00	102.50	–	449.56	nr	552.06
ref 1100 048; 3.3 m wide	408.34	5.00	102.50	–	466.95	nr	569.45
ref 1100 049; 3.6 m wide	422.28	5.00	102.50	–	480.89	nr	583.39

Q PAVING/PLANTING/FENCING/SITE FURNITURE

Item Excluding site overheads and profit	PC £	Labour hours	Labour £	Plant £	Material £	Unit	Total rate £
Hardwood field gate; five bar diamond braced; planed iroko; fixed to 150 × 150 mm softwood posts; inclusive of hinges and furniture							
ref 1100 100; 0.9 m wide	143.44	4.00	82.00	–	202.05	nr	**284.05**
ref 1100 101; 1.2 m wide	155.84	4.00	82.00	–	214.45	nr	**296.45**
ref 1100 102; 1.5 m wide	185.80	4.00	82.00	–	244.41	nr	**326.41**
ref 1100 103; 1.8 m wide	199.71	4.00	82.00	–	258.32	nr	**340.32**
ref 1100 104; 2.1 m wide	245.91	4.00	82.00	–	304.52	nr	**386.52**
ref 1100 105; 2.4 m wide	260.54	4.00	82.00	–	319.15	nr	**401.15**
ref 1100 106; 2.7 m wide	275.21	4.00	82.00	–	333.82	nr	**415.82**
ref 1100 108; 3.3 m wide	304.51	4.00	82.00	–	363.12	nr	**445.12**
ref 1100 109; 3.6 m wide	319.91	4.00	82.00	–	378.52	nr	**460.52**
Gates; softwood; Jacksons Fencing							
Timber field gates; treated softwood; including wrought iron ironmongery; five bar type; diamond braced; 1.80 m high; to 200 × 200 mm posts; setting 750 mm into firm ground							
width 2400 mm	81.63	10.00	205.00	–	202.90	nr	**407.90**
width 2700 mm	90.63	10.00	205.00	–	211.90	nr	**416.90**
width 3000 mm	97.56	10.00	205.00	–	218.83	nr	**423.83**
width 3300 mm	103.32	10.00	205.00	–	224.59	nr	**429.59**
Featherboard garden gates; treated softwood; including ironmongery; to 100 × 120 mm posts; 1 nr diagonal brace							
1.0 × 1.2 m high	25.11	3.00	61.50	–	108.02	nr	**169.52**
1.0 × 1.5 m high	32.94	3.00	61.50	–	118.91	nr	**180.41**
1.0 × 1.8 m high	36.90	3.00	61.50	–	129.49	nr	**190.99**
Picket garden gates; treated softwood; including ironmongery; to match picket fence; width 1000 mm; to 100 × 120 mm posts; 1 nr diagonal brace							
950 mm high	68.22	3.00	61.50	–	102.89	nr	**164.39**
1200 mm high	71.91	3.00	61.50	–	106.58	nr	**168.08**
1800 mm high	84.38	3.00	61.50	–	130.84	nr	**192.34**
Gates; tubular steel; Jacksons Fencing							
Tubular mild steel field gates; galvanized; including ironmongery; diamond braced; 1.80 m high; to tubular steel posts; setting in concrete							
width 3000 mm	110.25	5.00	102.50	–	208.32	nr	**310.82**
width 3300 mm	116.50	5.00	102.50	–	214.57	nr	**317.07**
width 3600 mm	122.89	5.00	102.50	–	220.96	nr	**323.46**
width 4200 mm	135.68	5.00	102.50	–	233.74	nr	**336.24**
Kissing gates							
Kissing gates; Jacksons Fencing; in galvanized metal bar; fixing to fencing posts (posts not included); 1.65 × 1.30 × 1.00 m high	221.40	5.00	102.50	–	221.40	nr	**323.90**
Stiles							
Stiles; 2 nr posts; setting into firm ground; 3 nr rails; 2 nr treads	72.00	3.00	61.50	–	83.56	nr	**145.06**

R DISPOSAL SYSTEMS

Item Excluding site overheads and profit	PC £	Labour hours	Labour £	Plant £	Material £	Unit	Total rate £
R12 DRAINAGE BELOW GROUND							
Silt pits and inspection chambers							
Excavating pits; starting from ground level; by machine							
maximum depth not exceeding 1.00 m	–	–	–	3.58	–	m³	3.58
maximum depth not exceeding 2.00 m	–	–	–	4.08	–	m³	4.08
Disposal of excavated material; depositing on site in permanent spoil heaps; average 50 m	–	–	–	6.96	–	m³	6.96
Surface treatments; compacting; bottoms of excavations	–	0.05	1.02	–	–	m²	1.02
Earthwork support; distance between opposing faces not exceeding 2.00 m							
maximum depth not exceeding 1.00 m	–	0.20	4.10	–	7.30	m³	11.40
maximum depth not exceeding 2.00 m	–	0.30	6.15	–	7.30	m³	13.45
maximum depth not exceeding 4.00 m	–	0.67	13.65	–	3.33	m³	16.98
Silt pits and inspection chambers; in situ concrete							
Beds; plain in situ concrete; 11.50 N/mm² – 40 mm aggregate							
thickness not exceeding 150 mm	–	1.00	20.50	–	118.07	m³	138.57
Benchings in bottoms; plain in situ concrete; 25.50 N/mm² – 20 mm aggregate							
thickness 150–450 mm	–	2.00	41.00	–	118.07	m³	159.07
Isolated cover slabs; reinforced in situ concrete; 21.00 N/mm² – 20 mm aggregate							
thickness not exceeding 150 mm	–	4.00	82.00	–	118.07	m³	200.07
Fabric reinforcement; A193 (3.02 kg/m²) in cover slabs	5.55	0.06	1.28	–	5.55	m²	6.83
Formwork to reinforced in situ concrete; isolated cover slabs							
soffits; horizontal	–	3.28	67.24	–	6.44	m²	73.68
height not exceeding 250 mm	–	0.97	19.89	–	3.58	m	23.47
Brickwork							
Walls to manholes; bricks; PC £300.00/1000; in cement mortar (1:3)							
one brick thick	37.80	3.00	61.50	–	46.36	m²	107.86
one and a half brick thick	56.70	4.00	82.00	–	69.54	m²	151.54
two brick thick projection of footing or the like	75.60	4.80	98.40	–	92.72	m²	191.12
Walls to manholes; engineering bricks; PC £260.00/1000; in cement mortar (1:3)							
one brick thick	36.54	3.00	61.50	–	45.10	m²	106.60
one and a half brick thick	54.81	3.00	61.50	–	63.37	m²	124.87
two brick thick projection of footing or the like	73.08	3.00	61.50	–	81.64	m²	143.14
Extra over common or engineering bricks in any mortar for fair face; flush pointing as work proceeds; English bond walls or the like	–	0.13	2.73	–	–	m²	2.73
In situ finishings; cement: sand mortar (1:3); steel trowelled; 13 mm one coat work to manhole walls; to brickwork or blockwork base; over 300 mm wide	–	0.80	16.40	–	2.85	m²	19.25

R DISPOSAL SYSTEMS

Item Excluding site overheads and profit	PC £	Labour hours	Labour £	Plant £	Material £	Unit	Total rate £
Building into brickwork; ends of pipes; making good facings or renderings							
small	–	0.20	4.10	–	–	nr	**4.10**
large	–	0.30	6.15	–	–	nr	**6.15**
extra large	–	0.40	8.20	–	–	nr	**8.20**
extra large; including forming ring arch cover	–	0.50	10.25	–	–	nr	**10.25**
Inspection chambers; polypropylene; Hepworth Plc							
Mini access chamber; up to 600 mm deep; including cover and frame							
300 mm diameter × 600 mm deep; three 100/110 mm inlets	142.78	3.00	61.50	–	145.49	nr	**206.99**
Up to 1200 mm deep; including polymer cover and frame with screw down lid							
475 mm diameter × 940 mm deep; five 100/110 mm inlets; supplied with four stoppers in inlets	211.09	4.00	82.00	–	296.93	nr	**378.93**
Extra over for square ductile iron cover and frame to 1 tonne load; screw down lid	–	–	–	–	35.77	nr	**35.77**
Step irons; Ashworth Ltd; drainage systems; malleable cast iron; galvanized; building into joints							
General purpose pattern; for one brick walls	4.39	0.17	3.48	–	4.39	nr	**7.87**
Best quality vitrified clay half section channels; Hepworth Plc; bedding and jointing in cement: mortar (1:2)							
Channels; straight							
100 mm	6.86	0.80	16.40	–	10.29	m	**26.69**
150 mm	11.43	1.00	20.50	–	14.86	m	**35.36**
225 mm	25.67	1.35	27.68	–	30.82	m	**58.50**
300 mm	52.70	1.80	36.90	–	56.13	m	**93.03**
Bends; 15, 30, 45 or 90°							
100 mm bends	6.18	0.75	15.38	–	7.90	nr	**23.28**
150 mm bends	10.68	0.90	18.45	–	14.11	nr	**32.56**
225 mm bends	41.42	1.20	24.60	–	44.85	nr	**69.45**
300 mm bends	84.46	1.10	22.55	–	88.75	nr	**111.30**
Intercepting traps; Hepworth Plc							
Vitrified clay; inspection arms; brass stoppers; iron levers; chains and staples; galvanized; staples cut and pinned to brickwork; cement: mortar (1:2) joints to vitrified clay pipes and channels; bedding and surrounding in concrete; 11.50 N/mm² – 40 mm aggregate; cutting and fitting brickwork; making good facings							
100 mm inlet; 100 mm outlet	84.25	3.00	61.50	–	100.68	nr	**162.18**
150 mm inlet; 150 mm outlet	121.47	2.00	41.00	–	139.85	nr	**180.85**

R DISPOSAL SYSTEMS

Item Excluding site overheads and profit	PC £	Labour hours	Labour £	Plant £	Material £	Unit	Total rate £
R12 DRAINAGE BELOW GROUND – CONT							
Excavating trenches; using 3 tonne tracked excavator; to receive pipes; grading bottoms; earthwork support; filling with excavated material to within 150 mm of finished surfaces and compacting; completing fill with topsoil; disposal of surplus soil							
Services not exceeding 200 mm nominal size							
average depth of run not exceeding 0.50 m	–	0.02	0.41	3.75	2.45	m	6.61
average depth of run not exceeding 0.75 m	–	0.03	0.68	5.19	2.45	m	8.32
average depth of run not exceeding 1.00 m	–	0.05	1.02	8.76	2.45	m	12.23
average depth of run not exceeding 1.25 m	–	0.10	2.05	11.54	2.45	m	16.04
Granular beds to trenches; lay granular material; to trenches excavated separately; to receive pipes (not included)							
300 mm wide × 100 mm thick							
reject sand	–	0.03	0.51	0.83	0.99	m	2.33
shingle 20 mm aggregate	–	0.03	0.51	0.83	1.22	m	2.56
sharp sand	–	0.03	0.51	0.83	1.13	m	2.47
300 mm wide × 150 mm thick							
reject sand	–	0.04	0.77	1.25	0.50	m	2.52
shingle 20 mm aggregate	–	0.04	0.77	1.25	0.61	m	2.63
sharp sand	–	0.04	0.77	1.25	0.57	m	2.59
Excavating trenches; using 3 tonne tracked excavator; to receive pipes; grading bottoms; earthwork support; filling with imported granular material type 2 and compacting; disposal of surplus soil							
Services not exceeding 200 mm nominal size							
average depth of run not exceeding 0.50 m	5.80	0.02	0.41	2.64	7.80	m	10.85
average depth of run not exceeding 0.75 m	8.70	0.03	0.51	3.09	11.70	m	15.30
average depth of run not exceeding 1.00 m	11.60	0.03	0.59	4.30	15.60	m	20.49
average depth of run not exceeding 1.25 m	14.50	0.03	0.59	7.27	19.50	m	27.36
Excavating trenches; using 3 tonne tracked excavator; to receive pipes; grading bottoms; earthwork support; filling with concrete ready mixed ST2; disposal of surplus soil							
Services not exceeding 200 mm nominal size							
average depth of run not exceeding 0.50 m	9.62	0.02	0.41	2.23	11.62	m	14.26
average depth of run not exceeding 0.75 m	14.42	0.03	0.51	3.09	17.42	m	21.02
average depth of run not exceeding 1.00 m	19.23	0.03	0.59	4.30	23.23	m	28.12
average depth of run not exceeding 1.25 m	24.04	0.03	0.59	7.27	29.04	m	36.90
Clay pipes and fittings; Hepworth Plc; Supersleve							
100 mm clay pipes; polypropylene slip coupling; in trenches (trenches not included)							
laid straight	3.18	0.25	5.13	–	5.53	m	10.66
short runs under 3.00 m	3.18	0.31	6.41	–	5.53	m	11.94

R DISPOSAL SYSTEMS

Item Excluding site overheads and profit	PC £	Labour hours	Labour £	Plant £	Material £	Unit	Total rate £
Extra over 100 mm clay pipes for							
bends; 15–90°; single socket	10.59	0.25	5.13	–	10.59	nr	15.72
junction; 45 or 90°; double socket	22.33	0.25	5.13	–	22.33	nr	27.46
slip couplings; polypropylene	3.76	0.08	1.71	–	3.76	nr	5.47
gully with P trap; 100 mm; 154 × 154 mm							
plastic grating	34.59	1.00	20.50	–	51.23	nr	71.73
PVC-u pipes and fittings; Wavin Plastics Ltd; OsmaDrain system							
110 mm PVC-u pipes; in trenches (trenches not included)							
laid straight	10.81	0.08	1.64	–	10.81	m	12.45
short runs under 3.00 m	10.81	0.12	2.46	–	10.81	m	13.27
Extra over 110 mm PVC-u pipes for							
bends; short radius	21.13	0.25	5.13	–	21.13	nr	26.26
bends; long radius	39.61	0.25	5.13	–	39.61	nr	44.74
junctions; equal; double socket	25.21	0.25	5.13	–	25.21	nr	30.34
slip couplings	12.23	0.10	2.05	–	12.23	nr	14.28
adaptors to clay	23.82	0.50	10.25	–	23.99	nr	34.24
Kerbs to gullies							
Kerbs to gullies; in one course Class B engineering bricks; to 4 nr sides; rendering in cement: mortar (1:3); dished to gully gratings	2.32	1.00	20.50	–	3.18	nr	23.68
Gullies; vitrified clay; Hepworth Plc; bedding in concrete; 11.50 N/mm² – 40 mm aggregate							
Yard gullies (mud); trapped; domestic duty (up to 1 tonne)							
100 mm outlet; 100 mm diameter; 225 mm internal width; 585 mm internal depth	102.25	3.50	71.75	–	102.87	nr	174.62
Yard gullies (mud); trapped; medium duty (up to 5 tonnes)							
100 mm outlet; 100 mm diameter; 225 mm internal width; 585 mm internal depth	102.25	3.50	71.75	–	102.87	nr	174.62
150 mm outlet; 100 mm diameter; 225 mm internal width; 585 mm internal depth	158.38	3.50	71.75	–	159.01	nr	230.76
Combined filter and silt bucket for yard gullies							
225 mm diameter	36.97	–	–	–	36.97	nr	36.97
Road gullies; trapped with rodding eye							
100 mm outlet; 300 mm internal diameter; 600 mm internal depth	98.35	3.50	71.75	–	98.97	nr	170.72
150 mm outlet; 300 mm internal diameter; 600 mm internal depth	100.72	3.50	71.75	–	101.34	nr	173.09
150 mm outlet; 400 mm internal diameter; 750 mm internal depth	116.80	3.50	71.75	–	117.42	nr	189.17
150 mm outlet; 450 mm internal diameter; 900 mm internal depth	158.03	3.50	71.75	–	158.65	nr	230.40
Hinged gratings and frames for gullies; alloy							
193 mm for 150 mm diameter gully	–	–	–	–	24.50	nr	24.50
120 × 120 mm	–	–	–	–	8.71	nr	8.71
150 × 150 mm	–	–	–	–	15.76	nr	15.76
230 × 230 mm	–	–	–	–	28.82	nr	28.82
316 × 316 mm	–	–	–	–	76.41	nr	76.41

R DISPOSAL SYSTEMS

Item Excluding site overheads and profit	PC £	Labour hours	Labour £	Plant £	Material £	Unit	Total rate £
R12 DRAINAGE BELOW GROUND – CONT							
Gullies – cont							
Hinged gratings and frames for gullies; cast iron							
265 mm for 225 mm diameter gully	–	–	–	–	48.75	nr	**48.75**
150 × 150 mm	–	–	–	–	15.76	nr	**15.76**
230 × 230 mm	–	–	–	–	28.82	nr	**28.82**
316 × 316 mm	–	–	–	–	76.41	nr	**76.41**
Universal gully trap; PVC-u; Wavin Plastics Ltd; OsmaDrain system; bedding in concrete; 11.50 N/mm^2 – 40 mm aggregate							
Universal gully fitting; comprising gully trap only							
110 mm outlet; 110 mm diameter; 205 mm							
internal depth	17.93	3.50	71.75	–	18.55	nr	**90.30**
Vertical inlet hopper c\w plastic grate							
272 × 183 mm	24.09	0.25	5.13	–	24.09	nr	**29.22**
Sealed access hopper							
110 × 110 mm	56.64	0.25	5.13	–	56.64	nr	**61.77**
Universal gully; PVC-u; Wavin Plastics Ltd; OsmaDrain system; accessories to universal gully trap							
Hoppers; backfilling with clean granular material; tamping; surrounding in lean mix concrete							
plain hopper; with 110 spigot 150 mm long	19.69	0.40	8.20	–	21.42	nr	**29.62**
vertical inlet hopper; with 110 spigot							
150 mm long	24.09	0.40	8.20	–	25.81	nr	**34.01**
sealed access hopper; with 110 spigot							
150 mm long	56.64	0.40	8.20	–	58.37	nr	**66.57**
plain hopper; solvent weld to trap	13.45	0.40	8.20	–	15.17	nr	**23.37**
vertical inlet hopper; solvent wed to trap	23.11	0.40	8.20	–	24.83	nr	**33.03**
sealed access cover; PVC-u	28.85	0.10	2.05	–	28.85	nr	**30.90**
Gullies PVC-u; Wavin Plastics Ltd; OsmaDrain system; bedding in concrete; 11.50 N/mm^2 – 40 mm aggregate							
Bottle gully; providing access to the drainage system for cleaning							
bottle gully; 228 × 228 × 317 mm deep	47.82	0.50	10.25	–	48.37	nr	**58.62**
sealed access cover; PVC-u; 217 × 217 mm	37.18	0.10	2.05	–	37.18	nr	**39.23**
grating; ductile iron; 215 × 215 mm	28.66	0.10	2.05	–	28.66	nr	**30.71**
bottle gully riser; 325 mm	6.16	0.50	10.25	–	7.33	nr	**17.58**
Yard gully; trapped; 300 mm diameter; 600 mm deep; including catchment bucket and ductile iron cover and frame; medium duty loading							
305 mm diameter; 600 mm deep	290.00	2.50	51.25	–	293.62	nr	**344.87**
Linear Drainage							
Loadings for slot drains							
A15 – 1.5 tonne; pedestrian use							
B125 – 12.5 tonne; domestic use							
C250 – 25 tonne; car parks, supermarkets, industrial units							
D400 – 40 tonne; highways							
E600 – 60 tonne; forklifts							

R DISPOSAL SYSTEMS

Item Excluding site overheads and profit	PC £	Labour hours	Labour £	Plant £	Material £	Unit	Total rate £
Linear drainage; channels; Aco Building Products; laid to concrete bed C25 on compacted granular base on 200 mm deep concrete bed; haunched with 200 mm concrete surround; all in 750 × 430 mm wide trench with compacted 200 mm granular base surround (excavation and subbase not included)							
HexDrain channel drain; recycled polypropylene; load class A15; 1000 × 125 × 80 mm							
with black plastic grating	16.35	1.20	24.60	–	26.18	ea	**50.78**
with galvanized steel grating	19.55	1.20	24.60	–	29.38	ea	**53.98**
with black plastic brickslot grating	19.55	1.20	24.60	–	29.38	ea	**53.98**
Accessories for HexDrain							
endcap; black plastic	3.30	–	–	–	3.30	ea	**3.30**
corner unit; with metallic effect plastic grating	17.90	0.50	10.25	–	22.42	ea	**32.67**
sump unit; black plastic; 250 mm depth	28.80	1.00	20.50	–	33.32	ea	**53.82**
corner unit; with black plastic grating; 125 mm vertical outlet	18.70	0.75	15.38	–	19.23	ea	**34.61**
corner unit; brickslot; black plastic; 125 mm vertical outlet	20.20	0.50	10.25	–	24.72	ea	**34.97**
Universal 820 drain union; PVC-u; 110 mm diameter	3.35	–	–	–	3.35	ea	**3.35**
RainDrain lightweight polymer concrete channel; load class A15; 1000 × 118 × 97 mm							
with galvanized steel grating	23.60	1.20	24.60	–	28.12	ea	**52.72**
Accessories for RainDrain							
foul air trap; horizontal; 110 mm diameter	15.40	–	–	–	15.40	ea	**15.40**
closing endcap	3.95	–	–	–	3.95	ea	**3.95**
outlet endcap; 110 mm diameter	8.65	–	–	–	8.65	ea	**8.65**
sump; with galvanized steel grating and sediment bucket; 500 mm	59.80	0.50	10.25	–	64.32	ea	**74.57**
RainDrain Pro drainage channel							
with heelguard ductile iron grating	72.20	1.20	24.60	–	76.72	ea	**101.32**
Accessories for RainDrain Pro							
Universal silver closing end cap	3.55	0.13	2.56	–	3.55	ea	**6.11**
step connector; 50 mm	8.15	0.13	2.56	–	8.15	ea	**10.71**
vertical outlet connector; 110 mm or 160 mm	13.30	0.13	2.56	–	13.30	ea	**15.86**
roddable foul air trap; 110 mm	91.60	0.13	2.56	–	91.60	ea	**94.16**
roddable foul air trap; 160 mm	113.05	0.13	2.56	–	113.05	ea	**115.61**
Linear drainage; channels; Aco Building Products; laid to concrete bed C25 on compacted granular base on 200 mm deep concrete bed; haunched with 200 mm concrete surround; all in 750 × 430 mm wide trench with compacted 200 mm granular base surround (excavation and subbase not included)							
Aco MultiDrain M100PPD; recycled polypropylene drainage channel; range of gratings to complement installations which require discreet slot drainage							
142 mm wide × 150 mm deep	59.50	1.20	24.60	–	69.33	m	**93.93**

R DISPOSAL SYSTEMS

Item Excluding site overheads and profit	PC £	Labour hours	Labour £	Plant £	Material £	Unit	Total rate £
R12 DRAINAGE BELOW GROUND – CONT							
Linear drainage – cont							
Accessories for M100PPD							
connectors; vertical outlet 110 mm	13.30	–	–	–	13.30	nr	**13.30**
connectors; vertical outlet 160 mm	13.30	–	–	–	13.30	nr	**13.30**
sump units; black plastic with plastic silt							
bucket; 110/160 mm inlet/outlets	105.25	1.50	30.75	–	123.31	nr	**154.06**
universal gully and bucket; 440 × 440 ×							
1315 mm deep	650.00	3.00	61.50	–	683.86	nr	**745.36**
Aco drainlock gratings for M100PPD system to							
loading A15							
slotted galvanized steel	17.15	–	–	–	17.15	m	**17.15**
perforated galvanized steel	25.04	–	–	–	25.04	m	**25.04**
Aco drainlock gratings for M100PPD system to							
loading C250							
Heelguard; composite black; 500 mm long							
with security locking	23.00	–	–	–	23.00	m	**23.00**
intercept; ductile iron; 500 mm long	27.00	–	–	–	27.00	m	**27.00**
slotted galvanized steel; 1.00 m long	44.00	–	–	–	44.00	m	**44.00**
perforated galvanized steel; 1.00 m long	51.90	–	–	–	51.90	m	**51.90**
mesh galvanized steel; 1.00 m long	35.26	–	–	–	35.26	m	**35.26**
Aco MultiDrain MD brickslot; offset galvanized							
slot drain grating for M100PPD; load class C250							
brickslot galvanized steel; 1.00 m	67.20	–	–	–	67.20	m	**67.20**
Aco MultiDrain MD brickslot; offset galvanized							
slot drain grating for M100PPD; load class							
C250 – 400							
Aco gratings for M100PPD brickslot;							
galvanized steel; load class C250/D400; 1.00 m							
brickslot; galvanized steel; 1.00 m	83.30	–	–	–	83.30	m	**83.30**
brickslot; stainless steel; 1.00 m	134.77	–	–	–	134.77	m	**134.77**
Channel footpath drain							
Aco channel drain to pedestrian footpaths;							
connection to existing kerbdrain; laying on							
150 mm concrete base and surround;							
excavation subbase not included	27.09	1.20	24.60	–	36.92	m	**61.52**
Channel drains; Aco Technologies Ltd; Aco							
MultiDrain MD polymer concrete channel							
drainage system; traditional channel and							
grate drainage solution							
Shallow depth channels M100D system;							
135 mm wide							
100 mm deep	57.90	1.20	24.60	–	67.73	m	**92.33**
Accessories for MultiDrain MD							
sump unit; complete with sediment bucket	104.25	1.50	30.75	–	104.25	nr	**135.00**
Access Covers and Frames							
FACTA (Fabricated Access Cover Trade							
Association) class:							
A – 0.5 tonne maximum slow moving wheel load							
AA – 1.5 tonne maximum slow moving wheel							
load							
AAA – 2.5 tonne maximum slow moving wheel							
load							
B – 5 tonne maximum slow moving wheel load							
C – 6.5 tonne maximum slow moving wheel load							
D – 11 tonne maximum slow moving wheel load							

R DISPOSAL SYSTEMS

Item Excluding site overheads and profit	PC £	Labour hours	Labour £	Plant £	Material £	Unit	Total rate £
Access covers and frames; solid top; galvanized; Steelway Brickhouse; Bristeel; bedding frame in cement mortar (1:3); cover in grease and sand; clear opening sizes; base size shown in brackets (50 mm depth)							
FACTA AA; single seal							
450 × 450 mm (520 × 520 mm)	75.52	1.50	30.75	–	85.72	nr	**116.47**
600 × 450 mm (670 × 520 mm)	84.91	1.80	36.90	–	96.24	nr	**133.14**
600 × 600 mm (670 × 670 mm)	91.52	2.00	41.00	–	105.11	nr	**146.11**
FACTA AA; double seal							
450 × 450 mm (560 × 560 mm)	129.99	1.50	30.75	–	140.18	nr	**170.93**
60 × 450 mm (710 × 560 mm)	144.04	1.80	36.90	–	155.36	nr	**192.26**
600 × 600 mm (710 × 710 mm)	158.82	2.00	41.00	–	172.41	nr	**213.41**
FACTA B; single seal							
450 × 450 mm (520 × 520 mm)	94.58	1.50	30.75	–	104.77	nr	**135.52**
600 × 450 mm (670 × 520 mm)	114.01	1.80	36.90	–	125.33	nr	**162.23**
600 × 600 mm (670 × 670 mm)	134.28	2.00	41.00	–	147.87	nr	**188.87**
FACTA B; double seal							
450 × 450 mm (560 × 560 mm)	151.08	1.50	30.75	–	161.27	nr	**192.02**
600 × 450 mm (710 × 560 mm)	169.58	1.80	36.90	–	180.91	nr	**217.81**
600 × 600 mm (710 × 710 mm)	202.84	2.00	41.00	–	216.43	nr	**257.43**
Access covers and frames; recessed; galvanized; Steelway Brickhouse; Bripave; bedding frame in cement mortar (1:3); cover in grease and sand; clear opening sizes; base size shown in brackets (for block depths 50 mm, 65 mm, 80 mm or 100 mm)							
FACTA AA							
450 × 450 mm (562 × 562 mm)	204.74	1.50	30.75	–	214.93	nr	**245.68**
600 × 450 mm (712 × 562 mm)	229.33	1.80	36.90	–	240.66	nr	**277.56**
600 × 600 mm (712 × 712 mm)	239.51	2.00	41.00	–	497.03	nr	**538.03**
750 × 600 mm (862 × 712 mm)	272.94	2.40	49.20	–	551.92	nr	**601.12**
FACTA B							
450 × 450 mm (562 × 562 mm)	205.18	1.50	30.75	–	215.37	nr	**246.12**
600 × 450 mm (712 × 562 mm)	246.50	1.80	36.90	–	257.83	nr	**294.73**
600 × 600 mm (712 × 712 mm)	251.15	2.00	41.00	–	264.74	nr	**305.74**
750 × 600 mm (862 × 712 mm)	303.52	2.40	49.20	–	318.24	nr	**367.44**
FACTA D							
450 × 450 mm (562 × 562 mm)	235.89	1.50	30.75	–	246.08	nr	**276.83**
600 × 450 mm (712 × 562 mm)	253.69	1.80	36.90	–	265.01	nr	**301.91**
600 × 600 mm (712 × 712 mm)	263.73	2.00	41.00	–	277.32	nr	**318.32**
750 × 600 mm (862 × 712 mm)	331.61	2.40	49.20	–	346.33	nr	**395.53**
Extra over manhole frames and covers for filling recessed manhole covers with brick paviors; PC £305.00/1000	13.42	1.00	20.50	–	14.05	m²	**34.55**
filling recessed manhole covers with vehicular paving blocks; PC £8.72/m²	8.89	0.75	15.38	–	9.52	m²	**24.90**
filling recessed manhole covers with concrete paving flags; PC £12.54/m²	12.54	0.35	7.17	–	12.92	m²	**20.09**

R DISPOSAL SYSTEMS

Item Excluding site overheads and profit	PC £	Labour hours	Labour £	Plant £	Material £	Unit	Total rate £
R13 LAND DRAINAGE							
Land drainage; excavating for drains; by 3 tonne tracked excavator; including disposing spoil to spoil heaps not exceeding 100 m							
Width 225 mm							
depth 450 mm	–	–	–	354.13	–	100 m	**354.13**
depth 600 mm	–	–	–	363.97	–	100 m	**363.97**
depth 700 mm	–	–	–	374.66	–	100 m	**374.66**
depth 900 mm	–	–	–	399.10	–	100 m	**399.10**
depth 1000 mm	–	–	–	413.15	–	100 m	**413.15**
depth 1200 mm	–	–	–	428.68	–	100 m	**428.68**
Land drainage; excavating for drains; by hand; including disposing spoil to spoil heaps not exceeding 100 m							
Width 150 mm							
depth 450 mm	–	22.03	451.62	–	–	100 m	**451.62**
depth 600 mm	–	29.38	602.29	–	–	100 m	**602.29**
depth 700 mm	–	34.27	702.53	–	–	100 m	**702.53**
depth 900 mm	–	44.06	903.23	–	–	100 m	**903.23**
Width 225 mm							
depth 450 mm	–	33.05	677.52	–	–	100 m	**677.52**
depth 600 mm	–	44.06	903.23	–	–	100 m	**903.23**
depth 700 mm	–	51.41	1053.90	–	–	100 m	**1053.90**
depth 900 mm	–	66.10	1355.05	–	–	100 m	**1355.05**
depth 1000 mm	–	146.88	3011.04	–	–	100 m	**3011.04**
Wavin Plastics Ltd; flexible plastic perforated pipes in trenches (not included); to a minimum depth of 450 mm (couplings not included)							
OsmaDrain; flexible plastic perforated pipes in trenches (not included); to a minimum depth of 450 mm (couplings not included)							
80 mm diameter; available in 100 m coil	100.10	2.00	41.00	–	100.10	100 m	**141.10**
100 mm diameter; available in 100 m coil	162.40	2.00	41.00	–	162.40	100 m	**203.40**
160 mm diameter; available in 35 m coil	403.24	2.00	41.00	–	403.24	100 m	**444.24**
WavinCoil; plastic pipe junctions							
80 × 80 mm	4.53	0.05	1.02	–	4.53	nr	**5.55**
100 × 100 mm	5.06	0.05	1.02	–	5.06	nr	**6.08**
100 × 60 mm	4.82	0.05	1.02	–	4.82	nr	**5.84**
100 × 80 mm	4.82	0.05	1.02	–	4.82	nr	**5.84**
160 × 160 mm	12.05	0.05	1.02	–	12.05	nr	**13.07**
WavinCoil; couplings for flexible pipes							
80 mm diameter	1.59	0.03	0.68	–	1.59	nr	**2.27**
100 mm diameter	1.76	0.03	0.68	–	1.76	nr	**2.44**
160 mm diameter	2.38	0.03	0.68	–	2.38	nr	**3.06**
Market prices of backfilling materials							
Sand	39.60	–	–	–	39.60	m³	**39.60**
Gravel rejects	34.65	–	–	–	34.65	m³	**34.65**
Topsoil; allowing for 20% settlement	31.20	–	–	–	31.20	m³	**31.20**

R DISPOSAL SYSTEMS

Item Excluding site overheads and profit	PC £	Labour hours	Labour £	Plant £	Material £	Unit	Total rate £
Land drainage; backfilling trench after laying pipes with gravel rejects or similar; blind filling with ash or sand; topping with 150 mm topsoil from dumps not exceeding 100 m; by machine							
Width 150 mm							
depth 450 mm	–	3.20	65.60	46.99	220.72	100 m	**333.31**
depth 600 mm	–	4.20	86.10	62.66	297.54	100 m	**446.30**
depth 750 mm	–	4.86	99.63	78.32	338.85	100 m	**516.80**
depth 900 mm	–	6.20	127.10	93.99	433.35	100 m	**654.44**
Width 225 mm							
depth 450 mm	–	4.80	98.40	72.06	300.00	100 m	**470.46**
depth 600 mm	–	6.30	129.15	93.99	446.48	100 m	**669.62**
depth 750 mm	–	7.80	159.90	117.45	300.00	100 m	**577.35**
depth 900 mm	–	9.30	190.65	144.11	300.00	100 m	**634.76**
Land drainage; backfilling trench after laying pipes with gravel rejects or similar; blind filling with ash or sand; topping with 150 mm topsoil from dumps not exceeding 100 m; by hand							
Width 150 mm							
depth 450 mm	–	18.63	381.92	–	194.99	100 m	**576.91**
depth 600 mm	–	24.84	509.22	–	283.50	100 m	**792.72**
depth 750 mm	–	31.05	636.52	–	266.31	100 m	**902.83**
depth 900 mm	–	37.26	763.83	–	419.31	100 m	**1183.14**

V ELECTRICAL SUPPLY/POWER/LIGHTING SYSTEMS

Item Excluding site overheads and profit	PC £	Labour hours	Labour £	Plant £	Material £	Unit	Total rate £
TRENCHING FOR ELECTRICAL SERVICES							
Excavating; mechanical; trenches for							
electrical services							
3 tonne excavator (bucket volume 0.13 m³);							
arisings laid alongside							
600 mm deep	–	–	–	1.59	–	m	**1.59**
800 mm deep	–	–	–	1.86	–	m	**1.86**
1.00 m deep	–	–	–	2.23	–	m	**2.23**
Extra over any types of excavating irrespective							
of depth for breaking out existing materials;							
heavy duty 110 volt breaker tool							
hard rock	–	5.00	102.50	59.69	–	m³	**162.19**
concrete	–	3.00	61.50	35.81	–	m³	**97.31**
reinforced concrete	–	4.00	82.00	60.94	–	m³	**142.94**
brickwork, blockwork or stonework	–	1.50	30.75	17.91	–	m³	**48.66**
By hand							
600 mm deep	–	0.24	4.93	–	–	m	**4.93**
800 mm deep	–	0.43	8.76	–	–	m	**8.76**
1.00 mm deep	–	0.67	13.67	–	–	m	**13.67**
Backfilling of trenches; including laying of							
electrical marker tape; compacting lightly							
as work proceeds; spreading any							
remaining material							
To trenches containing armoured cable							
by machine	0.13	–	–	1.67	0.13	m	**1.80**
by hand	0.13	0.20	4.10	–	0.13	m	**4.23**
To trenches containing ducted cable including							
150 mm sharp sand over the duct							
by machine	2.40	–	–	2.09	2.40	m	**4.49**
by hand	0.13	0.25	5.13	–	2.40	m	**7.53**
V90 ELECTRICAL CABLES							
Cable to trenches (trenching operations							
not included); laying only cable							
Twin core steel wire armoured cable; 50 m							
drums							
1.5 mm core	0.80	0.02	0.41	–	0.80	m	**1.21**
2.5 mm core	0.96	0.02	0.41	–	0.96	m	**1.37**
Twin core steel wire armoured cable; lengths							
less than 50 m runs							
1.5 mm core	0.88	0.02	0.41	–	0.88	m	**1.29**
2.5 mm core	1.04	0.02	0.41	–	1.04	m	**1.45**
Three core steel wire armoured cable; 50 m							
drums							
1.5 mm core	0.90	0.02	0.41	–	0.90	m	**1.31**
2.5 mm core	1.23	0.02	0.41	–	1.23	m	**1.64**
4.0 mm core	1.48	0.02	0.41	–	1.48	m	**1.89**
6.0 mm core	1.92	0.02	0.46	–	1.92	m	**2.38**
10.0 mm core	3.39	0.03	0.51	–	3.39	m	**3.90**
Three core steel wire armoured cable; lengths							
less than 50 m runs							
1.5 mm core	0.93	0.02	0.41	–	0.93	m	**1.34**
2.5 mm core	1.33	0.02	0.41	–	1.33	m	**1.74**

V ELECTRICAL SUPPLY/POWER/LIGHTING SYSTEMS

Item Excluding site overheads and profit	PC £	Labour hours	Labour £	Plant £	Material £	Unit	Total rate £
Ducts to trenches; twin wall flexible cable ducts with drawstrings; laid on 150 mm clean sharp sand							
Twin wall duct; laying to trenches							
63 mm × 50 m coils	1.00	0.02	0.41	–	1.00	m	**1.41**
110 mm × 50 m coils	1.50	0.02	0.41	–	1.50	m	**1.91**
Cable drawing through ducts							
straight runs up to 50 m lengths	–	1.00	20.50	–	–	nr	**20.50**
Terminations to armoured cables; cutting, coiling and taping length of cable to receive connection to light fitting, transformer and the like; (all not included) fixing to temporary posts							
Armoured cable	–	0.33	6.83	–	–	nr	**6.83**
Plain ducted cable	–	0.13	2.56	–	–	nr	**2.56**
Junction boxes; fixing to cables to receive connections to light fittings or transformers; to IP68							
Underground jointing box							
2 way	29.00	–	–	–	46.50	nr	**46.50**
3 way	34.00	–	–	–	54.00	nr	**54.00**
Above ground jointing box							
2 way	3.50	0.33	8.33	–	3.50	nr	**11.83**
3 way	4.90	–	–	–	18.90	nr	**18.90**
Power Connections							
Internal connections and switches for control of external lighting; connections of switches or switch bank to distribution board and internal cables to switches							
per circuit located within 1.00 m of the distribution board; chasing of cables to walls not included	–	–	–	–	–	nr	**35.00**
Power connections to switches							
power cable to switches including builders work; drilling through walls and making good. chasing of cables to walls not included; price per circuit for the first circuit	–	1.00	20.50	–	74.00	nr	**94.50**
additional circuits located in the same location	–	–	–	–	12.75	nr	**12.75**
V90 EXTERIOR LIGHTING TRANSFORMERS							
Outdoor transformers for 12 volt exterior lighting; connecting to junction box (not included); IP67							
Single light transformers; 100 × 68 × 72 mm; 50 vA							
fused	21.00	–	–	–	67.67	nr	**67.67**
unfused	20.34	–	–	–	20.34	nr	**20.34**

V ELECTRICAL SUPPLY/POWER/LIGHTING SYSTEMS

Item Excluding site overheads and profit	PC £	Labour hours	Labour £	Plant £	Material £	Unit	Total rate £
V90 EXTERIOR LIGHTING **TRANSFORMERS – CONT**							
Outdoor transformers for 12 volt exterior **lighting – cont**							
Two light transformer; 130 × 85 × 85 mm; 100 vA							
fused	34.81	–	–	–	34.81	nr	**34.81**
unfused	33.21	–	–	–	33.21	nr	**33.21**
Three light transformer; 130 × 85 × 85 mm; 150 vA							
fused	44.67	–	–	–	44.67	nr	**44.67**
unfused	42.21	–	–	–	42.21	nr	**42.21**
Four light transformer; 130 × 85 × 100 mm; 200 vA							
fused	52.56	–	–	–	52.56	nr	**52.56**
unfused	49.13	–	–	–	49.13	nr	**49.13**
V41 EXTERIOR LIGHTS – LOW VOLTAGE							
Low voltage lights; connecting to **transformers (transformers shown** **separately); fixing as described**							
Hunza Spike spotlight; 63.5 mm diameter × 75 mm long; c/w 20/35/50 w lamp							
stainless steel	107.16	–	–	–	118.83	nr	**118.83**
copper	66.99	–	–	–	78.66	nr	**78.66**
black	43.32	–	–	–	54.99	nr	**54.99**
Hunza Spike spotlight; adjustable; 63.5 mm diameter × 75 mm long; c/w 20/35/50 w lamp							
stainless steel	134.68	–	–	–	146.35	nr	**146.35**
copper	77.25	–	–	–	88.92	nr	**88.92**
black	50.92	–	–	–	62.59	nr	**62.59**
Hunza pole lights; single; fixing to concrete base 150 × 150 × 150 mm							
stainless steel	166.94	1.00	20.50	–	178.61	nr	**199.11**
copper	113.96	1.00	20.50	–	126.13	nr	**146.63**
black	73.32	1.00	20.50	–	85.49	nr	**105.99**
Hunza pole lights; double; fixing to concrete base 150 × 150 × 150 mm							
stainless steel	268.28	1.00	20.50	–	280.45	nr	**300.95**
copper	113.96	1.00	20.50	–	126.13	nr	**146.63**
black	73.32	1.00	20.50	–	85.49	nr	**105.99**
Recessed wall lights; eyelid steplights; fixing to brick or stone walls; inclusive of core drilling							
stainless steel	268.28	1.50	30.75	–	279.95	nr	**310.70**
copper	136.29	1.50	30.75	–	147.96	nr	**178.71**
black	92.04	1.50	30.75	–	103.71	nr	**134.46**
Hunza; Recessed deck path or lawn lights; inclusive of core drilling excavation and all making good							
stainless steel light installed to deck	250.00	0.66	16.50	–	250.00	nr	**266.50**
stainless steel light installed to path	250.00	1.08	23.63	–	250.00	nr	**273.63**
stainless steel light installed to lawn	197.00	0.20	4.10	–	208.55	nr	**212.65**
stainless steel driveway light	250.00	1.08	23.63	–	250.00	nr	**273.63**

Rethinking Landscape
A Critical Reader

Ian H. Thompson

This unusually wide-ranging critical tool in the field of landscape architecture provides extensive excerpted materials from, and detailed critical perspectives on, standard and neglected texts from the 18th century to the present day. Considering the aesthetic, social, cultural and environmental foundations of our thinking about landscape this book explores the key writings which shaped the field in its emergence and maturity. Uniquely the book also includes original materials drawn from philosophical, ethical and political writings.

Selected Contents:

Part 1: Pluralism

Part 2: Aesthetics

Part 3: The Social Mission

Part 4: Ecology

Part 5: Some other Perspectives

Part 6: Conclusions and Suggestions

2008: 246x174: 288pp
Hb: 978-0-415-42463-9 **£95.00**
Pb: 978-0-415-42464-6 **£24.99**

To Order: Tel: +44 (0) 1235 400524 **Fax:** +44 (0) 1235 400525
or Post: Taylor and Francis Customer Services,
Bookpoint Ltd, Unit T1, 200 Milton Park, Abingdon, Oxon, OX14 4TA UK
Email: book.orders@tandf.co.uk

For a complete listing of all our titles visit:
www.tandf.co.uk

Construction Delays
Extensions of Time and Prolongation Claims

Roger Gibson

Providing guidance on delay analysis, the author gives readers the information and practical details to be considered in formulating and resolving extension of time submissions and time-related prolongation claims. Useful guidance and recommended good practice is given on all the common delay analysis techniques. Worked examples of extension of time submissions and time-related prolongation claims are included.

Selected Contents:

1. Introduction
2. Programmes & Record Keeping
3. Contracts and Case Law
4. The 'Thorny Issues'
5. Extensions of Time
6. Prolongation Claims Summary

April 2008: 234x156: 374pp
Hb: 978-0-415-35486-6 **£80.00**

To Order: Tel: +44 (0) 1235 400524 **Fax:** +44 (0) 1235 400525
or Post: Taylor and Francis Customer Services,
Bookpoint Ltd, Unit T1, 200 Milton Park, Abingdon, Oxon, OX14 4TA UK
Email: book.orders@tandf.co.uk

For a complete listing of all our titles visit:
www.tandf.co.uk

Taylor & Francis
Taylor & Francis Group

PART 6

Fees for Professional Services

LANDSCAPE ARCHITECTS' FEES

The Landscape Institute no longer sets any fee scales for its members. A publication which offers guidance in determining fees for different types of project is available from:

The Landscape Institute
33 Great Portland Street
London
W1W 8QG

Telephone: 020 7299 4500

The publication is entitled 'Engaging a Landscape Consultant – Guidance for Clients on Fees 2002'

The document refers to suggested fee systems on the following basis:

- Time Charged Fee Basis
- Lump Sum Fee Basis
- Percentage Fee Basis
- Retainer Fee Basis

In regard to the 'Percentage Fee Basis' the publication lists various project types and relates them to complexity ratings for works valued at £22,500 and above. The suggested fee scales are dependent on the complexity rating of the project.

The following are samples of the Complexity Categories:

Category 1: Golf Courses, Country Parks and Estates, Planting Schemes

Category 2: Coastal, River and Agricultural Works, Roads, Rural Recreation Schemes

Category 3: Hospital Grounds, Sport Stadia, Urban Offices and Commercial Properties, Housing

Category 4: Domestic/Historic Garden Design, Urban Rehabilitation and Environmental Improvements

Table Showing Suggested Percentage Fees for Various Categories of Complexity of Landscape Design or Consultancy

Project Value £	Complexity			
	1	2	3	4
22,500	15.00%	16.50%	18.00%	21.00%
30,000	13.75	15.00	16.50	19.25
50,000	10.50	12.75	14.00	16.25
100,000	9.50	10.50	11.50	13.50
150,000	8.75	9.50	10.50	12.50
200,000	8.00	8.75	9.75	11.25
300,000	7.50	8.25	9.00	10.50
500,000	6.75	7.50	8.25	9.75
750,000	6.50	7.25	7.75	9.25
1,000,000	6.25	7.00	7.50	8.75
10,000,000	6.25	6.75	7.25	8.25

(extrapolated from the graph/curve chart in the aforementioned publication)

Guide to Stage Payments of Fees, Relevant Fee Basis and Proportion of Fee Applicable to Lump Sum and Percentage Fee Basis. *Details of Preliminary, Standard and Other Services are set out in detail in the Landscape Consultant's Appointment.*

Work Stage		Relevant Fee Basis			Proportion of Fee	
		Time	Lump	%	*Proportion of fee*	Total
Preliminary Services						
A	Inception	✓	✓	n/a	n/a	n/a
B	Feasibility	✓	✓	n/a	n/a	n/a
Standard Services						
C	Outline Proposals	✓	✓	✓	15%	15%
D	Sketch Scheme Proposals	✓	✓	✓	15%	30%
E	Detailed Proposals	✓	✓	✓	15%	45%
FG	Production Information	✓	✓	✓	20%	65%
HJ	Tender Action & Contract Preparation	✓	✓	✓	5%	70%
K	Operations on Site	✓	✓	✓	25%	95%
L	Completion	✓	✓	✓	5%	100%
Other Services		✓	✓	n/a		

Timing of Fee Payments
Percentage fees are normally paid at the end of each work stage. Time based fees are normally paid at monthly intervals. Lump sum fees are normally paid at intervals by agreement. Retainer or term commission fees are normally paid in advance, for predetermined periods of service.

WORKED EXAMPLES OF PERCENTAGE FEE CALCULATIONS

Worked Example 1

Project Type   Caravan Site
Services Required   To Detailed Proposals – Work Stages C to E
Budget   £120,000

Step 1	Decide on Work Type and therefore Complexity Rating – Complexity Rating 2
Step 2	Decide on Services required and Proportion of Fee – To Detailed Proposals, 45%
Step 3	Read off Graph, Complexity Rating 2, the % fee of £120,000 – Graph Fee 9.9%
Step 4	Multiply the Proportion of Fee (45%) by the Graph Fee (9.9%) – Adjusted Fee – 4.46%
Step 5	Calculate the Guide Fee (4.46% of £120,000) – Guide Fee – £5,352
Step 6	Agree fee with Client, complete Memorandum of Agreement & Schedule of Services & Fees

Worked Example 2

Project Type   New Housing
Services Required   Full Standard Services – Work Stages C to L
Budget   £350,000

Step 1	Decide on Work Type and therefore Complexity Rating – Complexity Rating 3
Step 2	Decide on Services required and Proportion of Fee – To Completion, 100%
Step 3	Read off Graph, Complexity Rating 3, the % fee of £350,000 – Graph Fee 8.8%
Step 4	Multiply the Proportion of Fee (100%) by the Graph Fee (8.8%) – Adjusted Fee – 8.8%
Step 5	Calculate the Guide Fee (8.8% of £350,000) – Guide Fee – £30,800
Step 6	Agree fee with Client, complete Memorandum of Agreement & Schedule of Services & Fees

Worked Example 3

Project Type   Urban Environmental Improvements
Services Required   To Production Information – Work Stages C to G
Budget   £1,250,000

Step 1	Decide on Work Type and therefore Complexity Rating – Complexity Rating 4
Step 2	Decide on Services required and Proportion of Fee – To Production Information, 65%
Step 3	Read off Graph, Complexity Rating 4, the % fee of £1,250,000 – Graph Fee 8.6%
Step 4	Multiply the Proportion of Fee (65%) by the Graph Fee (8.6%) – Adjusted Fee – 5.59%
Step 5	Calculate the Guide Fee (5.59% of £1,250,000) – Guide Fee – £69,875
Step 6	Agree fee with Client, complete Memorandum of Agreement & Schedule of Services & Fees

PART 7

Rates of Wages

BUILDING INDUSTRY WAGES

Talks are currently ongoing between the parties to the Construction Industry Joint Council (CIJC). It is expected that the deal will be announced towards the end of June 2010.

The Working Rule Agreement includes a pay structure with a general operative and additional skilled rates of pay as well as craft rate. Plus rates and additional payments will be consolidated into basic pay to provide the following rates (for a normal 39 hour week) which will come into effect from the following dates:

Effective from 30 June 2010 – assumed

The following basic rates of pay are assumed:

	Rate per 39-hour week (£)	Rate per hour (£)
Craft Rate	409.73	10.51
Skill Rate 1	390.64	10.02
Skill Rate 2	376.32	9.65
Skill Rate 3	352.05	9.03
Skill Rate 4	332.16	8.52
General operative	308.30	7.91

Example of Computation using National Wage Awards – Building Craft and General Operatives
(not used for rate calculations in this book)

			Craft Operatives		General Operatives	
			£	£	£	£
Wages at standard basic rate						
Productive time	44.30	weeks	409.73	18,151.04	308.30	13,657.69
Lost time allowance	0.9	weeks	409.73	368.76	308.30	277.47
Overtime	0	weeks	0	0	0	0
				18,519.80		13,935.16
Extra payments under National Working Rules	45.2	weeks				
Sick Pay	1	week				
CITB Allowance (0.50% of payroll)	1	year		104.48		78.62
Holiday pay	4.20	weeks	409.73	1,720.87	302.25	1,294.86
Public Holiday pay	1.60	weeks	409.73	655.57	302.25	493.28
Employer's contribution to:						
EasyBuild Stakeholder Pension	52	weeks	5.00	260.00	5.00	260.00
National Insurance (average weekly payment)	48	weeks	35.30	1,694.40	22.35	1,107.84
				22,955.12		17,169.76
Severance pay and sundry costs	Plus		1.5%	344.33	1.5%	257.55
				23,299.45		17,427.31
Employer's Liability and Third Party Insurance	Plus		2.0%	465.99	2.0%	348.55
Total cost per annum				£23,765.44		£17,775.86
Total cost per hour				£13.76		£10.29

Notes:

1. Absence due to sickness has been assumed to be for periods not exceeding 3 days for which no payment is due (Working Rule 20.7.3).
2. EasyBuild Stakeholder Pension effective from 1 July 2002. Death and accident benefit cover is provided free of charge. Taken as £5.00/week average as range increased for 2006/9 wage award.
3. All N.I. Payments are at not-contracted out rates applicable from April 2009. National Insurance is paid for 48 complete weeks (52 wks – 4.2 wks) based on employer making regular monthly payments into the Template holiday pay scheme and by doing so the employer achieves National Insurance savings on holiday wages.
4. At the time of preparing the book, wage rates for June 2010 are deemed to be assumed.

PART 8

Daywork and Prime Cost

When work is carried out which cannot be valued in any other way it is customary to assess the value on a cost basis with an allowance to cover overheads and profit. The basis of costing is a matter for agreement between the parties concerned, but definitions of prime cost for the building industry have been prepared and published jointly by the Royal Institution of Chartered Surveyors and the National Federation of Building Trades Employers (now the Construction Confederation) for the convenience of those who wish to use them. These documents are reproduced with the permission of the Royal Institution of Chartered Surveyors, which owns the copyright.

The daywork schedule published by the Civil Engineering Contractors Association is included in the External Works and Landscape Price Book's companion title, *Spon's Civil Engineering and Highway Works Price Book*.

For larger Prime Cost contracts the reader is referred to the form of contract issued by the Royal Institute of British Architects.

BUILDING INDUSTRY

DEFINITION OF PRIME COST OF DAYWORK CARRIED OUT UNDER A BUILDING CONTRACT (JUNE 2007 – THIRD EDITION)

This definition of Prime Cost is published by the Royal Institution of Chartered Surveyors and the Construction Confederation, for convenience and for use by people who choose to use it. Members of the Construction Confederation are not in any way debarred from defining Prime Cost and rendering their accounts for work carried out on that basis in any way they choose. Building owners are advised to reach agreement with contractors on the Definition of Prime Cost to be used prior to issuing instructions.

INTRODUCTION

This new edition of the Definition includes two options for dealing with the prime cost of labour:

Option 'A' – Percentage Addition, is based upon the traditional method of pricing labour in daywork, and allows for a percentage addition to be made for incidental costs, overheads and profit, to the prime cost of labour applicable at the time the daywork is carried out.

Option 'B' – All-inclusive Rates, includes not only the prime cost of labour but also includes an allowance for incidental costs, overheads and profit. The all-inclusive rates are deemed to be fixed for the period of the contract. However, where a fluctuating price contract is used, or where the rates in the contract are to be index-linked, the all-inclusive rates shall be adjusted by a suitable index in accordance with the contract conditions.

Model documentation, intended for inclusion in a building contract, is included in Appendix A, which illustrates how the Definition of Prime Cost may be applied in practice.

Example calculations of the Prime Cost of Labour in Daywork are given in Appendix B.

SECTION 1 – APPLICATION

1.1   This Definition provides a basis for the valuation of daywork executed under such building contracts as provide for its use.

1.2   It is not applicable in the case of daywork executed after the date of practical completion.

1.3   It is applicable to works carried out incidental to contract work but may not be deemed appropriate for use in 'daywork only' work or work carried out on an 'hourly' basis only, for which the 'Definition of Prime Cost of Building Works of a Jobbing or Maintenance Character' may be more suitable.

1.4   The terms 'contract' and 'contractor' herein shall be read as 'subcontract' and 'subcontractor' as applicable.

1.5   Dayworks are to be calculated by reference to the rate(s) current and prevailing on the day the work is carried out, except where Option 'B' for labour is used which may be adjusted by a suitable index in accordance with the contract conditions.

SECTION 2 – COMPOSITION OF TOTAL CHARGES

2.1   The prime cost of daywork comprises the sum of the following costs:

2.1.1   Labour as defined in Section 3.
2.1.2   Material and goods as defined in Section 4.
2.1.3   Plant as defined in Section 5.

2.2   Incidental costs, overheads and profit as defined in Section 6, as provided in the building contract and expressed therein as percentage adjustments are applicable to each of 2.1.1, (Option A for Labour – Section 3) – 2.1.3. NB: If using Option 'B' for the labour element of prime cost in Section 3, incidental costs, overheads and profit are deemed included.

SECTION 3 – LABOUR

Option A – Percentage Addition

3.1   The prime cost of labour is defined in 3.5. Incidental costs, overheads and profit should be added as defined in Section 6.

3.2   The standard wage rates, payments and expenses referred to below and the standard working hours referred to in 3.3 are those laid down for the time being in the rules or decisions of the Construction Industry Joint Council (CIJC) and the terms of the Building and Civil Engineering Benefits Scheme (managed by the Building and Civil Engineering Holidays Scheme Management Ltd) applicable to the works, or the rules or decisions or agreements of such body, other than the CIJC, as may be applicable relating to the grade and type of operative concerned at the time when and in the area where the daywork is executed.

3.3   Hourly base rates for labour are computed by dividing the annual prime cost of labour, based upon standard working hours and as defined in 3.5, by the number of standard working hours per annum (see Example 1 on page 476)

3.4   The hourly rates computed in accordance with 3.3 shall be applied in respect of the time spent by operatives directly engaged on daywork, including those operating mechanical plant and transport and erecting and dismantling other plant (unless otherwise expressly provided in the building contract) and handling and distributing the materials and goods used in the daywork.

3.5   The annual prime cost of labour comprises the following:

(a)   Standard or guaranteed minimum weekly earnings.*
(b)   All other guaranteed minimum payments (unless included in Section 6). *
(c)   Differentials or extra payments in respect of skill, responsibility, discomfort, inconvenience or risk (excluding those in respect of supervisory responsibility – see 3.6). *
(d)   Payments in respect of public holidays.
(e)   Any amounts which may become payable by the Contractor to or in respect of operatives arising from the operation of the rules or decisions referred to in 3.2 which are not provided for in 3.5 (a) to (d) or in Section 6. *

(f)   Employer's contributions to industry's annual holiday with pay scheme or payment in lieu thereof.

(g)   Employer's contributions to industry's welfare benefits scheme or payment in lieu thereof.

(h)   Employer's National Insurance contributions applicable to 3.5 (a) to (g).

(i)   Any contribution, levy or tax imposed by statute, payable by the contractor in his capacity as an employer, or compliance with any legislation which has a direct effect on the cost of labour. *

3.6   Differentials or extra payments in respect of supervisory responsibility are excluded from the annual prime cost (see Section 6). The time of supervisory staff such as principals, foremen, gangers, leading hands and the like, when working manually, is admissible under this Section only at the appropriate standard/normal rates for the grade of operative suitable for the operation concerned.

3.7   An example calculation of a typical standard hourly base rate is provided in Example 1 on page 476.

Non-productive Overtime

3.8   * The prime cost for non-productive overtime should be based only on the hourly payments for items marked with an asterisk in 3.5 #.

3.9   An example calculation of a typical non-productive overtime rate is précised in Example 2 on page 477.

Option B – All-inclusive Rates

3.10   The prime cost of labour is based on the all-inclusive rates for labour provided for in the building contract. The all-inclusive rates are to include all costs associated with employing the labour including all items listed in 3.5.

3.11   The all-inclusive hourly rates are also to include all costs, fixed and time-related charges, overheads and profit (as defined in Section 6) in connection with labour.

3.12   The all-inclusive hourly rates shall be applied in respect of the time actually spent by the operatives directly engaged on daywork, including those operating mechanical plant and transport and erecting and dismantling other plant (unless otherwise expressly provided in the building contract) and handling and distributing the materials and goods used in the daywork.

3.13   The time of supervisory staff, such as principals, foremen, gangers, leading hands and the like, when working manually, is admissible under this Section only at the appropriate all-inclusive hourly rates for the grade of operative suitable for the operations concerned. Any extra payment in respect of supervisory responsibility is not allowable.

3.14   The all-inclusive rates are deemed to be fixed for the period of the contract. However, where a fluctuating price contract is used, or where the rates in the contract are to be index-linked, the all-inclusive rates shall be adjusted by a suitable index in accordance with the contract conditions.

Non-productive Overtime

3.15   Allowance for non-productive overtime should be made in accordance with the Model Documentation included in Appendix A #.

SECTION 4 – MATERIALS AND GOODS

4.1   The prime cost of materials and goods obtained specifically for the daywork is the invoice cost after deducting all trade discounts and any portion of cash discounts in excess of 5%, plus any appropriate handling and delivery charges.

4.2   The prime cost of materials and goods supplied from the Contractor's stock is based upon the current market prices after deducting all trade discounts and any portion of cash discounts in excess of 5%, plus any appropriate handling charges.

4.3   Any Value Added Tax which is treated, or is capable of being treated, as input tax (as defined in the Finance Act, 1972, or any re-enactment or amendment thereof or substitution thereof) by the Contractor is excluded, for the purpose of calculations.

SECTION 5 - PLANT

5.1   Unless otherwise stated in the building contract, the prime cost of plant comprises the cost of the following:
       (a)   Use or hire of mechanical operated plant and transport for the time employed/engaged for the day-work.
       (b)   Use of non-mechanical plant (excluding non-mechanical hand tools) for the time employed/engaged for the daywork.
       (c)   Transport/delivery to and from site and erection and dismantling where applicable.
       (d)   Qualified professional operators (e.g. crane drivers) not employed by the contractor (see 5.5 below).

5.2   Where plant is hired, the prime cost of plant shall be the invoice cost after deducting all trade discounts and any portion of cash discount in excess of 5%.
5.3   Where plant is not hired, the prime cost of plant shall be calculated in accordance with the latest edition of the Royal Institution of Chartered Surveyor's (RICS) Schedule of Basic Plant Charges for Use in Connection with Daywork Under a Building Contract.
5.4   The use of non-mechanical hand tools and of erected scaffolding, staging, trestles or the like is excluded (see Section 6).
5.5   Where hired or other plant is operated by the Contractor's operatives, the operative's time is to be included under Section 3 unless otherwise provided in the contract.
5.6   Any Value Added Tax which is treated, or is capable of being treated, as input tax (as defined by the Finance Act, 1972, or any re-enactment or amendment thereof or substitution therefor) by the Contractor is excluded, for the purposes of calculation.

SECTION 6 – INCIDENTAL COSTS, OVERHEADS AND PROFIT

6.1   The percentage adjustments provided in the building contract, which are applicable to each of the totals of Sections 3 (Option A), 4 and 5, include the following: #
       (a)   Head Office charges.
       (b)   Site staff, including site supervision.
       (c)   The additional cost of overtime (other than that referred to in #).
       (d)   Time lost due to inclement weather.
       (e)   The additional cost of bonuses and all other incentive payments in excess of any guaranteed mini-mum included in 3.5 (a).
       (f)   Apprentices' study time.
       (g)   Subsistence, lodging and periodic allowances.
       (h)   Fares and travelling allowances.
       (i)   Sick pay or insurance in respect thereof.
       (j)   Third-party and employers' liability insurance.
       (k)   Liability in respect of redundancy payments to employees.
       (l)   Employers' National Insurance contributions not included in Section 3.5.
       (m)   Tool allowances.
       (n)   Use and maintenance of non-mechanical hand tools.
       (o)   Use of erected scaffolding, staging, trestles or the like.
       (p)   Use of tarpaulins, plastic sheeting or the like, all necessary protective clothing, artificial lighting, safety and welfare facilities, storage and the like that may be available on the site.
       (q)   Any variation to basic rates required by the Contractor in cases where the building contract provides for the use of a specified schedule of basic plant charges (to the extent that no other provision is made for such variation – see Section 5).
       (r)   All other liabilities and obligations whatsoever not specifically referred to in this Section nor charge-able under any other Section.
       (s)   Any variation in welfare/pension payments from industry standard.
       (t)   Profit (including main contractor's profit as appropriate).

Non-productive Overtime

6.2   When calculating the percentage adjustment for incidental costs, overheads and profit, if the Option A calculation of price cost of labour is prescribed in the contract, it should be borne in mind that not all items listed in 6.1 are necessarily applicable to non-productive overtime. When Option B is prescribed, non-productive overtime should be shown separately in the contract documents as detailed in the Model Documentation in Appendix A

The additional cost of non-productive overtime, where specifically ordered by the Architect/Supervising Officer/ Contract Administrator/Employer's Agent, shall only be chargeable on the terms of prior written agreement between the parties to the building contract.

APPENDIX A

Model Documentation for Inclusion in a Building Contract

This model document is included to illustrate how the Definition of Prime Cost may be applied in practice. It does not form part of the Definition. It is, however, in a form agreed between the RICS and the Construction Confederation and its use in this form amended only as required to suit the specific building contract is encouraged.

Where Using Option A for Labour

Dayworks

The Contractor will be paid as defined below for the cost of works carried out as daywork in accordance with the building contract.

For building works, the prime cost of daywork will be calculated in accordance with the latest *Definition of Prime Cost of Daywork carried out under a Building Contract, (State edition_____)*, published by the Royal Institution of Chartered Surveyors and the Construction Confederation.

For electrical works, the prime cost of daywork will be calculated in accordance with the latest *Definition of Prime Cost of Daywork carried out under an Electrical Contract, (State edition_____)*, published by the Royal Institution of Chartered Surveyors, the Electrical Contractors' Association and 'SELECT' the Electrical Contractors' Association of Scotland.

For heating and ventilating work etc, the prime cost of daywork will be calculated in accordance with the latest *Definition of Prime Cost of Daywork carried out under a Heating, Ventilating, Air Conditioning, Refrigeration, Pipework and/or Domestic Engineering Contract, (State edition_____)*, published by the Royal Institution of Chartered Surveyors and the Heating and Ventilating Contractors' Association.

For plumbing work, the prime cost of daywork will be calculated in accordance with the latest *Definition of Prime Cost of Daywork carried out under a Plumbing Contract, (State edition_____)*, published by the Royal Institution of Chartered Surveyors, the Association of Plumbing and Heating Contractors and the Scottish and Northern Ireland Plumbing Employers' Confederation.

Labour

| Building Operatives | Provisional Sum | £ |
| Add for Incidental Costs, Overheads and Profit | ....% | £ |

| Electrical Operatives | Provisional Sum | £ |
| Add for Incidental Costs, Overheads and Profit | ....% | £ |

| Heating and Ventilating Operatives | Provisional Sum | £ |
| Add for Incidental Costs, Overheads and Profit | ....% | £ |

| Plumbing Operatives | Provisional Sum | £ |
| Add for Incidental Costs, Overheads and Profit | ....% | £ |

Non-productive Overtime

| Building Operatives | Provisional Sum | £ |
| Add for Incidental Costs, Overheads and Profit | ....% | £ |

| Electrical Operatives | Provisional Sum | £ |
| Add for Incidental Costs, Overheads and Profit | ....% | £ |

| Heating and Ventilating Operatives | Provisional Sum | £ |
| Add for Incidental Costs, Overheads and Profit | ....% | £ |

| Plumbing Operatives | Provisional Sum | £ |
| Add for Incidental Costs, Overheads and Profit | ....% | £ |

Where using Option B for Labour

Dayworks

The Contractor will be paid as defined below for the cost of works carried out as daywork in accordance with the building contract.

For building works, the prime cost of daywork will be calculated in accordance with the latest *Definition of Prime Cost of Daywork carried out under a Building Contract, (State edition_____)*, published by the Royal Institution of Chartered Surveyors and the Construction Confederation.

For electrical works, the prime cost of daywork will be calculated in accordance with the latest *Definition of Prime Cost of Daywork carried out under an Electrical Contract, (State edition_____)*, published by the Royal Institution of Chartered Surveyors, the Electrical Contractors' Association and 'SELECT' the Electrical Contractors' Association of Scotland.

For heating and ventilating work etc, the prime cost of daywork will be calculated in accordance with the latest *Definition of Prime Cost of Daywork carried out under a Heating, Ventilating, Air Conditioning, Refrigeration, Pipe-work and/or Domestic Engineering Contract, (State edition_____), published by the Royal Institution of Chartered Surveyors and the Heating and Ventilating Contractors' Association.

For plumbing work, the prime cost of daywork will be calculated in accordance with the latest *Definition of Prime Cost of Daywork carried out under a Plumbing Contract, (State edition_____), published by the Royal Institution of Chartered Surveyors, the Association of Plumbing and Heating Contractors and the Scottish and Northern Ireland Plumbing Employers' Confederation.

Labour

The Contractor must state below the all-inclusive prime cost hourly rates required for labour as defined in Section 3 (Option B) and the core working ours to which they apply.

Core Hours

General Operatives	£............ per hour
Skilled Operatives (all grades)	£............ per hour
Craft Operatives	£............ per hour

Other Grades/Trades:

....	£............ per hour
....	£............ per hour
....	£............ per hour
....	£............ per hour
....	£............ per hour

Core hours are ____am to ____pm Monday to Friday (excluding statutory holidays)

Overtime specifically ordered by the Architect/Supervising Officer/Contract Administrator/ Employer's Agent

The non-productive element of overtime should be as defined in the relevant Working Rule Agreement. However, if different, please state below.

Trade	Day	Time	Non-Productive Element (hours)
....	....	.........to.........	....
....	....	.........to.........	....
....	....	.........to.........	....
....	....	.........to.........	....
....	....	.........to.........	....
....	....	.........to.........	....
....	....	.........to.........	....

Provide the all-inclusive prime cost of labour as defined in Section 3 (Option B)

Productive Hours

[ ] hours (Provisional) General Operatives   @ £.........per hour   £

[ ] hours (Provisional) General Operatives   @ £.........per hour   £

[ ] hours (Provisional) General Operatives   @ £.........per hour   £

Other Grades/Trades:

[ ] hours (Provisional) General Operatives   @ £.........per hour   £

[ ] hours (Provisional) General Operatives   @ £.........per hour   £

[ ] hours (Provisional) General Operatives   @ £.........per hour   £

Non-productive Hours

[ ] hours (Provisional) General Operatives   @ £.........per hour   £

[ ] hours (Provisional) General Operatives   @ £.........per hour   £

[ ] hours (Provisional) General Operatives   @ £.........per hour   £

Other Grades/Trades:

[ ] hours (Provisional) General Operatives   @ £.........per hour   £

[ ] hours (Provisional) General Operatives   @ £.........per hour   £

[ ] hours (Provisional) General Operatives   @ £.........per hour   £

Materials and Goods

Provide for the prime cost of materials and goods
as defined in Section 4 (Provisional)   £ [ ]

Add the percentage addition for incidental costs,
overheads and profit as defined in Section 6   _____%

Plant

Provide for the prime cost of plant hired by the
Contractor as defined in Section 5 (Provisional)   £ [ ]

Add the percentage addition for incidental costs,
overheads and profit as defined in Section 6   _____%

Rates for plant not hired by the Contractor shall be as set out in *The Schedule of Basic Plant Charges for Use in Connection with Daywork Under a Building Contract* published by the Royal Institution of Chartered Surveyors (_____ Edition dated _____)

Provide for the prime cost of plant not hired by the
Contractor, as defined in Section 5 (Provisional)   £ [ ]

Add the percentage addition for incidental costs,
overheads and profit as defined in Section 6   _____%

Daywork Rates – Building Operatives

APPENDIX B

Example Calculations of Prime Cost of Labour in Daywork

Example 1

Option A

Example of calculation of typical standard hourly base rate (as defined in Section 3) for CIJC Building Craft Operative and General Operative based upon rates applicable 30th June 2010 – assumed.

		Rate (£)	Craft Operative	Rate (£)	General Operative
Basic Wages:	46.2 weeks	409.73	£18,929.53	308.30	£14,243.46
Extra Payments:	Where applicable		0.00		0.00
Sub Total:			£18,929.53		£14,243.46
National Insurance:	12.80% above earnings threshold (ET) (46.2 wks @£110.01pw)		£1,724.94		£1,136.83
Holidays with Pay:	226 hours	10.51	£2,375.26	7.91	£1,787.66
Welfare Benefit:	52 weeks stamps	11.00	£572.00	11.00	£572.00
CITB Levy:	0.5% of payroll		£104.43		£78.58
Annual labour cost:			**£23,706.16**		**£17,818.53**
Hourly Base Rate:			**£13.16**		**£9.89**

For the convenience of readers, the example which appears on the previous page has been updated by the Editors for rates applicable 30th June 2010 – assumed

Note:

(1)   Standard working hours per annum calculated as follows:

52 weeks @ 39 hours	2028
Less \	
hours annual holiday	163
hours public holiday	63
Standard working hours per year	1802

(2)   It has been assumed that employers who follow the CIJC Working Rules Agreement will match the employee pension contributions (part of welfare benefit) between £3.00 and £10.00 per week. Furthermore it has been assumed that employees have contributed £10.00 per week to the pension scheme and £1.00 per week for life insurance.

(3)   It should be noted that all labour costs incurred by the Contractor in his capacity as an employer other than those contained in the hourly base rate, are to be taken into account under Section 6.

(4)   The above example is for the convenience of users only and does not form part of the Definition; all the basic costs are subject to re-examination according to the time when and in the area where the daywork is executed.

Example 2

Non-productive Overtime

Option A

Example of calculation of typical non-productive overtime rate (as defined in section 3) for CIJC Building Craft Operative and General Operative based upon rates applicable 6th April 2007.

		Rate (£)	Craft Operative	Rate (£)	General Operative
Basic Wages:	46.2 weeks	409.73	£18,929.53	308.30	£14,243.46
Extra Payments:	Where applicable		0.00		0.00
Sub Total:			£18,929.53		£14,243.46
National Insurance:	12.80% above earnings threshold (ET) (46.2 wks @£110.01pw)		£1,724.94		£1,136.83
CITB Levy:	0.5% of payroll		£92.79		£69.82
Annual labour cost:			**£20,747.26**		**£15,450.11**
Hourly Base Rate:			**£11.51**		**£8.57**

For the convenience of readers, the example which appears on the previous page has been updated by the Editors for rates applicable 30th June 2010 – assumed.

Note:

(1)   Standard working hours per annum calculated as follows:

52 weeks @ 39 hours	2028
Less \	
hours annual holiday	163
hours public holiday	63
Standard working hours per year	1802

(2)   It should be noted that all labour costs incurred by the Contractor in his capacity as an employer other than those contained in the hourly base rate, are to be taken into account under Section 6.

(3)   The above example is for the convenience of users only and does not form part of the Definition; all the basic costs are subject to re-examination according to the time when and in the area where the daywork is executed.

(4)   The above example is for the convenience of users only and does not form part of the Definition; all the basic costs are subject to re-examination according to the time when and in the area where the daywork is executed.

European Landscape Architecture

Best Practice in Detailing

Edited by **Ian Thompson**, **Torben Dam**, **Jens Balsby Nielsen**

Drawing together case studies from all over Europe, this text explores the relationship between the overall idea of the landscape architecture for a site and the design of details.

Examining concept sketches and design development drawings in relation to the details of the design, the book offers a more profound understanding of decision making through all stages of the design process.

The book includes the study of the choice of materials and techniques of construction, and explores the cultural and symbolic significance of such choices, as well as questions of environmental sustainability.

With projects analyzed and evaluated here that have won international acclaim, or have been awarded national prizes, *European Landscape Architecture* is a core book in the study and understanding of the subject.

2007: 250x250: 272pp
Hb: 978-0-415-30736-9: **£95.00**
Pb: 978-0-415-30737-6: **£35.99**

PART 9

Tables and Memoranda

This part of the book contains the following sections:

CONVERSION TABLES
QUICK REFERENCE CONVERSION TABLES

	Imperial		Metric	
1. LINEAR				
0.039	in	1	mm	25.4
3.281	ft	1	metre	0.305
1.094	yd	1	metre	0.914
2. WEIGHT				
0.020	cwt	1	kg	50.802
0.984	ton	1	tonne	1.016
2.205	lb	1	kg	0.454
3. CAPACITY				
1.760	pint	1	litre	0.568
0.220	gal	1	litre	4.546
4. AREA				
0.002	in²	1	mm²	645.16
10.764	ft²	1	m²	0.093
1.196	yd²	1	m²	0.836
2.471	acre	1	ha	0.405
0.386	mile²	1	km²	2.59
5. VOLUME				
0.061	in³	1	cm³	16.387
35.315	ft³	1	m³	0.028
1.308	yd³	1	m³	0.765
6. POWER				
1.310	HP	1	kW	0.746

CONVERSION FACTORS – METRIC TO IMPERIAL

Multiply Metric	Unit	By	To obtain Imperial	Unit
Length				
kilometre	km	0.6214	statute mile	ml
metre	m	1.0936	yard	yd
centimetre	cm	0.0328	foot	ft
millimetre	mm	0.0394	inch	in
Area				
hectare	ha	2.471	acre	
square kilometre	km²	0.3861	square mile	sq ml
square metre	m²	10.764	square foot	sq ft
square metre	m²	1550	square inch	sq in
square centimetre	cm²	0.155	square inch	sq in
Volume				
cubic metre	m³	35.336	cubic foot	cu ft
cubic metre	m³	1.308	cubic yard	cu yd
cubic centimetre	cm³	0.061	cubic inch	cu in
cubic centimetre	cm³	0.0338	fluid ounce	fl oz
Liquid volume				
litre	l	0.0013	cubic yard	cu yd
litre	l	61.02	cubic inch	cu in
litre	l	0.22	Imperial gallon	gal
litre	l	0.2642	US gallon	US gal
litre	l	1.7596	pint	pt
Mass				
metric tonne	t	0.984	long ton	lg ton
metric tonne	t	1.102	short ton	sh ton
kilogram	kg	2.205	pound, avoirdupois	lb
gram	g or gr	0.0353	ounce, avoirdupois	oz
Unit mass				
kilograms/cubic metre	kg/m³	0.062	pounds/cubic foot	lbs/cu ft
kilograms/cubic metre	kg/m³	1.686	pounds/cubic yard	lbs/cu yd
tonnes/cubic metre	t/m³	1692	pound/cubic yard	lbs/cu yd
kilograms/sq centimetre	kg/cm²	14.225	pounds/square inch	lbs/sq in
kilogram-metre	kg.m	7.233	foot-pound	ft-lb

Multiply Metric	Unit	By	To obtain Imperial	Unit
Force				
meganewton	MN	9.3197	tons force	tonf
kilonewton	kN	225	pounds force	lbf
newton	N	0.225	pounds force	lbf
Pressure and stress				
meganewton per square metre	MN/m²	9.3197	tons force/square foot	tonf/ft²
kilopascal	kPa	0.145	pounds/square inch	psi
bar		14.5	pounds/square inch	psi
kilograme metre	kgm	7.2307	foot-pound	ft-lb
Energy				
kilocalorie	kcal	3.968	British thermal unit	Btu
metric horsepower	CV	0.9863	horsepower	hp
kilowatt	kW	1.341	horsepower	hp
Speed				
kilometres/hour	km/h	0.621	miles/hour	mph

CONVERSION FACTORS – IMPERIAL TO METRIC

Multiply Imperial	Unit	By	To obtain metric	Unit
Length				
statute mile	ml	1.609	kilometre	km
yard	yd	0.9144	metre	m
foot	ft	0.3048	metre	m
inch	in	25.4	millimetre	mm
Area				
acre	acre	0.4047	hectare	ha
square mile	sq ml	2.59	square kilometre	km^2
square foot	sq ft	0.0929	square metre	m^2
square inch	sq in	0.0006	square metre	m^2
square inch	sq in	6.4516	square centimetre	cm^2
Volume				
cubic foot	cu ft	0.0283	cubic metre	m^3
cubic yard	cu yd	0.7645	cubic metre	m^3
cubic inch	cu in	16.387	cubic centimetre	cm^3
fluid ounce	fl oz	29.57	cubic centimetre	cm^3
Liquid volume				
cubic yard	cu yd	764.55	litre	l
cubic inch	cu in	0.0164	litre	l
Imperial gallon	gal	4.5464	litre	l
US gallon	US gal	3.785	litre	l
US gallon	US gal	0.833	Imperial gallon	gal
pint	pt	0.5683	litre	l
Mass				
long ton	lg ton	1.016	metric tonne	tonne
short ton	sh ton	0.907	metric tonne	tonne
pound	lb	0.4536	kilogram	kg
ounce	oz	28.35	gram	g
Unit mass				
pounds/cubic foot	lb/cu ft	16.018	kilograms/cubic metre	kg/m^3
pounds/cubic yard	lb/cu yd	0.5933	kilograms /cubic metre	kg/m^3
pounds/cubic yard	lb/cu yd	0.0006	tonnes/cubic metre	t/m^3
foot-pound	ft-lb	0.1383	kilogram-metre	kg.m

Multiply Imperial	Unit	By	To obtain metric	Unit
Force				
tons force	tonf	0.1073	meganewton	MN
pounds force	lbf	0.0045	kilonewton	kN
pounds force	lbf	4.45	newton	N
Pressure and stress				
pounds/square inch	psi	0.1073	kilogram/sq. centimetre	kg/cm²
pounds/square inch	psi	6.89	kilopascal	kPa
pounds/square inch	psi	0.0689	bar	
foot-pound	ft-lb	0.1383	kilogram metre	kgm
Energy				
British thermal unit	Btu	0.252	kilocalorie	kcal
horsepower (hp)	hp	1.014	metric horsepower	CV
horsepower (hp)	hp	0.7457	kilowatt	kW
Speed				
miles/hour	mph	1.61	kilometres/hour	km/h

Length

millimetre	mm	1 in	=	25.4 cm	1 mm	=	0.0394 in
centimetre	cm	1 in	=	2.54 cm	1 cm	=	0.3937 in
metre	m	1 ft	=	0.3048 m	1 m	=	3.2808 ft
		1 yd	=	0.9144 m	1 m	=	1.0936 yd
kilometre	km	1 mile	=	1.6093 km	1 km	=	0.6214 mile
	Note:	1 cm	=	10 mm	1 ft	=	12 in
		1 m	=	100 cm	1 yd	=	3 ft
		1 km	=	1,000 m	1 mile	=	1,760 yd

Area

square millimetre	mm²	1 in²	=	645.2 mm²	1 mm²	=	0.0016 in²
square centimetre	cm²	1 in²	=	6.4516 cm²	1 cm²	=	1.1550 in²
square metre	m²	1 ft²	=	0.0929 m²	1 m²	=	10.764 ft²
		1 yd²	=	0.8361 m²	1 m²	=	1.1960 yd²
square kilometre	km²	1 mile²	=	2.590 km²	1 km²	=	0.3861 mile²
	Note:	1 cm²	=	100 m²	1 ft²	=	144 in²
		1 m²	=	10,000 cm²	1 yd²	=	9 ft²
		1 km²	=	100 hectares	1 mile²	=	640 acres
					1 acre	=	4,840 yd²

Volume

cubic centimetre	cm³	1 cm³	=	0.0610 in³	1 in³	=	16.387 cm³
cubic decimetre	dm³	1 dm³	=	0.0353 ft³	1 ft³	=	28.329 dm³
cubic metre	m³	1 m³	=	35.3147 ft³	1 ft³	=	0.0283 m³
		1 m³	=	1.3080 yd³	1 yd³	=	0.7646 m³
litre	L	1 L	=	1.76 pint	1 pint	=	0.5683 L
		1 L	=	2.113 US pt	1 pint	=	0.4733 L
	Note:	1 dm³	=	1,000 cm³	1 ft³	=	1.728 in³
		1 m³	=	1,000 dm³	1 yd³	=	27 ft³
		1 L	=	1 dm³	1 pint	=	20 fl oz
		1 HL	=	100 L	1 gal	=	8 pints

Mass

milligram	mg	1 mg	= 0.0154 grain	1 grain	= 64.935 mg
gram	g	1 g	= 0.0353 oz	1 oz	= 28.35 g
kilogram	kg	1 kg	= 2.2046 lb	1 lb	= 0.4536 kg
tonne	t	1 t	= 0.9842 ton	1 ton	= 1.016 t

Note:

	1 g	= 1,000 mg	1 oz	= 437.5 grains
	1 kg	= 1000 g	1 lb	= 16 oz
	1 t	= 1,000 kg	1 stone	= 14 lb
			1 cwt	= 112 lb
			1 ton	= 20 cwt

Force

newton	N	1 lbf	= 4.448 N	1 kgf	= 9.807 N
kilonewton	kN	1 lbf	= 0.00448 kN	1 tonf	= 9.964 kN
meganewton	MN	100 tonf	= 0.9964 MN		

Pressure and stress

kilonewton per square metre		1 lbf/in²	= 6.895 kN/m²
	kN/m²	1 bar	= 100 kN/m²
meganewton per square metre		1 tonf/ft²	= 107.3 kN/m² = 0.1073 MN/m²
	MN/m²	1 kgf/cm²	= 98.07 kN/m²
		1 lbf/ft²	= 0.04788 kN/m²

Temperature

degree Celsius °C

$$^{\circ}C = \frac{5 \times (^{\circ}F - 32)}{9} \qquad ^{\circ}F = \frac{(9 \times {}^{\circ}C) + 32}{5}$$

Metric Equivalents

1 km	=	1000 m
1 m	=	100 cm
1 cm	=	10 mm
1 km²	=	100 ha
1 ha	=	10,000 m²
1 m²	=	10,000 cm²
1 cm²	=	100 mm²
1 m³	=	1,000 litres
1 litre	=	1,000 cm³
1 metric tonne	=	1,000 kg
1 quintal	=	100 kg
1 N	=	0.10197 kg
1 kg	=	1000 g
1 g	=	1000 mg
1 bar	=	14.504 psi
1 cal	=	427 kg.m
1 cal	=	0.0016 cv.h
torque unit	=	0.00116 kw.h
1 CV	=	75 kg.m/s
1 kg/cm²	=	0.97 atmosphere

English Unit Equivalents

1 mile	=	1760 yd
1 yd	=	3 ft
1 ft	=	12 in
1 sq mile	=	640 acres
1 acre	=	43,560 sq ft
1 sq ft	=	144 sq in
1 cu ft	=	7.48 gal liq
1 gal	=	231 cu in
	=	4 quarts liq
1 quart	=	32 fl oz
1 fl oz	=	1.80 cu in
	=	437.5 grains
1 stone	=	14 lb
1 cwt	=	112 lb
1 sh ton	=	2000 lb
1 lg ton	=	2240 lb
	=	20 cwt
1 lb	=	16 oz, avdp
1 Btu	=	778 ft lb
	=	0.000393 hph
	=	0.000293 kwh
1 hp	=	550 ft-lb/sec
1 atmosph	=	14.7 lb/in²

Power Units		
kW	=	Kilowatt
HP	=	Horsepower
CV	=	Cheval Vapeur (Steam Horsepower)
	=	French designation for Metric Horsepower
PS	=	Pferdestarke (Horsepower)
	=	German designation of Metric Horsepower
1 HP	=	1.014 CV = 1.014 PS
	=	0.7457 kW
1 PS	=	1 CV = 0.9863 HP
	=	0.7355 kW
1 kW	=	1.341 HP
	=	1.359 CV
	=	1.359 PS

SPEED CONVERSION

km/h	m/min	mph	fpm
1	16.7	0.6	54.7
2	33.3	1.2	109.4
3	50.0	1.9	164.0
4	66.7	2.5	218.7
5	83.3	3.1	273.4
6	100.0	3.7	328.1
7	116.7	4.3	382.8
8	133.3	5.0	437.4
9	150.0	5.6	492.1
10	166.7	6.2	546.8
11	183.3	6.8	601.5
12	200.0	7.5	656.2
13	216.7	8.1	710.8
14	233.3	8.7	765.5
15	250.0	9.3	820.2
16	266.7	9.9	874.9
17	283.3	10.6	929.6
18	300.0	11.2	984.3
19	316.7	11.8	1038.9
20	333.3	12.4	1093.6
21	350.0	13.0	1148.3
22	366.7	13.7	1203.0
23	383.3	14.3	1257.7
24	400.0	14.9	1312.3
25	416.7	15.5	1367.0
26	433.3	16.2	1421.7
27	450.0	16.8	1476.4
28	466.7	17.4	1531.1
29	483.3	18.0	1585.7
30	500.0	18.6	1640.4
31	516.7	19.3	1695.1
32	533.3	19.9	1749.8
33	550.0	20.5	1804.5
34	566.7	21.1	1859.1
35	583.3	21.7	1913.8
36	600.0	22.4	1968.5
37	616.7	23.0	2023.2
38	633.3	23.6	2077.9
39	650.0	24.2	2132.5
40	666.7	24.9	2187.2

km/h	m/min	mph	fpm
41	683.3	25.5	2241.9
42	700.0	26.1	2296.6
43	716.7	26.7	2351.3
44	733.3	27.3	2405.9
45	750.0	28.0	2460.6
46	766.7	28.6	2515.3
47	783.3	29.2	2570.0
48	800.0	29.8	2624.7
49	816.7	30.4	2679.4
50	833.3	31.1	2734.0

FORMULAE

Two dimensional figures

Figure	Diagram of figure	Surface area	Perimeter
Square		a^2	$4a$
Rectangle		ab	$2(a + b)$
Triangle		$\frac{1}{2}ch$	$a + b + c$
Circle		πr^2 $\frac{1}{4}\pi d^2$ where $2r = d$	$2\pi r$ πd
Parallelogram		ah	$2(a + b)$
Trapezium		$\frac{1}{2}h(a + b)$	$a + b + c + d$
Ellipse		Approximately πab	$\pi(a + b)$
Hexagon		$2.6 \times a^2$	

Figure	Diagram of figure	Surface area	Perimeter
Octagon		$4.83 \times a^2$	
Sector of a circle		$\frac{1}{2}rb$ or $\frac{q}{360}\pi r^2$ note b = angle $\frac{q}{360} \times \pi 2r$	
Segment of a circle		S – T where S = area of sector, T = area of triangle	
Bellmouth		$\frac{3}{14} \times r^2$	

Three dimensional figures

Figure	Diagram of figure	Surface area	Volume
Cube		$6a^2$	a^3
Cuboid/ rectangular block		$2(ab + ac + bc)$	abc
Prism/ triangular block		$bd + hc + dc + ad$	$\frac{1}{2}hcd$

Figure	Diagram of figure	Surface area	Volume
Cylinder		$2\pi r^2 + 2\pi h$	$\pi r^2 h$ $\frac{1}{4}\pi d^2 h$
Sphere		$4\pi r^2$	$\frac{4}{3}\pi r^3$
Segment of sphere		$2\pi R h$	$\frac{1}{6}\pi h(3r^2 + h^2)$ $\frac{1}{3}\pi h^2(3R - H)$
Pyramid		$(a + b)l + ab$	$\frac{1}{3}abh$
Frustum of a pyramid		$l(a + b + c + d) + \sqrt{(ab + cd)}$ [rectangular figure only]	$\frac{h}{3}(ab + cd + \sqrt{abcd})$

Figure	Diagram of figure	Surface area	Volume
Cone		πrl (excluding base) $\pi rl + \pi r^2$ (including base)	$\frac{1}{3}\pi r^2 h$ $\frac{1}{12}\pi d^2 h$
Frustum of a cone		$\pi r^2 + \pi R^2 + \pi l(R + r)$	$\frac{1}{3}\pi(R^2 + Rr + r^2)$

Formula	Description
Pythagoras theorem	$A^2 = B^2 + C^2$ where A is the hypotenuse of a right-angled triangle and B and C are the two adjacent sides
Simpsons Rule	The Area is divided into an even number of strips of equal width, and therefore has an odd number of ordinates at the division points area $= \dfrac{S(A + 2B + 4C)}{3}$ where S = common interval (strip width) A = sum of first and last ordinates B = sum of remaining odd ordinates C = sum of the even ordinates The Volume can be calculated by the same formula, but by substituting the area of each coordinate rather than its length.
Trapezoidal Rule	A given trench is divided into two equal sections, giving three ordinates, the first, the middle and the last. volume $= \dfrac{S \times (A + B + 2C)}{2}$ where S = width of the strips A = area of the first section B = area of the last section C = area of the rest of the sections
Prismoidal Rule	A given trench is divided into two equal sections, giving three ordinates, the first, the middle and the last. volume $= \dfrac{L \times (A + 4B + C)}{6}$ where L = total length of trench A = area of the first section B = area of the middle section C = area of the last section

EARTHWORK

Weights of Typical Materials Handled by Excavators

The weight of the material is that of the state in its natural bed and includes moisture. Adjustments should be made to allow for loose or compacted states.

Material	kg/m³	lb/cu yd
Adobe	1914	3230
Ashes	610	1030
Asphalt, rock	2400	4050
Basalt	2933	4950
Bauxite: alum ore	2619	4420
Borax	1730	2920
Caliche	1440	2430
Carnotite	2459	4150
Cement	1600	2700
Chalk (hard)	2406	4060
Cinders	759	1280
Clay: dry	1908	3220
Clay: wet	1985	3350
Coal: bituminous	1351	2280
Coke	510	860
Conglomerate	2204	3720
Dolomite	2886	4870
Earth: dry	1796	3030
Earth: moist	1997	3370
Earth: wet	1742	2940
Feldspar	2613	4410
Felsite	2495	4210
Fluorite	3093	5220
Gabbro	3093	5220
Gneiss	2696	4550
Granite	2690	4540
Gravel, dry	1790	3020
Gypsum	2418	4080
Hardcore (consolidated)	1928	120
Lignite broken	1244	2100
Limestone	2596	4380
Magnesite, magnesium ore	2993	5050
Marble	2679	4520
Marl	2216	3740
Peat	700	1180

Material	kg/m³	lb/cu yd
Potash	2193	3700
Pumice	640	1080
Quarry waste	1438	90
Quartz	2584	4360
Rhyolite	2400	4050
Sand: dry	1707	2880
Sand: wet	1831	3090
Sand and gravel – dry	1790	3020
– wet	2092	3530
Sandstone	2412	4070
Schist	2684	4530
Shale	2637	4450
Slag (blast)	2868	4840
Slate	2667	4500
Snow – dry	130	220
– wet	510	860
Taconite	3182	5370
Topsoil	1440	2430
Trachyte	2400	4050
Traprock	2791	4710
Water	1000	62

Transport Capacities

Type of vehicle	Capacity of vehicle	
	Payload	Heaped capacity
Wheelbarrow	150	0.10
1 tonne dumper	1250	1.00
2.5 tonne dumper	4000	2.50
Articulated dump truck (Volvo A20 6 × 4)	18500	11.00
Articulated dump truck (Volvo A35 6 × 6)	32000	19.00
Large capacity rear dumper (Euclid R35)	35000	22.00
Large capacity rear dumper (Euclid R85)	85000	50.00

Machine Volumes for Excavating and Filling

Machine type	Cycles per minute	Volume per minute (m^3)
1.5 tonne excavator	1	0.04
	2	0.08
	3	0.12
3 tonne excavator	1	0.13
	2	0.26
	3	0.39
5 tonne excavator	1	0.28
	2	0.56
	3	0.84
7 tonne excavator	1	0.28
	2	0.56
	3	0.84
21 tonne excavator	1	1.21
	2	2.42
	3	3.63
Backhoe loader JCB3CX excavator Rear bucket capacity 0.28 m^3	1	0.28
	2	0.56
	3	0.84
Backhoe loader JCB3CX loading Front bucket capacity 1.00 m^3	1	1.00
	2	2.00

Machine Volumes for Excavating and Filling

Machine type	Loads per hour	Volume per hour (m³)
1 tonne high tip skip loader	5	2.43
Volume 0.485 m³	7	3.40
	10	4.85
3 tonne dumper	4	7.60
Max. volume 2.40 m³	5	9.50
Available volume 1.9 m³	7	13.30
	10	19.00
6 tonne dumper	4	15.08
Max. volume 3.40 m³	5	18.85
Available volume 3.77 m³	7	26.39
	10	37.70

Bulkage of Soils (After excavation)

Type of soil	Approximate bulking of 1 m 3 after excavation
Vegetable soil and loam	25–30%
Soft clay	30–40%
Stiff clay	10–20%
Gravel	20–25%
Sand	40–50%
Chalk	40–50%
Rock, weathered	30–40%
Rock, unweathered	50–60%

Shrinkage of Materials (On being deposited)

Type of soil	Approximate bulking of 1 m 3 after excavation
Clay	10%
Gravel	8%
Gravel and sand	9%
Loam and light sandy soils	12%
Loose vegetable soils	15%

Voids in Material Used as Subbases or Beddings

Material	m³ of voids/m³
Alluvium	0.37
River grit	0.29
Quarry sand	0.24
Shingle	0.37
Gravel	0.39
Broken stone	0.45
Broken bricks	0.42

Angles of Repose

Type of soil		Degrees
Clay	– dry	30
	– damp, well drained	45
	– wet	15–20
Earth	– dry	30
	– damp	45
Gravel	– moist	48
Sand	– dry or moist	35
	– wet	25
Loam		40

Slopes and Angles

Ratio of base to height	Angle in degrees
5:1	11
4:1	14
3:1	18
2:1	27
1½:1	34
1:1	45
1:1½	56
1:2	63
1:3	72
1:4	76
1:5	79

Grades (In Degrees and Percents)

Degrees	Percent	Degrees	Percent
1	1.8	24	44.5
2	3.5	25	46.6
3	5.2	26	48.8
4	7.0	27	51.0
5	8.8	28	53.2
6	10.5	29	55.4
7	12.3	30	57.7
8	14.0	31	60.0
9	15.8	32	62.5
10	17.6	33	64.9
11	19.4	34	67.4
12	21.3	35	70.0
13	23.1	36	72.7
14	24.9	37	75.4
15	26.8	38	78.1
16	28.7	39	81.0
17	30.6	40	83.9
18	32.5	41	86.9
19	34.4	42	90.0
20	36.4	43	93.3
21	38.4	44	96.6
22	40.4	45	100.0

Bearing Powers

Ground conditions		Bearing power		
		kg/m^2	lb/in^2	Metric t/m^2
Rock	broken	483	70	50
	solid	2,415	350	240
Clay,	dry or hard	380	55	40
	medium dry	190	27	20
	soft or wet	100	14	10
Gravel,	cemented	760	110	80
Sand,	compacted	380	55	40
	clean dry	190	27	20
Swamp and alluvial soils		48	7	5

Earthwork Support

Maximum depth of excavation in various soils without the use of earthwork support

Ground conditions	Feet (ft)	Metres (m)
Compact soil	12	3.66
Drained loam	6	1.83
Dry sand	1	0.3
Gravelly earth	2	0.61
Ordinary earth	3	0.91
Stiff clay	10	3.05

It is important to note that the above table should only be used as a guide. Each case must be taken on its merits and, as the limited distances given above are approached, careful watch must be kept for the slightest signs of caving in.

CONCRETE WORK

Weights of Concrete and Concrete Elements

Type of material		kg/m³	lb/cu ft
Ordinary concrete (dense aggregates)			
Non-reinforced plain or mass concrete			
Nominal weight		2305	144
Aggregate	– limestone	2162 to 2407	135 to 150
	– gravel	2244 to 2407	140 to 150
	– broken brick	2000 (av)	125 (av)
	– other crushed stone	2326 to 2489	145 to 155
Reinforced concrete			
Nominal weight		2407	150
Reinforcement	– 1%	2305 to 2468	144 to 154
	– 2%	2356 to 2519	147 to 157
	– 4%	2448 to 2703	153 to 163
Special concretes			
Heavy concrete			
Aggregates	– barytes, magnetite	3210 (min)	200 (min)
	– steel shot, punchings	5280	330
Lean mixes			
Dry-lean (gravel aggregate)		2244	140
Soil-cement (normal mix)		1601	100

Weights of Concrete and Concrete Elements – cont

Type of material		kg/m² per mm thick	lb/sq ft per inch thick
Ordinary concrete (dense aggregates)			
Solid slabs (floors, walls etc.)			
Thickness:	75 mm or 3 in	184	37.5
	100 mm or 4 in	245	50
	150 mm or 6 in	378	75
	250 mm or 10 in	612	125
	300 mm or 12 in	734	150
Ribbed slabs			
Thickness:	125 mm or 5 in	204	42
	150 mm or 6 in	219	45
	225 mm or 9 in	281	57
	300 mm or 12 in	342	70
Special concretes			
Finishes etc.			
	Rendering, screed etc. Granolithic, terrazzo	1928 to 2401	10 to 12.5
	Glass-block (hollow) concrete	1734 (approx)	9 (approx)
Prestressed concrete		Weights as for reinforced concrete (upper limits)	
Air-entrained concrete		Weights as for plain or reinforced concrete	

Average Weight of Aggregates

Materials	Voids %	Weight kg/m³
Sand	39	1660
Gravel 10–20 mm	45	1440
Gravel 35–75 mm	42	1555
Crushed stone	50	1330
Crushed granite (over 15 mm)	50	1345
(n.e. 15 mm)	47	1440
'All-in' ballast	32	1800–2000

Material	kg/m³	lb/cu yd
Vermiculite (aggregate)	64–80	108–135
All-in aggregate	1999	125

Applications and mix design
Site mixed concrete

Recommended mix	Class of work suitable for	Cement (kg)	Sand (kg)	Coarse aggregate (kg)	Nr 25 kg bags cement per m³ of combined aggregate
1:3:6	Roughest type of mass concrete such as footings, road haunching over 300 mm thick	208	905	1509	8.30
1:2.5:5	Mass concrete of better class than 1:3:6 such as bases for machinery, walls below ground etc.	249	881	1474	10.00
1:2:4	Most ordinary uses of concrete, such as mass walls above ground, road slabs etc. and general reinforced concrete work	304	889	1431	12.20
1:1.5:3	Watertight floors, pavements and walls, tanks, pits, steps, paths, surface of 2 course roads, reinforced concrete where extra strength is required	371	801	1336	14.90
1:1:2	Works of thin section such as fence posts and small precast work	511	720	1206	20.40

Applications and mix design
Ready mixed concrete

Application	Designated concrete	Standardized prescribed concrete	Recommended consistence (nominal slump class)
Foundations			
Mass concrete fill or blinding	GEN 1	ST2	S3
Strip footings	GEN 1	ST2	S3
Mass concrete foundations			
Single storey buildings	GEN 1	ST2	S3
Double storey buildings	GEN 3	ST4	S3
Trench Fill foundations			
Single storey buildings	GEN 1	ST2	S4
Double storey buildings	GEN 3	ST4	S4
General applications			
Kerb bedding and haunching	GEN 0	ST1	S1
Drainage works – immediate support	GEN 1	ST2	S1
Other drainage works	GEN 1	ST2	S3
Oversite below suspended slabs	GEN 1	ST2	S3
Floors			
Garage and house floors with no embedded steel	GEN 3	ST4	S2
Wearing surface: Light foot and trolley traffic	RC30	ST4	S2
Wearing surface: General industrial	RC40	N/A	S2
Wearing surface: Heavy industrial	RC50	N/A	S2
Paving			
House drives, domestic parking and external parking	PAV 1	N/A	S2
Heavy-duty external paving	PAV 2	N/A	S2

Prescribed Mixes for Ordinary Structural Concrete

Weights of cement and total dry aggregates in kg to produce approximately one cubic metre of fully compacted concrete together with the percentages by weight of fine aggregate in total dry aggregates.

Conc. grade	Nominal max. size of aggregate (mm)	40		20		14		10	
	Workability	Med.	High	Med.	High	Med.	High	Med.	High
	Limits to slump that may be expected (mm)	50–100	100–150	25–75	75–125	10–50	50–100	10–25	25–50
7	Cement (kg)	180	200	210	230	–	–	–	–
	Total aggregate (kg)	1950	1850	1900	1800	–	–	–	–
	Fine aggregate (%)	30–45	30–45	35–50	35–50	–	–	–	–
10	Cement (kg)	210	230	240	260	–	–	–	–
	Total aggregate (kg)	1900	1850	1850	1800	–	–	–	–
	Fine aggregate (%)	30–45	30–45	35–50	35–50	–	–	–	–
15	Cement (kg)	250	270	280	310	–	–	–	–
	Total aggregate (kg)	1850	1800	1800	1750	–	–	–	–
	Fine aggregate (%)	30–45	30–45	35–50	35–50	–	–	–	–
20	Cement (kg)	300	320	320	350	340	380	360	410
	Total aggregate (kg)	1850	1750	1800	1750	1750	1700	1750	1650
	Sand								
	Zone 1 (%)	35	40	40	45	45	50	50	55
	Zone 2 (%)	30	35	35	40	40	45	45	50
	Zone 3 (%)	30	30	30	35	35	40	40	45
25	Cement (kg)	340	360	360	390	380	420	400	450
	Total aggregate (kg)	1800	1750	1750	1700	1700	1650	1700	1600
	Sand								
	Zone 1 (%)	35	40	40	45	45	50	50	55
	Zone 2 (%)	30	35	35	40	40	45	45	50
	Zone 3 (%)	30	30	30	35	35	40	40	45
30	Cement (kg)	370	390	400	430	430	470	460	510
	Total aggregate (kg)	1750	1700	1700	1650	1700	1600	1650	1550
	Sand								
	Zone 1 (%)	35	40	40	45	45	50	50	55
	Zone 2 (%)	30	35	35	40	40	45	45	50
	Zone 3 (%)	30	30	30	35	35	40	40	45

Weights of Bar Reinforcement

Nominal sizes (mm)	Cross-sectional area (mm²)	Mass kg/m	Length of bar m/tonne
6	28.27	0.222	4505
8	50.27	0.395	2534
10	78.54	0.617	1622
12	113.1	0.888	1126
16	201.06	1.578	634
20	314.16	2.466	405
25	490.87	3.853	260
32	804.25	6.313	158
40	1265.64	9.865	101
50	1963.5	15.413	65

Weights of Bars at Specific Spacings

Weights of metric bars in kilograms per square metre

Size (mm)	Spacing of bars in millimetres									
	75	100	125	150	175	200	225	250	275	300
6	2.96	2.220	1.776	1.480	1.27	1.110	0.99	0.89	0.81	0.74
8	5.26	3.95	3.16	2.63	2.26	1.97	1.75	1.58	1.44	1.32
10	8.22	6.17	4.93	4.11	3.52	3.08	2.74	2.47	2.24	2.06
12	11.84	8.88	7.10	5.92	5.07	4.44	3.95	3.55	3.23	2.96
16	21.04	15.78	12.63	10.52	9.02	7.89	7.02	6.31	5.74	5.26
20	32.88	24.66	19.73	16.44	14.09	12.33	10.96	9.87	8.97	8.22
25	51.38	38.53	30.83	25.69	22.02	19.27	17.13	15.41	14.01	12.84
32	84.18	63.13	50.51	42.09	36.08	31.57	28.06	25.25	22.96	21.04
40	131.53	98.65	78.92	65.76	56.37	49.32	43.84	39.46	35.87	32.88
50	205.51	154.13	123.31	102.76	88.08	77.07	68.50	61.65	56.05	51.38

Basic weight of steelwork taken as 7850 kg/m³
Basic weight of bar reinforcement per metre run = 0.00785 kg/mm²
The value of pi has been taken as 3.141592654

Fabric Reinforcement

Fabric reference	Longitudinal wires			Cross wires			Mass
	Nominal wire size (mm)	Pitch (mm)	Area (mm/m²)	Nominal wire size (mm)	Pitch (mm)	Area (mm/m²)	(kg/m²)
Square mesh							
A393	10	200	393	10	200	393	6.16
A252	8	200	252	8	200	252	3.95
A193	7	200	193	7	200	193	3.02
A142	6	200	142	6	200	142	2.22
A98	5	200	98	5	200	98	1.54
Structural mesh							
B1131	12	100	1131	8	200	252	10.90
B785	10	100	785	8	200	252	8.14
B503	8	100	503	8	200	252	5.93
B385	7	100	385	7	200	193	4.53
B283	6	100	283	7	200	193	3.73
B196	5	100	196	7	200	193	3.05
Long mesh							
C785	10	100	785	6	400	70.8	6.72
C636	9	100	636	6	400	70.8	5.55
C503	8	100	503	5	400	49.00	4.34
C385	7	100	385	5	400	49.00	3.41
C283	6	100	283	5	400	49.00	2.61
Wrapping mesh							
D98	5	200	98	5	200	98	1.54
D49	2.5	100	49	2.5	100	49	0.77
Stock sheet size	Length 4.8 m m		Width 2.4 m		Sheet area 11.52 m²		

Wire

SWG	6g	5g	4g	3g	2g	1g	1/0g	2/0g	3/0g	4/0g	5/0g
Diameter in mm	0.192	0.212	0.232	0.252	0.276	0.300	0.324	0.348	0.372	0.400	0.432
	4.9	5.4	5.9	6.4	7.0	7.6	8.2	8.8	9.5	0.2	1.0
Area in mm^2	0.029	0.035	0.042	0.050	0.060	0.071	0.082	0.095	0.109	0.126	0.146
	19	23	27	32	39	46	53	61	70	81	95

Average Weight (kg/m³) of Steelwork Reinforcement in Concrete for Various Building Elements

	kg/m³ concrete
Substructure	
Pile caps	110–150
Tie beams	130–170
Ground beams	230–330
Bases	90–130
Footings	70–110
Retaining walls	110–150
Superstructure	
Slabs – one way	75–125
Slabs – two way	65–135
Plate slab	95–135
Cantilevered slab	90–130
Ribbed floors	80–120
Columns	200–300
Beams	250–350
Stairs	130–170
Walls – normal	30–70
Walls – wind	50–90

Note: For exposed elements add the following %: Walls 50%, Beams 100%, Columns 15%

Formwork Stripping Times – Normal Curing Periods

Conditions under which concrete is maturing	Minimum periods of protection for different types of cement					
	Number of days (where the average surface temperature of the concrete exceeds 10°C during the whole period)			Equivalent maturity (degree hours) calculated as the age of the concrete in hours multiplied by the number of degrees Celsius by which the average surface temperature of the concrete exceeds 10°C		
	Other	SRPC	OPC or RHPC	Other	SRPC	OPC or RHPC
1. Hot weather or drying winds	7	4	3	3500	2000	1500
2. Conditions not covered by 1	4	3	2	2000	1500	1000

KEY
OPC – Ordinary Portland Cement
RHPC – Rapid-hardening Portland Cement
SRPC – Sulphate-resisting Portland Cement

Minimum period before striking formwork

	Minimum period before striking		
	Surface temperature of concrete		
	16°C	17°C	t°C (0–25)
Vertical formwork to columns, walls and large beams	12 hours	18 hours	300 hours t+10
Soffit formwork to slabs	4 days	6 days	100 days t+10
Props to slabs	10 days	15 days	250 days t+10
Soffit formwork to beams	9 days	14 days	230 days t+10
Props to beams	14 days	21 days	360 days t+10

MASONRY

Weights of Bricks and Blocks

Walls and components of walls	kg/m² per mm thick	lb/sq ft per inch thick
Blockwork		
Hollow clay blocks; average)	1.15	6
Common clay blocks	1.90	10
Brickwork		
Engineering clay bricks	2.30	12
Refractory bricks	1.15	6
Sand-lime (and similar) bricks	2.02	10.5

Weights of Stones

Type of stone	kg/m³	lb/cu ft
Natural stone (solid)		
Granite	2560 to 2927	160 to 183
Limestone – Bathstone	2081	130
– Marble	2723	170
– Portland stone	2244	140
Sandstone	2244 to 2407	140 to 150
Slate	2880	180
Stone rubble (packed)	2244	140

Quantities of Bricks and Mortar

Materials per m² of wall:		
Thickness	**No. of Bricks**	**Mortar m 3**
Half brick (112.5 mm)	58	0.022
One brick (225 mm)	116	0.055
Cavity, both skins (275 mm)	116	0.045
1.5 brick (337 mm)	174	0.074
Mass brickwork per m³	464	0.36

Mortar Mixes: Quantities of Dry Materials

Mix	Imperial cu yd			Metric m³		
	Cement cwts	Lime cwts	Sand cu yds	Cement tonnes	Lime tonnes	Sand cu m
1:3	7.0	–	1.04	0.54	–	1.10
1:4	6.3	–	1.10	0.40	–	1.20
1:1:6	3.9	1.6	1.10	0.27	0.13	1.10
1:2:9	2.6	2.1	1.10	0.20	0.15	1.20
0:1:3	–	3.3	1.10	–	0.27	1.00

Mortar Mixes for Various Uses

Mix	Use
1:3	Construction designed to withstand heavy loads in all seasons.
1:1:6	Normal construction not designed for heavy loads. Sheltered and moderate conditions in spring and summer. Work above d:p:c – sand, lime bricks, clay blocks, etc.
1:2:9	Internal partitions with blocks which have high drying shrinkage, pumice blocks, etc. any periods.
0:1:3	Hydraulic lime only should be used in this mix and may be used for construction not designed for heavy loads and above d: p: c spring and summer.

Quantities of Bricks and Mortar Required per m² of Walling

Description	Unit	Nr of bricks required	Mortar required (m³)		
			No frogs	Single frogs	Double frogs
Standard bricks					
Brick size					
215 × 102.5 × 50 mm					
half brick wall (103 mm) (103 mm)	m²	72	0.022	0.027	0.032
2 × half brick cavity wall (270 mm)	m²	144	0.044	0.054	0.064
one brick wall (215 mm)	m²	144	0.052	0.064	0.076
one and a half brick wall (328 mm)	m²	216	0.073	0.091	0.108
mass brickwork	m³	576	0.347	0.413	0.480
Brick size					
215 × 102.5 × 65 mm					
half brick wall (103 mm)	m²	58	0.019	0.022	0.026
2 × half brick cavity wall (270 mm)	m²	116	0.038	0.045	0.055
one brick wall (215 mm)	m²	116	0.046	0.055	0.064
one and a half brick wall (328 mm)	m²	174	0.063	0.074	0.088
mass brickwork	m³	464	0.307	0.360	0.413
Metric modular bricks – Perforated					
Brick coordinating size					
200 × 100 × 75 mm					
90 mm thick	m²	67	0.016	0.019	
190 mm thick	m²	133	0.042	0.048	
290 mm thick	m²	200	0.068	0.078	
Brick coordinating size					
200 × 100 × 100 mm					
90 mm thick	m²	50	0.013	0.016	
190 mm thick	m²	100	0.036	0.041	
290 mm thick	m²	150	0.059	0.067	
Brick coordinating size					
300 × 100 × 75 mm					
90 mm thick	m²	33	–	0.015	
Brick coordinating size					
300 × 100 × 100 mm					
90 mm thick	m²	44	0.015	0.018	

Note: Assuming 10 mm deep joints

Mortar Required per m² Blockwork (9.88 blocks/m²)

Wall thickness	75	90	100	125	140	190	215
Mortar m³/m²	0.005	0.006	0.007	0.008	0.009	0.013	0.014

Mortar Mixes

Mortar group	Cement: lime: sand	Masonry cement: sand	Cement: sand with plasticizer
1	1:0–0.25:3		
2	1:0.5:4–4.5	1:2.5–3.5	1:3–4
3	1:1:5–6	1:4–5	1:5–6
4	1:2:8–9	1:5.5–6.5	1:7–8
5	1:3:10–12	1:6.5–7	1:8

Group 1: strong inflexible mortar
Group 5: weak but flexible

All mixes within a group are of approximately similar strength.
Frost resistance increases with the use of plasticizers.
Cement: lime: sand mixes give the strongest bond and greatest resistance to rain penetration.
Masonry cement equals ordinary Portland cement plus a fine neutral mineral filler and an air entraining agent.

Calcium Silicate Bricks

Type	Strength	Location
Class 2 crushing strength	14.0 N/mm²	not suitable for walls
Class 3	20.5 N/mm²	walls above dpc
Class 4	27.5 N/mm²	cappings and copings
Class 5	34.5 N/mm²	retaining walls
Class 6	41.5 N/mm²	walls below ground
Class 7	48.5 N/mm²	walls below ground

The Class 7 calcium silicate bricks are therefore equal in strength to Class B bricks.
Calcium silicate bricks are not suitable for DPCs.

Durability of Bricks

FL	Frost resistant with low salt content
FN	Frost resistant with normal salt content
ML	Moderately frost resistant with low salt content
MN	Moderately frost resistant with normal salt content

Brickwork Dimensions

No. of horizontal bricks	Dimensions (mm)	No. of vertical courses	No. of vertical courses
½	112.5	1	75
1	225.0	2	150
1½	337.5	3	225
2	450.0	4	300
2½	562.5	5	375
3	675.0	6	450
3½	787.5	7	525
4	900.0	8	600
4½	1012.5	9	675
5	1125.0	10	750
5½	1237.5	11	825
6	1350.0	12	900
6½	1462.5	13	975
7	1575.0	14	1050
7½	1687.5	15	1125
8	1800.0	16	1200
8½	1912.5	17	1275
9	2025.0	18	1350
9½	2137.5	19	1425
10	2250.0	20	1500
20	4500.0	24	1575
40	9000.0	28	2100
50	11250.0	32	2400
60	13500.0	36	2700
75	16875.0	40	3000

Standard Available Block Sizes

Block	Coordinating size Length × height (mm)	Work size Length × height (mm)	Thicknesses (work size) (mm)
A	400 × 100	390 × 90	75, 90, 100, 140, 190
	400 × 200	440 × 190	75, 90, 100, 140, 190
	450 × 225	440 × 215	75, 90, 100, 140, 190, 215
B	400 × 100	390 × 90	75, 90, 100, 140, 190
	400 × 200	390 × 190	75, 90, 100, 140, 190
	450 × 200	440 × 190	75, 90, 100, 140, 190, 215
	450 × 225	440 × 215	75, 90, 100, 140, 190, 215
	450 × 300	440 × 290	75, 90, 100, 140, 190, 215
	600 × 200	590 × 190	75, 90, 100, 140, 190, 215
	600 × 225	590 × 215	75, 90, 100, 140, 190, 215
C	400 × 200	390 × 190	60, 75
	450 × 200	440 × 190	60, 75
	450 × 225	440 × 215	60, 75
	450 × 300	440 × 290	60, 75
	600 × 200	590 × 190	60, 75
	600 × 225	590 × 215	60, 75

TIMBER
Weights of Timber

Material	kg/m³	lb/cu ft
General	806 (avg)	50 (avg)
Douglas fir	479	30
Yellow pine, spruce	479	30
Pitch pine	673	42
Larch, elm	561	35
Oak (English)	724 to 959	45 to 60
Teak	643 to 877	40 to 55
Jarrah	959	60
Greenheart	1040 to 1204	65 to 75
Quebracho	1285	80
Material	kg/m² per mm thickness	lb/sq ft per inch thickness
Wooden boarding and blocks		
Softwood	0.48	2.5
Hardwood	0.76	4
Hardboard	1.06	5.5
Chipboard	0.76	4
Plywood	0.62	3.25
Blockboard	0.48	2.5
Fibreboard	0.29	1.5
Wood-wool	0.58	3
Plasterboard	0.96	5
Weather boarding	0.35	1.8

Conversion Tables (for sawn timber only)

Inches	>	Millimetres	Feet	>	Metres
1		25	1		0.300
2		50	2		0.600
3		75	3		0.900
4		100	4		1.200
5		125	5		1.500
6		150	6		1.800
7		175	7		2.100
8		200	8		2.400
9		225	9		2.700
10		250	10		3.000
11		275	11		3.300
12		300	12		3.600
13		325	13		3.900
14		350	14		4.200
15		375	15		4.500
16		400	16		4.800
17		425	17		5.100
18		450	18		5.400
19		475	19		5.700
20		500	20		6.000
21		525	21		6.300
22		550	22		6.600
23		575	23		6.900
24		600	24		7.200

Planed Softwood

The finished end section size of planed timber is usually 3/16" less than the original size from which it is produced. This however varies slightly dependent upon availability of material and origin of species used.

Standard (timber) to Cubic Metres and Cubic Metres to Standard (timber)

m³	m³/Standards	Standard
4.672	1	0.214
9.344	2	0.428
14.017	3	0.642
18.689	4	0.856
23.361	5	1.070
28.033	6	1.284
32.706	7	1.498
37.378	8	1.712
42.05	9	1.926
46.722	10	2.140
93.445	20	4.281
140.167	30	6.421
186.890	40	8.561
233.612	50	10.702
280.335	60	12.842
327.057	70	14.982
373.779	80	17.122
420.502	90	19.263
467.224	100	21.403

Standards (timber) to Cubic Metres and Cubic Metres to Standards (timber)

1 cu metre	=	35.3148 cu ft	=	0.21403 std
1 cu ft	=	0.028317 cu metres		
1 std	=	4.67227 cu metres		

Basic Sizes of Sawn Softwood Available (cross sectional areas)

Thickness (mm)	Width (mm)								
	75	100	125	150	175	200	225	250	300
16	X	X	X	X					
19	X	X	X	X					
22	X	X	X	X					
25	X	X	X	X	X	X	X	X	X
32	X	X	X	X	X	X	X	X	
36	X	X	X	X					
38	X	X	X	X	X	X	X		
44	X	X	X	X	X	X	X	X	X
47*	X	X	X	X	X	X	X	X	X
50	X	X	X	X	X	X	X	X	X
63	X	X	X	X	X	X	X		
75		X	X	X	X	X	X	X	X
100		X		X		X		X	X
150				X		X			X
200						X			
250								X	
300									X

* This range of widths for 47 mm thickness will usually be found to be available in construction quality only.
Note: The smaller sizes below 100 mm thick and 250 mm width are normally but not exclusively of European origin. Sizes beyond this are usually of North and South American origin.

Basic Lengths of Sawn Softwood Available (metres)

1.80	2.10	3.00	4.20	5.10	6.00	7.20
	2.40	3.30	4.50	5.40	6.30	
	2.70	3.60	4.80	5.70	6.60	
		3.90			6.90	

Note: Lengths of 6.00 m and over will generally only be available from North American species and may have to be re-cut from larger sizes.

Reductions From Basic Size to Finished Size of Timber By Planing of Two Opposed Faces

Purpose	15–35 mm	36–100 mm	101–150 mm	Over 150 mm
a) constructional timber	3 mm	3 mm	5 mm	6 mm
b) matching interlocking boards	4 mm	4 mm	6 mm	6 mm
c) wood trim not specified in BS 584	5 mm	7 mm	7 mm	9 mm
d) joinery and cabinet work	7 mm	9 mm	11 mm	13 mm

Note: The reduction of width or depth is overall the extreme size and is exclusive of any reduction of the face by the machining of a tongue or lap joints.

METAL

Weights of Metals

Material	kg/m³	lb/cu ft
Metals, steel construction, etc.		
Iron		
– cast	7207	450
– wrought	7687	480
– ore – general	2407	150
– (crushed) Swedish	3682	230
Steel	7854	490
Copper		
– cast	8731	545
– wrought	8945	558
Brass	8497	530
Bronze	8945	558
Aluminium	2774	173
Lead	11322	707
Zinc (rolled)	7140	446
	g/mm² per metre	**lb/sq ft per foot**
Steel bars	7.85	3.4
Structural steelwork	Net weight of member @ 7854 kg/m³	
riveted	+ 10% for cleats, rivets, bolts, etc.	
welded	+ 1.25% to 2.5% for welds, etc.	
Rolled sections		
beams	+ 2.5%	
stanchions	+ 5% (extra for caps and bases)	
Plate		
web girders	+ 10% for rivets or welds, stiffeners, etc.	
	kg/m	**lb/ft**
Steel stairs: industrial type 1 m or 3 ft wide	84	56
Steel tubes 50 mm or 2 in bore	5 to 6	3 to 4
Gas piping 20 mm or ¾ in	2	1¼

KERBS/EDGINGS/CHANNELS

Precast Concrete Kerbs to BS 7263
Straight kerb units: length from 450 to 915 mm

150 mm high × 125 mm thick		
bullnosed	type BN	
half battered	type HB3	
255 mm high × 125 mm thick		
45° splayed	type SP	
half battered	type HB2	
305 mm high × 150 mm thick		
half battered	type HB1	
Quadrant kerb units		
150 mm high × 305 and 455 mm radius to match	type BN	type QBN
150 mm high × 305 and 455 mm radius to match	type HB2, HB3	type QHB
150 mm high × 305 and 455 mm radius to match	type SP	type QSP
255 mm high × 305 and 455 mm radius to match	type BN	type QBN
255 mm high × 305 and 455 mm radius to match	type HB2, HB3	type QHB
225 mm high × 305 and 455 mm radius to match	type SP	type QSP
Angle kerb units		
305 × 305 × 225 mm high × 125 mm thick		
bullnosed external angle	type XA	
splayed external angle to match type SP	type XA	
bullnosed internal angle	type IA	
splayed internal angle to match type SP	type IA	
Channels		
255 mm wide × 125 mm high flat	type CS1	
150 mm wide × 125 mm high flat type	CS2	
255 mm wide × 125 mm high dished	type CD	
Transition kerb units		
from kerb type SP to HB	left handed	type TL
	right handed	type TR
from kerb type BN to HB	left handed	type DL1
	right handed	type DR1
from kerb type BN to SP	left handed	type DL2
	right handed	type DR2

Number of kerbs required per quarter circle (780mm kerb lengths)

Radius (m)	Number in quarter circle
12	24
10	20
8	16
6	12
5	10
4	8
3	6
2	4
1	2

Precast Concrete Edgings

Round top type ER	Flat top type EF	Bullnosed top type EBN
150 × 50 mm	150 × 50 mm	150 × 50 mm
200 × 50 mm	200 × 50 mm	200 × 50 mm
250 × 50 mm	250 × 50 mm	250 × 50 mm

BASES

Cement Bound Material for Bases and Subbases

CBM1:	very carefully graded aggregate from 37.5–75 ym, with a 7-day strength of 4.5 N/mm^2
CBM2:	same range of aggregate as CBM1 but with more tolerance in each size of aggregate with a 7-day strength of 7.0 N/mm^2
CBM3:	crushed natural aggregate or blast furnace slag, graded from 37.5 mm – 150 ym for 40 mm aggregate, and from 20–75 ym for 20 mm aggregate, with a 7-day strength of 10 N/mm^2
CBM4:	crushed natural aggregate or blast furnace slag, graded from 37.5 mm – 150 ym for 40 mm aggregate, and from 20–75 ym for 20 mm aggregate, with a 7-day strength of 15 N/mm^2

INTERLOCKING BRICK/BLOCK ROADS/PAVINGS

Sizes of Precast Concrete Paving Blocks

Type R blocks	**Type S**
200 × 100 × 60 mm	Any shape within a 295 mm space
200 × 100 × 65 mm	
200 × 100 × 80 mm	
200 × 100 × 100 mm	

Sizes of Clay Brick Pavers
200 × 100 × 50 mm
200 × 100 × 65 mm
210 × 105 × 50 mm
210 × 105 × 65 mm
215 × 102.5 × 50 mm
215 × 102.5 × 65 mm

Type PA: 3 kN
Footpaths and pedestrian areas, private driveways, car parks, light vehicle traffic and over-run.

Type PB: 7 kN
Residential roads, lorry parks, factory yards, docks, petrol station forecourts, hardstandings, bus stations.

PAVING AND SURFACING

Weights and Sizes of Paving and Surfacing

Description of item	Size	Quantity per tonne
Paving 50 mm thick	900 × 600 mm	15
Paving 50 mm thick	750 × 600 mm	18
Paving 50 mm thick	600 × 600 mm	23
Paving 50 mm thick	450 × 600 mm	30
Paving 38 mm thick	600 × 600 mm	30
Path edging	914 × 50 × 150 mm	60
Kerb (including radius and tapers)	125 × 254 × 914 mm	15
Kerb (including radius and tapers)	125 × 150 × 914 mm	25
Square channel	125 × 254 × 914 mm	15
Dished channel	125 × 254 × 914 mm	15
Quadrants	300 × 300 × 254 mm	19
Quadrants	450 × 450 × 254 mm	12
Quadrants	300 × 300 × 150 mm	30
Internal angles	300 × 300 × 254 mm	30
Fluted pavement channel	255 × 75 × 914 mm	25
Corner stones	300 × 300 mm	80
Corner stones	360 × 360 mm	60
Cable covers	914 × 175 mm	55
Gulley kerbs	220 × 220 × 150 mm	60
Gulley kerbs	220 × 200 × 75 mm	120

Weights and Sizes of Paving and Surfacing

Material	kg/m³	lb/cu yd
Tarmacadam	2306	3891
Macadam (waterbound)	2563	4325
Vermiculite (aggregate)	64–80	108–135
Terracotta	2114	3568
Cork – compressed	388	24
	kg/m²	lb/sq ft
Clay floor tiles, 12.7 mm	27.3	5.6
Pavement lights	122	25
Damp-proof course	5	1
	kg/m² per mm thickness	lb/sq ft per inch thickness
Paving Slabs (stone)	2.3	12
Granite setts	2.88	15
Asphalt	2.30	12
Rubber flooring	1.68	9
Poly-vinylchloride	1.94 (avg)	10 (avg)

Coverage (m²) Per Cubic Metre of Materials Used as Subbases or Capping Layers

Consolidated thickness laid in (mm)	Square metre coverage		
	Gravel	Sand	Hardcore
50	15.80	16.50	–
75	10.50	11.00	–
100	7.92	8.20	7.42
125	6.34	6.60	5.90
150	5.28	5.50	4.95
175	–	–	4.23
200	–	–	3.71
225	–	–	3.30
300	–	–	2.47

Approximate Rate of Spreads

Average thickness of course	Description	Approximate rate of spread			
		Open Textured		Dense, Medium & Fine Textured	
mm		kg/m²	m²/t	kg/m²	m²/t
35	14 mm open textured or dense wearing course	60–75	13–17	70–85	12–14
40	20 mm open textured or dense base course	70–85	12–14	80–100	10–12
45	20 mm open textured or dense base course	80–100	10–12	95–100	9–10
50	20 mm open textured or dense, or 28 mm dense base course	85–110	9–12	110–120	8–9
60	28 mm dense base course, 40 mm open textured of dense base course or 40 mm single course as base course		8–10	130–150	7–8
65	28 mm dense base course, 40 mm open textured or dense base course or 40 mm single course	100–135	7–10	140–160	6–7
75	40 mm single course, 40 mm open textured or dense base course, 40 mm dense roadbase	120–150	7–8	165–185	5–6
100	40 mm dense base course or roadbase	–	–	220–240	4–4.5

Surface Dressing Roads: Coverage (m²) per Tonne of Material

Size in mm	Sand	Granite chips	Gravel	Limestone chips
Sand	168	–	–	–
3	–	148	152	165
6	–	130	133	144
9	–	111	114	123
13	–	85	87	95
19	–	68	71	78

Sizes of Flags

Reference	Nominal size (mm)	Thickness (mm)
A	600 × 450	50 and 63
B	600 × 600	50 and 63
C	600 × 750	50 and 63
D	600 × 900	50 and 63
E	450 × 450	50 and 70 chamfered top surface
F	400 × 400	50 and 65 chamfered top surface
G	300 × 300	50 and 60 chamfered top surface

Sizes of Natural Stone Setts

Width (mm)		Length (mm)		Depth (mm)
100	×	100	×	100
75	×	150 to 250	×	125
75	×	150 to 250	×	150
100	×	150 to 250	×	100
100	×	150 to 250	×	150

SPORTS

Sizes of Sports Areas
Sizes in metres given include clearances

Association football	Senior	114 × 72
	Junior	108 × 58
	International	100–110 × 64–75
Football	American	Pitch 109.80 × 48.80
		Overall 118.94 × 57.94
	Australian Rules	Overall 135–185 × 110–155
	Canadian	Overall 145.74 × 59.47
	Gaelic	128–146.4 × 76.8–91.50
Handball		91–110 × 55–65
Hurling		137 × 82
Rugby	Union pitch	56 × 81
	League pitch	134 × 80
Hockey pitch		100.5 × 61
Men's lacrosse pitch		106 × 61
Women's lacrosse pitch		110 × 60
Target archery ground		150 × 50
Archery (Clout)		7.3 m firing area
		Range 109.728 (Women), 146.304 (Men).
		182.88 (Normal range)
400 m running track		115.61 bend length × 2
		84.39 straight length × 2
		Overall 176.91 long × 92.52 wide
Baseball		Overall 60 m × 70
Basketball		14.0 × 26.0
Badminton		6.10 × 13.40
Camogie		91–110 × 54–68
Discus and Hammer		Safety cage 2.74 m^2
		Landing area 45 arc (65° safety) 70 m radius
Javelin		Runway 36.5 × 4.27
		Landing area 80–95 × 48

Jump	High	Running area 38.8 × 19 Landing area 5 × 4
	Long	Runway 45 × 1.22 Landing area 9 × 2.750
	Triple	Runway 45 × 1.22 Landing area 7.3 × 2.75
Korfball		90 × 40
Netball		15.25 × 30.48
Pole Vault		Runway 45 × 1.22 Landing area 5 × 5
Polo		275 × 183
Rounders		Overall 19 × 17
Shot Put		Base 2.135 dia Landing area 65° arc, 25m radius from base
Shinty		128 –183 × 64–91.5
Tennis		Court 23.77 × 10.97 Overall minimum 36.27 × 18.29
Tug-of-war		46 × 5

SEEDING/TURFING AND PLANTING

Topsoil Quality

Topsoil grade	Properties
Premium	Natural topsoil, high fertility, loamy texture, good soil structure, suitable for intensive cultivation.
General Purpose	Natural or manufactured topsoil of lesser quality than Premium, suitable for agriculture or amenity landscape, may need fertilizer or soil structure improvement.
Economy	Selected subsoil, natural mineral deposit such as river silt or greensand. The grade comprises two subgrades; 'Low clay' and 'High clay' which is more liable to compaction in handling. This grade is suitable for low-production agricultural land and amenity woodland or conservation planting areas.

Forms of Trees

Standards:	shall be clear with substantially straight stems. Grafted and budded trees shall have no more than a slight bend at the union. Standards shall be designated as Half, Extra light, Light, Standard, Selected standard, Heavy, and Extra heavy.
Sizes of Standards	
Heavy standard	12–14 cm girth × 3.50 to 5.00 m high
Extra Heavy standard	14–16 cm girth × 4.25 to 5.00 m high
Extra Heavy standard	16–18 cm girth × 4.25 to 6.00 m high
Extra Heavy standard	18–20 cm girth × 5.00 to 6.00 m high
Semi-mature trees:	between 6.0 m and 12.0 m tall with a girth of 20 to 75 cm at 1.0 m above ground.
Feathered trees:	shall have a defined upright central leader, with stem furnished with evenly spread and balanced lateral shoots down to or near the ground.
Whips:	shall be without significant feather growth as determined by visual inspection.
Multi-stemmed trees:	shall have two or more main stems at, near, above or below ground.

Seedlings grown from seed and not transplanted shall be specified when ordered for sale as:

1+0 one year old seedling		
2+0 two year old seedling		
1+1 one year seed bed,	one year transplanted	= two year old seedling
1+2 one year seed bed,	two years transplanted	= three year old seedling
2+1 two year seed bed,	one year transplanted	= three year old seedling
1u1 two years seed bed,	undercut after 1 year	= two year old seedling
2u2 four years seed bed,	undercut after 2 years	= four year old seedling

Cuttings

The age of cuttings (plants grown from shoots, stems, or roots of the mother plant) shall be specified when ordered for sale. The height of transplants and undercut seedlings/cuttings (which have been transplanted or undercut at least once) shall be stated in centimetres. The number of growing seasons before and after transplanting or undercutting shall be stated.

0+1	one year cutting
0+2	two year cutting
0+1+1	one year cutting bed, one year transplanted = two year old seedling
0+1+2	one year cutting bed, two years transplanted = three year old seedling

Grass Cutting Capacities in m² per Hour

Speed mph	Width Of Cut in metres												
	0.5	0.7	1.0	1.2	1.5	1.7	2.0	2.0	2.1	2.5	2.8	3.0	3.4
1.0	724	1127	1529	1931	2334	2736	3138	3219	3380	4023	4506	4828	5472
1.5	1086	1690	2293	2897	3500	4104	4707	4828	5069	6035	6759	7242	8208
2.0	1448	2253	3058	3862	4667	5472	6276	6437	6759	8047	9012	9656	10944
2.5	1811	2816	3822	4828	5834	6840	7846	8047	8449	10058	11265	12070	13679
3.0	2173	3380	4587	5794	7001	8208	9415	9656	10139	12070	13518	14484	16415
3.5	2535	3943	5351	6759	8167	9576	10984	11265	11829	14082	15772	16898	19151
4.0	2897	4506	6115	7725	9334	10944	12553	12875	13518	16093	18025	19312	21887
4.5	3259	5069	6880	8690	10501	12311	14122	14484	15208	18105	20278	21726	24623
5.0	3621	5633	7644	9656	11668	13679	15691	16093	16898	20117	22531	24140	27359
5.5	3983	6196	8409	10622	12834	15047	17260	17703	18588	22128	24784	26554	30095
6.0	4345	6759	9173	11587	14001	16415	18829	19312	20278	24140	27037	28968	32831
6.5	4707	7322	9938	12553	15168	17783	20398	20921	21967	26152	29290	31382	35566
7.0	5069	7886	10702	13518	16335	19151	21967	22531	23657	28163	31543	33796	38302

Number of Plants per m²: For Plants Planted on an Evenly Spaced Grid

Planting distances

mm	0.10	0.15	0.20	0.25	0.35	0.40	0.45	0.50	0.60	0.75	0.90	1.00	1.20	1.50
0.10	100.00	66.67	50.00	40.00	28.57	25.00	22.22	20.00	16.67	13.33	11.11	10.00	8.33	6.67
0.15	66.67	44.44	33.33	26.67	19.05	16.67	14.81	13.33	11.11	8.89	7.41	6.67	5.56	4.44
0.20	50.00	33.33	25.00	20.00	14.29	12.50	11.11	10.00	8.33	6.67	5.56	5.00	4.17	3.33
0.25	40.00	26.67	20.00	16.00	11.43	10.00	8.89	8.00	6.67	5.33	4.44	4.00	3.33	2.67
0.35	28.57	19.05	14.29	11.43	8.16	7.14	6.35	5.71	4.76	3.81	3.17	2.86	2.38	1.90
0.40	25.00	16.67	12.50	10.00	7.14	6.25	5.56	5.00	4.17	3.33	2.78	2.50	2.08	1.67
0.45	22.22	14.81	11.11	8.89	6.35	5.56	4.94	4.44	3.70	2.96	2.47	2.22	1.85	1.48
0.50	20.00	13.33	10.00	8.00	5.71	5.00	4.44	4.00	3.33	2.67	2.22	2.00	1.67	1.33
0.60	16.67	11.11	8.33	6.67	4.76	4.17	3.70	3.33	2.78	2.22	1.85	1.67	1.39	1.11
0.75	13.33	8.89	6.67	5.33	3.81	3.33	2.96	2.67	2.22	1.78	1.48	1.33	1.11	0.89
0.90	11.11	7.41	5.56	4.44	3.17	2.78	2.47	2.22	1.85	1.48	1.23	1.11	0.93	0.74
1.00	10.00	6.67	5.00	4.00	2.86	2.50	2.22	2.00	1.67	1.33	1.11	1.00	0.83	0.67
1.20	8.33	5.56	4.17	3.33	2.38	2.08	1.85	1.67	1.39	1.11	0.93	0.83	0.69	0.56
1.50	6.67	4.44	3.33	2.67	1.90	1.67	1.48	1.33	1.11	0.89	0.74	0.67	0.56	0.44

Grass Clippings Wet: Based on 3.5 m³/tonne

Annual kg/100 m²	Average 20 cuts kg/100m²	m²/tonne	m²/m³
32.0	1.6	61162.1	214067.3

Nr of cuts	22	20	18	16	12	4
kg/cut	1.45	1.60	1.78	2.00	2.67	8.00
Area capacity of 3 tonne vehicle per load						
m²	206250	187500	168750	150000	112500	37500
Load m³	**100 m² units/m³ of vehicle space**					
1	196.4	178.6	160.7	142.9	107.1	35.7
2	392.9	357.1	321.4	285.7	214.3	71.4
3	589.3	535.7	482.1	428.6	321.4	107.1
4	785.7	714.3	642.9	571.4	428.6	142.9
5	982.1	892.9	803.6	714.3	535.7	178.6

Transportation of Trees

To unload large trees a machine with the necessary lifting strength is required. The weight of the trees must therefore be known in advance. The following table gives a rough overview. The additional columns with root ball dimensions and the number of plants per trailer provide additional information for example about preparing planting holes and calculating unloading times.

Girth in cm	Rootball diameter in cm	Ball height in cm	Weight in kg	Numbers of trees per trailer
16–18	50–60	40	150	100–120
18–20	60–70	40–50	200	80–100
20–25	60–70	40–50	270	50–70
25–30	80	50–60	350	50
30–35	90–100	60–70	500	12–18
35–40	100–110	60–70	650	10–15
40–45	110–120	60–70	850	8–12
45–50	110–120	60–70	1100	5–7
50–60	130–140	60–70	1600	1–3
60–70	150–160	60–70	2500	1
70–80	180–200	70	4000	1
80–90	200–220	70–80	5500	1
90–100	230–250	80–90	7500	1
100–120	250–270	80–90	9500	1

Data supplied by Lorenz von Ehren GmbH
The information in the table is approximate; deviations depend on soil type, genus and weather.

FENCING AND GATES

Types of Preservative

Creosote (tar oil) can be 'factory' applied	by pressure to BS 144: pts 1&2
	by immersion to BS 144: pt 1
	by hot and cold open tank to BS 144: pts 1&2
Copper/chromium/arsenic (CCA)	by full cell process to BS 4072 pts 1&2
Organic solvent (OS)	by double vacuum (vacvac) to BS 5707 pts 1&3
	by immersion to BS 5057 pts 1&3
Pentachlorophenol (PCP)	by heavy oil double vacuum to BS 5705 pts 2&3
Boron diffusion process (treated with disodium octaborate to BWPA Manual 1986.	

Note: Boron is used on green timber at source and the timber is supplied dry.

Cleft Chestnut Pale Fences

Pales	Pale spacing	Wire lines	
900 mm	75 mm	2	temporary protection
1050 mm	75 or 100 mm	2	light protective fences
1200 mm	75 mm	3	perimeter fences
1350 mm	75 mm	3	perimeter fences
1500 mm	50 mm	3	narrow perimeter fences
1800 mm	50 mm	3	light security fences

Close-boarded Fences

Close-boarded fences 1.05 to 1.8 m high
Type BCR (recessed) or BCM (morticed) with concrete posts 140 × 115 mm tapered and Type BW with timber posts.

Palisade Fences

Wooden palisade fences
Type WPC with concrete posts 140 × 115 mm tapered and Type WPW with timber posts.

For both types of fence:
Height of fence 1050 mm: two rails
Height of fence 1200 mm: two rails
Height of fence 1500 mm: three rails
Height of fence 1650 mm: three rails
Height of fence 1800 mm: three rails

Post and Rail Fences

Wooden post and rail fences
Type MPR 11/3 morticed rails and Type SPR 11/3 nailed rails
Height to top of rail 1100 mm
Rails: three rails 87 mm, 38 mm

Type MPR 11/4 morticed rails and Type SPR 11/4 nailed rails
Height to top of rail 1100 mm
Rails: four rails 87 mm, 38 mm.

Type MPR 13/4 morticed rails and Type SPR 13/4 nailed rails
Height to top of rail 1300 mm
Rail spacing 250 mm, 250 mm, and 225 mm from top
Rails: four rails 87 mm, 38 mm.

Steel Posts

Rolled steel angle iron posts for chain link fencing:

Posts	Fence height	Strut	Straining post
1500 × 40 × 40 × 5 mm	900 mm	1500 × 40 × 40 × 5 mm	1500 × 50 × 50 × 6 mm
1800 × 40 × 40 × 5 mm	1200 mm	1800 × 40 × 40 × 5 mm	1800 × 50 × 50 × 6 mm
2000 × 45 × 45 × 5 mm	1400 mm	2000 × 45 × 45 × 5 mm	2000 × 60 × 60 × 6 mm
2600 × 45 × 45 × 5 mm	1800 mm	2600 × 45 × 45 × 5 mm	2600 × 60 × 60 × 6 mm
3000 × 50 × 50 × 6 mm	1800 mm	2600 × 45 × 45 × 5 mm	3000 × 60 × 60 × 6 mm
with arms			

Concrete Posts

Concrete posts for chain link fencing:

Posts and straining posts	Fence height	Strut
1570 mm 100 × 100 mm	900 mm	1500 mm × 75 × 75 mm
1870 mm 125 × 125 mm	1200 mm	1830 mm × 100 × 75 mm
2070 mm 125 × 125 mm	1400 mm	1980 mm × 100 × 75 mm
2620 mm 125 × 125 mm	1800 mm	2590 mm × 100 × 85 mm
3040 mm 125 × 125 mm	1800 mm	2590 mm × 100 × 85 mm (with arms)

Rolled Steel Angle Posts

Rolled steel angle posts for rectangular wire mesh (field) fencing

Posts	Fence height	Strut	Straining post
1200 × 40 × 40 × 5 mm	600 mm	1200 × 75 × 75 mm	1350 × 100 × 100 mm
1400 × 40 × 40 × 5 mm	800 mm	1400 × 75 × 75 mm	1550 × 100 × 100 mm
1500 × 40 × 40 × 5 mm	900 mm	1500 × 75 × 75 mm	1650 × 100 × 100 mm
1600 × 40 × 40 × 5 mm	1000 mm	1600 × 75 × 75 mm	1750 × 100 × 100 mm
1750 × 40 × 40 × 5 mm	1150 mm	1750 × 75 × 100 mm	1900 × 125 × 125 mm

Concrete Posts

Concrete posts for rectangular wire mesh (field) fencing

Posts	Fence height	Strut	Straining post
1270 × 100 × 100 mm	600 mm	1200 × 75 × 75 mm	1420 × 100 × 100 mm
1470 × 100 × 100 mm	800 mm	1350 × 75 × 75 mm	1620 × 100 × 100 mm
1570 × 100 × 100 mm	900 mm	1500 × 75 × 75 mm	1720 × 100 × 100 mm
1670 × 100 × 100 mm	600 mm	1650 × 75 × 75 mm	1820 × 100 × 100 mm
1820 × 125 × 125 mm	1150 mm	1830 × 75 × 100 mm	1970 × 125 × 125 mm

Cleft Chestnut Pale Fences
Timber Posts

Timber posts for wire mesh and hexagonal wire netting fences
Round timber for general fences

Posts	Fence height	Strut	Straining post
1300 × 65 mm dia.	600 mm	1200 × 80 mm dia	1450 × 100 mm dia
1500 × 65 mm dia	800 mm	1400 × 80 mm dia.	1650 × 100 mm dia
1600 × 65 mm dia.	900 mm	1500 × 80 mm dia	1750 × 100 mm dia
1700 × 65 mm dia.	1050 mm	1600 × 80 mm dia	1850 × 100 mm dia
1800 × 65 mm dia.	1150 mm	1750 × 80 mm dia	2000 × 120 mm dia

Squared timber for general fences

Posts	Fence height	Strut	Straining post
1300 × 75 × 75 mm	600 mm	1200 × 75 × 75 mm	1450 × 100 × 100 mm
1500 × 75 × 75 mm	800 mm	1400 × 75 × 75 mm	1650 × 100 × 100 mm
1600 × 75 × 75 mm	900 mm	1500 × 75 × 75 mm	1750 × 100 × 100 mm
1700 × 75 × 75 mm	1050 mm	1600 × 75 × 75 mm	1850 × 100 × 100 mm
1800 × 75 × 75 mm	1150 mm	1750 × 75 × 75 mm	2000 × 125 × 100 mm

Steel Fences to BS 1722: Part 9: 1992

	Fence height	Top/bottom rails and flat posts	Vertical bars
Light	1000 mm	40 × 10 mm 450 mm in ground	12 mm dia. at 115 mm cs
	1200 mm	40 × 10 mm 550 mm in ground	12 mm dia. at 115 mm cs
	1400 mm	40 × 10 mm 550 mm in ground	12 mm dia. at 115 mm cs
Light	1000 mm	40 × 10 mm 450 mm in ground	16 mm dia. at 120 mm cs
	1200 mm	40 × 10 mm 550 mm in ground	16 mm dia. at 120 mm cs
	1400 mm	40 × 10 mm 550 mm in ground	16 mm dia. at 120 mm cs
Medium	1200 mm	50 × 10 mm 550 mm in ground	20 mm dia. at 125 mm cs
	1400 mm	50 × 10 mm 550 mm in ground	20 mm dia. at 125 mm cs
	1600 mm	50 × 10 mm 600 mm in ground	22 mm dia. at 145 mm cs
	1800 mm	50 × 10 mm 600 mm in ground	22 mm dia. at 145 mm cs
Heavy	1600 mm	50 × 10 mm 600 mm in ground	22 mm dia. at 145 mm cs
	1800 mm	50 × 10 mm 600 mm in ground	22 mm dia. at 145 mm cs
	2000 mm	50 × 10 mm 600 mm in ground	22 mm dia. at 145 mm cs
	2200 mm	50 × 10 mm 600 mm in ground	22 mm dia. at 145 mm cs

Notes: Mild steel fences: round or square verticals; flat standards and horizontals. Tops of vertical bars may be bow-top, blunt, or pointed. Round or square bar railings.

Timber Field Gates to BS 3470: 1975

Gates made to this standard are designed to open one way only.
All timber gates are 1100 mm high.
Width over stiles 2400, 2700, 3000, 3300, 3600, and 4200 mm.
Gates over 4200 mm should be made in two leaves.

Steel Field Gates to BS 3470: 1975

All steel gates are 1100 mm high.
Heavy duty: width over stiles 2400, 3000, 3600 and 4500 mm
Light duty: width over stiles 2400, 3000, and 3600 mm

Domestic Front Entrance Gates to BS 4092: Part 1: 1966

Metal gates:	Single gates are 900 mm high minimum, 900 mm, 1000 mm and 1100 mm wide

Domestic Front Entrance Gates to BS 4092: Part 2: 1966

Wooden gates:	All rails shall be tenoned into the stiles Single gates are 840 mm high minimum, 801 mm and 1020 mm wide Double gates are 840 mm high minimum, 2130, 2340 and 2640 mm wide

Timber Bridle Gates to BS 5709:1979 (Horse or Hunting Gates)

Gates open one way only	
Minimum width between posts	1525 mm
Minimum height	1100 mm

Timber Kissing Gates to BS 5709:1979

Minimum width	700 mm
Minimum height	1000 mm
Minimum distance between shutting posts	600 mm
Minimum clearance at mid-point	600 mm

Metal Kissing Gates to BS 5709:1979

Sizes are the same as those for timber kissing gates. Maximum gaps between rails 120 mm.

Categories of Pedestrian Guard Rail to BS 3049:1976

Class A for normal use. Class B where vandalism is expected. Class C where crowd pressure is likely.

DRAINAGE

Weights and Dimensions – Vitrified Clay Pipes

Product	Nominal diameter	Effective length	BS 65 limits of tolerance		Crushing strength	Weight	
			min	max			
	(mm)	(mm)	(mm)	(mm)	(kN/m)	kg/pipe	kg/m
Supersleve	100	1600	96	105	35.00	14.71	9.19
	150	1750	146	158	35.00	29.24	16.71
Hepsleve	225	1850	221	236	28.00	84.03	45.42
	300	2500	295	313	34.00	193.05	77.22
	150	1500	146	158	22.00	37.04	24.69
Hepseal	225	1750	221	236	28.00	85.47	48.84
	300	2500	295	313	34.00	204.08	81.63
	400	2500	394	414	44.00	357.14	142.86
	450	2500	444	464	44.00	454.55	181.63
	500	2500	494	514	48.00	555.56	222.22
	600	2500	591	615	57.00	796.23	307.69
	700	3000	689	719	67.00	1111.11	370.45
	800	3000	788	822	72.00	1351.35	450.45
Hepline	100	1600	95	107	22.00	14.71	9.19
	150	1750	145	160	22.00	29.24	16.71
	225	1850	219	239	28.00	84.03	45.42
	300	1850	292	317	34.00	142.86	77.22
Hepduct (conduit)	90	1500	–	–	28.00	12.05	8.03
	100	1600	–	–	28.00	14.71	9.19
	125	1750	–	–	28.00	20.73	11.84
	150	1750	–	–	28.00	29.24	16.71
	225	1850	–	–	28.00	84.03	45.42
	300	1850	–	–	34.00	142.86	77.22

Weights and Dimensions – Vitrified Clay Pipes

Nominal internal diameter (mm)	Nominal wall thickness (mm)	Approximate weight kg/m
150	25	45
225	29	71
300	32	122
375	35	162
450	38	191
600	48	317
750	54	454
900	60	616
1200	76	912
1500	89	1458
1800	102	1884
2100	127	2619

Wall thickness, weights and pipe lengths vary, depending on type of pipe required.
The particulars shown above represent a selection of available diameters and are applicable to strength class 1 pipes with flexible rubber ring joints.
Tubes with Ogee joints are also available.

Weights and Dimensions – PVC-u Pipes

	Nominal size	Mean outside diameter (mm)		Wall thickness	Weight
	min	max	(mm)	(kg/m)	
Standard pipes	82.4	82.4	82.7	3.2	1.2
	110.0	110.0	110.4	3.2	1.6
	160.0	160.0	160.6	4.1	3.0
	200.0	200.0	200.6	4.9	4.6
	250.0	250.0	250.7	6.1	7.2
Perforated pipes heavy grade	As above	As above	As above	As above	As above
thin wall	82.4	82.4	82.7	1.7	–
	110.0	110.0	110.4	2.2	–
	160.0	160.0	160.6	3.2	–

Width of Trenches Required for Various Diameters of Pipes

Pipe diameter (mm)	Trench n.e. 1.5 m deep (mm)	Trench over 1.5 m deep (mm)
n.e. 100	450	600
100–150	500	650
150–225	600	750
225–300	650	800
300–400	750	900
400–450	900	1050
450–600	1100	1300

DRAINAGE BELOW GROUND AND LAND DRAINAGE

Flow of Water Which Can Be Carried by Various Sizes of Pipe

Clay or concrete pipes

Pipe size	Gradient of pipeline							
	1:10	1:20	1:30	1:40	1:50	1:60	1:80	1:100
	Flow in litres per second							
DN 100 15.0	8.5	6.8	5.8	5.2	4.7	4.0	3.5	
DN 150 28.0	19.0	16.0	14.0	12.0	11.0	9.1	8.0	
DN 225 140.0	95.0	76.0	66.0	58.0	53.0	46.0	40.0	

Plastic pipes

Pipe size	Gradient of pipeline							
	1:10	1:20	1:30	1:40	1:50	1:60	1:80	1:100
	Flow in litres per second							
82.4 mm i/dia	12.0	8.5	6.8	5.8	5.2	4.7	4.0	3.5
110 mm i/dia	28.0	19.0	16.0	14.0	12.0	11.0	9.1	8.0
160 mm i/dia	76.0	53.0	43.0	37.0	33.0	29.0	25.0	22.0
200 mm i/dia	140.0	95.0	76.0	66.0	58.0	53.0	46.0	40.0

Vitrified (Perforated) Clay Pipes and Fittings to BS En 295-5 1994

Length not specified		
75 mm bore	**250 mm bore**	**600 mm bore**
100	300	700
125	350	800
150	400	1000
200	450	1200
225	500	

Precast Concrete Pipes: Prestressed Non-pressure Pipes and Fittings: Flexible Joints to BS 5911: Pt. 103: 1994

Rationalized metric nominal sizes: 450, 500	
Length:	500 – 1000 by 100 increments 1000 – 2200 by 200 increments 2200 – 2800 by 300 increments
Angles: length:	450–600 angles 45, 22.5, 11.25° 600 or more angles 22.5, 11.25°

Precast Concrete Pipes: Un-reinforced and Circular Manholes and Soakaways to BS 5911: Pt. 200: 1994

Nominal sizes:	
Shafts:	675, 900 mm
Chambers:	900, 1050, 1200, 1350, 1500, 1800, 2100, 2400, 2700, 3000 mm
Large chambers:	To have either tapered reducing rings or a flat reducing slab in order to accept the standard cover
Ring depths:	1. 300–1200 mm by 300 mm increments except for bottom slab and rings below cover slab, these are by 150 mm increments
	2. 250–1000 mm by 250 mm increments except for bottom slab and rings below cover slab, these are by 125 mm increments
Access hole:	750 × 750 mm for DN 1050 chamber
	1200 × 675 mm for DN 1350 chamber

Calculation of Soakaway Depth
The following formula determines the depth of concrete ring soakaway that would be required for draining given amounts of water.

$$h = \frac{4ar}{3\pi D^2}$$

h = depth of the chamber below the invert pipe
a = The area to be drained
r = The hourly rate of rainfall (50 mm per hour)
π = pi
D = internal diameter of the soakaway

This table shows the depth of chambers in each ring size which would be required to contain the volume of water specified. These allow a recommended storage capacity of ⅓ (one third of the hourly rainfall figure).

Table Showing Required Depth of Concrete Ring Chambers in Metres

Area m²	50	100	150	200	300	400	500
Ring size							
0.9	1.31	2.62	3.93	5.24	7.86	10.48	13.10
1.1	0.96	1.92	2.89	3.85	5.77	7.70	9.62
1.2	0.74	1.47	2.21	2.95	4.42	5.89	7.37
1.4	0.58	1.16	1.75	2.33	3.49	4.66	5.82
1.5	0.47	0.94	1.41	1.89	2.83	3.77	4.72
1.8	0.33	0.65	0.98	1.31	1.96	2.62	3.27
2.1	0.24	0.48	0.72	0.96	1.44	1.92	2.41
2.4	0.18	0.37	0.55	0.74	1.11	1.47	1.84
2.7	0.15	0.29	0.44	0.58	0.87	1.16	1.46
3.0	0.12	0.24	0.35	0.47	0.71	0.94	1.18

Precast Concrete Inspection Chambers and Gullies to BS 5911: Part 230: 1994

Nominal sizes:	375 diameter, 750, 900 mm deep
	450 diameter, 750, 900, 1050, 1200 mm deep
Depths:	from the top for trapped or un-trapped units:
	centre of outlet 300 mm
	invert (bottom) of the outlet pipe 400 mm
Depth of water seal for trapped gullies:	
	85 mm, rodding eye int. diam. 100 mm
Cover slab:	65 mm min.

Bedding Flexible Pipes: PVC-u Or Ductile Iron

Type 1 =	100 mm fill below pipe, 300 mm above pipe: single size material
Type 2 =	100 mm fill below pipe, 300 mm above pipe: single size or graded material
Type 3 =	100 mm fill below pipe, 75 mm above pipe with concrete protective slab over
Type 4 =	100 mm fill below pipe, fill laid level with top of pipe
Type 5 =	200 mm fill below pipe, fill laid level with top of pipe
Concrete =	25 mm sand blinding to bottom of trench, pipe supported on chocks, 100 mm concrete under the pipe, 150 mm concrete over the pipe.

Bedding Rigid Pipes: Clay or Concrete
(for vitrified clay pipes the manufacturer should be consulted)

Class D:	Pipe laid on natural ground with cut-outs for joints, soil screened to remove stones over 40 mm and returned over pipe to 150 m min depth. Suitable for firm ground with trenches trimmed by hand.
Class N:	Pipe laid on 50 mm granular material of graded aggregate to Table 4 of BS 882, or 10 mm aggregate to Table 6 of BS 882, or as dug light soil (not clay) screened to remove stones over 10 mm. Suitable for machine dug trenches.
Class B:	As Class N, but with granular bedding extending half way up the pipe diameter.
Class F:	Pipe laid on 100 mm granular fill to BS 882 below pipe, minimum 150 mm granular fill above pipe: single size material. Suitable for machine dug trenches.
Class A:	Concrete 100 mm thick under the pipe extending half way up the pipe, backfilled with the appropriate class of fill. Used where there is only a very shallow fall to the drain. Class A bedding allows the pipes to be laid to an exact gradient.
Concrete surround:	25 mm sand blinding to bottom of trench, pipe supported on chocks, 100 mm concrete under the pipe, 150 mm concrete over the pipe. It is preferable to bed pipes under slabs or wall in granular material.

PIPED SUPPLY SYSTEMS
Identification of Service Tubes From Utility to Dwellings

Utility	Colour	Size	Depth
British Telecom	grey	54 mm od	450 mm
Electricity	black	38 mm od	450 mm
Gas	yellow	42 mm od rigid 60 mm od convoluted	450 mm
Water	may be blue	(normally untubed)	750 mm

ELECTRICAL SUPPLY/POWER/LIGHTING SYSTEMS
Electrical Insulation Class En 60.598 BS 4533

Class 1: luminaires comply with class 1 (I) earthed electrical requirements
Class 2: luminaires comply with class 2 (II) double insulated electrical requirements
Class 3: luminaires comply with class 3 (III) electrical requirements

Protection to Light Fittings
BS EN 60529:1992 Classification for degrees of protection provided by enclosures.
(IP Code – International or ingress Protection)

1st characteristic: against ingress of solid foreign objects		
The figure	2	indicates that fingers cannot enter
	3	that a 2.5 mm diameter probe cannot enter
	4	that a 1.0 mm diameter probe cannot enter
	5	the fitting is dust proof (no dust around live parts)
	6	the fitting is dust tight (no dust entry)
2nd characteristic: ingress of water with harmful effects		
The figure	0	indicates unprotected
	1	vertically dripping water cannot enter
	2	water dripping 15° (tilt) cannot enter
	3	spraying water cannot enter
	4	splashing water cannot enter
	5	jetting water cannot enter
	6	powerful jetting water cannot enter
	7	proof against temporary immersion
	8	proof against continuous immersion
Optional additional codes:		A-D protects against access to hazardous parts
	H	High voltage apparatus
	M	fitting was in motion during water test
	S	fitting was static during water test
	W	protects against weather
Marking code arrangement:		(example) IPX5S = IP (International or Ingress Protection)
		X (denotes omission of first characteristic)
		5 = jetting
		S = static during water test

European Gardens
History, Philosophy and Design

By **Tom Turner**

EUROPEAN GARDENS
PHILOSOPHY AND DESIGN

Tom Turner   R

Garden design and usage has been a feature of human civilisation as far back as Neolithic times, when the first gardens began to be used for residential, horticultural and sacred tasks. Tom Turner follows the entire history of the European garden from its prehistoric roots right up to the present day in this beautifully illustrated book.

European Gardens is divided into ten periods of history and garden development, detailing the advancement of land usage for over 10,000 years. Some of the topics covered in this comprehensive book include the Egyptian gardens of the Pharaohs, the castle gardens of medieval times, eclectic gardens of the nineteenth century and abstract gardens of the last 100 years. The geographical scope of this book covers the whole of the European continent, and touches the garden designs of North Africa and the Middle East.

Tom Turner is a skilled landscape architect and garden historian, who supports his engaging writing with his own detailed plans and diagrams. *European Gardens* also features almost 1,000 colour photographs from across the continent allowing the reader to see for themselves how the design and structure of gardens has developed over time.

A companion to the *Asian Gardens* book, published by Routledge in 2010, *European Gardens* is a development of the original Garden History book from 2004.

December 2010: 250x250: 304pp
Hb: 978-0-415-49684-1: **£35.00**

To Order: Tel: +44 (0) 1235 400524 **Fax:** +44 (0) 1235 400525
or Post: Taylor and Francis Customer Services,
Bookpoint Ltd, Unit T1, 200 Milton Park, Abingdon, Oxon, OX14 4TA UK
Email: book.orders@tandf.co.uk

For a complete listing of all our titles visit:
www.tandf.co.uk

Taylor & Francis
Taylor & Francis Group

Index

Spon's Estimating Costs Guide to Minor Works,

Alterations and Repairs to Fire, Flood, Gale and Theft Damage

Fourth Edition

Bryan Spain

Specially written for contractors, quantity surveyors and clients carrying out small works, Spon's Estimating Costs Guide to Minor Works, Alterations and Repairs to Fire, Flood, Gale and Theft Damage contains accurate information on thousands of rates each broken down to labour, material overheads and profit.

Selected Contents: Introduction. Standard Method of Measurement/Trades Link. Part 1: Unit Rates. Part 2: Damage Repairs. Part 3: Approximate Estimating. Part 4: Plant and Tool Hire. Part 5: General Construction Data. Part 6: Business Matters

August 2008: 216x138: 320pp
Pb: 978-0-415-46906-7: **£31.99**

To Order: Tel: +44 (0) 1235 400524 **Fax:** +44 (0) 1235 400525
or Post: Taylor and Francis Customer Services,
Bookpoint Ltd, Unit T1, 200 Milton Park, Abingdon, Oxon, OX14 4TA UK
Email: book.orders@tandf.co.uk

For a complete listing of all our titles visit:
www.tandf.co.uk

Environmental Noise Barriers

2nd edition

Benz Kotzen and Colin English

Environmental Noise Barriers is a unique one-stop reference for practitioners, whether acoustical engineers, landscape architects or manufacturers, and for highways departments in local and central authorities. This extensively revised new edition is updated in line with UK and EU legislation and international provision of barriers.

Contents
Introduction;
Defining the Need for Barriers; Acoustic and Visual Impact Assessment – Environmental Impact Assessment Techniques and Mitigation;
Acoustic Performance of Barriers; Barrier Morphology and Design;
Types of Barrier and Barrier Materials; Planting and Bio-barriers;
Noise Amelioration and Roads; Tunnels;
Airports;
Environmental Noise Barriers in Other Locations – Industrial Uses

April 2009 246 x 189mm: 280pp
Hb: 978-0-415-43708-0 **£90.00**

To Order: Tel: +44 (0) 1235 400524 **Fax:** +44 (0) 1235 400525
or Post: Taylor and Francis Customer Services,
Bookpoint Ltd, Unit T1, 200 Milton Park, Abingdon, Oxon, OX14 4TA UK
Email: book.orders@tandf.co.uk

For a complete listing of all our titles visit:
www.tandf.co.uk

Taylor & Francis
Taylor & Francis Group

Software and eBook Single-User Licence Agreement

We welcome you as a user of this Spon Price Book Software and eBook and hope that you find it a useful and valuable tool. Please read this document carefully. **This is a legal agreement** between you (hereinafter referred to as the "Licensee") and Taylor and Francis Books Ltd. (the "Publisher"), which defines the terms under which you may use the Product. **By breaking the seal and opening the document inside the back cover of the book containing the access code you agree to these terms and conditions outlined herein. If you do not agree to these terms you must return the Product to your supplier intact, with the seal on the document unbroken.**

1.	**Definition of the Product**
	The product which is the subject of this Agreement, *Spon's External Works and Landscape Price Book 2011* Software and eBook (the "Product") consists of:
1.1	Underlying data comprised in the product (the "Data")
1.2	A compilation of the Data (the "Database")
1.3	Software (the "Software") for accessing and using the Database
1.4	An electronic book containing the data in the price book (the "eBook")
2.	**Commencement and licence**
2.1	This Agreement commences upon the breaking open of the document containing the access code by the Licensee (the "Commencement Date").
2.2	This is a licence agreement (the "Agreement") for the use of the Product by the Licensee, and not an agreement for sale.
2.3	The Publisher licenses the Licensee on a non-exclusive and non-transferable basis to use the Product on condition that the Licensee complies with this Agreement. The Licensee acknowledges that it is only permitted to use the Product in accordance with this Agreement.
3.	**Multiple use**
	For more than one user or for a wide area network or consortium, use is only permissible with the purchase from the Publisher of a multiple-user licence and adherence to the terms and conditions of that licence.
4.	**Installation and Use**
4.1	The Licensee may provide access to the Product for individual study in the following manner: The Licensee may install the Product on a secure local area network on a single site for use by one user.
4.2	The Licensee shall be responsible for installing the Product and for the effectiveness of such installation.
4.3	Text from the Product may be incorporated in a coursepack. Such use is only permissible with the express permission of the Publisher in writing and requires the payment of the appropriate fee as specified by the Publisher and signature of a separate licence agreement.
4.4	The Product is a free addition to the book and no technical support will be provided.
5.	**Permitted Activities**
5.1	The Licensee shall be entitled:
	5.1.1 to use the Product for its own internal purposes;
	5.1.2 to download onto electronic, magnetic, optical or similar storage medium reasonable portions of the Database provided that the purpose of the Licensee is to undertake internal research or study and provided that such storage is temporary;
5.2	The Licensee acknowledges that its rights to use the Product are strictly set out in this Agreement, and all other uses (whether expressly mentioned in Clause 6 below or not) are prohibited.
6.	**Prohibited Activities**
	The following are prohibited without the express permission of the Publisher:
6.1	The commercial exploitation of any part of the Product.
6.2	The rental, loan, (free or for money or money's worth) or hire purchase of this product, save with the express consent of the Publisher.
6.3	Any activity which raises the reasonable prospect of impeding the Publisher's ability or opportunities to market the Product.
6.4	Any networking, physical or electronic distribution or dissemination of the product save as expressly permitted by this Agreement.
6.5	Any reverse engineering, decompilation, disassembly or other alteration of the Product save in accordance with applicable national laws.
6.6	The right to create any derivative product or service from the Product save as expressly provided for in this Agreement.
6.7	Any alteration, amendment, modification or deletion from the Product, whether for the purposes of error correction or otherwise.

7. General Responsibilities of the License

7.1 The Licensee will take all reasonable steps to ensure that the Product is used in accordance with the terms and conditions of this Agreement.

7.2 The Licensee acknowledges that damages may not be a sufficient remedy for the Publisher in the event of breach of this Agreement by the Licensee, and that an injunction may be appropriate.

7.3 The Licensee undertakes to keep the Product safe and to use its best endeavours to ensure that the product does not fall into the hands of third parties, whether as a result of theft or otherwise.

7.4 Where information of a confidential nature relating to the product of the business affairs of the Publisher comes into the possession of the Licensee pursuant to this Agreement (or otherwise), the Licensee agrees to use such information solely for the purposes of this Agreement, and under no circumstances to disclose any element of the information to any third party save strictly as permitted under this Agreement. For the avoidance of doubt, the Licensee's obligations under this sub-clause 7.4 shall survive the termination of this Agreement.

8. Warrant and Liability

8.1 The Publisher warrants that it has the authority to enter into this agreement and that it has secured all rights and permissions necessary to enable the Licensee to use the Product in accordance with this Agreement.

8.2 The Publisher warrants that the Product as supplied on the Commencement Date shall be free of defects in materials and workmanship, and undertakes to replace any defective Product within 28 days of notice of such defect being received provided such notice is received within 30 days of such supply. As an alternative to replacement, the Publisher agrees fully to refund the Licensee in such circumstances, if the Licensee so requests, provided that the Licensee returns this copy of *Spon's External Works and Landscape Price Book 2011* to the Publisher. The provisions of this sub-clause 8.2 do not apply where the defect results from an accident or from misuse of the product by the Licensee.

8.3 Sub-clause 8.2 sets out the sole and exclusive remedy of the Licensee in relation to defects in the Product.

8.4 The Publisher and the Licensee acknowledge that the Publisher supplies the Product on an "as is" basis. The Publisher gives no warranties:

8.4.1 that the Product satisfies the individual requirements of the Licensee; or

8.4.2 that the Product is otherwise fit for the Licensee's purpose; or

8.4.3 that the Data are accurate or complete or free of errors or omissions; or

8.4.4 that the Product is compatible with the Licensee's hardware equipment and software operating environment.

8.5 The Publisher hereby disclaims all warranties and conditions, express or implied, which are not stated above.

8.6 Nothing in this Clause 8 limits the Publisher's liability to the Licensee in the event of death or personal injury resulting from the Publisher's negligence.

8.7 The Publisher hereby excludes liability for loss of revenue, reputation, business, profits, or for indirect or consequential losses, irrespective of whether the Publisher was advised by the Licensee of the potential of such losses.

8.8 The Licensee acknowledges the merit of independently verifying Data prior to taking any decisions of material significance (commercial or otherwise) based on such data. It is agreed that the Publisher shall not be liable for any losses which result from the Licensee placing reliance on the Data or on the Database, under any circumstances.

8.9 Subject to sub-clause 8.6 above, the Publisher's liability under this Agreement shall be limited to the purchase price.

9. Intellectual Property Rights

9.1 Nothing in this Agreement affects the ownership of copyright or other intellectual property rights in the Data, the Database of the Software.

9.2 The Licensee agrees to display the Publishers' copyright notice in the manner described in the Product.

9.3 The Licensee hereby agrees to abide by copyright and similar notice requirements required by the Publisher, details of which are as follows:

"© 2011 Taylor & Francis. All rights reserved. All materials in *Spon's External Works and Landscape Price Book 2011* are copyright protected. All rights reserved. No such materials may be used, displayed, modified, adapted, distributed, transmitted, transferred, published or otherwise reproduced in any form or by any means now or hereafter developed other than strictly in accordance with the terms of the licence agreement enclosed with *Spon's External Works and Landscape Price Book 2011*. However, text and images may be printed and copied for research and private study within the preset program limitations. Please note the copyright notice above, and that any text or images printed or copied must credit the source."

9.4 This Product contains material proprietary to and copyedited by the Publisher and others. Except for the licence granted herein, all rights, title and interest in the Product, in all languages, formats and media throughout the world, including copyrights therein, are and remain the property of the Publisher or other copyright holders identified in the Product.

10.   Non-assignment

This Agreement and the licence contained within it may not be assigned to any other person or entity without the written consent of the Publisher.

11.   Termination and Consequences of Termination.

11.1   The Publisher shall have the right to terminate this Agreement if:

      11.1.1 the Licensee is in material breach of this Agreement and fails to remedy such breach (where capable of remedy) within 14 days of a written notice from the Publisher requiring it to do so; or

      11.1.2 the Licensee becomes insolvent, becomes subject to receivership, liquidation or similar external administration; or

      11.1.3 the Licensee ceases to operate in business.

11.2   The Licensee shall have the right to terminate this Agreement for any reason upon two month's written notice. The Licensee shall not be entitled to any refund for payments made under this Agreement prior to termination under this sub-clause 11.2.

11.3   Termination by either of the parties is without prejudice to any other rights or remedies under the general law to which they may be entitled, or which survive such termination (including rights of the Publisher under sub-clause 7.4 above).

11.4   Upon termination of this Agreement, or expiry of its terms, the Licensee must destroy all copies and any back up copies of the product or part thereof.

12.   General

12.1   **Compliance with export provisions**

The Publisher hereby agrees to comply fully with all relevant export laws and regulations of the United Kingdom to ensure that the Product is not exported, directly or indirectly, in violation of English law.

12.2   **Force majeure**

The parties accept no responsibility for breaches of this Agreement occurring as a result of circumstances beyond their control.

12.3   **No waiver**

Any failure or delay by either party to exercise or enforce any right conferred by this Agreement shall not be deemed to be a waiver of such right.

12.4   **Entire agreement**

This Agreement represents the entire agreement between the Publisher and the Licensee concerning the Product. The terms of this Agreement supersede all prior purchase orders, written terms and conditions, written or verbal representations, advertising or statements relating in any way to the Product.

12.5   **Severability**

If any provision of this Agreement is found to be invalid or unenforceable by a court of law of competent jurisdiction, such a finding shall not affect the other provisions of this Agreement and all provisions of this Agreement unaffected by such a finding shall remain in full force and effect.

12.6   **Variations**

This agreement may only be varied in writing by means of variation signed in writing by both parties.

12.7   **Notices**

All notices to be delivered to: Spon's Price Books, Taylor & Francis Books Ltd., 2 Park Square, Milton Park, Abingdon, Oxfordshire, OX14 4RN, UK.

12.8   **Governing law**

This Agreement is governed by English law and the parties hereby agree that any dispute arising under this Agreement shall be subject to the jurisdiction of the English courts.

If you have any queries about the terms of this licence, please contact:

Spon's Price Books
Taylor & Francis Books Ltd.
2 Park Square, Milton Park, Abingdon, Oxfordshire, OX14 4RN
Tel: +44 (0) 20 7017 6672
Fax: +44 (0) 20 7017 6702
www.tandfbuiltenvironment.com

Multiple-user use of the Spon Press Software and eBook

To buy a licence to install your Spon Press Price Book Software and eBook on a secure local area network or a wide area network, and for the supply of network key files, for an agreed number of users please contact:

Spon's Price Books
Taylor & Francis Books Ltd.
2 Park Square, Milton Park, Abingdon, Oxfordshire, OX14 4RN
Tel: +44 (0) 207 017 6672
Fax: +44 (0) 207 017 6072
www.pricebooks.co.uk

Number of users	Licence cost
2–5	£450
6–10	£915
11–20	£1400
21–30	£2150
31–50	£4200
51–75	£5900
76–100	£7100
Over 100	Please contact Spon for details

Software Installation and Use Instructions

System requirements

Minimum
- Pentium processor
- 256 MB of RAM
- 20 MB available hard disk space
- Microsoft Windows 98/2000/NT/ME/XP/Vista
- SVGA screen
- Internet connection

Recommended
- Intel 466 MHz processor
- 512 MB of RAM (1,024MB for Vista)
- 100 MB available hard disk space
- Microsoft Windows XP/Vista
- XVGA screen
- Broadband Internet connection

Microsoft® is a registered trademark and Windows™ is a trademark of the Microsoft Corporation.

Installation

Spon's External Works and Landscape Price Book 2011 Electronic Version is supplied solely by internet download. No CD-ROM is supplied.

In your internet browser type in www.pricebooks.co.uk/downloads (or as a fallback www.ebookstore.tandf.co.uk/ supplements/pricebook) and follow the instructions on screen. Then type in the unique access code which is sealed inside the back cover of this book.

When the access code is successfully validated, click on the download links to download and then save the files.

Please note: you will only be allowed one download of these files and onto one computer.

A folder called *PriceBook* will be added to your desktop, which will need to be unzipped. Then click on the application called *install*.

Use
- The installation process will create a folder containing the price book program links as well as a program icon on your desktop.
- Double click the icon (from the folder or desktop) installed by the Setup program.
- Follow the instructions on screen.

Technical Support

Support for the installation is provided on www.ebookstore.tandf.co.uk/supplements/helpdesk.asp

The *Electronic Version* is a free addition to the book. For help with the running of the software please visit www. pricebooks.co.uk

The software used in *Spon's External Works and Landscape Price Book 2011 Electronic Version* is furnished under a single user licence agreement. The software may be used only in accordance with the terms of the licence agreement, unless the additional multi-user licence agreement is purchased from Spon Press, 2 Park Square, Milton Park, Abingdon, Oxon OX14 4RN Tel: +44 (0) 20 7017 6000

Free Updates

with three easy steps…

1.   Register today on www.pricebooks.co.uk/updates

2.   We'll alert you by email when new updates are posted on our website

3.   Then go to www.pricebooks.co.uk/updates
     and download the update.

All four Spon Price Books – *Architects' and Builders'*, *Civil Engineering and Highway Works*, *External Works and Landscape* and *Mechanical and Electrical Services* – are supported by an updating service. Three updates are loaded on our website during the year, in November, February and May. Each gives details of changes in prices of materials, wage rates and other significant items, with regional price level adjustments for Northern Ireland, Scotland and Wales and regions of England. The updates terminate with the publication of the next annual edition.

As a purchaser of a Spon Price Book you are entitled to this updating service for the 2011 edition – free of charge. Simply register via the website www.pricebooks.co.uk/updates and we will send you an email when each update becomes available.

If you haven't got internet access we can supply the updates by an alternative means. Please write to us for details: Spon Price Book Updates, Spon Press Marketing Department, 2 Park Square, Milton Park, Abingdon, Oxfordshire, OX14 4RN.

Find out more about Spon books
Visit www.sponpress.com for more details.